ENVIRONMENTAL ENGINEERING

WATER, SANITATION & SEWERAGE, DRAINAGE, LIQUID AND SOLID WASTE MANAGEMENT, ENERGY, POLLUTION CONTROL, ENVIRONMENTAL PLANNING

ENVIRONMENTAL ENGINEERING

WATER, SANITATION & SEWERAGE, DRAINAGE, LIQUID AND SOLID WASTE MANAGEMENT, ENERGY, POLLUTION CONTROL, ENVIRONMENTAL PLANNING

A.K. JAIN

KHANNA BOOK PUBLISHING CO. (P) LTD.

Publisher of Engineering and Computer Books

4C/4344, Ansari Road, Darya Ganj, New Delhi-110002

Phone : 011-23244447-48 **Mobile:** +91-9910909320

E-mail : contact@khannabooks.com

Website : www.khannabooks.com

Dear Readers,

To prevent the piracy, this book is secured with HIGH SECURITY HOLOGRAM on the front title cover. In case you don't find the hologram on the front cover title, please write us to at contact@khannabooks.com or whatsapp us at +91-99109 09320 and avail special gift voucher for yourself.

Specimen of Hologram on front Cover title:

Moreover, there is a SPECIAL DISCOUNT COUPON for you with EVERY HOLOGRAM.

How to avail this SPECIAL DISCOUNT:

Step 1: Scratch the hologram
Step 2: Under the scratch area, your "coupon code" is available
Step 3: Logon to www.khannabooks.com
Step 4: Use your "coupon code" in the shopping cart and get your copy at a special discount
Step 5: Enjoy your reading!

Thanks,
Khanna Publishing House

Price: ₹ 525.00

ISBN: 978-93-86173-63-8

ENVIRONMENTAL ENGINEERING
by **A. K. Jain**

Edition: 2018
Reprint: 2024

To view complete list of books, please scan the QR Code:

Published by:
Khanna Book Publishing Co. (P) Ltd.
CIN: U22110DL1998PTC095547

Visit us at: www.khannabooks.com
Write us at: contact@khannabooks.com

Printed in India.

Preface

The critical level of pollution, contamination and despoliation of th environment are the repugnant calamities of human blunders. There is a growing alarm at rapid deterioration of quality of air, water, depletion of natural resources and the ozone layer. The situation is gloomy, but the fact is that there are professionals, government and people who are concerned and interested to recognise shows a promise. This needs an integrated and preventive vision, rather than stand alone, remedial actions. Pollution abatement in its ramifications comprises a whole series of inter-related problems which must all be considered together, but tackled one at a time. It involves diverse concerns such as soil utilities, sanitation and drainage, waste creation noise, traffic and transport impacts, urban decay and blight, use of energy, water and other resources. Environmental engineering seeks to reverse the downward trend of environmental degradation and re-establish the ecological cycle and life support system.

The concept of environmental engineering is closely related with the five elements i.e. earth/ vegetable, water, fire, air and skash (sky) interlinked and interlocked with human and other forms of life. The five elements are linked with the life and have both personal, (material) and universal (spiritual) attributes.

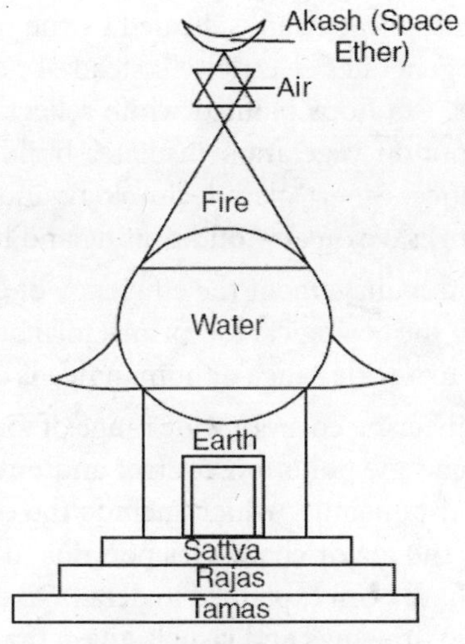

The Prakriti Stupa

The services of future have to be smart and sustainable, which integrate the built and natural environment. Time and energy are conserved and human productivity increases by availability of services, water, energy, etc. Urban services with circular metabolism can give as much to the environment as they take out, thus reducing the ecological impact. An ecological city and the buildings respect the nature so as to minimize the use of energy and carbon emissions. Cities and buildings can protect the environment using the minimum of mechanical systems and by maximizing sun, wind, water, earth and space.

To facilitate the efficient use of natural energy the form of cities and buildings has to be different. These are designed to incorporate both old and new ecological strategies, learning from traditional forms, as well as modern technology. For example, courtyard house and aerodynamic gliders which uses the minimum amount of material to maximum affect, both work in harmony with nature. Passive design and engineering need to maintain a harmonious balance. Sustainability can be achieved by the use of renewable energy and by eco-efficiency.

Indian cities are yet to explore the opportunity to create solar cities and solar buildings with new technology.

Sustainable development is giving birth to a new aesthetic, where the nature (sun, wind, water and earth) integrates with technology (nano-technology, renewables, biotechnology, eco-efficiency, etc.) to create an efficient, comfortable, peaceful and beautiful built environment.

Sensors for services, buildings and mobility will make the city energy efficient, smart and reduce emissions. Common Utility Ducts carrying electricity, water, cable television and broadband internet minimize damage from traffic, road repairs, rains, etc. Trigeneration energy systems combining power, cooling and heating, dual piping for recycled water and automated waste collection/utilization bundle "green infrastructure" together. Water saving toilets with recycled wastewater cisterns and showerheads will save water. Satellite controlled micro-irrigation system cuts water and power consumption.

Solid waste extracted from sewage is burned to make electricity. Three bin recycling is adopted with separate bins for trash, recyclable and compost which is transported and processed by an underground pneumatic system. Bio-technology, enzyme based STP, bio-remedial treatment, sludge gas/energy recovery, vermi-culture, fossilization and composting options are explored for waste treatment. Swales, porous paving, bio-drainage and storm surge gates in river, drains and canals are proposed for zero run off drainage and flood control. Rooftop water tanks, heated by the solar energy (photovoltaics) provide cooking and hot water. Solar panel generates electricity instead of power plants and also shade rooftops to lower a building's cooling needs. Rooftops painted white reflect heat, lowering a building's cooling cost and a city's heat build-up. Rooftop vegetation insulates buildings and super-insulated windows quadruple their thermal performance. Smart glass technology and cool air draughts in clay pipes, 4 meters below ground (PEDC system) save on air-conditioning and high energy cost.

By proper planning, design and management the efficiency of environmental engineering services can be ensured. This helps obviating the ecological footprints, formation of heat islands and outdoor and indoor pollution, while addressing to a wide range of human needs and sustainability issues.

The book 'Environmental Engineering covers a wide range of topics- water, sanitation and drainage, solid and liquid waste treatment, energy, pollution control and environmental planning. The basis of environmental engineering is the sustainability which include the ecology, climate, geology, hydrology and infrastructure services. One of the major challenges pertains to making the cities more sustainable by environmental planning, which have been explained in detail. Several examples, explained in a simple, flowing language and more than 200 drawings and visuals give a practical focus to the book.

The preparation of such a book involves drawing from the many fountains. I am grateful to each author of various works and have tried to acknowledge them, wherever known. Although my name appears on the cover of the book, behind the scene many people, both known and unknown, have contributed a great deal to make it possible. Saying thank you to them seems a poor reward for all their support and help. The Khanna Books, especially Mr. Mukul Seth, deserves a special mention for converting an unwieldy manuscript into this presentable book.

A.K. Jain

Contents

1 Environmental Engineering and Infrastructure Services: An Introduction

According to U.N. estimates the current population of India is 1,380,233,788 (as on July 8, 2020). Out of this 35% is urban population. The population density in India is 460 persons per square kilometer. It is expected that by 2030 India's population will be 1.512 billion and will become most populous country in the world.

As per the United Nations Millennium Development Goals (MDG) programme 88 million people roughly equal to 6.5% of India's population, are living below the poverty line of $1.25 (2018-19).

As of 2019, there are 6,64,369 villages in India. Considerable urbanization has occurred during last decade. Presently (2020) there are 739 districts. (Govt. of India web Directory), up from 640 in the 2011 census and 593 recorded in the 2001 census of India.

As per the 2011 census, there were three mega cities (having population more than 10 million). According to latest (2020) U.N. estimates there are 5 megacities in India : Mumbai (20.4 million), Kolkata (14.8 million), Bengaluru (12.3 million), Chennai (10.9 million) and Hderabad (10.0 million).

The current housing shortage in urban areas is around 10 million units and most of the shortage lies in Economically Weaker Sections (EWS) and Low Income Group Segments (ET-Realty 2019). According to the report, 8.36 million houses have been sanctioned under the governments' "Housing for all by 2022" initiative.

Even as India is the fastest growing economy in the world to-day with a growth rate of 8.2% in the first quarter of 2018-19 - 63.67 million urban and rural households across India do not have adequate housing. U.N. defined 'adequate housing' as "adequate privacy, adequate space, adequate security, adequate lighting and ventilation, adequate basic infrastructure and adequate location with regard to work and basic facilities- all at a reasonable cost."

On an average about 24% of the population mainly the poor are living in the Slums termed as Jhuggi-Jhopris or bastis. However, there is a trend of declining number of Slums in India. It has declined from 42% (in 2000) to 29% (in 2010) according to a UN report.

The indiscriminate expansion of urban footprints on the landscape of India is putting serious pressure on its ever-dwindling water resources. India is the home to 16% of world's population and has only 2.5% of world's land area and 4% of world's water resources at its disposal.

Most of the municipal water supply in India comes from the groundwater resource. According to the Composite Water Management Index (CWMI) report released by Niti

Ayog (2018), 21 major cities-including Delhi, Bengaluru, Chennai, Hyderabad and others are racing to reach zero groundwater levels by 2020 affecting access for 100 million people.

Government of India recently formed a new Jal Shakti (Translation : Water Power) ministry which aims at tackling water issues with a holistic and integrated perspective on the subject. The ministry has announced an ambitious plan to provide piped water connections to every household in India by 2024. According to National Sample Survey 76th round only 21.4% of households in India have piped drinking water connections.

The World Health Organisation (WHO) has stated that an individual requires around 25 litres of water daily for meeting basic hygiene and food needs. The rest is used for non-potable purposes like mopping and cleaning. This indicates that for most of non-potable uses, a quality lower than drinking water is required. Thus, for economic efficiency and environment sustainability, water must be treated and supplied according to usage.

Other serious issues are the leakages losses, water pricing and metering of water. It is said that 40% of piped water in urban areas is lost by leakage.

Most important cause of concern is the ever-increasing pollution of our waters, both surface and underground waters. It is said, 'Modernization is always accompanied by various types of pollutions'. According to United Nations Environmental Programme (UNEP), 1.2 billion people in the world are being affected by polluted water. Dirty water contributes to 15 million child deaths every year.

WASTEWATER AND SURFACE WATER POLLUTION

The coverage of underground sewerage system varies between 24% in Nagpur to 85% in Pune with a weighted average of 60%. In the metropolitan cities 100% of the sewage is not subject to treatment before disposal. Sewage treatment facility is available for a total flow of 960 mld. Which is 22% of the flow rate generated. Although nearly 300 urban centres have sewerage system only 70 towns / cities have sewage treatment facilities.

Although 90% of urban population is reported to have access to safe drinking water, there are severe deficiencies in quantity and quality of water available to urban residents. As per the Central Pollution Control Board (CPCB) only 5 metropolitan cities have proper wastewater collection system. Among the metropolitan cities, eight cities have partial primary and partial secondary treatment facilities while two have secondary treatment facilities only. A survey conducted by CPCB for 142 class-I cities reveals that in about 30% of the class-1 cities mode of disposal is on agricultural land, another 30% discharge into rivers either directly or through drains. About 22% cities followed mode of disposal both on agricultural land and into rivers while rest discharged to other recipient system such as sea, lakes, ponds etc. Similar situation for mode of disposal of wastewater is prevalent in Class-II towns. In none of the mega cities 100% wastewater is treated. In Delhi about 78% of wastewater is treated. While in Bangalore, Chennai and Kolkata the proportion of wastewater treated is about 75, 63 and 50 percent respectively.

India is blessed with good rainfall. The average annual rainfall in the country is 1170 mm. A reasonably uniform distribution of 14 major, 44 medium and 55 minor rivers, makes India a water-rich country. The total available sweet water in the country is estimated at 4000 billion cubic meters per annum. Of this usable water is only 1123 billion cubic meters.

In this the share of surface water is 728 billion cubic meters and 395 billion cubic meters groundwater. The per capita availability of water which was 2464 cubic meters in 1991 has been decreasing steadily due to population explosion and pollution of water bodies including groundwater. At present the per capita availability of fresh water is about 1400 cubic meters. It is feared that by 2050 it will fall to 1000 cubic meters, at which India will fall into the category of water stressed country. This would further hamper food security, as the scarcity of water will directly suppress agricultural production.

Many important cities are situated on the banks of rivers like : Ahmadabad (Sabarmati), Allahabad now Pryagraj (Ganga - Yamuna), Varanasi -Patna- Kanpur (Ganga), Guwahati (Brahmputra), Kolkata (Hugli) etc. Most of these carry high social, cultural and religious value. However, with indiscriminate urbanization these rivers have become highly polluted. The Ganga most important river of India. is now among the most polluted rivers in the world ranked fourth (2019). Ganga supports over 500 million people covering 26% of countries land mass spread in 11 States.

Each year 2.9 billion litres of waste water from sewage, domestic and industrial sources is dumped directly into the river. Most of it comes from big cities like Pryagraj (Allahabad), Kanpur and Varanasi. Even few thousand bodies are thrown into the river each year.

The National Mission for Clean Ganga under its Namami Gange programme has initiated several projects to clean and rejuvenate Ganga by sewage management, construction of new ghats ad crematoria and spreading public awareness about keeping the river pollution-free under Namami Gange Programme, 46 STPs are being made on Ganga-Yamuna rivers.

Water Supply and Sanitation in India: In India the water supply and sanitation has improved considerably. Still large number of people lack access to clean water, toilets and sewage infrastructure. At the Central Government, State Government and community levels various programmes have been launched to improve sanitation and drinking water supply. In 1980 rural sanitation coverage was only 1% which reached 95% in 2018.

The basic sanitation was available to 59.5% Indians (2017). During 2014-2019, the Government claims to have built about 110 million toilets across India, thus providing sanitation coverage to 93.3% (2019) households. If Government figures are correct, it would be a huge achievement.

According to a U.N. report, India has made rapid progress in increasing access to sanitation in schools. The Joint Monitoring Programme (JMP) of WHO and UNICEF has found that Indian schools have achieved a major feat as almost all schools (2018) in the country have some type of sanitation facility compared to half the schools having no sanitation facility a decade ago.

About half of India is facing drinking water crisis. A Neeti Ayog report released last year predicts Day Zero for 21 Indian cities by 2020. Day zero refers to the day when a place is likely to have no drinking water of its own. According to Niti Ayog's composite Water Management Index (CWMI), Bengaluru, Chennai, Delhi and Hyderabad are among the most susceptible.

Sewage Generation and Treatment : According to the Ministry of Forest and Climate Change the sewage generated in Urban areas (2018) is 61,948 million litres per day (mld) with only 23,277 mld being treated. Main sources of wastewater and effluents are the industries, refineries, dairies, power stations etc. Maharashtra being the first at 8143 mld followed by Uttar Pradesh at 7124 mld and Tamil Nadu at 5599 mld. Delhi, West Bangal and Gujrat are also major contributors.

As per the data given by CPCB (2016), there are 920 Sewage Treatment Plants in India with Punjab (86), Maharashtra (78), Tamilnadu (73) and UP (73) leading the table.

There is a large gap between generation and treatment of wastewater in India. Even the treatment capacity existing is also not effectively utilized due to operation and maintenance problem. More than one third plants are not conforming to the general standards as per the CPCBs survey report.

Drainage Systems : The drainage systems are in place to remove excess rainwater. Drainage systems are also in place to remove wastewater effectively, and this is referred to as a Sewer System. Not having a proper drainage system in place will result in flooding of low-lying areas thereby causing property damage and health risk.

Drainage systems can be separate for storm water (rainwater) and wastewater (sewers) but sometimes these inputs drain into the same system mostly underground drainage system. The advantage of having separate drainage systems is that storm water can be directly discharged into water bodies. Recently, defective drainage systems have caused urban floods in many cities of the world including India.

Overburdened drainage, frenzied and unregulated construction, no regard to the natural topography and hydrogeomorphology all make urban flooding a man-made disaster. Frequent and large scale flooding in Indian cities disrupt normal life and cause extensive damage to the property and environment.

The recent floods in Bihar (2019) have taken by surprise both the government and people. Urban flooding has a long history in India- Chennai (2004, 2015), Mumbai (2005, 2017), Delhi (2002, 2003, 2009, 2010), Patna (2019) are most notable urban floods.

Due to open drainage channels and poorly drained areas there is serious threat of spreading of pathogens and thus increase in water borne diseases. Polluted water gets entry into low-pressure water supply system and underground aquifers, contaminating the drinking water.

Solid Waste: India is one of the mega underdeveloped countries of the world with respect to population status and dynamics. The growing rate of population has direct impact on enhancement of municipal solid waste generations. Press Information Bureau estimates put total annual solid waste generated at 62 million tonnes per year (2016). Various estimates of total municipal solid waste quantity are quite confusing. One of these estimates put the total MSW at 96 million tonnes per annum (2019). Considering World Bank estimate of 34% Indian population living in Urban areas, this works out to be solid waste generation per person per day to be 550 grams. It seems that by 2040 the MSW will be of the order of 160 million tonnes.

Presently, only half of the MSW is collected, processed or recycled. This is a major cause of insanitation and poor hygiene, which degrades and pollutes the environment. According to an estimate about 50% of the recyclables material is collected by Rag Pickers, who get exposed to various hazards.

Energy: Cities account for three quarters of the total energy demand including fossil fuel used for transport and produce almost 80% of CO_2 emissions. Over 95,000 villages and 60 percent of the households, still have no access to electricity. The concept of energy efficiency, renewable energy and Zero-fossil Energy Development (ZED) are the pillars of sustainable energy policy which can reduce the level of energy demand and slow down the rate at which resources are depleted. Energy efficiency involves a synergy of the various levels of planning, design, construction and maintenance, leading to a sustainable a regime. The renewable energy not only helps in energy generation, but also in a pollution-free environment. It is estimated that India has a potential of generating more than 100,000 MW from the renewable sources of energy.

The installed generating capacity increased nearly one hundred fifty-folds from 1362 MW in 1947 to 2,00,000 MW in 2017. The per capita consumption of electricity, which was less than 15 units in 1951, increased to 1000 units (2.74 kW per capita) by 2017. Despite these achievements, the power sector has not been able to meet growing aspirations of the people in the country. There is severe shortage of power supply vis-a-vis demand both in urban and rural areas and there are frequent power cuts, low voltage, low frequency and high incidences of breakdowns.

Transport: The urban transport scenario in Indian cities is quite depressing. Vehicle population in India increased 80 folds in the last 40 years, but road area increased by only 5 per cent. Only 17 of the largest cities have organized bus services, and only six cities – Mumbai, Kolkata, Delhi, Hyderabad, Bangalore and Chennai – have an intra-urban rail system. In metropolitan cities, 40 to 50 percent people walk and use non-motorised transport, while 30% people use public transport Private vehicles are growing three to four times the population growth in urban areas. It is estimated that 2.5 billion square meter of roads, 7,400 km of metro rail network and subways are needed to be built by 2031. India with more than 151,000 (2018) road deaths and more than 500,000 major road accidents per year is one of the most unsafe countries to travel in the world. 81 % of the accidents happen because of the poor, rash driving, of which mostly the pedestrians are victims. In Indian cities about one-third to two-thirds of trips are solely by walk. Still 90% of city roads do not have footpaths. Except few highways and major roads there are no signage, signboards, road marking and basic information for drivers and travellers. About half of urban roads have potholes and are in poor repair. By and large there is no accountable system of road maintenance. The intensive use of fossil fuels for urban transport (private vehicles, cars, 2 stroke engines, two wheelers, etc.) has made urban air very unhealthy.

Disasters: India in her 3.3 million sq. km. area contains 29 States, 7 Union Territories, 739 districts, 7936 cities and 6,49,481 villages (2018). The nature has bestowed upon India all welcome diversity in the form of hills, plains and long coastline. However, each of these regions face regular natural disasters also, such as : droughts, floods, cyclones, earthquakes, landslides, avalanches and forest fires.

An area of over 40 million hectare (12.5% of land) is prone to floods, out of the 7516 km long coastline 5,700 km is prone to cyclones and tsunamis, about 68% of cultivable land is vulnerable to drought. The four seismic zones cover about 57% of our land area. This shows that all the regions experience disasters causing huge economic, human lives and environment related loss. There exists no definite trend in these losses.

India has suffered economic losses due to natural hazards of 79.5 billion U.S. dollars in the last two decades according to the United Nations. India proposes to spend 1 Trillion U.S. dollars in infrastructure to combat disasters.

The United Nations global assessment report on disaster risks (2015) estimates that India's average annual economic loss due to disasters is likely to be $9.8 billion, which includes more than $7 billion loss on account of floods. The report warns that without adequate State investments in disaster risk reduction (DRR) development would be hindered. India has projected a $1 trillion investment in infrastructure in the next five years and unless adequate steps are taken to make them resilient to floods and other natural calamities, the investment runs the risk of going waste.

Climate Change: India with its diverse agro-climatic and morphological zones is particularly affected by climate change. Climate change can bring considerable changes in the hazard profile of the region by its type intensity, magnitude and frequency. Many of these impacts may not be uniform across spatial and temporal scales. Devastating floods, typhoons and a series of hurricanes are being associated with climate change. According to the initial report of the United Nations Framework on the Convention on Climate Change, the predicted impacts of climate change in India include a surface air temperature rise up to 4° Celsius by 2100, up to 30% decline in yield in rain-fed areas for some crops and an increase in incidences of extreme events, such as droughts, floods and cyclones. Urban livelihoods, property, quality of life and future prosperity are threatened by the risks of flooding, landslides, heat waves and droughts and adverse events due to climate change. While large population densities in urban areas create increased vulnerability, they also create the potential for changes that can mitigate human impacts on climate. Since these impacts will strongly vary in its extent and form throughout the country, they require customised interventions in different states threatened population.

Urban Transformation (Smart Cities): With the advent of the 20th century, India embarked into the era of rapid urbanization. In 100 years between 1901-2001, its share of urban population increased from 11% to over 28% and total urban agglomeration increased from 1827 to 4386. According to 2011 census the urban population was 31.16%. Worldometers estimates India's urbanization in 2019 as 33.6% and by 2030 it will be 40%.

In order to improve the quality of life of urban residents, Government has designed the Smart City Mission. In the approach to the smart city mission, the objective is to promote cities that provide core infrastructure and give a decent quality of life to its citizens, a clean and sustainable environment and application of 'smart' solutions. The focus is on sustainable and inclusive development and the idea is to look at compact areas.

Urban Transformation: To make the cities more liveable and inclusive, the Government of India in 2015 has embarked on Smart Cities programme for 100 cities and the Atal Mission for Rejuvenation and Urban Transformation (AMRUT) for 500 cities. The focus is on core infrastructure services like adequate and clean water supply, sanitation and solid waste management, efficient urban mobility and public transportation, affordable housing for the poor, power supply, IT connectivity, e-governance and citizen participation along with municipal reforms and providing aid to the state governments and the ULBs.

With the commitment that by the time the Nation completes 75 years of Independence, every family will have a pucca house with water connection, toilet facilities, 24×7 electricity supply and access. The Central Government has launched (2015) "Prime Minister's Housing for All by 2022" mission. The mission seeks to provide 20 million housing and take up the Slum rehabilitation project. The program provides interest subsidy of 6.5 per cent on housing loans with tenure of up to 15 years for EWS and LIG, which works to nearly Rs 100,000 to Rs 230,000 per unit. It promotes slum free cities and mandates house in the name of women, or joint ownership.

For conservation of urban heritage and cultural resources, the Government of India (2015) has launched the Historic City Development Augmentation Yojana (HRIDAY), which will initially cover 12 cities viz. Ajmer, Amrawati, Amritsar, Badami, Dwarka, Gaya, Kanchipuram, Mathura, Puri, Varanasi, Velankanni and Warangal.

Swachh Bharat Mission (SBM) is another major initiative of the Government of India, covering 4,041 statutory towns, to clean the streets, roads and infrastructure of the country. SBM includes focus on provisioning of household toilets, community toilets, public toilets, solid waste management and bringing in behavioural change in people regarding healthy sanitation practices. The following key initiatives aim to make cities and villages clean and hygienic:

- Every house with a toilet
- Public toilets and total sanitation
- Sewerage and decentralised STP
- Solid Waste Management, Circular Metabolism
- Drainage, wastewater recycling
- Integrated Sewerage-Sanitation- Water-Wastewater- Drainage
- Pollution control
- Public health and hygiene focus

The restoration of rivers is a high priority on the agenda of Government of India. National Gange Plan aims to integrate the efforts to clean the river Ganga, with an outlay of Rs 200,000 million for the next five years. It focuses on pollution abatement, interception, diversion and treatment of wastewater flowing in the open drain through bio-remediation in-situ treatment, use of innovative technologies, sewage treatment plants (STPs), effluent treatment plants (ETPs) rehabilitation and augmentation of existing STPs. Short term measures include-arresting the pollution at exit points on riverfront to prevent inflow of sewage, effluents and solid waste into the river.

All these missions focus on urban transformation of India. However, their success is largely contingent upon rigourous, sustained efforts of the urban local bodies and meticulous planning, together with legal, institutional and financial reforms, which facilitate and mobilise public, private and peoples' resources.

These involve integration of public, private and community resources, Central Government grants and funding and adoption of smart systems. Capacity building and implementation focus are necessary to achieve the results. Planning is a pre-requisite which should be based on a comprehensive Information System, Digitized Mapping and Geo-portal. Development of Service Level Benchmarks based maintenance and monitoring of public services, especially public toilets, water supply, sanitation, sewerage and drainage, solid waste management, wastewater recycling, pollution control, etc. are necessary for effective implementation of the plans and programmes.

SERVICE LEVEL BENCHMARKING FOR INFRASTRUCTURE SERVICES

The MoUD has initiated an exercise to develop Service Level Benchmarks (SLBs). Measuring service levels of civic agencies implies measuring outcomes, and indirectly reflects on institutional capacity, financial performance and other parameters with respect to water supply, sewerage, drainage and solid waste management for efficient functioning.

These SLB should be considered as benchmark for measuring, reporting and monitoring the performance and comparing inter and intra city level infrastructure. The SLBs for different utility services in an urban area are mentioned in table below:

Table 1.1: Service Level Benchmarks

Sr. No.	Indicators	Benchmark Levels
Water Supply		
1.	Coverage of WS connections (Population)	100%
2.	Per capita availability of WS at consumer end	135 Lpcd
3.	Extent of metering of WS connections	100%
4.	Extent of Non-Revenue Water	20%
5.	Continuity of Water Supply	24×7
6.	Efficiency of redress of Customer Complaints	80%
7.	Quality of Water Supplied	100%
8.	Cost recovery in Water Supply Service	100%
9.	Efficiency in collection of Water Supply Charge	90%
Sewerage		
1.	Coverage of Wastewater network service	100%
2.	Collection effciency of Wastewater network	100%
3.	Adequacy of Wastewater treatment capacity	100%
4.	Quality of Wastewater treatment	100%
5.	Extent of reuse & recycling of treated Wastewater	20%
6.	Extent of cost recovery in Wastewater management	100%
7.	Efficiency of redqress of Customer Complaints	80%
8.	Efficiency in collection of sewerage charges	90%
9.	Coverage of toilets	100%

Storm Water Drainage		
1.	Coverage of storm Water Drainage network	100%
2.	Incidence of water logging/flooding	0
Solid Waste Management		
1.	Household level Coverage of Solid Waste Management service	100%
2.	Efficiency of Collection of Municipal Solid Waste	100%
3.	Extent of segregation of Municipal Solid Waste	100%
4.	Extent of Municipal Solid Waste recovered/recycled	80%
5.	Extent of scientific disposal of Municipal Solid Waste	100%
6.	Extent of cost recovery in Solid Waste management service	100%
7.	Efficiency of redress of Customer Complaints	80%
8.	Efficiency of collection of user charges	90%

Source: Advisory note on Improving Urban Water Supply and Sanitation Services, MoUD, 2012

URBAN GOVERNANCE

The Indian federal has a three-tier institutional system with the Central government at the top, state governments at the second level and local governments at the grassroots level. The responsibility for providing basic services and implementing development programs is mainly of the local government. India has the distinction of having the largest number of democratically elected local governments in the world—more than 250,000 rural and urban local bodies with almost 3 million elected representatives.

The form of governance prevalent in the country at all levels is generally bureaucratic and involves conservative controls. With the 74th Constitutional Amendment Act (1992), the urban local bodies have been assigned greater role in urban planning and governance. However, effectively this has not happened in most of the States, as the governments are reluctant to part with their powers. Poor municipal capacity and finances are acting as barriers in the area of decentralization.

The process of urban governance, linked with the participatory democratic decentralization has direct bearing on the issues of livelihoods, poverty, infrastructure/basic services, equity and capacity building. The local bodies have a catalyst role to act as a bridge between the people, NGOs, CBOs, political apparatus and the government.

Information and communication technology (ICT) and e-governance are emerging as the tool of community empowerment, efficient service delivery and reduced corruption. The local municipal governments, police, service agencies, banks, etc. are gradually adopting digital planning, web sites, e-banking and e-governance, replacing time consuming and corruption generating paper work. This is the fundamental building block of urban governance. The key areas to achieve better results are sustainable environment planning and engineering practices, enhanced efficiency, employment generation, social inclusion, gender equity.

DEBRIEFING QUESTIONS

1. What is the purpose and scope of Environment Engineering?
2. Give a brief of status of water, sewerage, drainage, energy and transport in India and why an environmental approach is necessary in their planning and development?
3. How the problems of natural disasters in India can be mitigated?
4. Explain the phenomenon of climate change and its impact on sustainability.
5. What are the approaches towards containing air pollution?
6. How the water pollution can be reduced?
7. What is the role of urban governance in environmental sustainability?

2 Water

India is considered to be a water rich country with a fairly uniform network of rivers. These rivers are classified into major, medium and minor river basins according the area of basin. India has a total of 113 river basins of which 14 are major with catchment area ≥ 20,000 km^2, medium with area between 20000 and 2000 km2 and minor ≤ 200 km^2. The major river basins form about 84% of the total drainage area of the country. The three major river systems are: Ganga, Indus and the Brahmputra. Ganga (25%) and Brahmputra (40%) together contribute 65% water. Over 90% of water is drained into the Bay of Bengal and rest in Arabian sea.

India is blessed with good rainfall. The average annual rainfall in the country is 1170 mm. The total available sweet water in the country is about 4000 billion cubic meters per year. There is large variation in rainfall both temporally and spatially. About 90% of rainfall falls during 3 monsoon months out of this usable water is only 1123 billion cubic meters. Out of this total usable water 728 bcm is contribution of surface water and 395 bcm from groundwater.

The availability of water, both surface and groundwater vary extremely in various regions of the country. Rajasthan has only 1% of the total water resources of the country, Maghalya and North-Eastern state get heavy rainfall of the order of 11,000 mm, almost 10 times of all India average.

Due to rapidly increasing population, urbanization and industrialization the water demand is rapidly increasing while our freshwater resources are declining due to man-made pollution of our water resources.

Water-stress condition:

Water stress occurs when the demand for water exceeds the available amount during a certain period or when poor quality restricts its use. Based on the average requirement of water for various purposes, the situation is considered of Water Stress condition when per capita water availability ranges from 1000 to 1700 cubic meter per capita per year, and it is considered Water Scarcity condition when availability reduces to below 1000 cubic meters per year. If present trend continues most of India will turn into Water-Stress condition very soon and water scarcity condition by mid 21st century.

The Figure 2.1 shows the observed and projected decline in per capita average annual freshwater availability and growth of population from 1951 to 2050. This clearly indicates the two-sided effect on water resources. As a result, availability of water is bound to reach critical levels sooner or later.

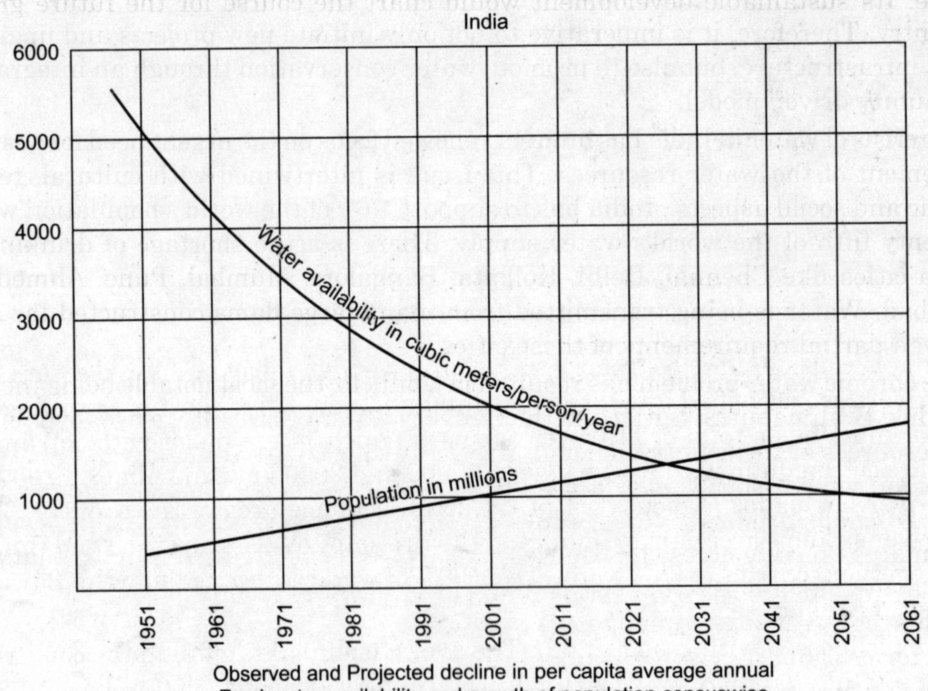

Observed and Projected decline in per capita average annual Freshwater availability and growth of population censuswise

Fig.2.1

Groundwater, which is an important source for domestic water supply is reported to be declining due to overdraft. The total annual groundwater extraction of the entire country for the year 2017 has been estimated as 248.69 bcm. About 89% of this is used by agriculture sector, only 27.24 bcm is used for domestic and industrial use. Annual groundwater recharge is 431.86 bcm. Though the overall stage of groundwater development in the country is 63% the stage of extraction is very high in the states of Delhi, Haryana, Punjab and Rajasthan where it is more 100% which implies decline in groundwater table.

Groundwater was considered to be pure and fit for even drinking. However there is growing concern on deterioration of groundwater quality mainly due to anthropogenic activities. The groundwater quality has undergone a change to an extent that its use can be hazardous. Increase in overall salinity and presence of high concentration of fluoride, nitrate, iron, arsenic and traces of toxic metals have been noticed in large areas of several states.

As India grows and urbanizes, its surface water bodies are getting toxic. It is estimated (2019) that around 70% of surface water in India is unfit for consumption. Everyday almost 40 million liters of wastewater enters rivers and other water bodies with only a tiny fraction adequately treated.

CRITICAL ISSUES

Fast increasing population is leading to indiscriminate expansion of urban footprints on the landscape of India. This is putting unbearable pressures on the ever-dwindling water resource. Its sustainable development would chart the course for the future growth of the country. Therefore, it is imperative to not only initiate new projects and upgrade our present infrastructure, but also to promote water conservation through an integrated and a community driven model.

The crisis of water in India has brought renewed focus on the urgent need for sustainable management of the water resources. This issue is intertwined with cultural, technical, economic and social aspects. India has to support 1/6th of the world's population with only one-twenty fifth of the world's water supply. There is acute shortage of drinking water in mega cities like Chennai, Delhi, Kolkata, Bangalore, Mumbai, Pune, Ahmedabad & Hyderabad. Water is being transmitted from many large dams constructed far away to cater even partial requirements of these cities.

The chronic water problem has resulted in conflicts, the most notable being the dispute over Indus, Brahmaputra, Sutlej and the Cauvery waters. Agriculture uses over 80% of all water resources of the country. Though the quantity needed for drinking water supply is much less in comparison, we still face acute shortage of drinking water both in quantity and quality. Official statistics indicate a high level of service coverage of 92% of people living in urban areas having access to safe water (1999). But the ground reality is that water is usually available in most cities for about two hours per day. Most of the cities are facing acute shortages of water. Among metro cities, Chennai and Delhi are the worst affected cities in terms of hours of water availability. Supply varies from 3 hours in Delhi, 4 hours in Chennai, 5 hours in Mumbai and 10 hours in Kolkata. Frequency of water supply is just for half an hour once or twice a week in Surendranagar of Gujarat. In Attur and Gudivattam towns of Tamil Nadu, water is available only twice a week. In cities like Bangalore, Bhilwara and Barmer in several areas water comes on alternate days.

Other causes for concern are enormous waste and pollution of water. Only about three-fifths of the waste water collected is treated. In many cities, underground systems only collect waste water and discharge it into land, lakes or rivers without treatment. Indiscriminate discharge of domestic and industrial waste water has seriously polluted most water resources. Groundwater is under continuous threat of over exploitation as well as seepage of surface pollutants. Cities as a result, are reaching out to remote sources of water supply.

Urban Water Supply

Indiscriminate and rapid urbanization significantly alter the hydrology and quality of surface water. Incompatible land uses, together with encroachments and pollution of the water bodies are killing the water bodies, and no amount of effluent treatment or legislation can reverse the damage. It is a paradox that, while on one hand we face the water crisis, on the other there is flooding in the same area, during the monsoons. This is largely due to lack of water based planning and development. Excessive hard paving of open areas, disturbing the natural drainage pattern, building of structures and roads, the natural depressions, silting of drains, water bodies and rivers, disposal of high amount of solid waste in water channels and bodies and their pollution are largely responsible for such a situation.

The over-exploitation of groundwater for irrigation, industry and urban uses has caused rapid Groundwater depletion. As a result large areas in various States have been declared 'dark' and 'Grey' zones, and there has been no let-up in the depletion of Groundwater aquifers. In many cases Groundwater depletion has aggravated water quality problems due to excess fluoride, arsenic and brackishness. Though India has an annual average precipitation of 1150 mm and drains a total volume of 400 million hectare metres of rain water, hardly 100 million hectare meters is retained in the soil. The remaining water runs off, carrying with it fertile soil.

The rapid runoff of precipitation and groundwater prevents natural percolation into the earth and water table. This is demonstrated by the frequent droughts in many parts of the country. In view of the fast growing population and urbanization, coupled with diminishing water resources, it is necessary to conserve and harvest rain water to the maximum extent possible. Conserving the scarce water and recycling of used water is a major challenge for the urban India. The wasteful practices of "drill, pump, and spill" still prevail and the efficient methods of water conservation, use, and recycling are not generally applied. Water pollution is a serious problem in India as almost 70% of its surface water resources and a growing number of its Groundwater reserves are already contaminated by biological, toxic organic and inorganic pollutants. In many cases, these sources have been rendered unsafe for human consumption, and even for irrigation and industrial needs, aggravating further to water scarcity. In many cities, about half of the water that is treated and distributed at public expense is not accounted for. This is due to unrecorded usage or illegal taps, and water connections are not registered. Reducing such losses is cheaper than reducing leakages which can be possible by adoption of differential cost recovery. In view of critical issues there is a need is evolve a holistic and a comprehensive approach to develop and manage the water resource. A blueprint for blue networks comprises the following:

- Comprehensive water supply standards
- Reviving indigenous skills and community participation
- Blue Network Planning
- Conserving Groundwater
- Rainwater harvesting and groundwater recharge
- Water Recycling
- Regulate Freshwater Consumption
- Protecting the quality of water
- Cost recovery, checking of leakages and unaccounted flow of water.

WATER SUPPLY STANDARDS

The objective of a public protected water supply system is to supply safe and clean water in adequate quantity, conveniently and as economically as possible. Rising demand of water due to rapid urbanisation is putting enormous stress. While planning the water supply system for an area, it is evident to consider water conservation aspects, which may be possible through optimal use of available water resources, prevention and control of wastage of water and effective demand management.

Rural Water Supply

- Planning Commission (1996): 55 lpcd and one HP/PSP for 150 persons and a distance norm of 0.5 km. Additional 30 lpcd of water to cater to drinking water requirements of cattle in DDP areas.

Sources of Drinking Water

Table 2.1: Sources of Drinking Water

Springs	Flow from underground formations to the surface of earth
Lakes, Rivers, Reservoirs, Well	From surface/underground aquifers through wells dug or bored reservoir, well
Sterile water	Purified as per the requirement of pharmacopoeia
Artesian wells	Tapped from a confined aquifer underground
Groundwater	From a subsurface saturated zone under atmospheric or higher pressure.

A large proportion of urban population living in informal settlements, usually do not have access to municipal water supply. They largely depend on the following sources of water:

Table 2.2: Comparison of options for existing and potential water services in informal settlements

Typical existing water sources	*Potential improved service options*
• Unregulated kiosks	• Utility supported private water kiosks
• Handcart vendors (expensive)	• Community managed kiosks
• Unauthorised connections	• Community managed local water distribution pipes
• Contaminated pools or rivers	• Shared water connections
• Distant springs or boreholes	• Prepaid metered kiosks
• Seasonal dug wells	• Water kiosks with storage tanks

Besides taps, tube-wells and hand pumps are common sources of water supply in urban settlements.

Table 2.3: Source of Drinking Water for Sample Households

Sector	Tap	Tube well	Handpump	Well	Tank, pond excl-usive	Other Tank, Pond	River, Canal, Lake	Spring	Other	All Sources
Rural	189	166	279	317	13	8	17	9	3	1000
	18.9%	16.6%	27.9%	31.7%	1.3%	0.8%	1.7%	0.9%	0.3%	100.0%
Urban	704	99	86	86	4	4	1	1	14	1000
	70.4%	9.9%	8.6%	8.6%	0.4%	0.4%	0.1%	0.1%	1.4%	100.0%

Source: NSSO (1999).

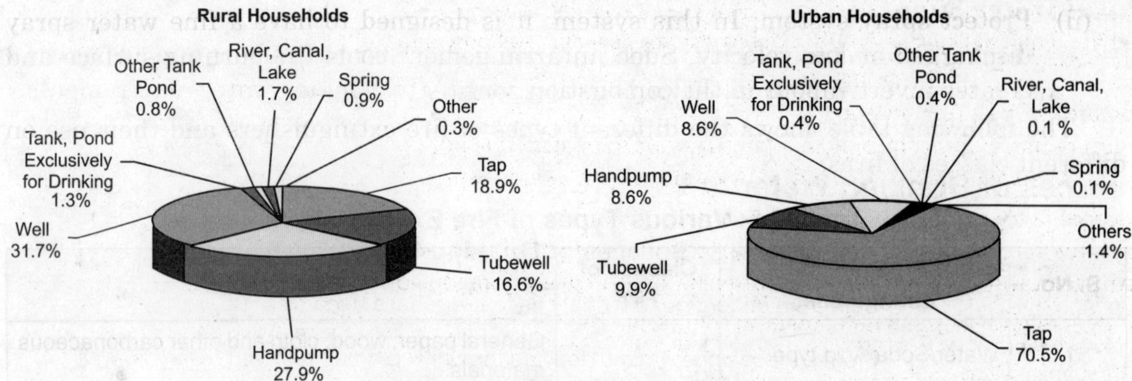

Fig.2.2: Sources of Drinking Water

In an already overstrained water scenario, it is projected that by the year 2050 the water requirement shall be almost twice of the present. This poses huge challenge for the country

Table 2.4: Future Drinking Water demand

Year	Total Water Demand		BCM/Year		BCM/Year
	Based on Past Census	Based on Un Projection	Based on Past Census	Based on UN Projection	Total water Requirement for Urban and Rural
2011	54810	63555	20	23.20	42
2021	66555	83375	24.29	30.20	NA
2025	71340	91350	26.04	33.34	55
2050	100755	1706050	36.77	51.33	90

Note: The water demand has been worked out @ 170 lpcd for 65% of the urban population presumed to be living in class-I cities and @ 100 lpcd for balance 35% of the urban population living in class-I cities.

Source: Report of the National Commission for Integrated Water Resources Development Plan (NCIWRDP).

FIRE HYDRANTS

The fire hydrants are located at suitable points and they can be operated either manually or automatically. In manual operation, a pump has to be started manually after opening the hydrant valve. In automatic operation, the pump will start automatically when the hydrant valve is opened manually.

The hydrant valve should be designed for the prescribed discharge under pressure such that a jet throw of water is obtained from the nozzle of hose pipe. This system has wide applications and in one sense, it is the backbone of all fire protection systems. It can be installed for multi-storeyed buildings, power plants, textile mills, hospitals, paper mills, airports, warehouses printing presses, etc.

Water spray is used to extinguish fire and depending on the type of spray, the system can be grouped into the following two categories:

(i) Multi-fire system: In this system, it is designed to have a coarse water spray discharged at high velocity. Such an arrangement gives the water spray to reach up to the base of flame zone.

(ii) Protect spray system: In this system, it is designed to have a fine water spray discharged at low velocity. Such an arrangement cools the burning surface and creates invert vapour in the combustion zone.

The following table shows the different types of fire extinguishers and their use on different classes of fires.

Table 2.5: Various Types of Fire Extinguishers

S. No.	Type of Fire Extinguisher	Classes of Fire	Description
1.	Water/Soda Acid type	A	General paper, wood, cloth and other carbonaceous materials.
2.	Dry Chemical Powder	B	Inflammable liquid, oil, gasoline chemicals, paints and solvents.
		C	Electrical installations, motors, switchgears and compressed gases.
		D	Metals i.e. Magnesium, Sodium, Potassium.
3.	Carbon dioxide	B and C	As above in B and C
4.	Halon	B and C	As above in B and C
5.	Chemical Foam	B	As above in B and C

The CPHEEO Manual recommends fire-fighting water demand as a function of population, i.e. water demand for fire-fighting purpose = $100\sqrt{P}$, where P stands for forecasted population may be adopted for communities larger than 50,000. It is desirable that one-third of fire-fighting requirements from part of the service storage. The balance requirement may be distributed in several static tanks at strategic points. These static tanks may be filled from nearby ponds, streams or canals by water tankers wherever feasible. The pressure required for fire-fighting would have to be boosted by fire engines.

Sectorwise Water Consumption in India

On account of rapid economic growth and demographic changes, the water demand in all sectors is showing increasing trend. The following table indicates the estimated water demand in India for various sectors:

Table 2.6: Projected Water Demand in India (By Different Use)

Sector	Water Demand in Billon Cubic Meter (BCM)								
	Standing Sub-Committee (MOWR)			National Commission on Integrated Water Resources Development (NCIWRD)					
	2010	2025	2050	2010		2025		2050	
				Low	High	Low	High	Low	High
Irrigation	688 (84.6)	910 (83.26)	1072 (74.1)	543 (78.24)	557 (78.45)	561 (71.6)	611 (72.50)	628 (64.55)	807 (68.4)
Drinking Water	56 (6.9)	73 (6.68)	102 (7.0)	42 (6.05)	43 (6.06)	55 (7.0)	62 (7.35)	90 (9.25)	111 (9.4)

Industry	12 (1.5)	23 (2.10)	63 (4.4)	37 (5.30)	37 (5.21)	67 (8.5)	67 (7.95)	81 (8.30)	81 (6.9)
Energy	5 (0.6)	15 (1.37)	130 (9.0)	18 (2.60)	19 (2.67)	31 (4.0)	33 (3.90)	63 (6.50)	70 (5.9)
Others	52 (6.4)	72 (6.59)	80 (5.5)	54 (7.81)	54 (7.61)	70 (8.9)	70 (8.30)	111 (11.40)	111 (9.4)
Total	**813 (100)**	**1093 (100)**	**1447 (100)**	**694 (100)**	**710 (100)**	**784 (100)**	**843 (100)**	**973 (100)**	**1180 (100)**

1 Billion = 100 Crore;
MOWR : Ministry of Water Resources;
% in brackets.

Source : *Basin Planning Directorate, CWC, XI Plan document.*

Table 2.7: Water Use by Different Industries in India

Industrial Sector	Annual Water Consumption (Million Cubic meters)	Proportion of total Water Consumed in Industry (%)
Thermal Power Plants	35,157.4	87.87
Engineering	2019.9	5.05
Pulp and Paper	905.8	2.265
Textiles	829.8	2.075
Steel	516.6	1.29
Sugar	194.9	0.49
Fertilizer	73.5	0.18
Others	314.2	0.78
Total	**40,012.1**	**100**

Source : *Centre for Science and Environment (CSE) 2004*

Railway Station water requirements

Water requirement for the Station for various uses (per day):

- 15 liters per passengers (Metcalf Eddy)
- 455 liters per employee (ref: CPHEEO Manual)
- Apron washing – 10 liters per sq. meters as per Indian Railways Work Manual (IRWM)
- Cleaning/mopping of platform and floor – 5 liters per sqm.
- Meters and greeters – 5 liters per visitor
- Catering – 45 liters per passengers
- Gardening/ Horticulture – 22500 liters per hectares (IRWM)
- Cleaning of trains on platform – 500 liters per coach (IRWM)
- Cleaning of trains in washing lanes – 3600 liters per coach (IRWM)
- Watering of trains – 1800 liters/ coach (IRWM)

- Fire-fighting (CPHEEO Norms)
- Commercial – 45000 liters per hectares per day (IRWM)
- UFW – 15% of total demand for new system (CPHEEO)
- 75% of total capacity shall be filled at the station for long haul trains (passing trains)

Water requirement for metro station could be calculated based on the type of station (underground or elevated). In absence of any specific norms to calculate water demand for metro station, the water estimations can be calculated from the above mentioned information for stations. Parameters such as staff requirement, HVAC requirement if station is underground, toilet requirement and passenger requirement to be considered based on the facilities provided at metro stations.

DISTRICT METER AREA PLANNING (BULK METERING)

The term 'district metering' is used to describe the method whereby flow meters are installed on all major supply lines and strategic points within the distribution system. The meters should be used to monitor the overall performance of the system establishing average daily flows into various districts. District meter areas ideally consist of 2000 to 5000 properties.

Size of the district meter should be such that it is capable of recording night flow without loss of accuracy and also must be capable of supplying peak flow without introducing serious head loss.

The District Meters should be read at weekly intervals at the same time of day as previous readings of the meter be calculated from the above mentioned information for stations. Parameters such as staff requirement, HVAC requirement if station is underground, toilet requirement and passenger requirement to be considered based on the facilities provided at metro stations.

Billing and Collection

Revenue management system is an important aspect of any Water Supply System as it governs the financial aspect. Besides fixing a tariff structure, billing and collection of revenue play an important part. The water charges to be fixed taking into account the ability of the system to meet the expenditure on the following heads.

- Operating Cost (excluding establishment cost),
- Establishment Cost,
- Depreciation,
- Debt Services & Doubtful Charges,
- Asset replacement fund.

Tariff structure should be fixed and revised periodically. Automatic increase of tariff periodically on index basis can also be adopted. Where the same authority also provides sewerage system, charges for this can also supply through Public stand post, may be charged and also be included as a percentage of the water charges.

There are various methods by which Water Billing can be done. The user is advised to refer 'Manual on Operation and Maintenance of Water Supply Systems' of CPHEEO while fixing the tariff, billing process and collection system.

Leakage Control

The overall objective of leakage control is to diagnose how water loss is caused and to formulate and implement action to reduce it, to technically and economically acceptable minimal. The water losses can be termed into two categories i.e. Physical losses and Non-Physical Losses. The Physical Losses is mainly due to leakage of water in the network and comprises of physical losses from pipes, joints & fittings, reservoirs & overflows of reservoirs & sumps. The Non-Physical Losses is due to theft of water through illegal, already disconnected connections, under-billing either deliberately or through defective meters, water wasted by consumer through open or leaky taps, errors in estimating flat rate consumption, public stand posts and hydrants. The major activities which should be taken up for efficient monitoring and leakage control are:

- Preliminary data collection and planning – The water distribution drawings are to be studied and updated. The number of service connections is to be obtained and in the drawings of the roads the exact locations of service connections marked. The district and sub-district boundaries are suitably fixed taking into consideration the number of service connections, length of mains, and pressure points in the main. The exact locations of valves, hydrants with their sizes should be noted on the drawings.
- Pipe location and survey – Electronic pipe locators can be used during survey. These instruments work on the principle of Electromagnetic signal propagation. It consists of a battery operated transmitter and a cordless receiver unit to pick up the signals of preset frequency. There are various models to choose from.
- Assessment of pressure and flows – Data loggers are used to record the pressure and flows. It is an instrument which stores the raw data electronically so as to be able to transfer it to the computer with a data cable link. Two types of portable data loggers are used either with a single channel or dual channel. In the absence of electronic equipment, the pressures can be ascertained by tapping and providing a pressure gauge. Flows can be assessed by using meters on a bypass line.
- Locating the leaks – Walking and sounding are the two general methods or their combination can be used to detect possible location of leakages. Leakage can be detected by walking over the main looking for tell-tale signs of presence of water. Whereas sounding is the cheapest and an effective method of detecting leaks in a medium-sized water supply system. The equipment used for detecting leakages is described in in 'Manual on Operation and Maintenance of Water Supply Systems' of CPHEEO.
- Assessment of leakage – To conduct tests for assessment of leak the following equipment are needed:
- Road measurer
- Pipe locator
- Valve locator
- Listening sticks or sounding rods
- Electronic sounding rods

Each method mentioned above is described in 'Manual on Operation and Maintenance of Water Supply Systems' of CPHEEO and can be referred during formulation of the strategy.

Water Quality Standards

The water quality standards as prescribed by the Bureau of Indian Standards are tabulated in Table 2.8.

Table 2.8: Organoleptic and Physical Parameters of Drinking Water

S.No.	Characteristics	Requirement (Acceptable limit)	Permissible limit in the absence of alternate source	Method of Test, ref to part of IS 3025	Remarks
1.	Colour, Hazen units, Max	5	15	Part 4	Extended to 15 only, if toxic substances are not suspected in absence of alterante sources
2.	Odour	Agreeable	Agreeable	Part 5	Test cold and when heated Test at several dilutions
3.	pH value	6.5-8.5	No relaxation	Part II	–
4.	Taste	Agreeable	Agreeable	Part 7 & 8	Test to be conducted only after safety has been established
5.	Turbidity, NTU, Max	1	5	Part 10	–
6.	Total dissolved	500	2000	Part 16	–

Source: IS 10500: 2012.

Note: It is recommended that the acceptable limit is to be implemented. Values in excess of those mentioned under 'acceptable' render the water not suitable, but still may be tolerated in the absence of an alternative source but up to the limits indicated under 'permissible limit in the absence of alternate source' in col. 4, above which the sources will have to be rejected.

Table 2.9: General Parameters Concerning Substances Undesirable in Excessive Amounts in Drinking Water

S.No.	Characteristics	Requirement (Acceptable limit)	Permissible limit in the absence of alternate source	Method of Ref. No.	Remarks
1.	Aluminium (as Al) mg/L. Max	0.03	0.2	IS 3025 (part 55)	–
2.	Ammonia (as total ammonia-N) mg/L. Max	0.5	No relaxation	IS 3025 (Part 34)	–
3.	Anionic, detetgents (as MBAS) mg/L., Max	0.2	1.0	Annex K of IS 13428	–
4.	Barium (as Ba), mg/L., Max	0.7	No relaxation	Annex F of IS 13428 or IS15302	–
5.	Boron (as B), mg/L. Max	0.5	1.0	IS 3025 (Part 57)	–

6.	Calcium (as Ca), mg/L., Max	75	200	IS 3025 (Part 40)	–
7.	Chlormines Cl_2), mg/L., Max	4.0	No relaxation	IS 3025 (Part 26 or ALPHA 4500–Cl.G)	–
8.	Chloride as (Cl)/ mg/L., Max	250	1000	IS 3025 (Part 32	–
9.	Copper (as Cu),mg/L., Max	0.05	1.5	IS 3025 (Part 42)	–
10.	Fluoride (as F) mg/L., Max	1.0	1.5	IS 3025 (Part 60)	
11.	Free residual	0.2	1	IS 3025 (Part 26)	To be applicable only when water is chlorinated. Tested at consumer end. When protection against viral infection is required, it should be minimum 0.5 mg/L.
12.	Iron (as Fe), mg/L. Max	0.3	No relaxation	IS 3025 (Part 53)	Total concentration of manganese (as Mn) and iron (as Fe) shall not exceed 0.3 mg/L.
13.	Magesium (as Mg.), mg/L., Max	30	100	IS 3025 (Part 46)	–
14.	Manganese (as Ms), mg'l, Max	0.1	0.3	IS 3025 (Part 59)	Total concentration of Manganese (as Mn) and iron (as Fe) shall not exceed 0.3 mg/L.
15.	Mineral Oil, Mg/L. Max	0.5	No relaxation	Clause 6 of IS 3025 (Part 39) Inframed partitioned method	–
16.	Nitrate (as NO_2) mg/L., Max	45	No relaxation	IS 3025 (Part 34)	–
17.	Phenolic compounds (as C_6H_3OH)	0.001	0.002	IS 3025 (Part 43)	–
18.	Selenium (as Se), mg/L., Max.	0.01	No relaxation	IS 3025 (Part 56) or IS 15303	–
19.	Silver (as Ag), mg/L., Max	0.1	No relaxation	Annex J of IS 13428	–
20.	Sulphate mg/L., Max	200	400	IS 3025 (Part 24)	May be extended to 400 provided that Magnesium does not exceed 30.

21.	Sulphide (as H_2S), mg/L., Max	0.05	No relaxation	IS 3025 (Part 29)	–
22.	Total alkalinity as Calcium, Carbonate, mg/L., Max	200	600	IS 3025 (Part 21)	–
23.	Total hardness (as $CaCO_3$), mg/L., Max	200	600	IS 3025 (Part 21	–
24.	Zinc (as Zn), mg/L., Max	5	15	IS 3025 (Part 49)	–

Source: IS 10500: 2012.

Table 2.10: Permissible Limits of Toxic Substances

S.No.	Characteristics	Requirement (Acceptable limit)	Permissible limit in the absence of source	Method of Ref. No.	Remarks
1.	Calcium (as Ca) mg/L. Max	0.003	No relaxation	IS 3025 (Part 41)	–
2.	Cyanide (as CN) mg/L. Max	0.05	No relaxation	IS 3025 (Part 27)	–
3.	Lead (as Pb), mg/L. Max	0.01	No relaxation	IS 3025 (Part 47)	–
4.	Mercury (as Hg), mg/L. Max	0.001	No relaxation	IS 3025 (Part 47) Mercury analyser	–
5.	Molybdenum (as Mo), mg/L. Max	0.02	No relaxation	IS 3025 (Part 2)	–
6.	Nickel (as Ni), mg/L. Max	0.07	No relaxation	IS 3025 (Part 54)	–
7.	Pesticides mg/L. Max	As given above	No relaxation	As given above	–
8.	Polychlorinated biphecnls, mg/L, Max	0.0005	No relaxation	ASTM 5175	–
9.	Polynuclear aromatic hydro-carbons (as PAH), mg/L. Max	0.0001	No relaxation	APHA 6440	–
10.	Total arsenic mg/L. Max	0.01	0.05	IS 3025 (Part 37)	–
11.	Tribalomethane:				
(a)	Bromoform, mg/L. Max	0.1	No relaxation	ASTM D 3973-85 or APHA 6232	
(b)	Dibromochloromethane mg/L. Max				
(c)	Bromodichloron ethane, mg/L. Max	0.06	No relaxation APHA 6232	ASTM D 3973-85 or APHA 6232	
(d)	Chloroform, mg/L. Max	0.2	No relaxation	ASTM D 3973.85 or APHA 6232	

Source: IS 10500: 2012.

Table 2.11: Pesticide Limits of Radioactive Substance

S. No.	Characteristics	Requirement (Acceptable limit)	Permissible limit in the absence of source	Method of test Ref. No.	Remarks
1.	Alpha emitters Bq/L, Max	0.1	No relaxation	Part 2	–
2.	Beta emitters Bq/L, Max	1.0	No relaxation	Part 1	–

Source: IS 10500: 2012.

Table 2.12: Pesticide Residues Limits of Drinking Water

S.No.	Pesticide	Limit ug/L	USEPA	AOAC/ISO
1.	Alachlor	20	525.2.507	–
2.	Atrazine	2	525.1.8141A	–
3.	Aidrin/Dieldin	0.03	508	–
4.	Alpha HCH	0.01	508	–
5.	Beta HCH	0.04	508	–
6.	Butachlor	125	525.18141A	–
7.	Chlorpyriphos	30	525.18141A	–
8.	Delta HCH	0.04	508	–
9.	24-Dichlorophe noxyacetic acid	30	515.1	–
10.	DDT (o.p and pp- Isomers of DDT, DDE and DDD)	1	508	AOAC 990.06
11.	Endosulfan (alpha, beta and sulphate)	0.4	508	AOAC 990.06
12.	Ethion	3	1657 A	–
13	Gamma- HCH (Lindane)	2	508	AOAC 900.06
14.	Isoproturon	2	508	–
15.	Malathion	190	532	–
16.	Methyl parathion	0.3	8141A	–
17.	Monocrotophos	1	8141 A	–
18.	Phorate	2		–

Source: IS 10500: 2012.

Table 2.13: Bacteriological Quality of Drinking Water

S.No.	Organisms	Requirements
1.	All water intended drinking:	
(a)	*E. Coli* or thermotolerant coliform bacteria	Shall not be detectable in any 100 mL. sample
2.	Treated water entering the distribution system	
(a)	*E. Coli* or thermotolerant coliform bacteria	Shall not be detectable in any 100 mL. sample
(b)	Total coliform bacteria	Shall not be detectable in any 100 mL. sample
3.	Treated water in the distribution system	
(a)	*E. Coli* or thermotolerant coliform bacteria	Shall not be detectable in any 100 mL. sample
(b)	Total coliform bacteria	Shall not be detectable in any 100 mL. sample

Source: IS 10500: 2012.

Table 2.14: Health Effects of Chemical Parameters

S. N.	Parameter	BIS Guideline value (maximum allowable)	General and health effect
1.	Total dissolved solids	2000 mg/l	Undesirable taste; gastro intestinal irritations; corrosion or incrustation
2.	pH	6.5-8.5	Affects mucous membrane; bitter taste; corrosion; affects aquatic life
3.	Alkalinity	600	Boiled rice turns yellowish
4.	Hardness	600	Poor lathering with soap; deterioration of the quality of clothes; scale forming; skin irritation; boiled meat and food become poor in quality
5.	Calcium	200 mg/l	Poor lathering and deterioration of the quality of clothes; incrustation in pipes; scale formation
6.	Magnesium	100 mg/l	Poor lathering and deterioration of clothes; with sulfate laxative
7.	Iron	0.3mg/l	Poor or sometimes bitter taste, color and turbidity; staining of clothes materials; iron bacteria causing slime
8.	Manganese	0.3 mg/l	Poor lathering and deterioration of clothes; with sulfate laxative
9.	Aluminium	0.2 mg/l	Neurological disorders; Alzheimer's disease
10.	Copper	1.5 mg/l	Liver damage; mucosal irritation, renal damage and depression; restricts growth of aquatic plants
11.	Zinc	15 mg/l	Astringent taste; opalescence in water; gastro intestinal irritation; vomiting, dehydration, abdominal pain, nausea and dizziness
12.	Ammonia	0.5 mg/l	Indicates pollution; growth of algae
13.	Nitrite		Forms nitrosoamines which are carcinogenic
14.	Nitrate		Blue baby disease (methemoglobineamia); algal growth
15.	Sulfate		Taste affected; laxative effect; gastro intestinal irritation

16.	Chloride		Taste affected; corrosive
17.	Fluoride		Dental and skeletal fluorosis; non-skeletal
18.	Phosphate		Algal growth
19.	Arsenic	0.05 mg/l	Toxic; bio-accumulation; central nervous system affected; carcinogenic
20.	Mercury	0.001 mg/l	Highly toxic; causes 'minamata' disease- neurological impairment and renal disturbances; mutagenic
21.	Cadmium	0.003 mg/l	Highly toxic; causes 'itai-itai' disease-painful rheumatic condition; cardio vascular system affected; gastro intestinal upsets and hyper tension
22.	Lead	0.01 mg/l	Causes plumbism-tiredness, lassitudes, abdominal discomfort, irritability, anaemia; bio-accumulation; impaired neurological and motor development, and damage to kidneys
23.	Chromium	0.05 mg/l	Carcinogenic; ulcerations, respiratory problems and skin complaints
24.	Pesticide		Affects central nervous system
25.	Detergent		Undesirable foaming

LAND REQUIREMENT FOR WATER SUPPLY SYSTEM

Identification of land for water infrastructure is an essential parameter. It is necessary to earmark land for Water Treatment Plant (WTP) in Development Plans. Also land for overhead reservoirs, pumping stations should be marked at the zonal level or local area plan level. However, land requirement may vary based on the capacity of WTPs and up?gradation of technology. Below is the table, which recommends land requirement based on different capacity.

Table 2.15: Recommended Land Requirement Based on Capacities

S.No.	Identified Capacities	Land Requirement (Hectares)
1.	5 MLD	0.10
2.	10 MLD	0.19
3.	50 MLD	0.93
4.	100 MLD	1.87
5.	200 MLD	3.73
6.	500 MLD	9.34

Source: CPHEEO.

Note: Additional 0.63 ha of land can also be considered for staff quarters

Table 2.16: Design Period for Water Supply Components

S.No.		Components	Design period (in years)
1.		Storage by Dams	50
2.		Infiltration works	30
3.		Pumping	
	i.	Pump house (civil works)	30
	ii.	Electric motors and pumps	15
4.		Water treatment units	15
5.		Pipe connection to several treatment units and other small appurtenances	30
6.		Raw water and clear water conveying mains	30
7.		Clear water reservoirs at the head works, balancing tanks and service reservoirs (overhead or ground level)	15
8.		Distribution system	30

Source: Manual on Water Supply and Treatment, 1999, CPHEEO.

Water Quality Index (WQI)

A Water Quality Index provides a single number (like a grade) that expresses overall water quality at a certain location and time based on certain quality parameters. The objective of an index is to turn complex water quality data into information that is easily understandable and useable by the public.

The use of such an index to grade water quality is a controversial issue among water quality scientists. This is because a single number cannot tell the whole story of water quality. However, a water index based on some very important parameters can provide a simple indicator of water quality. It gives the public a general idea of the possible problems with water in the region.

The purpose of finding Water Quality Index (WQI) is to decide whether a water body is healthy or not. The parameters are selected on the basis of their impact on overall quality of water and health effects. Guidance is taken from IS 10500 : 2012.

Calculating Water Quality Index

For calculating WQI, Nine (9) important parameters are chosen and weight (importance) was attached to each, as mentioned below:

Table 2.17: Water Quality Index

S.N.	Parameter	Unit	Weight
1.	Dissolved Oxygen (DO)	% Saturation	0.17
2.	Faecal coliform	Colonies per 100 ml	0.16
3.	pH	—	0.11
4.	Biological Oxygen Demand (BOD)	mg/L	0.11

5.	Temperature	°C	0.10
6.	Total Phosphate	mg/L	0.10
7.	Nitrates	mg/L	0.10
8.	Turbidity	NTU	0.08
9.	Total Solids	mg/L	0.07
	Total		**1.00**

* *NTU- Nephelometric Turbidity Unit*

After completion of water quality tests for all the nine parameters, the results are recorded and transferred to a weighting chart or curve where a numerical value called Q-value is obtained for each parameter.

Now the Q-value is multiplied by the weighting factor. The nine resulting values are then added to arrive at an overall WQI-value.

The Water Quality Index is evaluated on a scale of 0 to 100 (zero to hundred). The 100-point index is divided into several ranges corresponding to the general descriptive terms shown in the following table.

Table 2.18: Water Quality Index Legend

Range	Quality
90 to 100	Excellent
70 to 90	Good
50 to 70	Medium
25 to 50	Bad
0 to 25	Very Bad

The Q-Values: A Q-value is a way of standardizing all the different water quality test results so that they can be combined and used in calculating WQI.

A Q-value is a way of standardizing all the different water quality test results so that they can be combined and used to find an overall water quality value for the river. You can think of the Q-value like a score on a test. Less than 50 is like a failing grade, whereas 90 or more is like an "A."

Use the following charts to determine the Q-value for each of the water quality tests you perform. Find your water quality value on the x-axis. Determine where that value intersects the curved line on the graph, then follow it horizontally across to the y-axis and read the Q-value.

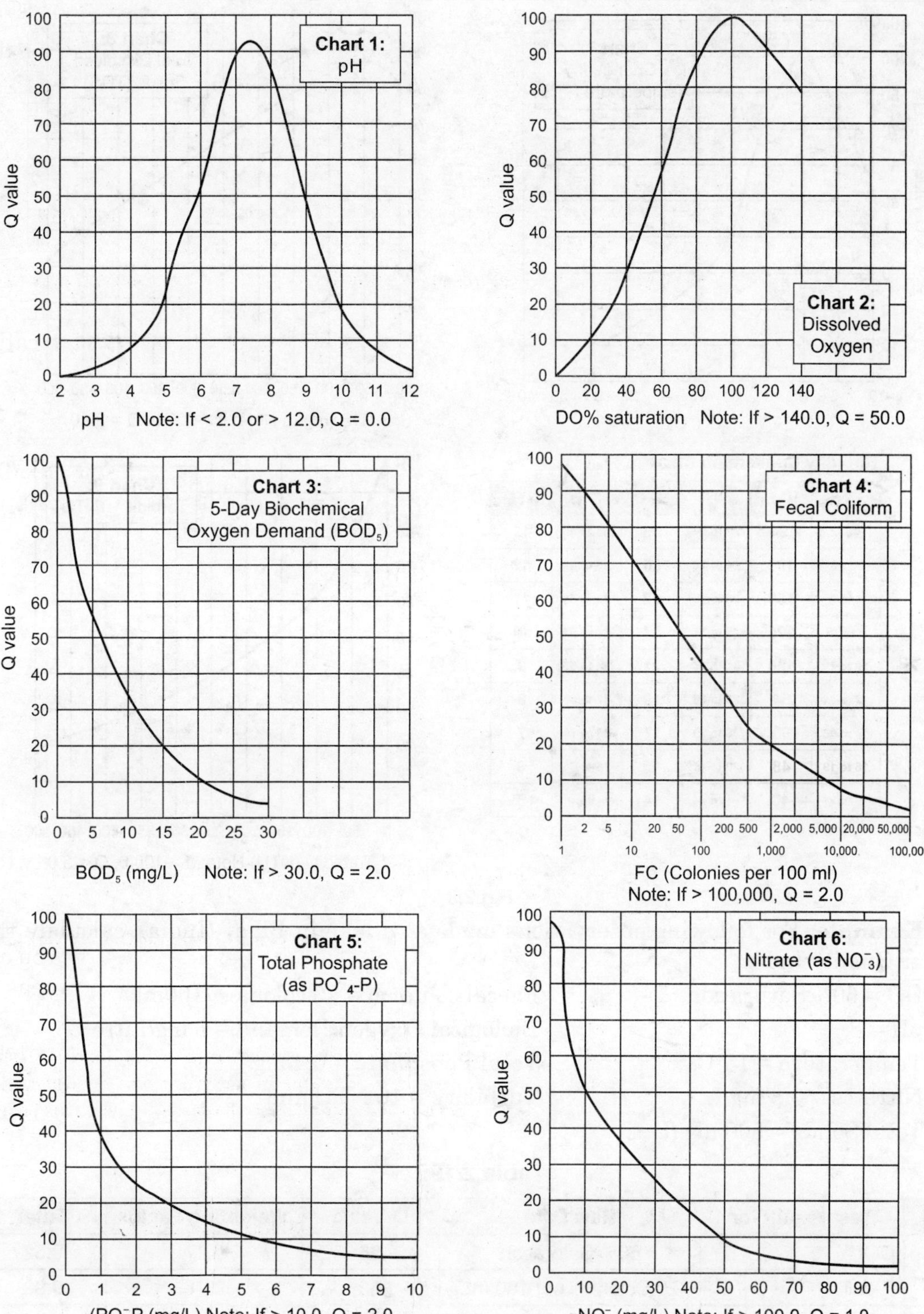
Chart 1:
pH
Q value
pH Note: If < 2.0 or > 12.0, Q = 0.0
Chart 2:
Dissolved
Oxygen
DO% saturation Note: If > 140.0, Q = 50.0
Chart 3:
5-Day Biochemical
Oxygen Demand (BOD5)
BOD5 (mg/L) Note: If > 30.0, Q = 2.0
Chart 4:
Fecal Coliform
FC (Colonies per 100 ml)
Note: If > 100,000, Q = 2.0
Chart 5:
Total Phosphate
(as PO-4-P)
(PO-4 P (mg/L) Note: If > 10.0, Q = 2.0
Chart 6:
Nitrate (as NO-3)
NO-3 (mg/L) Note: If > 100.0, Q = 1.0

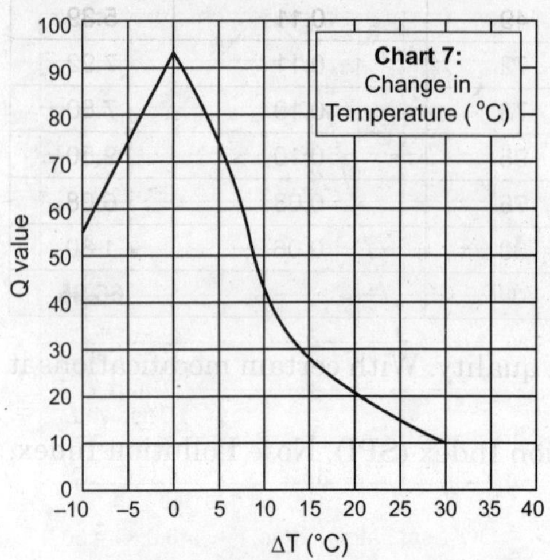

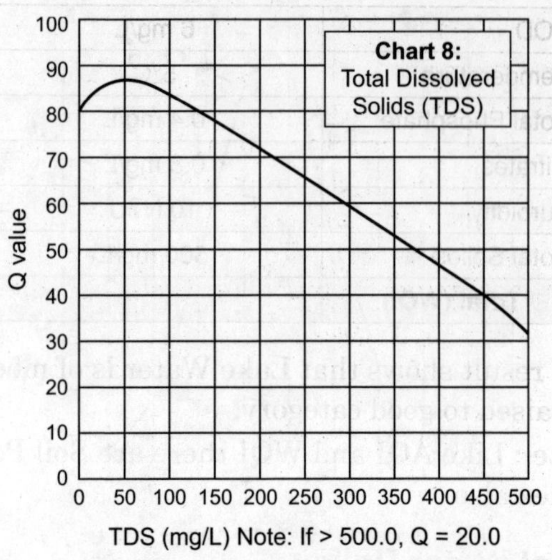

Turbidity Conversion Chart

Cm	NTU	Cm	NTU	Cm	NTU
< 7	> 240	21 to 24	35	44 to 46	13
7 to 8	185	24 to 26	30	46 to 49	12
8 to 9	150	26 to 29	27	49 to 51	11
9 to 10	120	29 to 31	24	51 to 54	10
10 to 12	100	31 to 34	21	54 to 57	9
12 to 14	84	34 to 36	19	57 to 60	8
14 to 16	60	36 to 39	17	60 to 70	7
16 to 19	48	39 to 41	15	70 to 85	6
19 to 21	40	41 to 44	14	> 85	< 5

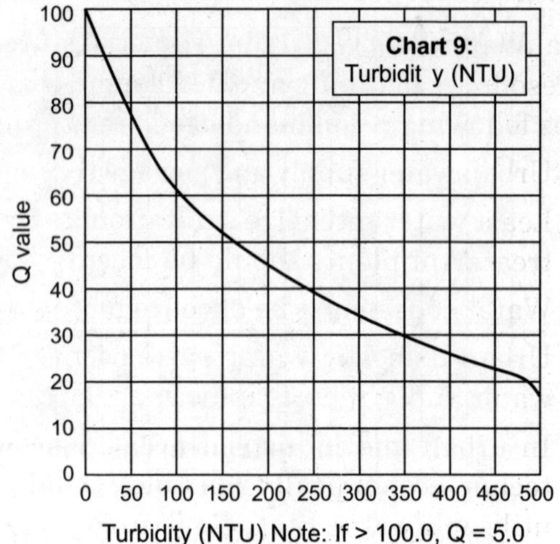

Fig.2.3

Example: For following observations made at a lake in Bihar (India), calculate the Water Quality Index.

DO = 60% Saturation,
pH = 8,
Temperature = 15°C,
Nitrates = 0.8 mg/L,
Total Solids = 500 mg /L
Faecal coliform = 20 colonies /100ml
Biological Oxygen Demand = 6 mg/Litre
Total Phosphate = 0.4 mg/L
Turbidity = 10 NTU and

Table 2.19:

S.N.	Test Result for	Raw Data	Q-Value	Weighting Factor	Total
1.	DO	60% Saturation	56	0.17	9.52
2.	E-coli	20 Colonies per 100 mL	62	0.16	9.92
3.	pH	8	82	0.11	9.02

4.	BOD	6 mg/L	49	0.11	5.39
5.	Temperature	5°C	72	0.11	7.92
6.	Total Phosphate	0.4 mg/L	78	0.10	7.80
7.	Nitrates	0.8 mg/L	95	0.10	9.50
8.	Turbidity	10 NTU	76	0.08	6.08
9.	Total Solids	500 mg/L	30	0.06	1.80
	Total (WQI)				**66.95**

The result shows that Lake Water is of medium quality. With certain modifications it can be raised to good category.

Note : Like AQI and WQI there are Soil Pollution Index (SPI), Nose Pollution Index, U-V Index also.

National Water Policy

National Water Policy, 2012 by Ministry of Water Resources was formulated to govern the water resource's planning and development along with its optimum utilisation. The Policy provides following recommendations regarding urban settlements:

- Urban water supply and sewage treatment must be integrated along with its billing,
- Least water intensive sanitation and sewerage systems with decentralized sewage treatment plants should be incentivized.
- Water reuse must be encouraged in urban settlements,
- Urban domestic water systems need to collect and publish water accounts and water audit reports indicating leakages and pilferages, which should be reduced.
- In urban and industrial areas, rain water harvesting and desalinisation where techno-economically feasible, should be encouraged to increase availability of utilizable water. State Governments shall draft State Water Policy for decentralised or micro level water management to yield better results.

Rain Water Harvesting

The term water harvesting refers to collection and storage of natural precipitation and also other activities, aimed at harvesting surface and groundwater, prevention of losses through evaporation and seepage, and all other hydrological studies and engineering interventions, aimed at conservation and efficient utilization of the limited water endowment of a physiographic unit, such as a watershed. The various options, which can be covered under water harvesting are the following:

- Construction of storage structures for "in situ" collection of rain water.
- Farm ponds for collection of runoff, either for supplemental irrigation or for augmentation of groundwater.
- Check dams for storing runoff in first order (small) streams.
- Percolation tanks at appropriate sites for augmentation of groundwater.

- Construction of ponds and reclamation/revitalization of traditional water arresting structures.
- Artificial recharge through wells.
- Reduction of evaporation from surface water bodies and soils.
- Prevention of seepage losses in appropriate situations.
- Enhancement of runoff through mechanical arid chemical treatment in catchmentareas.
- Sub-surface dams to arrest base flow of groundwater.
- Soil and water conservation practices comprising contour and terrace bunding.
- Control of seawater incursion in coastal aquifers/prevention of sub-surface dischargeof groundwater to sea.
- Control of transpiration without affecting normal plant growth.

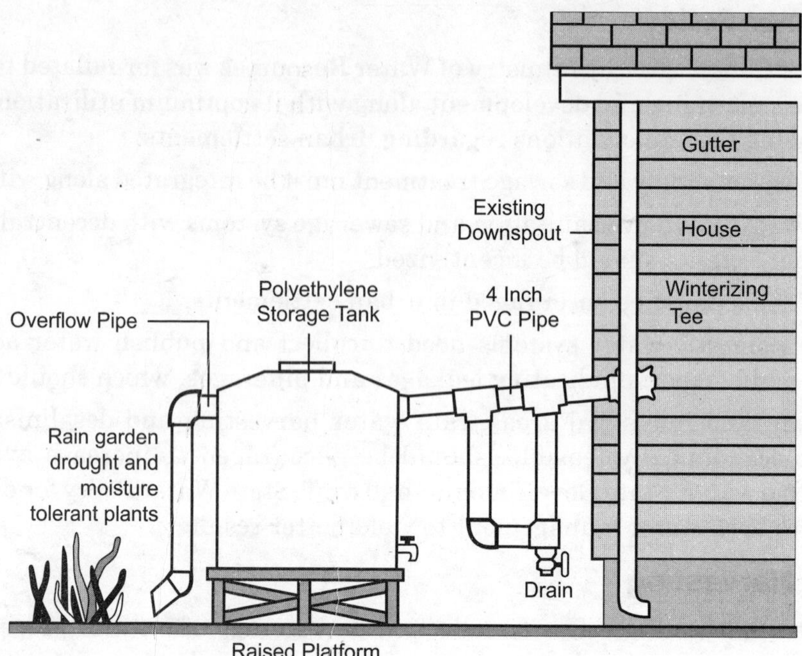

Fig.2.4: Rainwater Storage Tank Detail

Artificial Recharge Techniques

A variety of methods have been developed to recharge Groundwater. Most of the artificial recharge techniques are briefly described below:

1. Direct surface techniques, by:
   - Flooding
   - Basins or percolation tanks
   - Stream augmentation

- Ditch and furrow system
- Over irrigation

2. Direct sub surface techniques, by:
   - Injection wells or recharge wells
   - Recharge pits and shafts
   - Dug well recharge
   - Bore hole flooding
   - Natural openings, cavity fillings.
3. Combination surface – sub-surface techniques, by:
   - Basin or percolation tanks with pit shaft or wells.
4. Indirect Techniques, by:
   - Induced recharge from surface water source.
   - Aquifer modification

The 'Manual on Artificial Recharge of Groundwater', prepared by Ministry of Water Resources, Central Groundwater Board (CGWB) gives the following options for artificial groundwater recharge:

1. Ditch and Furrow Method
2. Lateral Ditch Pattern
3. Dendritic Pattern
4. Contour Pattern
5. Spreading Basin or Percolation Tanks
6. Gully Plug / Check Dam / Nala Bund / Gabbion Structures
7. Dug Well Recharge
8. Recharge Shafts / Pits / Trenches
9. Artificial Recharge through Injection Well
10. Induced Recharge from Surface Water Sources
11. Subsurface Dykes / Underground Bandharas

Source: www.cgwb.gov.in

While the various options given above differ in terms of their connotations, the common conceptual thread is that they describe water management on a small or micro-scale, which is different from construction of major dams or river diversions and canal networks for irrigation of command area.

Fig.2.5: Rainwater Harvesting and Grey Water Recycling: In most of the regions in India, water is a critical problem. By rainwater harvesting, wastewater recycling, primary treatment and checking of leakages, the problem can be litigated to a great extent.

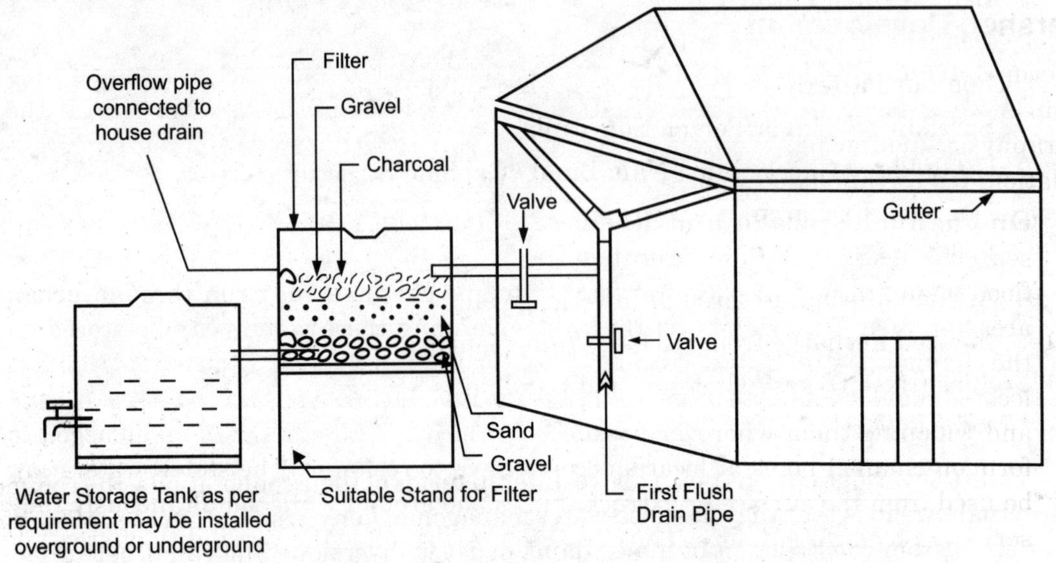

Fig.2.6: Rainwater can be Stored in Tanks for Drinking and Cooking

When untreated effluents are discharged into the ground, these contaminated waters are absorbed by the soil and thus the Groundwater is polluted. Large-scale exploitation of water lowers the water level and dries the soil. This makes the maximum absorption of the domestic sewage and industrial effluents including pesticides causing the groundwater pollution.

No other method will be efficient to decrease the solid content except the membrane separation technique. The groundwater may be recharged with the rainwater with the help of percolation wells.

The high-level contents of the parameters observed may be minimised if the groundwater is recharged with the available storm water in the rainy season. This not only dilutes the constituents of the groundwater but also raises the groundwater level that depletes due to large-scale exploitation.

Consider a building with flat terrace area of 125 sq.m The average annual rain fall is say 750 mm. Suppose there is no loss of water from terrace floor, then in one year, there will be rainwater on the terrace floor to a height of 750 mm.

Area of Terrace = 125 sq. m

Height of Rainfall = 750 mm-0.75 m

Volume of Rainfall = 125×0.75

= 93.75 cum or 93.750 litres.

Assuming that only 60% of the total rainfall is effectively harvested.

Volume of Water harvested = $93.750 \times 0.60 = 56{,}250$ litres.

A family of six needs about 21,900 litres of water per year.

Thus, harvested water of 56,250 litres is more than the double required by the family of six.

Watershed Development

The term watershed refers to a topographically distinctive area, which is drained by a stream. A watershed can be of a few hundred to few thousand hectares in size. If the topography is undulating, the area of the watershed is small and it increases as the undulation in a terrain is reduced.

- **On-channel Storage and Recharge of Storm Water Channels:** Several seasonal streams become activated during the monsoons providing the outlet to flood waters as well as local area runoff. The channels may run through urban areas or their hinterlands. With careful planning these waters can be stored on the channel itself. The waters can be detained through construction of appropriate located regulators. The storage capacity of channels can be enhanced by deepening and widening them wherever possible. Storm water channels can he enlarged to form on-channel lakes at locations conductive to recharge. The stored waters can be used from the surface or extracted from the ground while simultaneously they serve to recharge aquifers. These reservoirs also offer the possibility of creating rich urban bio-diversity habitats.
- **Off-channel Storage for Floodwaters:** As all storm water cannot be stored on channel, possibilities exist for the creation of off-channel reservoirs linked to the main channel. The creation of off-channel reservoirs, especially wherever a suitable recharge area exists in the neighbourhood of a channel, is a useful device for storing water for the lean season. These reservoirs also offer the possibility of creating rich urban bio-diversity habitats.

- **Storage in Lakes and Depressions:** In urban areas, lakes have often been misused for dumping of solid waste and sewage disposal. The storage capacity of these can be increased by desilting and widening. The water quality would need to be improved by replicating the principles of wetlands by the systematic induction of aquatic plants and fish. This is a very low cost treatment and subsequently only clean water would enter the underlying aquifer. The groundwater table is the most economical form of storage. In that there is no expense in construction of reservoirs and water can thereafter be extracted from the aquifer by battery of shallow tubewells for a long duration extending into the lean season. The yet - untapped, potentials of the wetland habitats comprising of oxbow lakes, marshes, low lying lands on the floodplains, lakes, estuarine lands, abandoned quarries, artificial lakes over good recharge zones for the sustenance of not only the ecology but for other activities such as recreation, flood prevention, water storage, water extraction and natural water treatment.
- **Floodplain Reservoirs for Conjunctive Extraction:** Very often the most copious fresh water aquifers underline the floodplains. In urban areas, there is no place to store the monsoon discharge of the river due to lack of topographical relief. To overcome this difficulty simple reservoirs on the floodplain can be created by scooping out earth at appropriate locations and letting these reservoirs to be filled up by the expanded monsoon season flow of the river.

Watershed development, bio-drainage, conservation/revival of village ponds, check dams, water, gully control structures, contour bunding, ridges and furrows, rainwater harvesting, micro-irrigation, etc. are essential for sustainable water and drainage management in rural areas.

In order to adopt appropriate measures for augmenting ground recharge, it is necessary to spread water harvesting techniques by construction of check dams arid percolation tanks, besides direct injection methods utilising the abandoned structures available in large numbers in the hard rock region. Before undertaking direct recharge into such abandoned structures, suitability of the structure for receiving recharge and also the quality of source water should be thoroughly examined to confirm the technical feasibility. Rain water harvesting and groundwater recharge can be suitably dovetailed with watershed development programmes aimed at soil and water conservation by arresting the run-off. This involves several actions as outlined below:

- Arresting groundwater decline and improving its accessibility
- Benefiting water quality in aquifers
- Preventing surface water run-off during monsoons
- Enhancing availability of surface water at a specific place and time
- Reducing power consumption
- Conserving urban waste water.

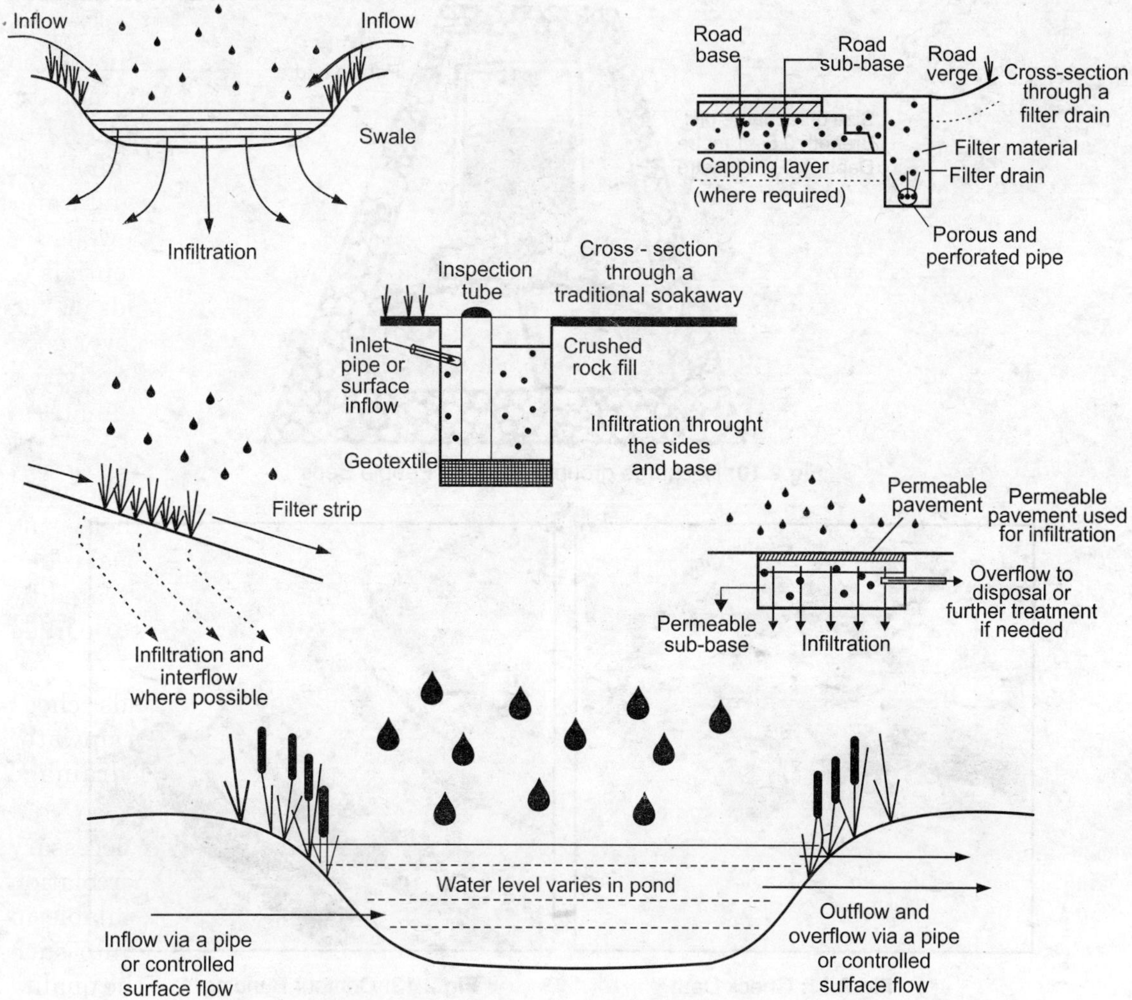

Fig.2.7: Components of Sustainable Drainage System

Source: www.environment-agency.gov.uk/suds/

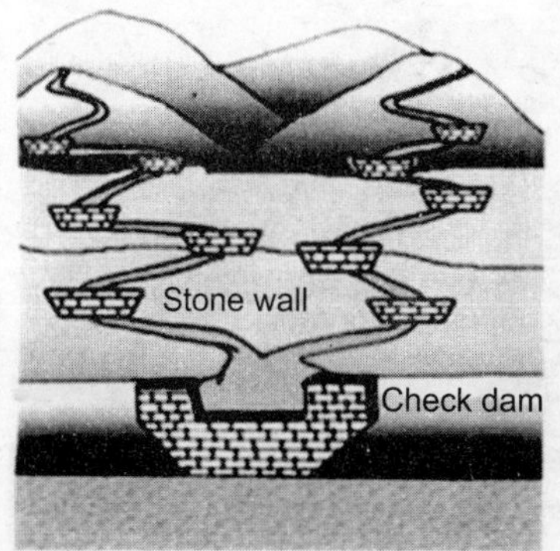

Fig.2.8: Rainwater harvesting through watershed management

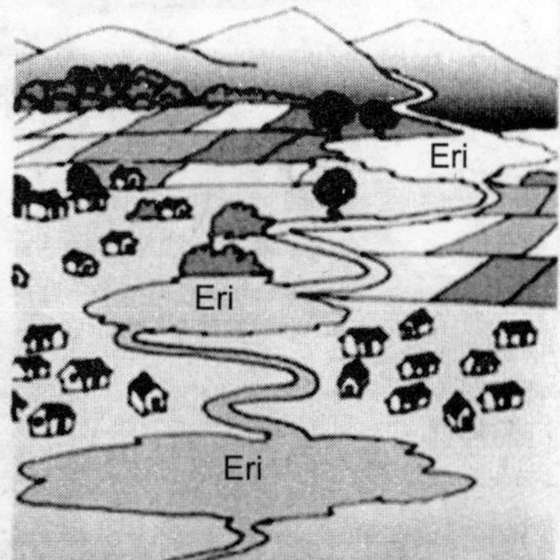

Fig.2.9: Traditional water harvesting through Eris (lakes)

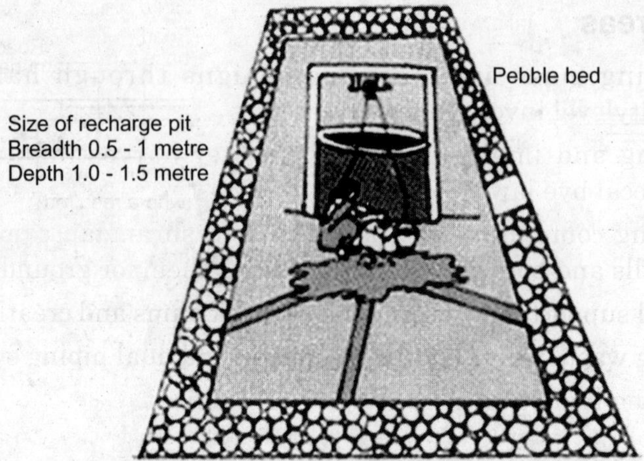

Fig.2.10: Recharge ground Water by Pebble Beds

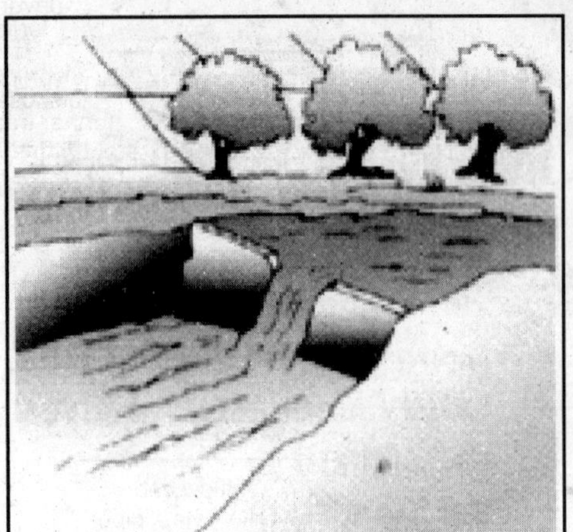

Fig.2.11: Check Dam

Fig.2.12: Contour Bunding

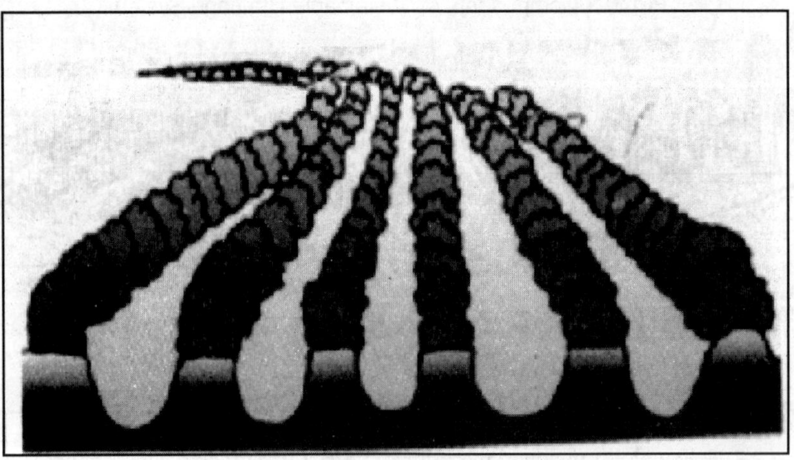

Fig.2.13: Ridge Sand Furrows Gully Control Structure and Trenching

Key Action Areas

- Organising mass awareness campaigns through hand-outs, posters and communitylevel involvement.
- Practising and implementing rain water harvesting and ground recharging throughlocal bye laws.
- Motivating communities to participate in shramdan camps for cleaning ponds, tanks,wells and borewells; and reutilising them for groundwater storage.
- Financial support for construction of check dams and creation of catchment ponds.
- Recycling waste water subject to quality and dual piping system.
- Recharging industrial effluents after treatment.

Guidelines for Water Conservation

- Redesigning cisterns/tanks/reservoirs for conservation of water, e.g. traditional 3.5 gallon toilet cistern can be replaced by a 5 to 7 litres one with equal effect.
- Using urban waste water effluents after suitable treatment for horticulture etc.
- Recycling of waste water.
- Groundwater recharge by bio-swales, rainwater harvesting, porous paving and artificial recharge techniques
- Conserving monsoon water in Johads, check dams, percolation tanks, village ponds, etc. and their regular desilting.
- Using existing or defunct wells and tube wells for purpose of recharging and rainwater harvesting.

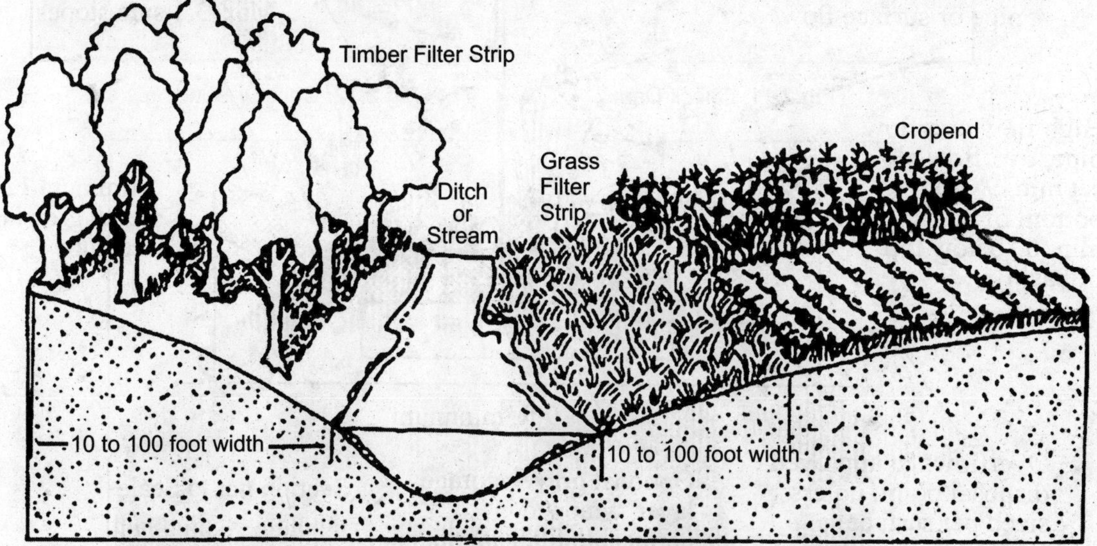

Fig.2.14: Vegetatives Filter Strips.

Stormwater inflow
Basin with grassed surface
Geotextile fabric
Overflow to street drain
Gravel
Subsoil
Infiltration to subsoil
Water table

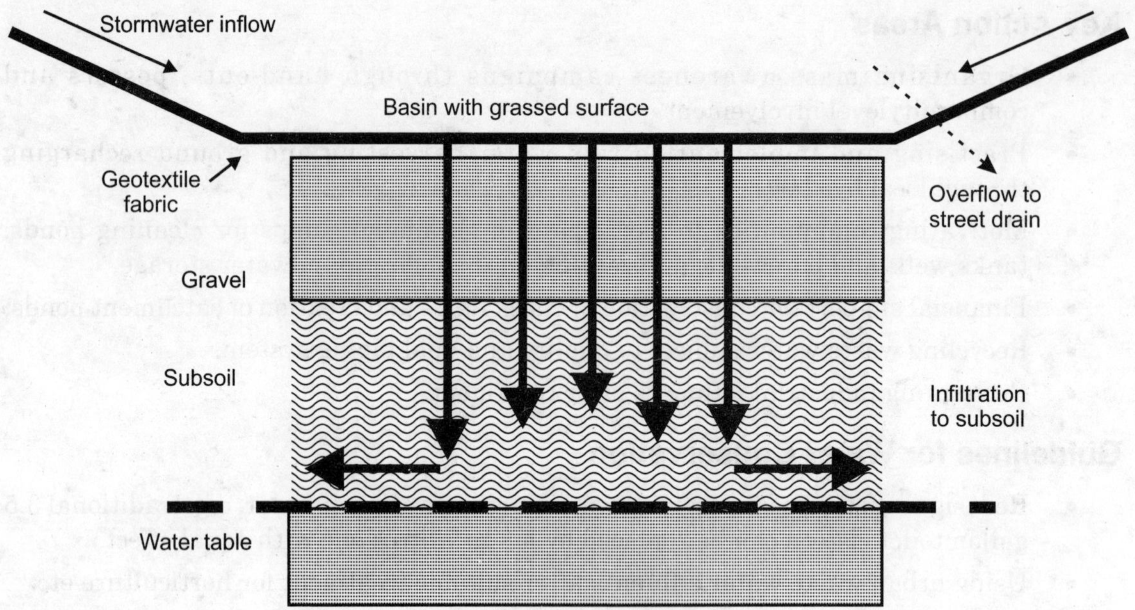

Fig.2.15: Detail-Filter Strip.

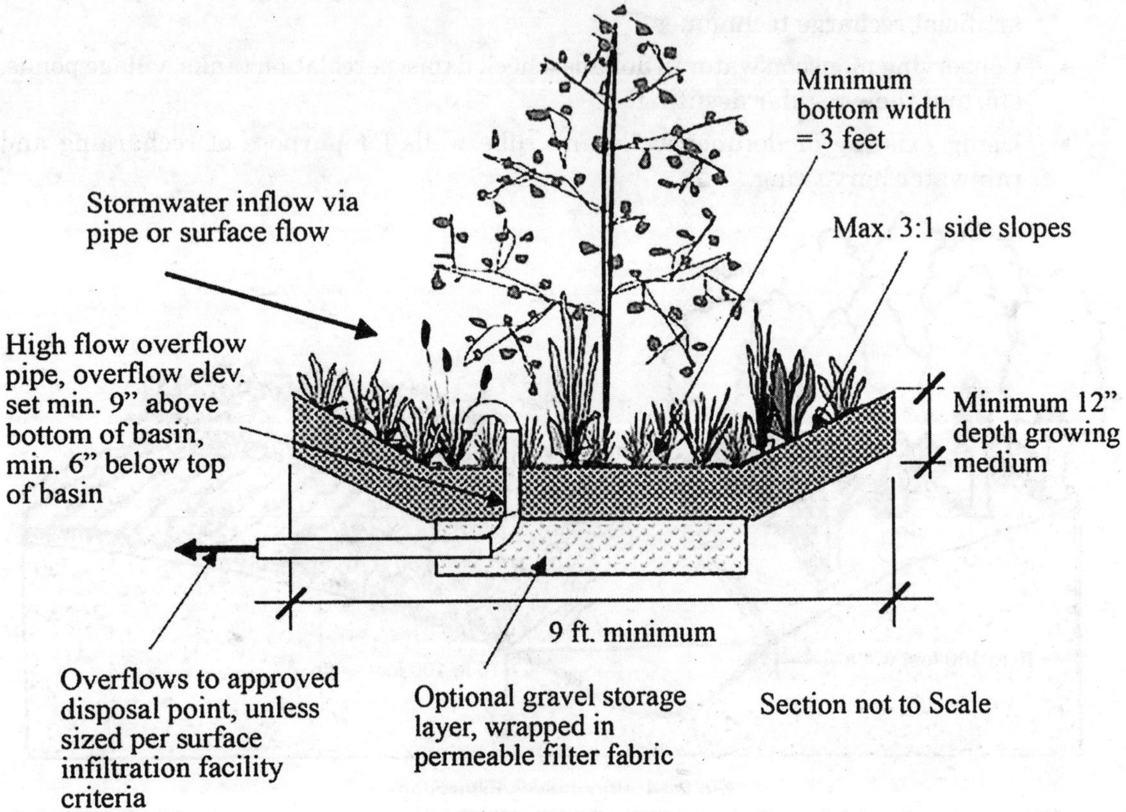

Minimum bottom width = 3 feet
Stormwater inflow via pipe or surface flow
Max. 3:1 side slopes
High flow overflow pipe, overflow elev. set min. 9" above bottom of basin, min. 6" below top of basin
Minimum 12" depth growing medium
9 ft. minimum
Overflows to approved disposal point, unless sized per surface infiltration facility criteria
Optional gravel storage layer, wrapped in permeable filter fabric
Section not to Scale

Fig.2.16: Infiltration Basin.

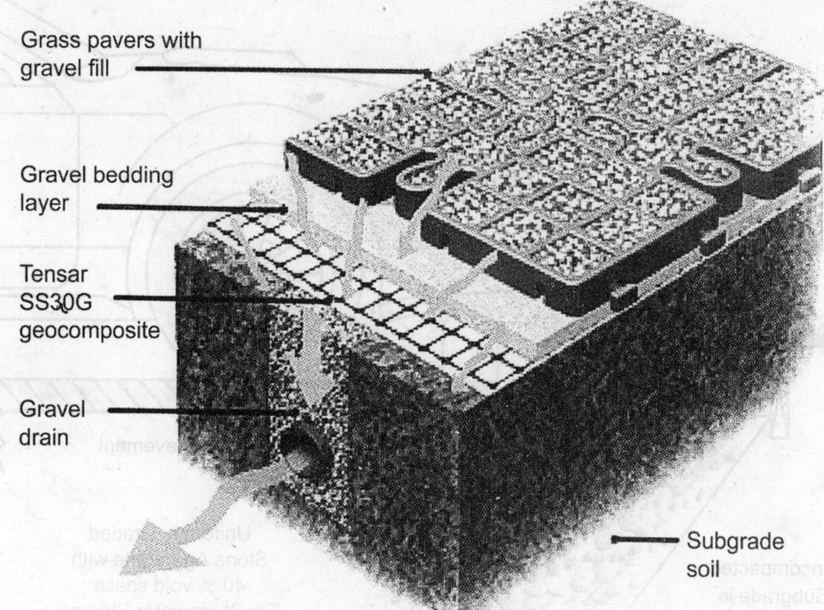

Fig.2.17: On-site Retention of Water (porous paving)

Fig.2.18: Rainwater Harvesting in an Infiltration Trench

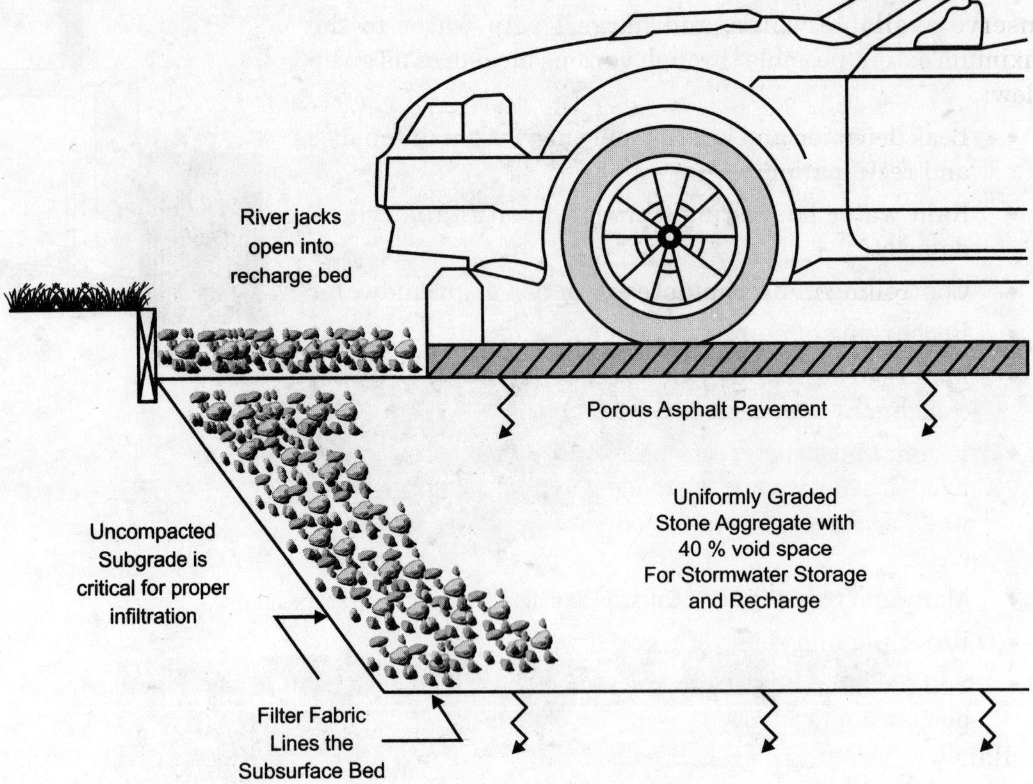

River jacks
open into
recharge bed
Porous Asphalt Pavement
Uniformly Graded
Stone Aggregate with
40 % void space
For Stormwater Storage
and Recharge
Uncompacted
Subgrade is
critical for proper
infiltration
Filter Fabric
Lines the
Subsurface Bed

Fig.2.19: Roadside Recharge Bed

Surface Runoff
Clean stone
1 to 2 metres
(50 mm diameter)
150 mm
Non-woven
Filter Fabric
150 mm - 300 mm
Sand Filter

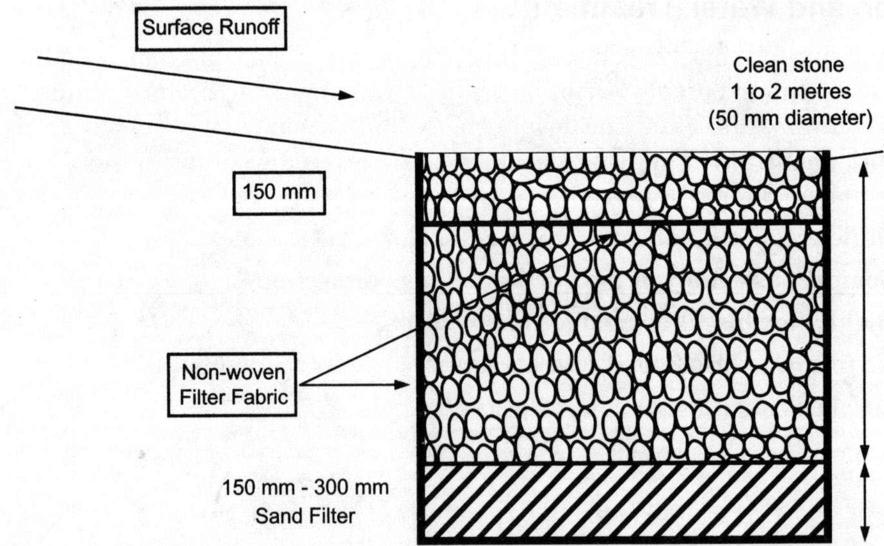

Fig.2.20: Section of Infiltration Trench

In view of the fast growing population and urbanisation, coupled with diminishing water resources, it is necessary to conserve available water, and harvest rain water to the maximum extent possible through various measures as given below:

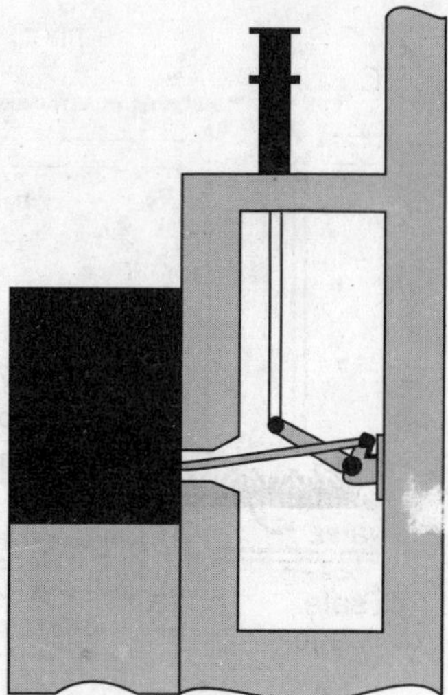

Fig.2.21: Schematic Diagram of the Flap Valve.

- Leak detection and water supply network maintenance and rectification.
- Rain water harvesting, through in-situ tanks check dams etc.
- Controlling indiscriminate extraction of groundwater
- Recharging of aquifers.
- Meters to be installed along trunk mains and for bulk consumers and tariff review.
- Percolation structures, wherever suited, be encouraged in open grounds and roadside will help recharge bore-wells and provide water for gardening, car washing etc.
- Mandatory system of roof-top rain water harvesting.
- Recycling and reuse of treated waste water
- Soft parking, landscape and pavements filled with pebbles and coarse sand to allow percolation.

Rain water falling on roof is collected and injected into groundwater reservoir through abandoned dug well, had pump or in a water storage tank.

Filtration and Water Treatment

Surface and sub-soil water is generally impure which need primary and secondary treatment to make it useful or potable. Drinking water must meet the highest quality, purity and hygiene requirements. Thus, the development of filtration technology has always been one of our most important. The filter design constitutes an optimum solution for both single and multiplayer filters:

- Outlet regulation: best conditions for direct filtration.
- Compact assembly of the back wash ing connections.
- Highly flexible filter media combinations.
- No filter media losses.
- Low backwash water consumption.

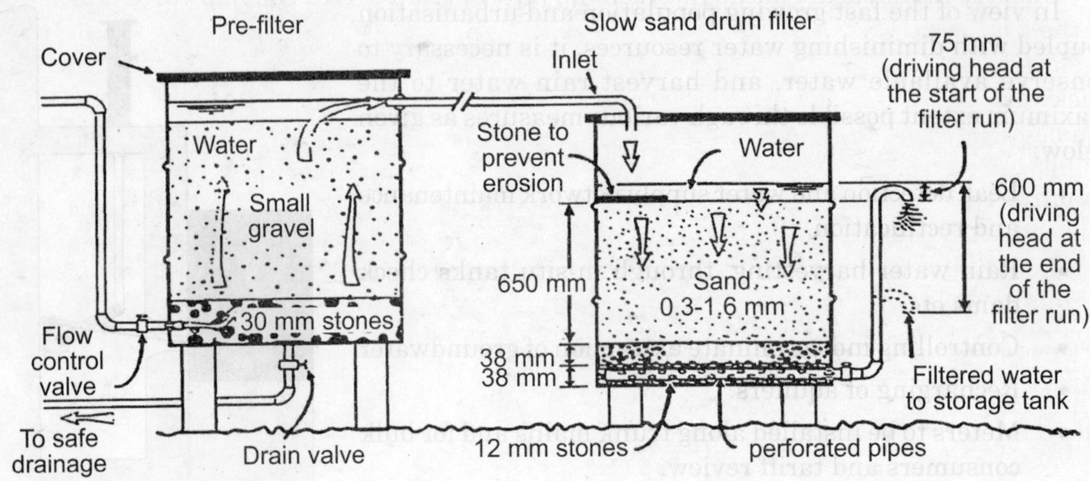

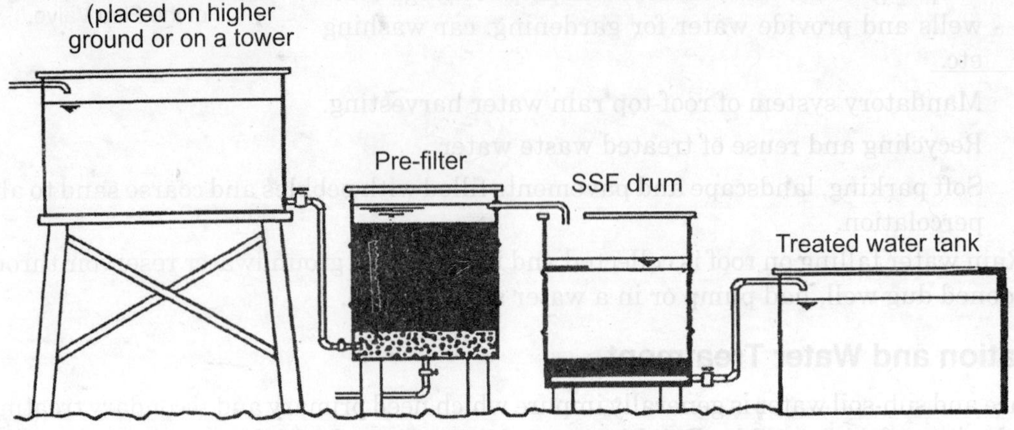

Fig.2.22: Slow Sand Drum Filter and Pre-filter.

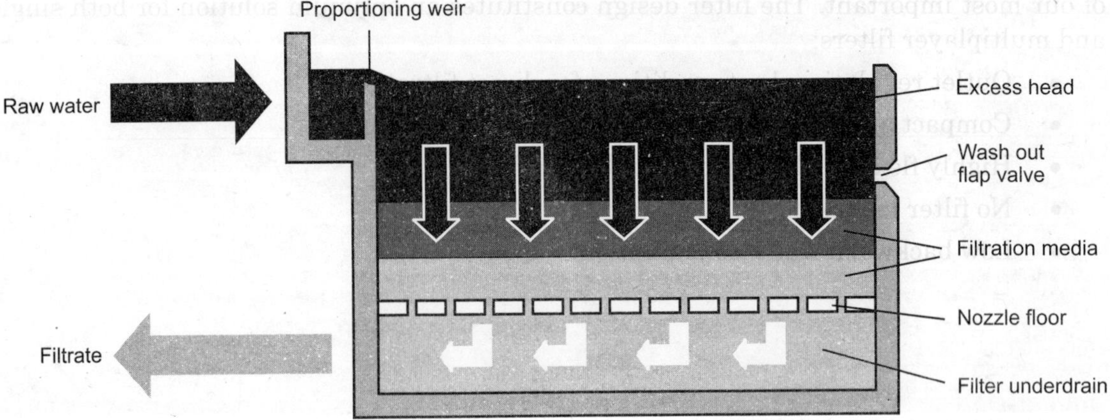

Fig.2.23: Schematic Diagram of the Filter Cell

One essential feature of the filters is the internal flap valve. The flap valve is situated at the front of the filter cell, just above the top level of the filter medium. The pneumatic or electric actuation is equipped with a positioner for the phased opening of the flap valve. Backwashing is based on the excess head principle, which means that the rising process can be precisely matched to individual operating conditions.

There are several packaged systems such as small, high rate, pressurized sand filter with backwash. A carbon fitler may also be incorporated.

Table 2.20: Comparison of Slow Sand Filters and Rapid Filters

	Slow sand filter (SSF)	Rapid filter
Filter medium	Fine	Coarse
Filtration rate	Low: 0.1-0.3 m/h	High: 5-15 m/h
Mechanisms for the removal of impurities	Sedimentation, adsorption, straining, chemical and microbiological processes	Sedimentation, adsorption, straining, chemical and microbiological processes
Main mechanism for removal of impurities	Microbiological	Physical mechanisms, especially adsorption
Principal application	Treatment of low turbidity water (<20 NTU) for distribution as potable water-post-chlorination may still be carried out	Treatment of low turbidity water-chlorination is required before distribution. Treatment of high turbidity raw water and following flocculation and sedimentation.
Principal advantages	Substantially reduces pathogenic bacteria, viruses and cysts, to produce a potable water without further treatment. No machinery required	Relatively small and compact
Principal disadvantages	Can only effectively treat low turbidity water	Cannot produce a potable water without further treatment. Backwashing water required to clean filter-this usually involves pumps.

Table 2.21: Advantages and Disadvantages of Chlorine Use as a Disinfectant

Advantages	Disadvantages
It comes in several forms: powder, granules, liquid, gas	It is powerful oxidizing agent which must be handled carefully do not breathe chlorine fumes
It is usually readily available and relatively cheap in one form or another	It does not effectively penetrate particulate matter
It dissolves easily Residual chlorine remaining in treated water provides some protection against further contamination	It can give an unpleasant taste when slightly overdosed, which could dissuade people from drinking a safe water.
It is effective against a wide range of pathogens	Its effectiveness against some pathogens-cysts, ova, viruses-requires higher chlorine concentrations and a longer contact time

REVIVING INDIGENOUS SKILLS AND COMMUNITY PARTICIPATION

Indians over centuries developed a range of techniques to harvest every possible form of water, from rainwater, streams, rivers and flood water. They have tapped water from hill streams of springs known as kuhls ranging in lengths from one to 15 kms and carrying a discharge of 15-100 liters per second. In Meghlaya, a 200-year-old system of tapping stream and spring water for irrigating plants by using bamboos still exists. The "kundis" of Rajasthan are unique structures which look like huge concrete saucers and are used for collecting rain water to meet the needs of local villagers and their animals. In Rajasthan, particularly Jodhpur, traditional water system still exists and village people maintain their traditional system even after they started getting piped water. Villages that neglected their traditional systems and relied solely on piped water sources faced scarcity under drought conditions, particularly when the canals dried up. An ironic contrast of water management are Jaisalmer and Cherrapunji - Jaisalmer gets 100 mm of annual rainfall while Cherrapunji gets 15000 mm - yet Jaisalmer was able to collect enough water for itself until recent years, while Cherrapunji, the wettest place on earth faces a drinking water shortage. Societies must learn to perform their roles in water harvesting, and this is where the local administration must take responsibility and ensure that villagers are fully involved in such projects.

In the desert towns of Rajasthan roof top water harvesting was a common practice. The sloping roofs of the houses were used to collect the rain water which was taken through a pipe into underground 'tanks' (tank) sometimes as big as a room built in the main house or the courtyard. This would meet the drinking water needs of the household. Sometimes rooms would be built next to the tank, which used to be coolest in the summer. Another traditional rainwater harvesting systems in Rajasthan was the kund. It was a simple saucer shaped structure, having a circular depression, of about 15 mt. diameter, whose sides were plastered with lime and ash. As water would naturally flow down to the centre along all the sides, a tank was built at the centre to collect and store the water. A dome shaped cover was given on top known as "bhinda" made from phog plant and plastered with mud.

The kund had two or three openings from which rainwater could enter. The tanks and ponds were constructed in lowest terrains so as store rain water. Apart from the tanks, there were Augor or Agar which handled water. The Aav or Payatams were the channels that fed the reservoir. They were wide and shallow so that water velocity was low to avoid or reduce silting of the pond. Khurra or stone pieces placed strategically within the Aav reduced the velocity further. The tanks were stepped to receive the maximum water. The lower portions were known as Neshta and were constructed in lime and stone to withstand water pressure. This portion received the first rainfall. The shallower portions were encircled with soil embankment, sometimes reinforced and were referred to as 'Pal'. The embankments were constructed or repaired before the rainfall. The Neshta collected the initial water providing additional time for the Pal to consolidate.

Wells, baoli or chuher were sometimes dug within the tanks or lakes to supplement the surface accumulation. The reservoir was planned and constructed taking into account the actual and projected demands of the community. Various sections of varying heights ensured supply throughout the year. Measuring rods with different coding indicated the quantum of water available at different points within the reservoir. Usually, the tanks were constructed 20 Basis or Purush deep, one Purush being roughly 1.8 mtrs. For large tanks small islands were created in the middle to absorb the wave energy and protect the mud embankments. Shrines were constructed on these islands to contribute to the waterbody's status and minimize its misuse. The catchment areas, the channels feeding the reservoir, the reserves and the outlets for irrigation or drainage were integral parts of the water harvesting system. Pani panchayat or water societies created by electing members from the community were responsible for its management together with that of arid agriculture. Special festival days were earmarked for repairs and desilting when the entire community participated.

Particular communities having indigenous skills were engaged to plan and construct water systems. The Gajdhar was the key person who was responsible to estimate the quantum of water that could be availed and plan and implement the system. The Gajdhars were held in high esteem and communities awarded Saropas or recognition for their efforts. The Sirbhavs or the Jalsungha could 'smell' or identify locations where Groundwater was available and wells could be dug. The Sonkar or Matkut actually constructed the tank by producing the soil bricks required for this purpose. The Chunker or Lunias helped them in this. The Agaria manufacture and supplied the tools required to construct the tanks. The Odhya and Gond tribes (Orrisa) were famous for their skills in constructing such tanks and were much sought after.

Having forgotten the indigenous skills of water management, today there is almost a desperate situation in Gujarat, Rajasthan and elsewhere. The water table has failed by 150-200 ft. Groundnut and cotton crops in over 4 lakh ha. in Gujarat were ruined (1999-2000) and a bucket of water costs between Rs. 4 to Rs.10. The fall in normal precipitation has reduced grain production by 60-70% and large number of families adversely affected due to depleting water table and excessive run-off. A study of Gujarat reveals that the areas where the water depth has gone below 10m. increased from 3% in 1973 to 25% in 1990 and 46% by 1994. The gap between water use and sustainable yield of the aquifer in Punjab will be alarming by 2025. It is feared that if water table goes below 15m. all the tube wells will stop functioning.

In order to bring about the type of sustainable water management that is able to address poverty, employment creation, the role of women and the protection of the environment, there is a need to move beyond conventional integrated approaches towards retrospection into the future. Rediscovering the traditional water wisdom would enable sustainable community based management of water, small-scale irrigation and catchment management.

BLUE NETWORK PLANNING

All along the history water has always been one of the basic determinants of human settlements - their location, growth and development. Today it is more so because of the impending water crisis, as well as the environmental threat, which our cities are facing. This calls for a review of the ongoing policies, concepts and strategies of urban development and' evolving a concerted approach towards land use planning, health, environment, legal framework and management.

Often the land use plans propose the development of industries, sanitary landfill sites, power stations, housing, etc. along with wetlands, water bodies and rivers, which destroy the quantity and quality of water and make it unfit for human use. Indiscriminate urbanisation and land use changes significantly alter the hydrology and quality of surface water. Incompatible land uses, together with encroachments along the water bodies are the built-in mechanism to kill the water bodies, and no amount of effluent treatment or legislation can reverse the damage. Therefore, as the preventive measures, the land use pattern should be carefully worked out so as not to precipitate such a situation. It is equally essential to provide proper sanitation, sewerage and solid waste, disposal system to avoid hazards to water. It is a paradox that, while on one hand we face the water crisis, on the other there is flooding in the same area, during the monsoons. This is largely due to lack of water based planning and development. Excessive hard paving of open areas, disturbing the natural drainage pattern, building of structures and roads, the natural depressions, silting of drains, water bodies and rivers, disposal of high amount of solid waste in water channels and bodies and lack of greenery are largely responsible for such a situation. It is necessary to work out planning and development strategies, which should be based on a detailed analysis of hydrology, topography, aquifers, rainfall, flooding, drainage, vegetation cover, water harvesting potential, together with other technical and financial considerations. Local skills and community often prove to be valuable resources in evolving realistic planning and sustainable development strategies, which should be identified and incorporated in the planning framework.

The starting point of planning should be to arrest the run-off, the conservation of natural valleys, water bodies and aquifers, where no roads or structures should be built. The natural drainage system should be meticulously studied and taken into consideration for allowing maximum Groundwater recharge and rainwater harvesting in the natural or man-made catchment areas.

The concept of 'zero run-off' and bio-drainage with retentive ponds, sediment traps and balancing lakes should be adopted for rainwater conservation and drainage, with a segregated wastewater disposal system so as not to pollute the rainwater. In this way, blue network and wastewater drainage network should be planned separately. A green network surrounding the blue network would protect the ecology of aquifers, as well as provide a pleasant environment. Simple methods of site planning, which incorporate porous/semi permeable paving drop inlet/down pipe. Sediment trap, storm sewer, retention ponds, etc. can immensely contribute in maintaining Groundwater table. This has been the mainstay of planning of Dwarka Sub-City for a population of 10 million, covering 56 sq.km. of area.

In order to maintain an equilibrium between the elements of natural eco-system of vegetation, land or water on the one hand and human activities on the other hand, the following are the necessary components of a blue network blue prints:

- Zero run-off drainage for rainwater harvesting
- Conservation of low lying lands for water harvesting, storage and channels
- Community participation
- Sustainable livelihood support system

- Decentralised planning and decision making
- Ridge to Valley treatment approach
- Adoption of bio-drainage

The basic principles of sound Watershed Management include the following considerations:

- Confine urban development to the uplands
- Protect the wet-lands, streams and water bodies by leaving adequate fringes of vegetation and giving them sacred or legal shields.
- Preserve and utilise the natural drain ways and maximise conservation of water by various methods, such as, underground rainwater storage, contour trenching, pits, deep holes and wells, stepwells, tanks and baolis, ponds, balancing lakes, etc.
- Promote on-channel storage and groundwater recharge.
- To improve the quality of river water and to secure its continuous flow and encourage aquatic life.
- Installation of water purification facilities.
- Improvement of drainage, waste water treatment and sewage.
- Adoption of porous paving for service roads, footpaths, plazas and parking areas; avoid pucca pavements and instead fill with pebbles and coarse sand to allow fast percolation.
- Leak detection and rectification works
- Rain water harvesting, through in-situ tanks, check dams etc.
- Controlling indiscriminate extraction of Groundwater
- Recharging of aquifers
- Meters to be installed along trunk mains and for bulk consumers
- System of roof-top rain water harvesting be made mandatory
- Recycling and reuse of treated waste water
- Percolation structures in open grounds and road side will help re-charge bore-wells and provide water for gardening, car washing, etc.

Rivers and canals can serve to give agreeable form and structure to the regions through which they pass. With indiscriminate growth and dumping of wastes, the water bodies have become polluted and foul. This needs appropriate and strict pollution control measures and eco-sensitive river-edge land use controls.

Water flow needs to be controlled at the tributaries and riverbanks need to be stabilized and marked at each kilometre station. The valleys should be zoned so that the rivers are flanked with greenery, farmlands and forests, so that each village and city has its attractive water portal. The river basin should be administered by a joint authority, with uniform signals and regulations.

With the reduction in water in the water bodies, channels and ponds, these are continuously getting converted into urban uses. It is necessary to revive and rehabilitate the old canal systems. Some of the most delightful water-edge living is to be found where homes and shops are clustered along the edges of such man-made waterways. Many cities can rediscover and refurbish old canals, locks, and channels as prime real estate and frontage. One of the prime objectives of urban development in Indian cities should be to improve the quality of river-water, to secure its continuous flow and to encourage the return of aquatic life. This needs installation of water purification facilities, improvement of drainage, wastewater treatment and pollution abatement by sewage treatment. This would lead to attractive and pleasant riverfront development besides improving the quality and quantity of water. The surplus water during the monsoons should be retained in swells or balancing ponds along the riverbed, rather than allowing to the downstream areas. Such ponds and reservoirs in the catchment zones in low-lying areas offer several advantages, such as a quicker, efficient and economical process of drainage, continuous replanishment and enrichment of underGroundwater table, availability of rolling greens and recreational areas, linking the population with the waterfront, improving the micro-climate, and interconnected parkways. There is no need for elaborate gardening of the greenways, but there wild, simple and natural stretch by itself would be ecologically important. Such trails could be one of the cheapest form of drainage and recreation.

No matter how littered or foul the stream, preserve the edges for public use or private development, together with restoration of water quality. Seal off the discharge of existing mines which are no longer working. Control by regulation the flow of acid and other wastes from active mining operations. Uniform laws and enforcement are needed to reduce the economic disadvantage of certain localities, but under no circumstances can the commonplace corruption of the watersheds be allowed. Since the costs of corrective programs are usually too heavy to be borne wholly by districts plagued with abandoned strip and deepmine workings, state and federal funds will be required.

CONSERVING GROUNDWATER

85% of the drinking water needs are met from Groundwater, although only 5% of total groundwater extraction is needed for domestic water supply, irrigation accounts for 90% of all groundwater extraction and industry takes the remainder. The rapid development in groundwater based irrigation and its overexploitation for urban use has caused Groundwater depletion, because of which the life of drinking water supply source becomes short. Highly subsidized irrigation electricity tariffs have led to an indiscriminate and disproportionate level of groundwater extraction. Although significant areas in States, such as Punjab, Haryana, Gujarat, Karnataka, Maharashtra, Rajasthan, Tamil Nadu and Uttar Pradesh (in all 144 districts in 10 States) have been declared 'dark' and 'Grey' zones, there has been no let up in the depletion of Groundwater aquifers. In many cases, Groundwater depletion has aggravated water quality problems due to excess fluoride, arsenic and brackishness. In certain areas, Public Health Engineering Departments were forced to abandon low-cost hand pump based systems and to undertake costly and complicated piped water supply schemes. The need for regulating the extraction of groundwater arises from the following considerations:

- Protection of water resource against over exploitation (particularly for irrigation).
- Protection against quality degradation
- To ensure social equity and to guarantee minimum provision to all sections of the society.

A number of steps need to be taken urgently in order to manage groundwater in a more scientific manner especially in dark and grey zones. There must be a Central legislation against subsidies on electricity tariffs for irrigation. Although the World Bank has approved a drinking water project in principle for Punjab, further processing was withheld by the World Bank because of the Punjab Government's policy of providing free electricity and power for irrigation sector. In fact farmers do not stand to benefit from this due to undependable power supply and consequent disruptions in the supply of water. Many of them are forced to operate diesel power tariff with an assured power supply is beneficial and acceptable to the community. The Model Bill framed by the Government of India and circulated to the States in 1996 should be adopted by the States and implement it seriously. Also the user groups should be formed on the lines of Joint forest Management Groups, who should do Groundwater monitoring and ensure that it is managed as common property resource, rather than allow it to be over exploited as an open access resource.

A two pronged strategy comprising of to regulating indiscriminate groundwater withdrawal and to adopt appropriate measures for augmenting its recharge through spreading techniques in alluvial areas, check dams and percolation tanks besides direct injection methods, utilizing the abandoned structure available in large numbers in the hard rock region. Before undertaking direct recharge into such abandoned structures, their suitability for receiving recharge and also the quality of source water should be thoroughly examined to confirm the technical feasibility. Rainwater harvesting is another attractive feasible option for conservation of water. Besides, Groundwater recharge can be suitably dovetailed with watershed development programs aimed at soil and water conservation by arresting the run-off.

WATER RECYCLING

Conserving the scarce water and recycling of used water is a major challenge for the urban India. For this, a new technology, known as root zone cleaning system is feasible. It consists of a receiving tank into which the toilet, kitchen and bath water flows by gravity; a mud area where suitable reeds, rushes and other plants are grown; and an outlet tank in which the water cleaned by the natural process is accumulated. The processed water is clean and without any smell. Fish and amphibians living and growing within it show that the water is clean and the system is successful. Chlorination, which removes germs and bacteria, leaves chemical residues that are unfavourable for human consumption.

Removal of forest cover and the addition of impervious surfaces will increase frequency of flooding and change stream characteristics. However, the use of a natural drainage system increases water-table recharge and increases lag time for run-off entering streams. Protection of flood plains and drainage swales is imperative for flood control, for regulating stream control and for maintaining water quality.

Indian Vedic knowledge (vastu) prescribes locating a city along a serpentine bend of a river, and towards its west so that the morning sun rays and breeze purify the homes (which should have lower height and vegetation towards the north and east). The reason for choosing site along the curving river are its drainage potential as well as self-purification of river water by oxidation and the gradient. The research at Banaras Hindu University reveals that the speed of the flow of the water along the concave bank of Varanasi (that is the city and the ghats) is much slower than the speed of the river along the opposite bank. The level of the opposite bank is much lower than on the side of the ghats. The sandy bed on the opposite side has a far greater capacity to absorb the effluents than is ppssible on the city side. The gradients and the basic asymmetries that nature has provided to cleanse the effluents generated by the city as the velocity of the river is markedly higher along the opposite (convex) bank. The capacity of the river to dilute, diffuse, disperse and digest the effluents and organic water matter is much higher than on the city side of the river.

Regulate Freshwater Consumption

The time has come when the unlimited use of this essential commodity can no longer be permitted. In many areas its use for irrigation or as an industrial coolant, may have to be reduced or even banned. In some cases these functions may be served as well by treated or even untreated fluid wastes. Otherwise, if new sources of fresh water cannot be tapped, the bulk consumers may have to relocate to other areas where suitable water is in adequate supply. In any event, it is incumbent upon planning authorities to run a check upon the quantity needs and uses of all existing or potential consumers, with an eye to the future regulation of freshwater resource as related to land use

Drinking Water

According to Central Public Health Environmental Engineering Organisation (CPHEEO) estimates, 90% of urban population has access to potable water supply. The average water availability in large cities is 5 to 6 hrs/day with low water pressure. Another concern is depletion of Groundwater that supplies to 80% of rural water and 50% of urban and industrial usage. Death tolls have increased recently to 1.5 million children every year by drinking of contaminated water in rural area. 1.51 lakh people suffered from water quality problem such as excess arsenic, fluoride, salinity and iron.

State Governments have the responsibility to provide safe drinking water and sanitation in the rural areas. Government of India supports and supplements the efforts of the State Governments. Programme for Drinking Water Supply and Sanitation have been under implementation ever since inception of the first five-year Plan. The Accelerated Rural Water Supply Programme (ARWSP) was introduced by the Government of India in 1972 -73. A Technology Mission on Drinking Water called the National Drinking Water Mission (NDWM) was launched in 1986, which subsequently was renamed as the Rajiv Gandhi National Drinking Water Mission (RGNDWM) in 1991. Upto end of November 2001, 87.89% habitations in the country is having access to adequate water (fully covered) with 40 litres per capita per day (lpcd). About 10.85% are partially covered and remaining about 20000 habitations are still not covered. States are also being supported from 2000-01, under the rural drinking water component of Prime Minister's Gramodaya Yojna (PMGY-RDW).

Sector Reform has been introduced in selected 63 districts in the country on a pilot basis on demand driven, community participation, and cost sharing approach by the user groups. Other support programme like social mobilization, capacity building, IEC, Human Resource Development (HRD) and MIS etc. supplement the main ARWSP and PMGY (RDW).

Providing safe drinking water to the poor is an ethical and moral responsibility of the government, industry and the society. No one should be denied access to safe household water. However, widespread use of unsafe water by the poor leads to diseases that harm the users and the nation as a whole. Although, about 83% coverage of rural habitations with drinking water supply facilities has been achieved in physical terms, there are wide gaps in quality term.

In spite of the high coverage as indicated above, people are not fully satisfied with the service provided as the overall ground realities are different and there are alarming incidence of quality problems. There are about 2,17,000 quality-affected habitations as per reports.

The main factors that can be attributed for the above situation are:

a. Quantity oriented and supply driven centralized government managed programme.
b. Absence of people's involvement in implementation of rural water supply schemes leading to lack of sense of ownership among them, thereby rendering the schemes unsustainable.
c. Unsustainable safe water sources due to lack of population control, improper planning and uncoordinated Groundwater management.
d. Lack of awareness and skill among the rural population to plan, implement and manage their own rural water supply schemes and safe water sources.

According to World Health Organisation (1992) the most immediate environment problems in the world are the ill health and premature deaths caused by biological agents in water. They contributed to the premature death of millions of people (mostly infants and children) and to the ill health or disability of hundreds of millions more. The problems are most acute in India.

Water Pollution

Water pollution is caused mainly by (a) discharge of waste water into natural water courses which results in depletion of life sustaining oxygen in the water bodies (b) introduction of all types of materials not normally found in water in its natural state, and (c) discharge of toxic materials which would affect health and life of the aquatic flora and fauna and also of human being. There are two major source of water pollution, namely industries and human habitations. Industrial waste from all types of industries contains polluting substances, which can be broadly defined as oxygen depleting substances, chemicals and toxic materials.

We are dumping pollutants into our water bodies in a reckless manner. Once inside the body, some of these live on their own terms. Cadmium, for example, never leaves the body. It keeps on accumulating inside and is difficult to detect. And one does not know how it reacts.

Nickel, which is released by electroplating industries, causes cancer of the lung's and is a known allergen. An electroplating unit can be established with less than five people, which does not being the unit under the Factory Act. The workers themselves are prone to getting cancer at the factory. Nitrate, which has been found in Delhi's groundwater in large quantities, is carcinogenic and can cause the blue, baby syndrome and hamper the transport of oxygen in the blood. Experts add that high concentrations of nitrate in drinking water can be fatal for infants, particularly in the first three months. Higher levels of nitrate can lead to gastric cancer and have an adverse effect on the cardiovascular system. Fluoride which is present in large quantities in our water can cause dental fluorosis (staining of tooth enamel and damage to teeth) and lead to neurological problems. Long term intake of water contaminated with fluoride can cause stiffness of the spine, bone deformities, thyroid and kidney changes. Some metals are difficult to trace in the body. A person can suffer from lead poisoning, which is extremely difficult to detect.

The most common chemical contaminants found in water sources are iron, dissolved salts, fluoride, arsenic and chromium. Biological contamination is brought about by bacteria, viruses, parasites and other pathogens caused by the farces of infected animals or people. This is likely to occur when public and private drinking water supply is driven from surface waters (ponds, creeks, rivers, lakes etc.), which can be contaminated by domestic pets, livestock, poultry, wild animals or humans who defecate in or near the water bodies. Sources of water pollution could be industrial, domestic or agricultural and pollutants in water could damage our health, agriculture or property.

The WHO estimates that contaminated water causes about 25 million deaths and other hundreds of million people suffering from various diseases. In India, major industries responsible for water pollution are fertilizers, sugar, textiles and chemicals, mines and minerals, pulp and paper, lather tanneries, and process industries. The problem is compounded by the presence of a large number of small scale industries. Water contamination situation in India is so severe that about 70% of all diseases in the country are water borne. About 73 million workdays are lost each year due to water borne diseases.

The records show that on an average about 50% of the cases are related to water borne diseases like diarrhoea (Gastro), Enteric Fever (Typhoid) and Hepatitis B (Jaundice). The effects of these diseases are more prevalent in the children below 12 years of age. Seasonal variation of incidence of diarrhoea remains constant over the years or stable during the pre-monsoon and post-monsoon periods, but there is a sharp increase in the cases during the monsoon season. This indicates possibility of mixing of sewage with water supply sources and mains. Male children are recorded more vulnerable to the diseases. This suggests possibility of a gender-biasness as if male child is sick, he will be given more attention and more likely to be taken to the hospital for treatment, whereas for the female child, she could be neglected or the disease could be overlooked and some home-remedy could be offered to her. Further more prevalence of diarrhoea for children under one year suggests possibility of dependency on external feeds to infants.

A survey carried out by Public Health Engineering Department in Rajasthan for status of water supply in villages and small rural colonies indicates that many areas were found to be affected by excess fluoride 2-3 PPM. Millions of people in some of the districts of Rajasthan are consuming water with fluoride concentrations even up to 18 mg per litre. The safe level is 1.0 to 1.5 mg per litre as prescribed by in Manual on Water Supply and

Treatments, 1991 in India and the WHO. Lack of awareness among people and availability and access to drinking water sources were found the major problems in rural areas.

An integrated river action plan in Maharashtra shows that the major source of river water pollution is the discharge of untreated or partially treated municipal sewage from the municipal corporations or councils situated on the bank of rivers. Industries are also responsible for about 20% of the river water pollution. Furthermore, major towns situated on the bank of rivers do not have sufficient number of sanitation facilities. Water pollution due to industries is being effectively checked and monitored by State Pollution Control Board but the pollution caused by urbanized local bodies remained unchecked and unmonitored despite the prevailing law.

The quality of water in a polluted stream not only depends on the pollution material but also on the abstraction of water from the stream, particularly for irrigation. In India, all rivers are interstate. An upper reparian state if over exploits water or pollutes river, it will affect the quality and quantity of water flowing to the lower reparian state. Further, pollution is a technical problem and treating the trade effluents could be a costly affair. The choice of method depends upon the volume of the effluents and its strength in terms of various pollution parameters.

Norms of Water Quality

Drinking water is a cause of several infections and diseases:

- Experts estimate that over 63% of water borne infection, worldwide are directly caused by Giardia and Cryptosporidium.
- Cryptosporidium - major concern since found in over 87% of world water supply systems.
- Bacteria, parasites cause wide range of diseases in humans ranging from Mild Diarrhoea to Acute Aseptic Meningitis and Paralysis can be fatal. High risk patients include very young, the elderly, pregnant woman and immuno compromised patients like Cancer and AIDS, recent transplant, Kidney Dialysis.
- 40-50% mortality rate in case of 'Immunocompromised'.
- Under inhospitable conditions presence of chlorine, Giardia and Crypto can go Cystic.
- Cystic totally resistant to Chlorination.

To avoid the infections and diseases the drinking water should have compliance of IS: 10500-1991 and packaged drinking water IS: 14543-1998. Natural waters are never completely pure. During their precipitation and their passage over or through ground they acquire a wide variety of dissolved and suspended impurities. The concentration of these impurities is seldom large in ordinary chemical sense but they modify the chemical behaviour of water or its usefulness. It is estimated that 70-80% Indian population do not get safe drinking water.

In India even bottled drinking water is not safe to drink. In analysis of various brands of bottled drinking water sold in and around many as five different residues in the samples. Taking a serious view of the report on the alleged presence of deadly pesticides in various brands of drinking water, the government ordered a high-level investigation into the matter. These pesticide residues reportedly cause long-term damage such as cancer, liver and kidney damage, birth defects and disruption of the immune system. The BIS has effected a change in its code IS: 14543-1998. The amended code says "packaged water will have to be tested for 32 pesticides; individually, the quantity of these pesticides residue should be less than 0.5 micro milligram per litre."

In water some impurities are toxic, some may affect health, some affect the portability while others indicate pollution. A list of such impurities allowable limit is given below. As per IS: 10500-1991 the drinking water should have pH value 6.5 to 8.5 and Maximum turbidity NTU:5.

Table 2.22: Toxic Substances in water

	Allowed Limit WHO Std. Maximum	IS 10500-1991 required desirable limit Maximum
Lead	0.05 mg/L	0.05 mg/L
Arsenic	0.05 mg/L	0.05 mg/L
Selenium	0.01 mg/L	0.01 mg/L
Chromium	0.05 mg/L	0.05 mg/L
Cyanide	0.20 mg/L	0.05 mg/L
Cadmium	0.01 mg/L	0.01 mg/L
Radioactive	1000	–
Compound gross beta activity		
Mercury	–	0.001 mg/L
Pesticides	–	Absesnt

Table 2.23: Components Hazardous to Health

Flourides	1.5 mg/L	1.0mg/L
Nitrates as NO_3	45 mg/L	45 mg/L
Sulphate as SO_4	–	200 mg/L

Table 2.24: Components Affecting Potability

Total dissolved solid	1500 mg/L	500 mg/L
Iron	5.0 mg/L	0.3 mg/L
Managanese	5.0 mg/L	0.1 mg/L
Copper	1500 mg/L	500 mg/L
Zinc	1500 mg/L	500 mg/L
Magnesium plus sodium sulphate	1000 mg/L	–
Surfactants ABS	0.5 mg/L	

Table 2.25: Chemical Indicator of Pollutant

Biochemical Oxygen Demand—BOD	6.0 mg/L	– –
Chemical Oxygen Demand—COD	10.0 mg/L	–
Total nitrogen exclusive of NO	1.0	mg/L
Ammoniacal Nitrogen	0.05 mg/L	–
Carbon Chloroform Extract—CCE	0.05 mg/L	–
Oil and Grease	1.0 mg/L	–
Dissolved Oxygen	40% saturation	–

Table 2.26: Different Estimates of the Number of Urban Dwellers Lacking Provision for Water and Sanitation in 2000

Region	**Number and proportion of urban dwellers without "improved" provision for:**		**Indicative estimates for the number (and proportion) or urban dwellers without "adequate" provision for:**	
	Water	**Sanitation**	**Water**	***Sanitation***
Africa	44 million (15%)	46 million (16%)	100-150million (35-50%)	150-180 million (50-60%)
Asia	98 million (7%)	297 million (22%)	500-700 million (35-50%)	600-800 million (45-60%)
Latin America and the Caribbean	29 million (7%)	51 million (13%)	80-120 million (20-30%)	100-150 million (25-40%)

Source: WHO and UNICEF (2000), Global Water Supply and Sanitation Assessment, 2000 Report, World Health Organization, UNICEF and Water Supply and Sanitation Collaborative Council. UN-Habitat (2003), Water and Sanitation in the World's Cities; Local Action for Global Goals, Earthscan, London.

Table 2.27: Standards of Water Quality

Designated best use	**Class of water**	***Criteria***	
Drinking Water Source without Conventional Treatment but after Disinfection	A	1.	Total Coliforms Organism. MPN/ 100 mL. shall be 50 or less.
		2.	pH between 6.5 & 8.5
		3.	Dissolved Oxygen 6 mg/L or more
		4.	Biochemical Oxygen Demand 5 days 20° C 2 mg/L or less
Outdoor bathing (organized)	B	1.	Total Coliforms Organism MPN/ 100 mL. shall be 500 or less
		2.	pH between 6.5 & 8.5
		3.	Dissolved Oxygen 5 mg/L or more
		4.	Biochemical Oxygen Demand 5 days 20° C 3 mg/L or less

Drinking Water Source	C	1.	Total Coliforms Organism MPN/ 100 mL. shall be 5000 or less
		2.	pH between 5 to 9
		3.	Dissolved Oxygen 4 mg/L or more
		4.	Biochemical Oxygen Demand 5 days 20° C 3 mg/L or less
Propagation of Wild Life	D	1.	pH between 6.5 & 8.5 fisheries
		2.	Dissolved Oxygen 4 mg/L or more
		3.	Free Ammonia (as N) 1.2 mg/L or less
Irrigation, Industrial Cooling Controlled Water	E	1.	pH between 6.0 or 8.5
		2.	Electrical conductivity at 25° C micro mhos/ cm max 2250
		3.	Sodium Absorption ratio max 26
		4.	Boron, max 2 mg/Litre

Source: Central Statistical Organisation, Compendium of Environment Statistics, 1997, Department of Statistics, M/o Planning & Programme Implementation, Govt. of India, New Delhi.

NEED FOR A SAFE WATER SYSTEM (SWS)

Existing water treatment and distribution networks in India have been largely ineffective in controlling water-borne diseases. At present, individual households, particularly in low-income areas, need information on water treatment and sanitation options within their reach, education on the use of such options and on safe water use practices, and access to the means to exercise such options.

The National Institute of Communicable Diseases estimates morbidity in India due to diarrhoea in excess of 600,000 deaths per year, most being children under five years of age. Despite investments in water and sanitation infrastructure, many low-income communities in India continue to lack access to safe drinking water, improved sanitation facilities, and information and education on hygiene.

As such low cost and rapid interventions should be promoted to mitigate the health effects arising from deficiencies in water supply and sanitation services and poor hygienic practices. Currently most available strategies and technologies to make water safe to drink are unaffordable and inaccessible to most low-income households, particularly those without piped water supply. The Indian National Family Health Survey (1998/ 99) reports that 68.4% of Indian households do not purify water by any means (rural, 75.3%), and the most common measure used for some purification (filtering of water through cloth) is one of the least effective.

While access to drinking water has increased over the past decade, to 82% of India's rural population and 87% of the urban population, the tremendous adverse impact of unsafe water on health continues. Twenty one per cent of communicable diseases are related to water.

A water disinfection and storage system, made accessible and affordable to lower income populations, has been shown to be an effective form of intervention to reduce the incidence of water-borne diseases at the household level. A point-of-use safe water system would empower underserved populations in India to ensure consumption of safe drinking water. Treatment of contaminated water before consumption, combined with targeted behaviour change messages, has the potential to greatly reduce waterborne diseases. The introduction of a household drinking water disinfectant system, along with education on safe health practices, has been conservatively estimated to reduce diarrhoea cases by 20%, and improved water and saitation can reduce water related disease Morbidity and mortality rates by 20% to 80%. These gains translate into better health profiles, improved productivity, and lower spending on curative health care in underserved populations.

The basis of the SWS intervention is:

- Point-of-use treatment of contaminated water with locally produced sodium hypochlorite solution, packaged in locally available bottles with a 5 to 10 mL. cap that serves as a dosing device.
- Safe water storage in plastic containers with a covered narrow mouth and a spigot to prevent recontamination.
- Behaviour change techniques, including social marketing, motivational interviewing, community mobilization, communication and education.

State Governments have the responsibility to provide safe drinking water and sanitation in the rural areas. Government of India support and supplements efforts of the State Governments. Programme for Drinking Water Supply and Sanitation have been under implementation ever since inception of the First Five-Year Plan. The Accelerated Rural Water Supply Programme (ARWSP) was introduced by the Government of India in 1972-73. A Technology Mission on Drinking Water called the "National Drinking Water Mission (NDWM) was launched in 1986, which subsequently was renamed as the Rajiv Gandhi National Drinking Water Mission (RGNDWM) in 1991. Upto end of November 2001, 87.89% habitations in the country is having access to adequate water (fully covered) with 40 litres per capita per day (lpcd). About 10.85% are partially covered and remaining about 20000 habitations are still not covered habitations. States are also being supported from 2000-01, under the rural drinking water component of Prime Minister Gramodaya Yojna (PMGY-RDW). Sector Reform has been introduced in selected 63 districts in the country on a pilot basis on demand driven, community participation, and cost sharing approach by the user groups. Other support programme like social mobilization, capacity building, IEC, Human Resource Development (HRD) and MIS etc. supplement the main ARWSP and PMGY (RDW).

Water Supply- Discovering the Unexploited Opportunities

Water stress is a growing problem in most parts of India. Delhi is no exception where the problem is reaching a crisis stage, particularly due to excessive water consumption and population growth. The Master Plan for Delhi-2021, which was notified in 2007, gave a close look to the issue of water supply and discovered that there are numerous unexploited opportunities for saving water and augmenting its supply. It gives the following suggestions in this regard.

All measures are to be taken to reduce unaccounted flow of water (UFW), production losses at existing plants. The existing drainage basins shall be made self-sustainable in water management by integrating water-sewerage-drainage systems. It is imperative to not only to initiate new projects and upgrade present infrastructure, but also to promote water conservation through an integrated and a community driven model, comprising of complimentary short term and long term measures.

To improve the water supply in accordance with the projected requirement, all measures are to be taken to reduce unaccounted flow of water (UFW), production losses at existing plants. The existing drainage basins shall be made self-sustainable in water management by integrating water-seweragedrainage systems. It is imperative to not only initiate new projects and upgrade present infrastructure, but also to promote water conservation through an integrated and a community driven model, comprising of complimentary short term and long term measures as given below:

(a) Recycling of treated wastewater with separate lines for potable water and recycled water. For this, dual pipe supply system has to be introduced in a phased manner in all the areas.

(b) Groundwater recharging through rain water harvesting, conserving water bodies and controlling groundwater extraction. Groundwater extraction is to be controlled through registering boreholes and recharging according to test yields. Groundwater management is to be enforced by concerned agency.

(c) Focused planning and action will be required to be taken to prepare and implement rain water as roof water harvesting schemes both with the aim of optimizing water use and Groundwater recharge. For this suitable mandatory provision to be made for planning and construction of various schemes.

(d) The planning should dovetail watershed management, and arrest the run-off. It should ensure the conservation of natural valleys, water bodies and aquifers. The concepts of 'zero run-off drainage', with retention ponds, sediments traps and balancing lakes should be adopted, with a segregated wastewater disposal system. A green network overlapping the blue network would protect the ecology of aquifers, and also provide a pleasant environment. Simple methods of site planning, which incorporate porous/semi permeable paving, drop inlet/down pipe, sediment trap, retention ponds, etc. will contribute in maintaining Groundwater table.

(e) Rivers, major drains and canals, with indiscriminate dumping of wastes, have become polluted and foul. These need strict pollution control measures and eco-sensitive land use controls. Water flow needs to be controlled and stabilized and marked at each kilometer station. The valleys should be zoned as water portals, so that these are flanked with greenery, farmlands and forests.

(f) One of the prime objectives of development should be to improve the quality of river-water, to secure its continuous flow and to encourage the return of aquatic life. This needs improvement of drainage, waste water treatment and pollution abatement by sewerage improvement. The surplus water during the monsoons should be retained in balancing ponds along the riverbed rather than allowing it to the downstream areas.

(g) The drains and waterfront can be landscaped in the form of interconnected parkways. There is no need for elaborate gardening of the greenways, but wild, simple and natural stretch by itself would be ecologically important. Such trails could be one of the cheapest forms of drainage and recreation.

(h) Water supply in new areas should incorporate separate lines – one for washing, water coolers and garden taps, the second for supplying potable water. All non-residential buildings having a discharge of over 10,000 liters a day should incorporate a wastewater recycling.

(i) The wasteful practice of 'drill, pump, and spill' has to be replaced by efficient methods of water conservation, use, and recycling as standard and mandatory procedures. There is a need to incorporate the mandatory stipulation of water saving/waterless flushing system in the Building Byelaws.

(j) Where the hazard of pollution exists, the minimum charge for operating permits should cover the expenses of adequate policing and controls. Mandatory performance bonds and liability insurance. should pay for all damages plus any corrective measures, which might be needed. As a governing rule, no new development, manufacturing, process or operation of any polluting activity should be permitted, which may result in the significant degradation of any water resource.

(k) About half of the water that is treated and distributed at public expense is non-revenue water. This is due to unrecorded usage or illegal taps and water connections. Reducing water losses is cheaper than augmenting water capacity for such losses

Source: *Delhi Development Authority (2007), Master Plan of Delhi, 2021, New Delhi,*

Drinking Water Conservation

Ensure 24 hours supply in residential areas:

(a) Psychologic satisfaction to the consumer.

(b) No hoarding of water resulting in:

- No underground/overhead tanks in individual flats.
- No booster pumps.
- No filtration device in the house.
- With availability of water constantly at the tap/outlet point the rusting due to intermittent supply attracting air in the system shall be reduced considerably resulting in longer life to geysers and other appliances.
- Saving in electrical energy.
- No disputes, no litigation if the consumer is charged the right amount for the actual consumption made.

Approach

To achieve the above, it is essential to:

- Optimise distribution network design.
- Ensure no leakage in the distribution network from the source to pumping station

as well as from pumping station to the consumer point (sometimes goes even up to 50%).

- Conservation in treated water by avoiding wastage.
- The treated water which is wasted needs complete retreatment. Thereby the cost of the water authority is not merely the cost to the treated water wasted but the cost of the retreatment for this quantity (almost double the initial investment).

Pricing

- Bare essential quantity (cooking, essential washing etc.): Very less charges or even free.
- Average requirement actual production cost of water recovered.
- Luxurious/Wasteful expenditure of water (such as consuming treated water for irrigating lawns/ washing cars etc. - compounded cost to be released on a telescopic scale as a deterrent to waste.

Rural Water Supply

In India's 6 lakh odd villages, water supply is by and large dependent upon surface and underground sources. There is hardly any municipal water supply system in the villages:

Table 2.28: Sources of Drinking Water in Inhabited Villages in India

Source Village	(Percent)
Well	69.8
Hand Pump	55.9
Tubewell	21.1
Tap	18.2
Tank	14.3
River	10.0
Nala	3.6
Canal	3.5
Fountain	2.6
Spring	1.7
Lake	0.1
Other	4.5

Source: Ministry of Rural Development, Government of India..

The strategies of Rural Water Supply Programme revolve around the basic premise that provision of safe drinking water is the responsibility of the Government. Increased outlay by the Government, particularly in the last one decade and, a change in technology focus to handpumps fitted on tube wells and bore wells, had resulted in an impressive increase in the total rural water supply coverage. However, the availability of potable drinking water in rural areas, especially during the summer months, is still not satisfactory. Event though about 1 lakh habitations are covered every year, the number of problem habitations has not declined proportionately. The reasons identified for uncovered villages were:

- Fast depletion of Groundwater level, which also increases incidence of quality problems of Arsenic and Fluoride etc.
- Sources go dry and defunct due to deforestation with consequent reduced recharge and lack of protection.
- Heavy emphasis on new construction and poor attention to maintenance.
- Non-involvement of people in operations and maintenance.
- Neglect of traditional water management practices/systems.

The revised guidelines envisage to energise the system towards overcoming the above problems and to achieve the goal of providing safe and sustainable drinking water to all rural habitations of the country by the following strategies:

- Control on over-extraction of groundwater
- More funds for repairs and rehabilitation
- Increasing people's participation
- Reserve 20% funds for states promoting sector reforms
- Water to be treated as a socio-economic good
- Stronger links with watershed development programmes

The prime objectives of these guidelines are as under:

- To ensure coverage of all rural habitations especially to reach the un-reached with access to safe drinking water.
- To ensure Sustainability of the systems and sources.
- To preserve quality of water by institutionalising water quality monitoring and surveillance through a Catchment Area approach.

Rajiv Gandhi National Drinking Water Mission

Funds are provided to the States by the Rajiv Gandhi National Drinking Water Mission under the following programmes:

- Accelerated Rural Water Supply Programme (ARWSP): To supplement the efforts of the State Governments in providing access to safe drinking water to all rural habitations or the country.
- Sector reform programme: Funds would be provided for institutionalising community participation in capital cost sharing, Operation & Maintenance and Water Quality Monitoring & Surveillance in identified pilot districts.
- Sub Missions: Five Sub Missions on problems of water quality and sources drying up needing construction of water conservation and recharge structures and other measures with the States planning and approving them on their own.
- Human Resource Development (HRD).
- Research &. Development (R&D).

- Information. Education and Communication (IEC).
- Management Information System (MIS).
- Provision of water supply in rural schools.
- Monitoring and investigation units, purchase of rigs, water quality monitoring and surveillance, monitoring and evaluation, solar pumps and innovative projects.

Accelerated Rural Water Supply Programme (ARWSP)

The Accelerated Rural Water Supply Programme (ARWSP) aims at providing safe anti adequate drinking water facilities to the rural population by supplementing the efforts made by the State Governments/UTs under the State Sector Minimum Needs Programme (MNP).

Objectives

The prime objectives of the ARWSP are:

(a) To ensure coverage of all rural habitations especially to reach the un-reached with access to safe drinking water.

(b) To ensure sustainability of the systems and sources; and

(c) To takley the problem of water quality in affected habitations and to preserve quality of water by institutionalising water quality monitoring and surveillance through a Catchment Area Approach.

The following norms are being adopted for providing safe drinking water to rural population in the habitations:

- 40 litres of safe drinking water per capita per day (Ipcd) for human beings.
- 30 lpcd additionally for cattle in the Desert Development Programme (DDP) areas.
- One handpump or standpost for every 250 persons.
- The water source should exist within the habitation /within 1.6 km in the plains and within 100 metres elevation in the hilly areas.

Water is defined as safe if it is free from biological contamination (guineaworm, cholera, typhoid, etc.) and chemical contamination (excess fluoride, brackishness, excess iron, arsenic, nitrates, etc.). The pointiest of rural drinking wale supply are following:

- To cover not covered (NC) habitations and to fully cover partially covered (PC) habitations getting less than 10 Ipcd. Among them, priority may be given to the ones inhabited exclusively by SC/ST or having larger SC/ST population enumerated in the Status Report of 1994 (Survey) and resurveyed in 1996-97.
- Coverage of quality affected habitations with acute toxicity first and the others, later.
- Upgradation of source level of safe source habitations which get less than 40 Ipcd water to the level of 40 Ipcd.
- Coverage of schools and Anganwadis where safe drinking water sources could not be provided under the outlays allocated by the Tenth Finance Commission.

Once the task of providing every habitation with safe drinking water source is completed as per the norms and priorities indicated above, in the entire State, the State Government may consider relaxation of norms with the prior approval of the Government of India, subject to the condition that beneficiaries of the relaxed norms are willing to share a part (which should not be less than 20%) of the capital cost and shoulder full responsibilities of subsequent O & M and replacement so as to meet their enhanced service expectations.

The criteria for allocation of ARWSP funds to the States, as approved by the Government of India requires the States/UTs have to earmark 25% as minimum outlay for SCs and 10% as minimum outlay for STs, for taking up Water Supply Schemes exclusively for them. Allocation of funds to the States, based on the above criteria, would be subject to the matching provision/ expenditure by the States under the respective State Sector MNP. Releases under the ARWSP would not exceed the provision for Rural Water Supply made by the State Governments under their MNP.

Government of India's Recommendations for Rural Water Supply

To ensure institutional sustainability, GOI recommends the following five action areas in its reform package:

- Enable Panchayat institutions and user community groups to assume lead roles in decision making for RWSS.
- Promote and facilitate participation of the NGOs and the private sector; define appropriety roles of the NGOs/ private sector as well as external support agencies.
- Strengthen existing institutional capabilities to undertake consumer orientation, policy development, planning, implementation, O & M, monitoring and evaluation, and promotion of health and sanitation.
- Restructure public sector institutions.
- Streamline operational guidelines and procedures for RWSS agencies.
- Accommodate adoption of a demand-driven approach and participatory processes based on empowerment of villagers.
- Focus on long-term sustainability of investments including good quality of construction.
- Ensure integrated service delivery for water, sanitation and hygiene education.
- Ensure stable and reliable back-up support to communities in O & M.

Key Strategies

- Ensure an enabling environment for reforms (policy framework).
- Establish institutional sustainability (clarifying and redefining roles of institutions and stakeholders).
- Establish financial viability and sustainability (by capital cost sharing, full O & M, cost recovery and reductions in operating costs).

Key Factors

- Quality/Quantity: Attention must be given to the quantity of water that is available as well as its quality. Until minimum standards for both quantity and quality are met, the priority should be to provide equitable access to an adequate quantity of water of intermediate quality, rather than to provide an inadequate quantity of water which meets the minimum standard for quality. If there are serious doubts about the microbiological quality of the water, it should be treated with a residual disinfectant as a first measure to improve quality.

- Access and Equity: Even if a sufficient quantity of water is available to meet minimum needs, additional measures may be needed to ensure that access is equitable. Unless water points are sufficiently close to their dwellings, people will not be able to collect enough water for their needs. In urban situations, it may be necessary to have water supplied into individual buildings to ensure that toilets continue to function. Water may need to be rationed to ensure that everyone's basic needs are met. If water is rationed or pumped at given times, this should be at times that are convenient to women and others who have responsibility for collecting water. Women and men from the affected population should be informed about their entielements, and should also be involved in monitoring the equitable distribution of water.
- Water Collection and Storage: People need vessels to collect water, to store it and to use it for washing, cooking and bathing. These vessels should be hygienic and appropriate to local needs and habits, in terms of size, shape and design.
- Communal Washing and Bathing Facilities: People may need a space where they can be the in privacy. If this is not possible at the family shelter, some central facilities may be needed. Washing clothes is an essential activity for hygiene, particularly for children, and cooking and eating utensils need washing. It is not possible to define universal standards relating to these activities, but if some facilities are needed for them to be carried out then they should be available. The design, numbers and location of these facilities should be decided in consultation with the intended users, especially women. Among the essential factors to consider are the safety, appropriateness and convenience of facilities for the users, especially women and girls, whose views on siting and design should be sought. As with latrines, facilities that are remote from the centre of a settlement are likely to pose additional risk of attack to female users.

To meet the need for improved Rural Water and Sanitation Services, the Government of India has revamped the Rural Water Supply Programme which aims to usher in reforms by institutionalising community participation in the rural water supply sector. 20% of the annual ARWSP outlay would be given to those State Governments who undertake community-based Rural Water Supply Programmes by adopting the demand-driven approach based on empowerment of villagers to ensure their full participation in the project, through, a decision-making role in the choice of key design and management arrangements, and part of capital cost and 100% O & M cost to be borne by the users.

The reforms stipulate taking up pilot projects which are mainly lEC-based with start-up activities and HRD campaigns. The IEC campaign is aimed at educating the rural people about the need for them to actively participate in the programme, various technologies available, it's capital cost, O & M cost and requirement of replacement costs, etc. of each of the possible technology options and the benefits of the people planning, funding, implementing, owning, operating, maintaining and replacing the water supply Schemes of their choice. The HRD campaigns are aimed at giving the rural population the requisite training to carry out the responsibilities indicated above. The physical work under the project would commence only subsequently on the basis of the awareness and demand generated by the campaigns in the first stage. Even though, the requirement of funds is likely to be slightly higher, it is expected that, it would be a onetime investment as by the time the project is completed the beneficiaries would be sufficiently equipped to operate and manage their rural water supply Schemes and to plan, fund and implement the replacement Schemes after the expiry of the life of the existing Schemes.

INVOLVEMENT OF WOMEN AND COMMUNITY

Since women are the principal beneficiaries of this programme, it is of crucial importance that women are involved at all the stages of Rural Water Supply Schemes, particularly while making decisions on the location of the standpost/spot sources in the villages/habitations. Information about sources of water should be obtained from women by the surveyors of the Schemes. They should also be involved in the following manner:

(a) At least 30% of handpump mistries under NHRD, TRYSEM and other training Schemes, should be women of the local areas/habitations as they can take better care of the operation and maintenance of the handpump Schemes than others.

(b) There should be women caretakers for handpumps in the habitations.

(c) Certificate about satisfactory completion of the Schemes may be obtained from women groups in the habitations.

(d) Prominent women from the habitation should be represented on the Village Level Water Monitoring Committees.

Experience has shown that where women are involved in such consultation and in the maintenance of the spot sources, the performance of the water supply system has been effective. This aspect should be kept in view and ensured by the implementing agencies at the time of formulation, implementation and maintenance of the Schemes. Each Scheme/ project of Rural Water Supply should mention the extent of the involvement of women.

Service Delivery Process

To overcome the severe water scarcity, there is a need to invoke new technology for the needs of rural India in terms of appropriateness, cost, operation and maintenance.

To ensure sustainability of solutions an integrated collaborative approach, compromising the following components needs to be adopted:

- Implementation through NGOs and agencies that have long-term community presence for overall project coordination, monitoring and financial arrangements.
- Community involvement - villagers encouraged to "own, use and care for" the technology.

- Training of community members to handle operation and maintenance as well as minor trouble shooting.
- Feedback from field to technology for need-based modifications.
- Awareness building programmes on the need for water conservation and treatment.

Village Water and Sanitation Committees (VWSCs)

The village water and sanitation committee is a formal body, with legal status as a sub-committee of the statutory gram panchayat (local government). The committee consists of 7 to 12 democratically elected and representative members, with at least 20 per cent representation from the socially and economically backward sections of the community. At least 30 per cent of the members are women. Unlike the gram panchayat, which is responsible for multi-sectorial development in the village, the village water and sanitation committee is a user association set up specifically for the project. Each committee has a Chairperson and Treasurer who jointly operate the bank account. Over 50 per cent of the treasures are women. While all the elected members work in an honorary capacity, the village committee usually employs a local technician, who is known as a village maintenance worker, to operate and maintain the water supply scheme after construction is completed.

Table 2.29: Community Contracting Models in Water Supply and Sanitation Technologies

Technology	Average Value (US$)	Type of Contract	Type of Contractor	Selection Process
Mechanized borewell with overhead tank and distribution network	30,000 (mid-sized scheme)	Written	Private/public sector corporation	Bidding/ market survey by VWSC/SO
Boreholes with handpumps (8 handpumps per village on an average)	4,000	Verbal	Private/government corporation	Market survey by VWSC/SO
Hand dug wells	4,000	Verbal	Private/government corporation	Market survey by VWSC/SO
Gravity-flow system	20,000	Verbal	Community technician	Community consensus
Individual household latrines	80	Verbal	Local mason	Household chooses
Drainage soakpit	50	Verbal	Community technician	Household chooses

The following priorities have been set for achieving the objective of providing safe drinking water supply to all rural habitations:

(i) Highest priority to be given to ensuring that the 'not covered' habitations are provided with sustainable and stipulated supply of drinking water.

(ii) It will be equally important to ensure that all the 'partially covered' habitations having a supply level of less than 10 litres per capita per day (Ipcd) and those habitations facing a severe water quality problem are fully covered with safe drinking water facilities on a sustainable basis.

(iii) Thereafter, other 'partially covered' and 'quality affected' habitations are to be covered.

(iv) Once drinking water supply facility is provided to all rural habitations, the efforts will focus on consolidation and the habitations which have slipped back to 'partially covered' or 'not covered' status due to a variety of reasons.

(v) Simultaneous action is needed to identify and tackle habitations where water quality problems have emerged recently.

(vi) It should be ensured that scheduled caste/scheduled tribe (SC/ST) population and other poor and weaker sections are covered fully on a priority basis. A systematic survey of all such identified habitations will be undertaken.

Table 2.30: Triggers for Disbursing Funds in the Swajal Project

Fund Release in Implementation Phase	Activities / Triggers
Instalment 1 (50%)	• IPTA signing • Market survey • Rates, quality and quantity endorsed • Proforma invoice obtained and submitted to PMU (v) Rate, quality and quality verified
Instalment 2 (40%)	• 80% utilization of Installment 1 funds • Achievement of construction and community development milestones as per IPTA • Physical and financial audit
Installment 3 (10%)	(i) Implementation Phase Completion Report (ii) Physical and financial audit.

The stipulated norms of supply would be 40 lpcd of safe drinking water within a walking distance of 0.5 km or elevation difference of 100 metres in hilly areas, to be relaxed as per field conditions persons is to be provided. Additional water is to be provided in the Desert Development Programme (DDP) areas for cattle, based on the cattle population. The water requirements for cattle need not necessarily be met through piped water supply and could be made through rain water harvesting structures/spot sources.

DEBRIEFING QUESTIONS

1. Give a brief status of water in India.
2. What are major issues of urban water supply?
3. What are the standards of water requirement and supply?
4. How the quality of water can be ensured and water sources can be protected from pollution?
5. What are the parameters of water for drinking?
6. Write a brief on National Water Policy?
7. What are the main options for rainwater harvesting?

8. What are the methods of artificial recharge of Groundwater?
9. What is understood by watershed development?
10. Write Short Notes on
    (a) Vegetative Filter Strip
    (b) Porous Paving
    (c) Swales
    (d) Recharge Bed
    (e) Infiltration Trench
11. Explain the major methods of water purification, filtration, disinfection and control of waterborne diseases.
12. What is understood by safe water system?

3 Water Supply in Buildings

The planning and design of water to be supplied in a city depends on many factors like population and their characteristics, climatic conditions, the type of sewerage system. In addition, it depends upon policy regarding metered supply, firefighting requirements, the number and types of industries and area under public gardens, and efficiency of management. It also depends upon the evaluation of potential sources of water supply.

Characteristics of population is as important as the magnitude of population for projecting the quantity of water supply. These include the economic status of the population (higher the status more the water use), living habits (number of baths taken per season), and average living and or garden space per family. The population in any urban area is heterogeneous and hence it is important to arrive to weighted average of these factor loading to assess the quantum of water required.

The climatic conditions not only affect the per capita consumption but also affect the seasonal demand of water. Cooler the climate less water is used and vice versa. Industries depending upon their products, consume varying amount of water. If there are more public gardens, the consumption would be more.

Table 3.1: Water Requirements of Institutions and Commercial Establishments

S.N.	Type of Unit/Building		Domestic Use (Litres per head per day)	Flushing (Litres per head per day)	Total Consumption (Litres per head per day)
1.	Factories including Canteen and bathrooms		30	15	45
2.	Factories with Canteen but no bathrooms		20	10	30
3.	Hospital (excluding laundry and kitchen)				
	(a)	Number of beds ≤ 100	230	110	340
	(b)	Number of beds > 100	300	150	450
	(c)	Out Patient Department (OPD)	10	5	15
4.	Nurses Homes and Medical Quarters		90	45	135
5.	Hostels		90	45	135
6.	Hotels (upto 3 Star) excluding laundry, kitchen, staff, water bodies		120	60	180
7.	Hotels (4 Star and above) excluding laundry, kitchen, staff, water bodies		260	60	320

8.	Offices (including canteen)		25	20	45
9.	Restaurants and Food Courts including Kitchen				
	(a)	Restaurants	55 per seat	15 per seat	70 per seat
	(b)	Food Courts	25 per seat	10 per seat	35 per seat
10.	Club House		25	20	45
11.	Stadiums		4	6	10
12.	Cinemas, Concert Halls, Theatres and Multiplex		5 per seat	10 per seat	15
13.	Schools/ Educational Institutions				
	(a)	Without boarding facilities	25	20	45
	(b)	With boarding facilities	90	45	135
14.	Shopping and Retail (Mall)				
	(a)	Staff	25	20	45
	(b)	Visitors	5	10	15
15.	Traffic Terminal Stations				
	(a)	Airports	40	30	70
	(b)	Railway Stations (Junction) with bathing facility	40	30	70
	(c)	Railway Stations (Junction) without bathing facility	25	20	45
	(d)	Railway Stations (Intermediate) with bathing facility	25	20	45
	(e)	Railway Stations (Intermediate) without bathing facility	15	10	25
	(f)	Interstate Bus Terminals	25	20	45
	(g)	Interstate Bus Terminals/Metro Stations	10	5	15

Source : National Building Code 2016, BIS.

WATER SAVING IN BUILDINGS

The mains water supply in buildings is used in thousands of ways. However, the choices are in fact up to us and we can reduce consumption if we wish to use our resources wisely. In Japan local authorities encourage recycling. For example, in Tokyo, for a building with a floor area of over 30,000 m^2 to get planning permission it must recycle rainwater and have an in-building grey water treatment system.

Definitions vary somewhat but often rainwater is considered as one category, grey water as another and black water as a third. In this definition, grey water is considered to be

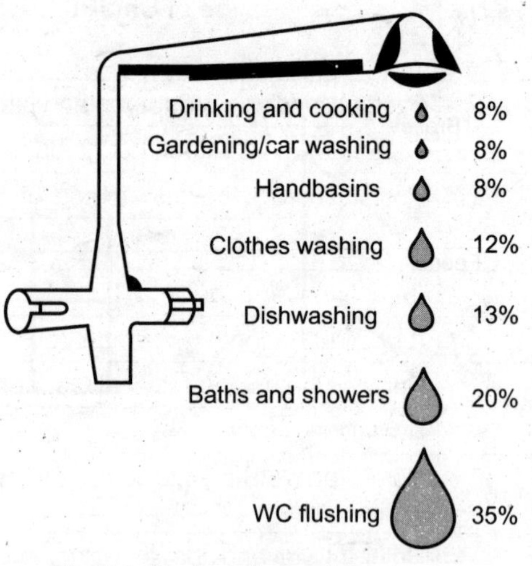

Fig.3.1: Typical domestic water use

all waste water from domestic appliances with the exception of toilets. Thus, it includes discharges from kitchen sinks, washroom basins, baths, showers, washing machines and dishwashers. Black water is all water that combines into the foul drain and then into the sewer system. It thus includes all grey water and the waste water from toilets.

Only 6 per cent or so is used for drinking and cooking even though all domestic water is to international standards for potable water. It is a bit like using our highest source of energy, electricity, for all power and heating in the home when one has a lower-grade source, gas, available.

About one-third is used for WC flushing and that clearly this water could be of a lower quality.

As mentioned before, the starting point is to reduce demand. This can be done in a number of ways, including the use of more efficient dishwashers and washing machines, low-water-volume WCs, taps that aerate the water stream and low-flow showerheads. By actively managing water use it should be possible to reduce consumption to a figure closer to 100 litres/ person/day. More radical alternatives are also available, such as composting toilets and waterless urinals.

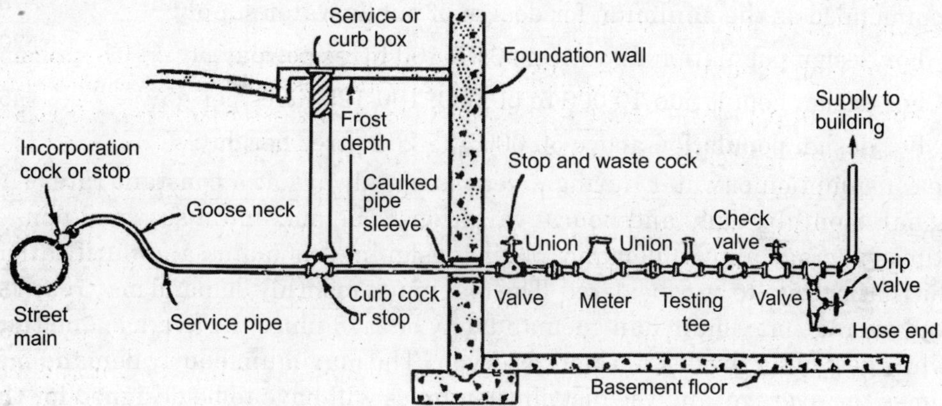

Fig.3.2: Typical Water Service Connection

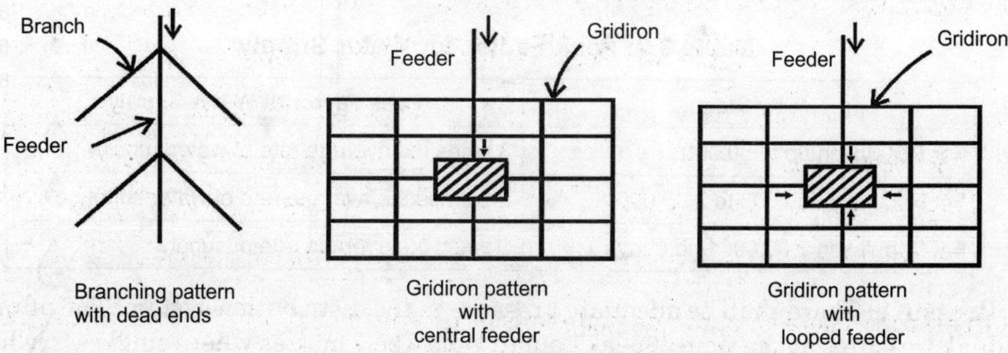

Fig.3.3: Water Supply Distribution System Pattern

WATER SUPPLY PATTERNS

Water should be available for fire fighting in any part of the town. Compared to the total requirement of water the actual amount of water used to extinguish fires in a year is very small, but the rate of demand during a fire is very great. Large quantity of water has to be rushed to the spot as quickly as possible. This has got a bearing on size of distribution mains and distribution reservoirs.

There are several types of sanitation technologies available in Indian cities demanding differing water requirements. Dry latrines require the least amount of water while the conventional sewerage require the maximum amount of water for flushing. Between the two extremes there are various levels of water use. The composition of sanitation technology mix, therefore is another factor to project the demand of water supply.

Design Period: For water supply projects, normally a 30 years of design period is assumed. Population forecast for this period take in to consideration the potential of various sectors of urban economy and regional physical and economic linkages.

Per Capita Rate of Water Supply: As a general rule, the following per capita rates are recommended as the minimum for design of public water supply:

- For design population up to 10,000;70-100 litres per day.
- For design population 10,000 to 50,000; 100-125 litres per day.
- For design population above 50,000;125-200 litres per day.

The consumption of water during a year is not always at a constant rate. There will be seasonal, monthly, daily and hourly variations. The maximum seasonal demand may be 1.3 times average annual demand. While designing the mains and purification works these fluctuations are to be considered. The maximum monthly demand may reach up to 1.5 times and average maximum daily demand may be 1.75 times the average and this will be noted while designing the distribution reservoir. The maximum hourly demand may go up to 2.5 times the average, and the distribution pipes will have to be designed for this load.

Peak Factor: The distribution system shall be designed to satisfy the peak demand of the population. The recommended Peak Factors for India are:

Table 3.2: Peak Factor for Water Supply

City Size	Peak Factor of Water Supply
For Population up to 50,000	3 times the average rate of water supply
For Population 50,000 to 2,00,000	2.5 times the average rate of water supply
For Population 2,00,000 and above	2 times the average rate of supply

Pressure: There shall be adequate pressure in the distribution system. The minimum residual pressures at the time of peak demand shall be 7 metres where single sorry houses are common. 12 metres where two storey houses are common. 17 metres where three storey houses are common, and 22 metres where four storey houses are common. The multistorey apartments should have booster pumps and individual underground and storage.

Water Meters: The policy whether meters will be installed is a critical decision. This involve expenditure on salary of municipal employees who read meters, consumers who by water meter as well as revenue from consumers and some control on water use. Therefore policy for metering involves balancing municipal revenue with expenditure and advantage accrued to the municipality for the conservation of water use. It has been observed that metering reduces the water consumption. However private gardens are likely to suffer due to metering which may ultimately affect the appearance of the town. It has been further observed that in not metered conditions whenever intermittent water supply is provided, it increases the consumption contrary to ones expectations.

Phasing: The factors given above need to be considered for different bench mark years corresponding to six five year plan (30 years) periods may be required to project water requirement. Further, the perspective of national, state and local government regarding the investment in water supply need to be realistically assessed for the projection of water requirement.

Efficiency: The efficiency of the water works management will reduce losses and waste. In a management system where leakages of pipes and taps are detected without delay, and stopped, a considerable amount of waste is avoided. This efficiency quantified should be another factor for the demand of water.

SOURCES OF WATER SUPPLY

The search for water sources that meet the water requirement for the next thirty years for a town is an important task. Once the sources are identified, and the quantum of water supply is assessed, it is necessary to protect the water quality and water field by proper environmental protection measures. This should be considered as an important component of the town development plan.

Ground and surface water are the two sources of urban water supply. The surface of the water under the ground level is called the water table. The water table moves up and down according to the fluctuation in the rainfall, temperature, tides and withdrawal of water from the near by wells. A water bearing strata is known as an aquifer. The ground water is under constant movement. It's velocity which is a function of speed and direction can be studied with the use of dyes (colours). Dyes commonly used for the test are fluorescing, uranin and eosin. Powdered fluorescein is reddish brown in colour, but when dissolved in water it becomes green. One part of fluorescein per 40 million part can be easily identified by unaided eye. Test wells are dug at equal intervals in all direction. Dyes in proper quantity is injected in the central well. The water in these test wells are continuously observed for change in colour. The well in which the colour is first detected and the time at which it is observed are noted. The distance between the well divided by the time interval between the dropping of the dye and its reappearance in the next well gives us the speed of the water movement. The position of the test well with respect to the central well, where the colour first appears gives us an idea of direction of flow.

The reappearance of ground water at the surface is either as natural spring or as an artesian spring, the first being the outflow of ground-water in a sloping ground and the second is the result of the coming out of percolated water held under pressure on account of typical ground formation. Natural filtration takes place underground and hence groundwater is generally devoid of suspended impurities, but the risk of dissolved impurities are more in ground-water supply. Sodium chloride if dissolved make water brackish, while carbonates, silicates or chlorides of magnesium and calcium imparts hardness to ground-water.

Yield of a Well

It is always desirable to estimate the quantity of water available from any well. In this context the term yield of a well is defined as the rate of withdrawal or pumping of water from wells without causing failure or drying of wells.

The yield of a well depends on porosity and permeability of aquifer, quantity of water present in aquifer, interference from nearby wells, rate of pumping and water table position.

When water is drawn out from a well, the water in the aquifer from around enters the well under the action of $(H_0 - H_w)$ as shown in the figure 3.4.

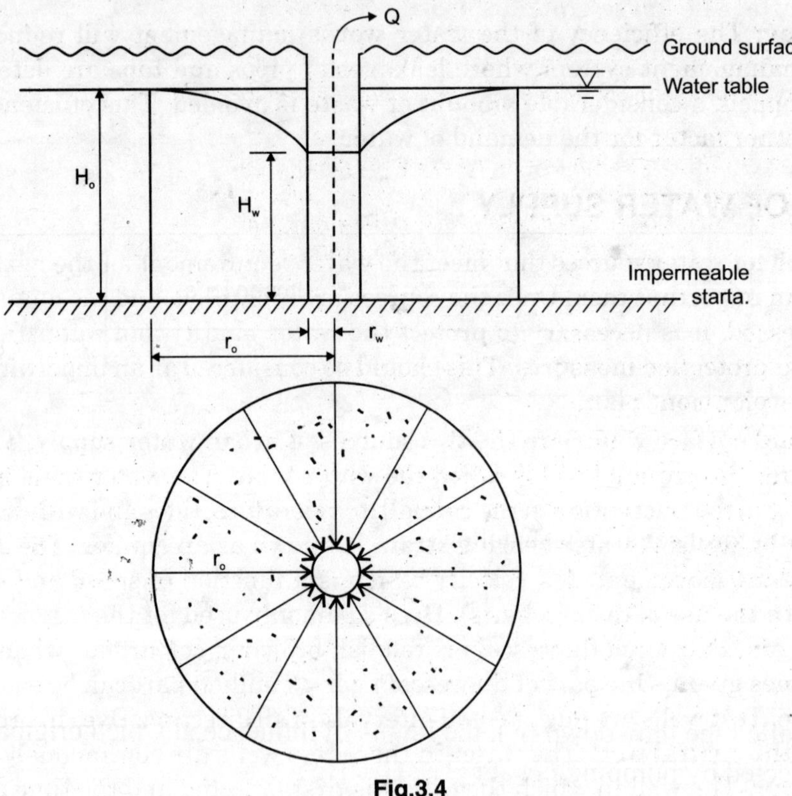

Fig.3.4

By applying Darcy's law for **unconfined** flow, Discharge formula can be developed as given below,

$$Q = \pi K \frac{H_o^2 - H_w^2}{\text{In } {}^{r_o}\!/\!{}_{r_w}}$$

Where, K, is Darey's coefficient of permeably
H_o is Aquifer Thickness
H_w is Depth of water in the well above impervious strata
r_0 is radius of influence.
r_w is radius of the well

Similarly, for homogenous confined aquifer flow equation for discharge through the well can be developed for a fully penetrating well.

Confined aquifers also known as artesian or pressure aquifers, occur where groundwater is confined under pressure greater than atmospheric pressure by overlying, relatively impermeable strata.

The piezometric surface of a confined aquifer is an imaginary surface coinciding with the hydrostatic level of the water in the aquifer. The water level in a well penetrating a confined aquifer defines the elevation of the piezometric surface at that point. Should the piezometric surface lie above ground surface, a flowing well results.

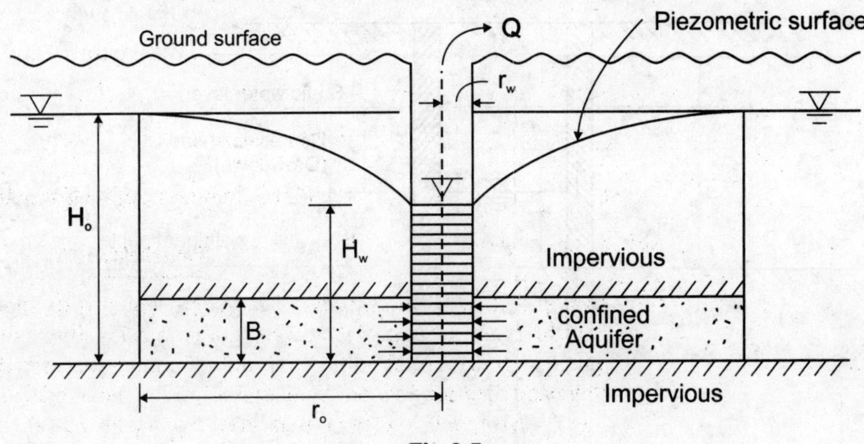

Fig.3.5

The formula for discharge is as follow,

$$Q = 2\pi KB\left[\frac{Ho - H_w}{\ln {r_0}/{r_w}}\right]$$

$H_o - H_w$ is called the drawdown (r_o), the radius of influence at which original water table remains unaffected by pumping *i.e.* at $r = r_o$, $H = H_o$

B = Thickness of the confined strata.

The radius of influence (r_o) value can be approximately found out by sichardt formula,

$$r_o = 3000\, s\sqrt{K}$$

Where, s is maximum drawdown = $H_o - H_w$
K is permeability in m/s.

However, suggested value of radius of influence fall in range of 160m to 320m.

Testing Wells for Yield

After a well is constructed, it should be tested to determine its yield and drawdown. This information provides a basis for determining the water supply available from the well, for selecting the type of pump and for estimating the cost of pumping,

For determining the yield of a well following are two tests,

1. Pumping test or constant level test
2. Recuperation test

1. Pumping test or constant level test

In this test water is pumped from the open well so that the water level in the well is depressed by an amount h known as **depression head** or **drawdown** as shown in the figure...

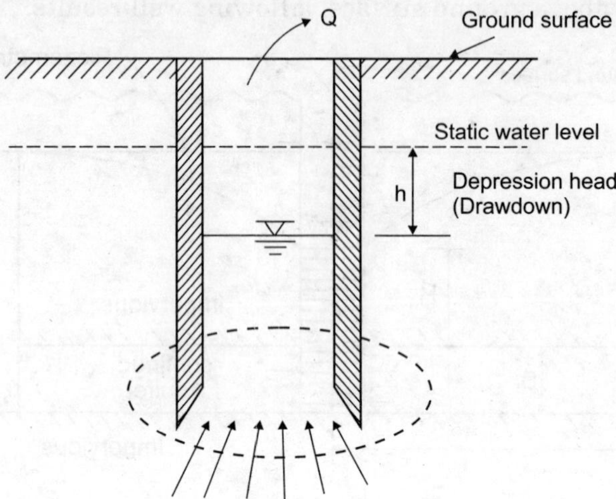

The rate of pumping is then so adjusted that the water level in the well remains constant. In this condition the rate of pumping will be equal to the rate of yield from the well at this drawdown. The rate of discharge can be measured by a suitable discharge measuring device.

The quantity pumped out in one hour gives the yield of the well per hour. Following formula is used to determine yield Q.

$$Q = A \times v \text{ or, } Q = A \times C \times h$$

Where, Q = Discharge in m^3/s

A = Cross – sectional area of the base.

V = Mean velocity of water percolating into the well in m/s.

h is depression head in meters

C is called Percolation Intensity Coefficient.

C depends upon aquifer characteristics. It has greater value for coarser soil. It is expressed in m/s under unit head.

Due to cavity formation the effective area of cross is taken as 4/3 times the actual cross sectional area of the bottom of the well.

The percolation head cannot be increased beyond a certain critical value because beyond this critical value percolation velocity will exceed and the soil particle will be dislodged. This critical value of head is called **critical depression head.** In practice the depression head is kept equal to 1/3 of the critical head. The yield or discharge at this working head is taken to be **maximum safe yield.**

2. Recuperation Test

This is not as accurate as the constant level pumping test, but is resorted to when it is difficult to regulate the pump to maintain a constant level in the well. In the recuperation test, water level is depressed to some level (h_1) below the normal water table level and the pumping is stopped. Now the time (T) taken by water to come back to some other level (h_2) is recorded. From this yield and specific yield can be calculated by the formula,

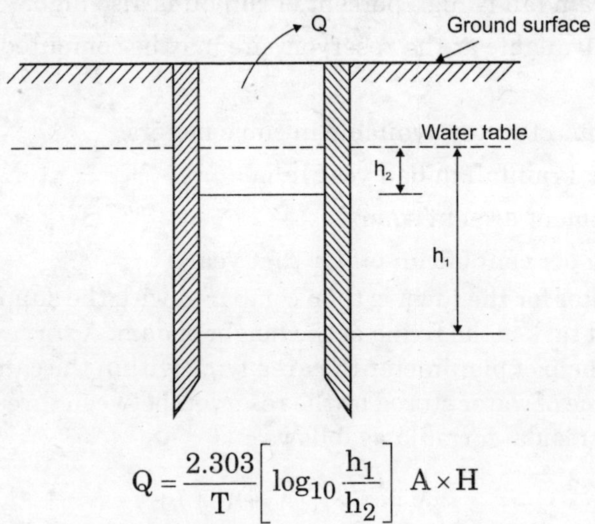

$$Q = \frac{2.303}{T}\left[\log_{10}\frac{h_1}{h_2}\right] A \times H$$

here H is working head

T is time in hours

Q is yield in m^3/hour

$\frac{Q}{AH}$ is called specific Yield.

Ground water supply most often requires costly pumping, but if water is available as surface flow nearby as of a river or stream, surface supply would be preferred as a source of supply. The yield of stream or river can be determined by studying catchment area, average annual rainfall, and the coefficient of run-off.

The catchment area can be determine by studying contour maps of the river on it's upstream. From the study of contour map, the ridge line can be marked. The area, contained within this ridge line up to the point of storage on the river is then measured by a planimeter, and it is the catchment area of the site.

Annual rainfall figures expressed as centimetres or millimetres are recorded in many places. The mean annual rainfall at a place, for 35 years, is called the average annual rainfall. There can be low or high rainfall. The maximum annual rainfall for any two consecutive good years may be 1.5 times the average rainfall and the minimum annual rainfall may be generally taken as 0.7 times the average annual rainfall. If the actual figures are worse than these, it should be considered for the design of storage capacity of the reservoir.

The percent of run-off over a catchment depends on several factors:

- greater the rain fall, greater is the percent of run-off and vice versa.
- in hot months, run-off is low percent and in wet months run-of is high percent.
- the more porous the soil greater is the absorption of water and less percent of run-off.
- the more presence vegetation and trees, there is more of run.
- the steeper the ground, run-off per cent higher.
- if the rate of rain fall is high percent of run-off is also high.

The amount of water available at the reservoir site may be computed as follows:

$W = R \times A \times P$

W = The amount of water available in cubic meters.

R = The annual rainfall in bad year in m.

A = The catchment area in sq.m.

P = The likely per cent of run-off for that year.

The demand of water for the town is then compared with the supply of water available in a bad year. The next task is the fixing of the height of dam. A survey map of 5 m contour is prepared. With the help of planimeter the area lying within the catchment and the dam is computed. The volume of water stored by the reservoir between three successive contours would be given by prismoidal formula as follows:

$$V = \frac{H}{6} \times (A_1 + A_2 + 4A_3)$$

Where H is the contour intervals (5 m) and A_1, A_2, A_3 are the area contained in three successive contours. Using this approach, starting from the lowest contour in the area, the amount of water contained between the successive contours is calculated. A sum of the amounts so calculated is made for every three consecutive areas starting from bottom till we reach a figure greater than the required storage capacity of the reservoir. The contour at which this capacity is reached gives the idea of the height of the proposed dam. Fixing of the dam site is an engineering decision and so also the type of dam.

Distribution of Water

Water supply distribution system is designed to satisfy the demand placed by different land uses, densities and fire-fighting requirement all the time. Piping systems, pumping stations, storage facilities, fire hydrants, house connections, water meters and other appurtenance constitute the distribution system. The distribution system normally costs sixty to seventy percent of the total cost. The cost of pipes and valves normally amount to eighty percent of the cost of distribution system.

Distribution reservoir should be located strategically for maximum benefit. Normally the reservoir should be located near the centre of the use. It will then reduce the friction loss. Reservoir providing service storage must be higher enough to develop adequate pressure in the system.

Storage, pumping and re-pumping regulates pressure. Sizes of pies and valves regulates flow of water and allows for by-pass for repairs, uncontrolled surges, water hammer, leakage and cross connection. Transmission pipes operates under high pressure and have limited number of branches and pressure reducing branches. A good distribution system is one in which adequate pressure is available at different points in the system at the time of maximum demand. The pressure or residual head should be sufficient for fire demand or the consumer.

Configuration of Distribution System: The configuration of the system is dictated primarily by (a) street patterns, (b) topography, (c) degree and type of development of the area and (d) location of treatment and storage work.

Topography and Distribution: If there is changes in topography the distribution system is divided into zones. This will preclude extreme high pressure at lower areas and maintain reasonable pressure at higher elevation.

Grid and Branching System: Distribution system may be generally classified as grid system, branching system or some other combination of these. The grid system is usually preferred since it can furnish supply to any point from at least two direction. While the branching system cannot permit such circulation. Grid system incorporates loop feeder which can distribute water at several directions.

Peak Demand for Rate of Supply: Ideally intermittent supply of water through the distribution system is not advisable and therefore 24-hour supply is suggested as far as possible. The distribution system is to be designed for the peak demand of the population. The recommended peak factors given in Table 3.2 should be followed.

Pipe Specification: Distribution pipes shall not be less than 150 mm in diameter in the Indian metropolitan cities to meet fire-fighting requirements. In other urban areas the minimum pipe diameter shall be 100 mm.

Water Reservoir and Operating Storage

Purified water is available from water works at constant rate, but the demand of water changes with the hour and at peak hour the rate of demand is very high. So, in order to get a balance between supply and demand, a balancing reservoir or a distribution reservoir becomes necessary. Care has to be taken to see that enough storage is available for fire protection, so that fire demand can be continuously met for six hours.

The operating storage can be determined from mass diagram or hydro graph (time / discharge graph). Mass diagram is a plot between cumulative demand of water and time.

For illustrating the computation for operating storage for a hypothetical city the following tabulation of hourly demand is given:

Table 3.3: Hydrograph (Time Discharge Graph)

Time in Hrs.	Av. Hrly. Demand 1000 g pm	Hrly. Demand 1000 g	Cu'tive Demand Hrly. 1000 g	Demand % of av.	Av. Hrly. minus demand	Demand Hrly.
0	0.00	0	0	0.00		
1	2.17	130	130	45.41		156.25
2	2.10	126	126	44.02		160.25
3	2.02	121	377	42.27		165.25
4	1.97	118	495	41.22		168.25
5	1.98	119	614	41.57		167.25
6	2.08	125	739	43.67		161.25
7	3.63	218	957	76.16		68.25
8	5.19	312	1269	109.00	-25.75	
9	5.62	337	1606	117.73	-50.75	
10	5.90	354	1960	123.67	-67.75	
11	6.04	363	2323	126.81	-76.75	
12	6.32	379	2702	132.40	-92.75	
13	6.44	387	3089	135.20	-100.75	
14	6.37	382	3471	133.45	-95.75	
15	6.32	379	3850	132.40	-92.75	
16	6.34	381	4231	133.10	-94.75	
17	6.64	399	4630	139.39	-112.75	
18	7.32	439	5069	153.36	-152.75	
19	9.33	560	5629	195.63	-273.75	
20	8.32	499	6128	174.32	-212.75	
21	5.05	303	6431	105.85	-16.75	
22	2.57	154	6585	53.80		132.25
23	2.47	148	6733	51.70		138.25
24	2.29	137	6870	47.86		149.25

Total		6870			-1466.50	1466.50

Average hourly demand = 6,870,000/24 = 2,86,250 gallons. *1 gallon = 4.546 litres*

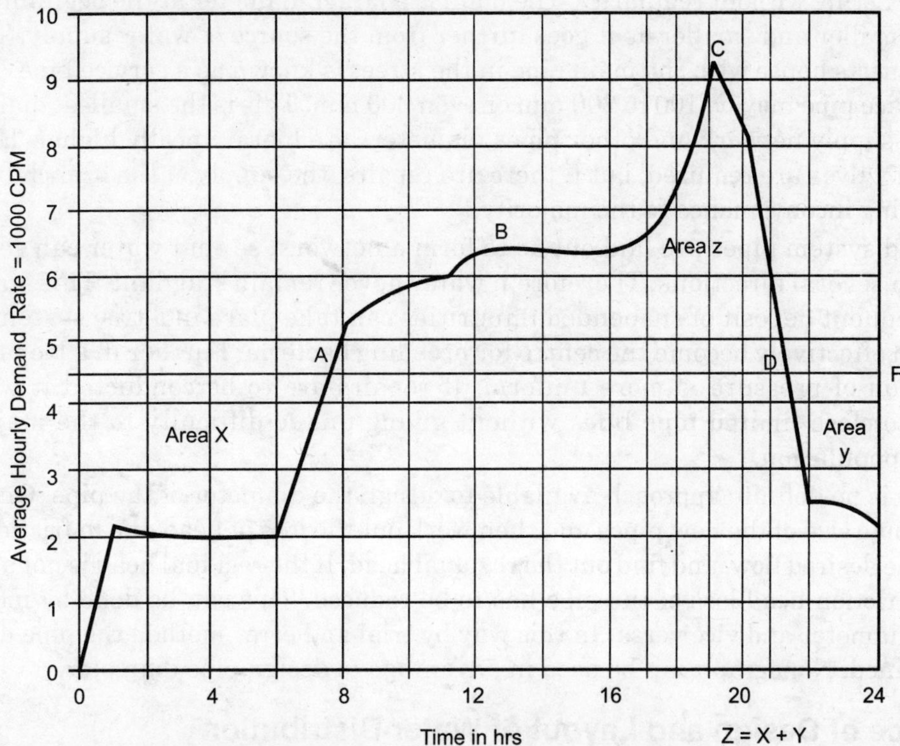

Fig.3.6: Graphical Determination of Equalising Water Storage.

Source: Institute of Town Planners, India, 1990, Network and Services—A Reader, New Delhi

Above figure plots hours of water supply with rate of demand. If the inflow is continuous for 24 hours, the inflow curve will be a straight line parallel to X axis. The rate of demand is however fluctuating as shown. When the demand flow goes above the inflow line there is an excess of demand which must be met with water stored in the reservoir. Thus ABCD shows excess of demand. The area rate of demand hours gives the demand in thousands of gallons/ litres. If the area above inflow line is made equal to area below inflow line then excess of inflow is equal excess of inflow.

Distribution System and Pipe Diameter

The spatial distribution of water consumption can be estimated by studying population densities, and commercial and industrial water use patterns. Predicted hydrograph for each zone is required for assessing the combined peak flow requirement.

The design of water distribution networks involves the selection of a system of pipes so that the predicted design flow can be carried with head losses which do not exceed those deemed necessary for adequate operations of the system.

Water supply networks are laid out as a tree system (also known as dead end system), grid iron system (also known as loop system) or a combination of tree grid iron system. Tree system is like a tree with branches and is suitable for towns that have one central road and by-lanes existing without regularity. The main is a larger diameter at the beginning and it becomes smaller and smaller as it goes further from the source of water supply. The pipe which connects house with the main pipe in the street is known as a service pipe. The size of the service pipe may be 100 to 200 mm or even 300 mm. This is the smallest diameter in the water supply network and other pipes diameters are hierarchically higher. Here less number of valves are required, but if there are repairs, the supply of the branch has to be cut off giving inconvenience to the majority.

In grid system pipes are laid out as to form a network, so that water can reach one point from several directions. Therefore,n water never remain stagnant. This stagnancy and consequent deposit of suspended impurities can take place in a tree system. These points can effectively become the centre for breeding bacteria. Further in a loop network distribution of pressure is more uniform. If repairs are to be conducted it would be possible to close limited pipe lines without giving undue difficulty to the majority of the town population.

There is no definite approach available to design the diameter of the pipe. One has to assume some size of the new pipes and then work out the loss of head due to friction in the pipe for the desired flow, and find out the residual head. If the residual head is not adequate then the friction head loss of the pipe has to be reduced. This can be done by increasing the pipe diameter and vice versa. In this way by trial and error method the pipe diameter is determined. Nonograph can be used in advantage to design pipe diameter.

Sequence of Design and Layout of Water Distribution

(a) On a development plan of the area to be serviced sketch the tentative location of all water mains which will be required to supply the area. The sketch should distinguish the proposed feeder mains and small service mains by colour coding.

(b) The pipe lines shall be interconnected at the interval of 400 m (1200 feet) or less. Looped feeder system shall be used wherever possible. Two small feeder line running parallel several blocks apart are preferable to a single large main with an equal or slightly larger capacity pipe than the two mains combined.

(c) Using estimated values of anticipated design flow select appropriate pipe size by assuming velocities of 1 to 1.5 m/ s (3 to 5 ft/ sec) from a monograph. Anticipated water flow is determined from demand of the district and peak factor.

(d) Mark the position of building service connections, fire hydrants and valves. Fire hydrants should not be further than 170 m (500 feet) apart. It is ideal to be located at the intersection of roads.

(e) Service connections from the link between the distribution system and individual consumer normally are one customer per service pipe.

(f) Using Hazen Williams formula or other methods given below, compute pressure at key points in the distribution system. From this information construct a map of pressure contours and or hydraulic gradient line (HGL).

Hazen Williams equation is as given below:

$$V = 0.849\ C\ R^{0.63}\ S^{0.54}$$

V = Average Velocity (m/s)

R = Hydraulic Radius $= \dfrac{A}{P}$ (m)

A = Area of cross-section (m^2)

P = Wetted perimeter $= \dfrac{h_f}{L}$

S = Energy line slope

h_f = Loss of head (m) due to friction in (m)

L = length of pipe (m).

By continuity equation.

$$Q = A \times V$$

Table 3.4: Value of 'C' in Hazen William Equation

Type of Pipe	Value of C
Asbestos	140
Cast Iron Pipe (5 years old)	120
Cast Iron Pipes (10 years old)	110
Cast Iron Pipe (30 years old)	80-90
Concrete Pipe	120-140
Steel pipe riveted or welded	110
Copper	130-140
Plastic	140-150

Selection of Pipe

Formula used is of Darcy-Weisbach formula.

$$h_f = \frac{f\,Lv^2}{2gD}$$

where f is Coefficient of friction.

d = Diameter of pipe (m)

value of (f) can be had from Moody's diagram based on Reynold's number.

Use of Nomograms

Nomograms are in use for various purposes. A pipe sizing nomogram cab also be used for a given Q and velocity.

Use the nomogram below to determine Hose ID Size based on Flow (Q) and Velocity.

Conversely, it can be used to determine Velocity, based on Flow (Q) and Hose ID,or Flow (Q) based on Velocity and Hose ID.

To use the nomogram, draw a straight line across the entire nomogram, intersecting the scale for the two known quantities.

In the example shown by the red line below, the desired Velocity of 25 feet per second and the desired Flow of 10 Gallons per minute were known. The straight line intersecting the two known quantities intersects the Hose ID scale at 1/2" Therefore, a hose with 1/2" ID (-08) would be selected.

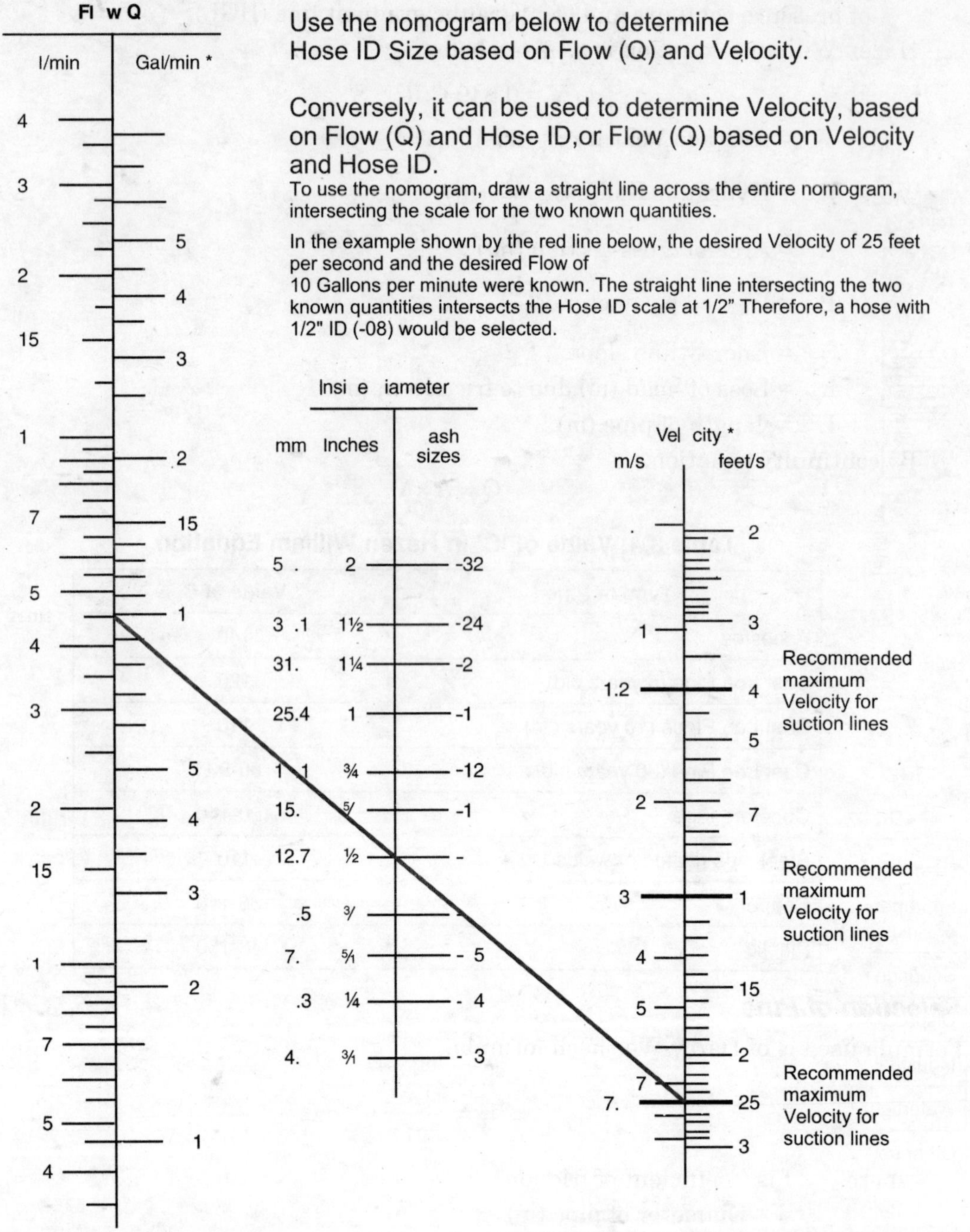

Fl w Q

l/min | Gal/min *

Insi e iameter

mm | Inches | ash sizes

5 . | 2 | -32
3 .1 | 1½ | -24
31. | 1¼ | -2
25.4 | 1 | -1
1 .1 | ¾ | -12
15. | 5/ | -1
12.7 | ½ | -
.5 | 3/ |
7. | 5/1 | - 5
.3 | ¼ | - 4
4. | 3/1 | - 3

Vel city *

m/s | feet/s

Recommended maximum Velocity for suction lines

Recommended maximum Velocity for suction lines

Recommended maximum Velocity for suction lines

* Gallons shown are Imperial (British) Gallons. To convert to US Liquid Gallons, multiply by 1.2009.

Other Conversions: Gal/min (Brit) x 4.546 = l/min
feet/s x 0.3048 = m/s

* Recommended Velocities shown are based on hydraulic fluids with maximum viscosity of 315 SSU at 100° F (38° C) working at room temperatures within the range 65° F (18° C) and 155° F (68° C).

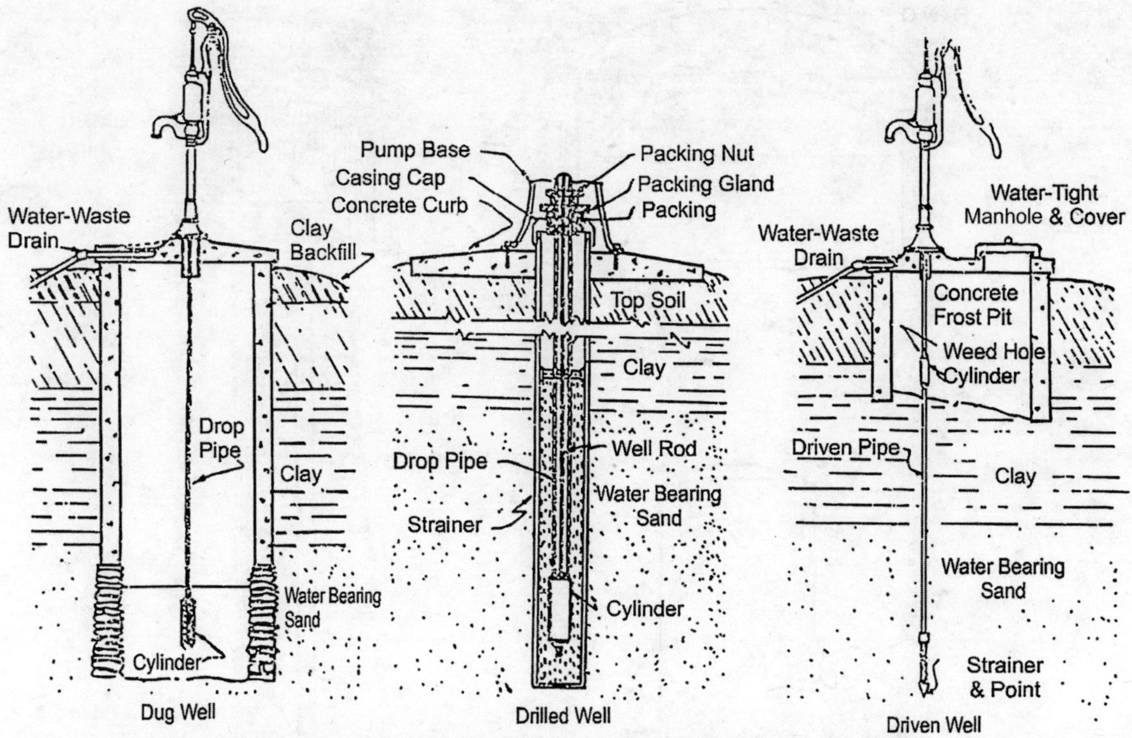

Fig.3.7: Sectional drawing of Dug well, Drilled Well and Driven Well

Table 3.5: Minimum Distances of Well from Source of Pollution

Distance From	Dug Well	Bored Well	Driven Well	Drilled Well
Property line	30m (100')	15m (50')	15m (50')	6m (20')
Improperly abandoned well or sinkhole		Unacceptable		60m (200')
Seepage pit	60m (200')	60m (200')	60m (200')	30m (100')
Disposal field of bed	60m (200')	60m (200')	60m (200')	30m (100')
Industrial lagoon		Unacceptable		(2')
Watertight sewer lines	15m (50')	15m (50')	15m (50')	3m (10')
Other sewer lines	30m (100')	30m (100')	30m (100')	15m (50')

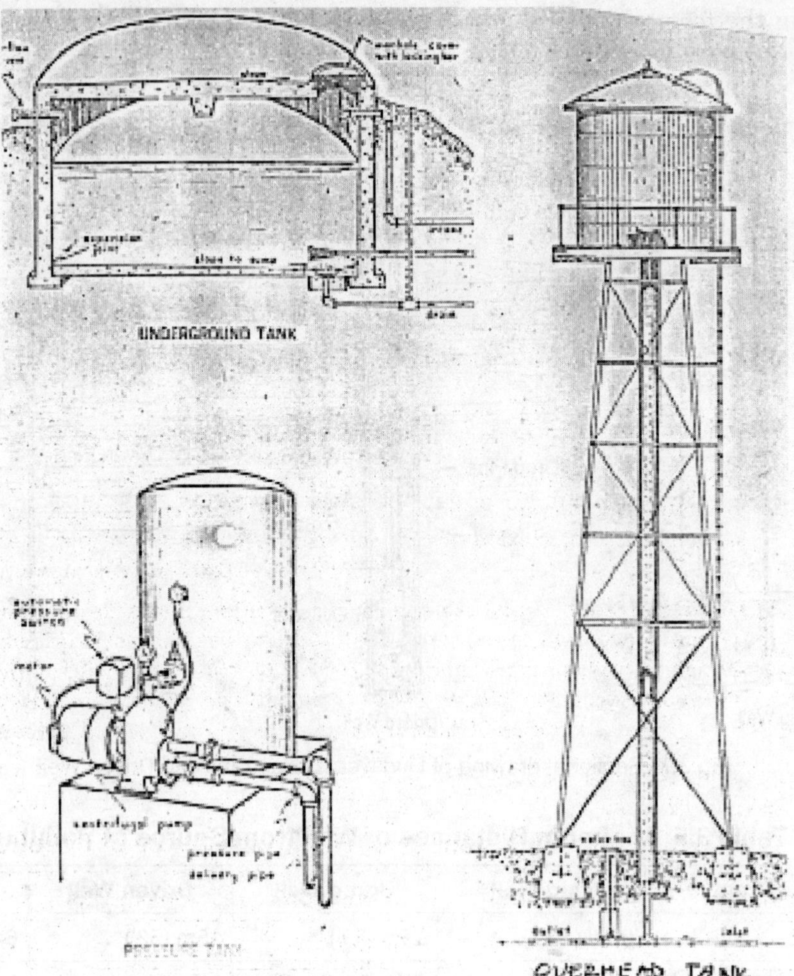

Fig.3.8: Water Storage Tanks.

INFILTRATION WELL

Most rural areas in the central and western Himalayas suffer an acute shortage of drinking water. The traditional systems of harvesting drinking water. from shallow sub-surface capillaries into *annuals/bauris* are unable to meet the needs of the ever-growing population. Moreover, the environmental degradation of critical catchment areas has affected the sustainability of primary water sources of communities.

To tackle this problem, the government has been investing large funds for the creation of public corporations, which are responsible for planning, implementing, operating and maintaining piped water supply systems in remote villages. These water supply schemes are mainly based on secondary or tertiary water sources, which are not within the jurisdiction of the concerned villages. As a result, appraisals reveal that for a variety of reasons, the water crisis in the area is far from resolved. Grassroots, a Kumaon based NGO, has been demonstrating the feasibility of an appropriate technology option in the drinking water sector through the construction of infiltration wells.

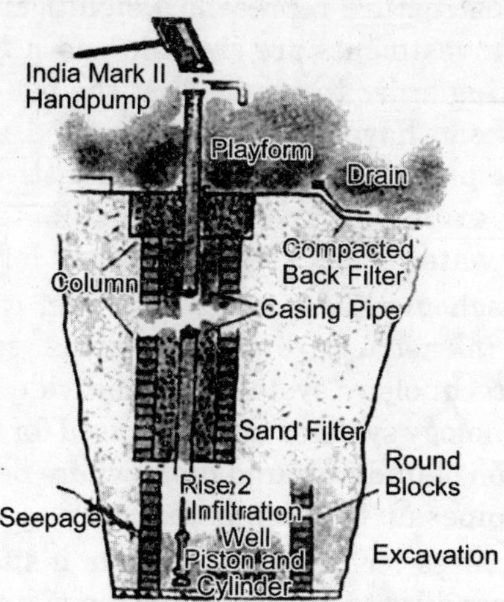

Fig.3.9: Standard Infiltration Well with India Mark II Handpump.

Source: Pan Himalayan Grassroot Development Foundation, Ranikhet

Infiltration well technology is merely an extension of the traditional system; instead of *annuals/bauris* which are only 3-6 feet deep, infiltration wells are located about 25 feet below ground level. This allows additional quantities of water to be harvested. Curved concrete well blocks made with a simple well block mould are laid at the base of the well to form a circle. The annular ring is backfired with clean sand and the cap sections laid. Cast column blocks are used to form a column around the casting pipe, which is inserted through a hole in the cap section. Sand backfilling continues till the cap is covered to a depth of 30 cm, after which the excavated material is used.

A block column is built to reach the desired elevation at which the feet of the handpump will be set. A platform is constructed, usually with the pump at the centre. The pump discharge faces downhill so that it does not seep back into the infiltration well. The wells are capped and fitted with a handpump, which prevents pollution. The construction of water seal toilets is encouraged to break the cycle of the spread of waterborne disease. Since infiltration wells are based on primary water resources, user groups are motivated to protect and conserve the environment of the immediate catchment area, providing sustainability to the source. A major advantage of these infiltration wells is that the drudgery of women collecting water from distant sources is considerably reduced.

Infiltration wells are easy to construct, and the operation and maintenance of handpumps/submersible pumps are within the capacities of local communities, albeit with some initial training.

User groups have the option to recover a small fee each month from each household to pay for the operation and maintenance of these decentralized water supply systems. Over the last few years, this appropriate technology option has spread to nearly 200 hamlets/ villages in seven districts in Uttaranchal and Himachal Pradesh.

TRENCHLESS TECHNOLOGIES

Water and wastewater infrastructure represent a significant investment on the part of most municipalities. These investments are sizeable and it is critical that infrastructure be maintained in good working order to ensure that the benefits from these investments are fully realized. Where systems have not been maintained, and in some cases where land subsidence has occurred, the problem of leaking water distribution and sewage collection pipes can be extensive. For example, experts have estimated that in the industrialized world, 20-40% of all treated water is lost as "unaccounted for" water.

Conventional open trench methods, while effective, can be costly and disruptive in areas where significant infrastructure already exists, such as buildings and roads. In many cases, trenchless technology systems can provide an innovative, cost-effective alternative. Trenchless technology systems are often used for the installation, replacement and renovation/ rehabilitation of underground pipes where personnel entry is not possible. This is usually the case for pipes up to 900 mm diameter.

Before the type of trenchless technology is chosen, a site visit is made to determine the soil and groundwater conditions surrounding pipes and ducts. Following this, an inspection is conducted to ascertain the present condition of pipes. Finally, a location survey is performed to identify any surrounding pipes or services that may inhibit work in the immediate area. There are several methods of trenchless construction used to install new pipeline infrastructure. The most common methods are horizontal directional drilling and micro-tunnelling.

Horizontal Directional Drilling

Horizontal directional drilling is a trenchless surface-launched method for installing relatively small diameter pipes. This method can be used to install various types of pipe, including polyethylene, polyvinyl chloride, steel and copper. Applications include water, gas and cable installations, usually in soft ground. Gravel-heavy soils and rock can present difficulties when using this technique.

Micro-tunnelling is used when constructing pipelines where close tolerance for line and grade is required. This method involves the use of a remote-controlled, laser-guided pipejacking system, for which personnel entry is not required. The process can be used for a range of pipeline diameters.

Micro-tunnelling can be cost-effective compared to open-cut construction when pipelines are to be installed in congested urban or environmentally sensitive areas, in unstable ground, or below the water table. Micro tunnelling has been used to install concrete, steel, centrifugally spun-cast polyester resin, fibreglass-reinforced and vitrified clay pipes. PVC pipe has also been installed using this technique.

Replacement

There are a variety of methods for pipe replacement, whereby rigid replacement pipes are pushed through an existing pipe or hole in the ground.

Pipe Bursting

Pipe bursting is a replacement method commonly used in urban centres where the area immediately below the paved surface is often congested with existing services, allowing little space for replacement of a defective service with a new line. The existing pipe/ hole in the ground thus becomes the guide for replacement.

Replacement occurs by splitting the defective pipe and displacing the fragments to enable a new pipeline of the same diameter, usually polyethylene, to be drawn in. Where increased capacity is needed, an expanding device, either hydronic or pneumatic, is introduced into the defective pipeline to shatter the pipe; then the new line is pulled through.

Pipe bursting is not appropriate in all sub-surface conditions. The forces exerted by the bursting head may adversely affect other pipelines near the pipeline being rehabilitated.

Rehabilitation

In recent years, rehabilitation of underground pipes by relining the deteriorated host pipe with an inner pipe or lining material has gained acceptance. The most common methods are cured-in-place pipe and fold-and-formed pipe.

Cured-in-Place Pipe

Cured-in-place pipe systems enable pipelines to be repaired from within by inserting a lining material through existing manholes or other entry points. The liner is composed of a fabric reconstruction tube which is impregnated with a thermosetting resin. After insertion, the liner is exposed to hot circulating water or steam and hardens into a structurally sound, joint less pipe. The rehabilitation liner not only serves to repair the deteriorated structure of the existing pipe (sometimes referred to as the host pipe or casing), but reduces inflow, infiltration and leakage, and usually improves the flow characteristics of the system.

Installation methods for cured-in-place pipe systems are as varied as the types of materials used. In some cases, the tube is inverted through the pipe using water or air pressure; in other cases, a winch is used to pull the tube through the pipe. The curing process usually requires hot water or steam to be applied to the liner for several hours after it has been formed within the host pipe.

Fold-and-Formed Pipe

More recently, some companies have introduced systems that can install folded plastic pipe relining products through a manhole entry. These systems, referred to as fold-and-formed pipe systems, utilize thermoplastic materials that have been folded from a circular shape to produce a smaller net cross-section that can be fed into an existing pipeline. These fold-and-formed pipe products are usually extruded PVC or high density polyethylene (HDPE) pipe that is flattened and folded longitudinally. The plastic pipe is fed from a spool into an existing pipe, then hot water or steam is applied until the liner reaches a temperature elevated enough for rounding. After rounding, the installed liner is allowed to cool, producing a snug-fitting liner in the host pipe.

Slip Lining

Slip lining is another rehabilitation method where a new pipeline of a smaller diameter is inserted into the defective pipe and the annulus (the space between the existing and inserted pipes) is grouted. It has the benefit of simplicity and is relatively inexpensive; however, there can be a significant loss of hydraulic capacity.

A modified version of slip lining, often called close-fit lining, makes use of the properties of polyethylene or PVC to allow temporary reduction in diameter or change in shape prior to insertion in the defective pipe. The inserted pipe is subsequently expanded to form a tight fit against the wall of the original pipe, thus avoiding the need for annulus grouting as in conventional slip lining. Temporary reduction in diameter is achieved either by mechanical rolling or drawing through a reduction die.

Table 3.6: Delivery from Different Sizes of Taps

Diameter of pipe (in mm)	15	20	25	32	40	50
Flow (in litres/minute)	10	20	30	45	70	110

Table 3.7: Recommended Sizes of Service Mains for Housing

S.N. Class of Building	Size of service main (mm)
1. Ordinary single family dwelling 2 to 3 storeys and not more than 10 rooms	32
2. Larger dwelling with about 16 rooms	40
3. Four apartment building with about 24 rooms	50
4. Twenty five apartment building with about 100 rooms	50

Notes:

- Service main should not be normally less than 25 mm dia. For very smaller houses it can be reduced to 200 mm and if the pressure is very high then it can be reduced to 15 mm.
- Where it is necessary for a pipe to pass through a wall or floor, a sleeve shall be fixed therein through which pipe should be passed and the gap later should be filled by a sealant.
- Tables above are on the assumption of having that much pressure in the system so as to have the exit velocity (at the taps) of about 1 m/s.

Table 3.8: Relative Discharge Capacities of Full Pipes

Relative Diameter of Pipe	Relative Discharge Capacity	Relative Diameter of Pipe	Relative discharge Capacity
1	1.00	15	871
1.5	2.76	16	1024
2	5.66	17	1192
2.5	9.90	18	1375
3	15.60	20	1789
4	32.00	22	2270
5	56.00	24	2822
6	88.00	26	3447
7	130.00	28	4149
8	181.00	30	4930
9	243.00	32	5793
10	316.00	34	6741
11	401.00	36	7776
12	498.00	38	8870
13	609.00	40	10150
14	733.00		

Note: Discharge varies as $\sqrt{d^5}$ *approx. (head remaining the same).*

Table 3.9: Number of Connections of Various Sizes that can be Fed from a Main Line (that will together have the same discharge as the main line)

Dia. of Delivery very Main (mm)	Diameter of Branch Pipe (mm)														
	15	20	25	32	40	50	65	80	100	125	150	200	250	300	350
	Number of Branch Connections														
15	1														
20	3	1													
25	6	2	1												
32	10	3	1.7	1											
40	16	6	2.7	1.6	1										
50	32	12	5.6	3.2	2	1									
65	56	20	10	5.6	3.6	1.7	1								
80	88	32	16	8.9	5.6	2.7	1.6	1							
100	181	66	32	18	11	5.6	3.2	2	1						

125		115	56	32	20	10	5.6	3.2	1.7	1					
150		181	88	50	32	15	8,9	5.6	2.7	1.6	1				
200			181		66	32	18	11	5.6	3.2	2	1			
250			316		115	56	32	20	10	5.6	3.6	1.7	1		
300						88	50	32	16	8.9	5.6	2.7	1.6	1	
350						130	74	47	23	13	8.3	4.0	2.3	1.5	1
400								66	32	18	11	5.6	3.2	2.0	1.4
450								88	43	25	16	7.6	4.3	2.7	1.9
500								115	56	32	20	10	5.6	3.6	2.4
600									88	50	32	16	8.9	5.6	3.8
700										74	47	23	13	8.3	5.6
750										88	56	27	16	10	6.5

$$\text{Number of branch connections} \simeq \sqrt{\left(\frac{\text{Dia. of larger pipe}}{\text{Dia. of smaller pipe}}\right)^5}$$

Table 3.10: Recommended Velocities for Water Pipes

Pipe dia.	Velocity in m/sec.	Pipe dia.	Velocity in m/sec.
100 mm	0.75 to 0.91	500 mm	1.40
150 mm	0.81 to 1.22	650 mm	1.69
200 mm	0.91 to 1.32	800 mm	1.92
250 mm	0.99 to 1.52	900 mm	1.95
320 mm	1.09 to 1.67	1000 mm	2.00
400 mm	1.27 to 1.83	Service pipes	2.13

Notes:

- Velocity in pipes should not be less than 0.75 m/s to prevent silting in pipes.
- Velocity in pipes shouldn't exceed 3 m/s because of the following reasons:
- It causes noise.
- It causes wear and tear of pipe.
- It causes heavy frictional losses thus reducing effective head. Frictional losses in pipe increase as the square of velocity.
- As the diameter increases, recommended velocity is also increased because with the larger diameter, same velocity causes less friction.

Table 3.11: Some Thumb rules for Water Supply and Sewerage Works

S.No.	Items	Dimension
1.	**Manhole sizes**	
	(a) For depth up to 3' (0.9 m)	900 × 800 mm rect. manhole
	(b) For depth 3'–5.5' (0.9 – 1.65 m)	1200 × 900 mm rect. manhole Or 3' (0.9 m) dia circular manhole
	(c) For depth 5.5' – 14' (1.65 m – 2.25 m)	4' (1.2 m) dia circular manhole
	(d) For depth 7.5' – 14' (2.25 m – 4.2 m)	5' (1.5 m) dia circular manhole
	Notes:	
	(i) All dimensions are clear internal dimensions.	
	(ii) Min. depth of sewer manhole (clear = 0.45 m.	
	(iii) Circular manholes will taper with height and their top dia (clear) will be only 0.56 m.	
	(iv) Min. dia bottom dia. of circular manhole = dia. of pipe +2'.	
2.	**Spacing of manholes**	
	Pipe dia (mm)	Max. Spacing (m)
	Up to 200 20 200–300	30
	300–500	45
	500–900	75
3.	**Floor depressions (in toilet sunken slabs)**	
	for I.W.C.	500 mm
	for E.W.C.	250 mm
4.	**Filling material for toilet sunken slabs**	
		(i) Cinder concrete 1:15 (1 cement, 15 cinders of 15.20 mm nominal size)
		(ii) Brick bat coba is mixture of Lime+ Surkhi + Brickbats
5.	**Slope of sanitary pipes in building**	1 : 10 to 1 : 50
6.	**Normal sizes of soil and waste to pipe inside the bldg.**	
	(i) Antisiphonage or vent pipe	2" C.I. Pipe
	(ii) W.C. soil pipe	4" C.I. Pipe
	(iii) Wash basin and bath waste pipe (below floor trap)	3" C.I. Pipe
	(iv) Wash basin, sink outlet (waste pipe above the floor) G.I. Pipe (can be PVC, Brass or Lead pipe etc.	11/4", 1.5" or 2" dia 32, 40 or 50 mm .
	(v) Min. dia. of a sewer line outside the building	-150 mm
	Min. dia. of a trunk sewer or public sewer	-250 mm
7a.	**Rungs to be provided in all manholes greater than 0.8 m depth**	
	Top rung	45 cm below manhole cover
	Lowest rung 30 cm above benching Intermediate Rungs	30 cm C/C

7b.	**1 Standard Box of Ceramic tiles contains**	
	150 tiles of size 4" × 4"	to cover an area of
	100 tiles of size 6" × 4"	1.5 m^2 on floor
8.	**Water seals**	
	Gully trap	75 mm
	W.C. trap	50 mm
	Floor (or Nahni) trap	35 mm
9.	**Height of vent pipes above roof parapet**	60 cm
	above topmost window	2 m
10.	**Clearance required between building walls and soil/waste pipes**	5 cm
11	**Maximum distance between manhole and gully chamber**	
12	**I.S. Codes for C.I. Pipe**	6 m
	(i) Specn. for centrifugally cast (spun) iron pipes	IS : 1536
	(ii) Specn. for C.I. fittings	IS : 1538
	(iii) Code of practice for laying C.I. pipes	IS : 3114
	(iv) Specn. for lead for caulking	IS : 782
13.	**Grades of PVC pipes**	**Test Pr.**
	Low Pr.	4 kg/cm^2
	Medium tr.	6 kg/cm^2
	High Pr.	8 kg/cm^2
14.	**Grades of A.C. Pipe**	**Test Pr.**
	Class1	5 kg/cm^2
	2	10 kg/cm^2
	3	15 kg/cm^2
	4	20 kg/cm^2
	5	25 kg/cm^2
15.	**Grades of cast iron centrifugally cast (spun) Pr. pipes**	**Test pr.**
	Class LA	12 kg/cm^2
	A	18 kg/cm^2
	B	24 kg/cm^2

Notes:

- Class A pipes have 10% more wall thickness than class LA.
- Class B pipes have 10% more wall thickness than class A.
- Cl pipes centrifugally cast are available both as flanged pipe and to S & S pipes.

16.	**Trenches for C.I. Pipes**	
	Depth of trench	1 meter under road; 0.75 meter elsewhere
	Width of trench	Nominal dia. of pipe + 40 cms with min of 55 cms in case of soil and 1 metre in case of rock.

Note: For exact details of width of trenches for laying underground. pipes etc.

17. **Trenches (or excavation) for G.I. Pipe**

Dia of Pipe	Width of trench	Depth of trench
15 mm to 50 mm	30 cm	60 cm
65 mm to 100 mm	45 cm	75 cm

- The depth of G.I. pipe below road should be minimum of 3'.
- The pipes shall be laid on a layer of 7.5 cm thickness sand and sand will be filled up to height 15 cms above and around the pipes. Remaining portion above the pipe can be filled with excavated earth.

18 **Width of bed concrete for underground pipe**
(C.I. or S.W. or C.C. pipes) — dia. of pipe + 30 cms (15cm both sides of barrel)

19 **Manholes size for storm water drainage**

1. For depth < 1 m — 800 × 800 mm
2. For depth 1 m to 1.5 m — 1200 × 900 mm
3. For depth > 1.5 m — 1.4 m internal dia.

20 **Identification of hot & cold water taps in W.B. or Sink or Bath Area**
Hot water tap (Red mark) - on left side
Cold water tap (Blue mark) - on right side

21. **Manholes to be provided at** — **Every change of alignment.**
22. **Every junction of sewers.**
23. **Every change in size of sewers.**
24. **At beginning and end of all sewers.**
25. **Type of solids in water** — Suspended solids-
Colloidal matter like clay, silt in suspension cause turbudity in water, can be removed by fine filter and by coagulation. Clay particles are -ve charged.
Dissolved solids-
Can be removed by evaporating or boiling water.
Settleable solids-
Heavy solids can be removed in sedimentation tanks by reducing the velocity of water and keeping it still for certain time. More bigger solids can also be removed by screens.

23. **Nature of stale sewage** - Acidic
Nature of fresh and treated sewage - Alkaline

24. **Volume and discharge in a pipe** at a point in a network of water system fed through a pump or water tank will depend upon:
    1. Size of pipe.
    2. Elevation of tank/capacity of pump.
    3. Distance or length of pipe up to that point.

		4. Type of pipe. 5. Number of fittings. 6. Branches in pipes.
25.	**G.I. Pipes normally manufactured in size**	8 mm to 100 mm
26.	**Flushing Cisterns for Urinals.**	• Automatic flushing cistern: Here as soon as cistern fills, it automatically empties. If rate of inflow higher it will empty at closer intervals. So stop cock can be adjusted accordingly. • Constant flow cistern: Here water flows out constantly from cistern to urinal bowls. Here if inlet is opened more so that rate of inflow rate of outflow, then water will start overflowing from cistern.

Maintenance and Installation of Pumps

Maintenance of Existing Pumps

- The worn-out impeller should be replaced with a new one so that it may suck water efficiently.
- If the discharge is sand mixed, the filter should be checked and replaced if necessary.
- If the pump does not lift water, the leakage on suction side should be checked. The reflux valve (N.R.V) flap may have some foreign material or rubber therein may be damaged/ worn out. Remedial measures may be taken accordingly.
- If the discharge is not regular, check the joints in suction pipe and plug the leakage.
- In case of cavity bores the impellers of pump should be checked regularly to see that there is no foreign material.
- The worn- out bushes, bearings and shaft should be got repaired/replaced during annual maintenance of pump.
- Whenever the shaft gets over-heated, the bearing and shaft should be greased after checking the same.
- Gland packing should be of proper quality. Never tighten a gland more than necessary. The water from the gland should drip at the rate of 15-30 drops per minute.

Installation of New Pumps

- The pump should be selected keeping in view the required discharge and total head. Every pump manufacturer has design charts for their various types of pumps. In these design charts, you can choose the pump of the requirement which gives you the required discharge against the stipulated head. One such design chart of a manufacturer (M/ s Best & Crompton) is enclosed herewith for your ready reference.

- ISI marked monoblock set should be installed, because it is more efficient.
- Compare the total head marked on name plate of pump with the required total head. The required head should be less than the head marked on the name plate by one/ two metres.
- High end suction pump may be used, if necessary. High end suction pumps give 25% more discharge of water.
- The pump set should be installed at a height not more than 10' from water level.
- The pump set should be securely fixed with nuts and bolts on levelled and strong foundation.
- Good quality joints should be used so that there is no air leakage.

The reflux valve should be of standard quality/ISI marked whose flap should open fully.

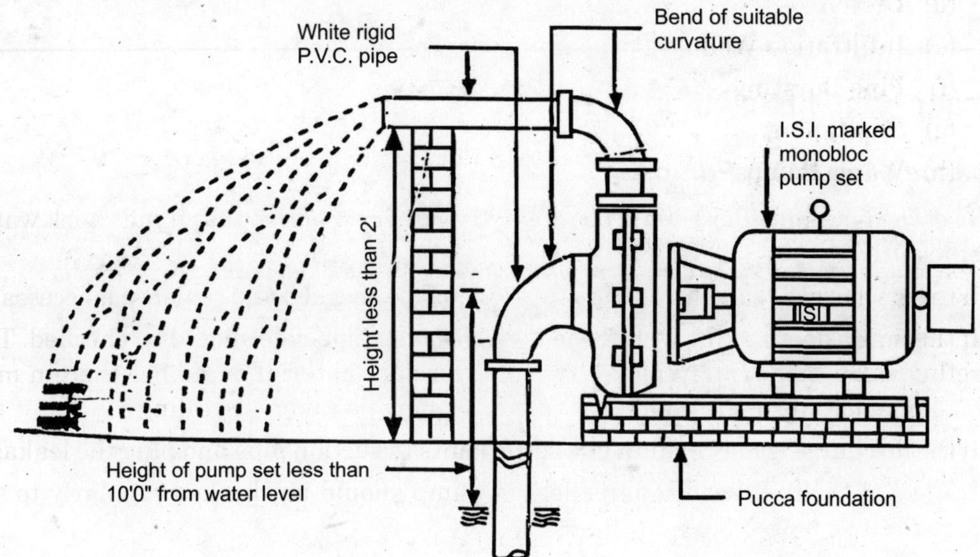

Fig.3.10: Diagram Showing Requisites of an Efficient Monoblock Pump.

- The pipe should be selected according to discharge.
- The size of suction pipe should be one number greater than that of delivery pipe.
- There should be no sharp bends in the delivery pipe. The bends should have large curvature.
- The delivery pipe should be near the ground and not more than 2' in height from ground level/ water channel.
- ISI marked LT shunt capacitors of requisite capacity should be installed on electric motors of the pump sets.

DEBRIEFING QUESTIONS

3. It is said that health and hygiene of the family depend upon availability of pure water in the premises. What do you understand with this statement from engineering point of view?
4. What are the norms of water supply for domestic and public buildings?
5. Explain the approaches and main considerations of water supply plan of a colony and building.
6. What are ways to achieve water efficiency in a building?
7. What is understood by dual piping system?
8. Write short notes on:
   (a) Distribution System
   (b) Grid and Branching System
   (c) Water supply pipe materials and specifications
   (d) Driven Well
   (e) Infiltration Well
   (f) Pipe Bursting
   (g) Slip Lining
   (h) Water Pumps

4 Rainwater Harvesting

The term rainwater harvesting refers to collection and storage of natural precipitation and also other activities, aimed at harvesting surface and groundwater, prevention of losses through evaporation and seepage, and all other hydrological studies and engineering interventions, aimed at conservation and efficient utilization of the limited water endowment of a physiographic unit, such as a watershed.

India has an annual average precipitation of 1150 mm and drains a total volume of 400 million hectare metres of rain water, of which hary 100 million hectare metres is retained in the soil. The remaining water runs off, carrying with it fertile soil. The amount of water available per person in India is decreasing steadily - from 3450 cu.m in 1951, it fell to 1250 cu.m in 1999. This according to the Ministry of Water Resources, is expected to decrease further to 760 cu.m per person in 2050.

Some 200 million Indians i.e. about 20% of population do not have access to safe and clean water. There is an urgent need for action to maintain the health of our water systems. Shortage of fresh water and its pollution threaten the quality of life of many Indians - nothing demonstrates this fact more than the frequent droughts in main parts of the country.

This gravity of water crisis has been realised in recent years and the government/ non-government agencies and people have taken up various measures to harvest rain water by various methods such as given below:

- Capturing runoff from rooftops
- Capturing runoff from local catchments
- Capturing seasonal floodwaters from local streams
- Conserving water through watershed management.

A very large section of the population uses raw water from the surface, river and lakes and groundwater bodies for human consumption. Rainfall is highly seasonal and occurs over a short rainy season with a very large dry period. As a result, there is a progressive decrease in the groundwater level. The main cause of groundwater pollution is direct contamination due to lateral travel of polluted water. The contaminated water requires permanent remedy otherwise it will affect the health of the consumer.

When untreated effluents are discharged into the ground, these contaminated waters are absorbed by the soil and thus the groundwater is polluted. Large-scale exploitation of water lowers the water level and dries the soil. This makes the maximum absorption of the domestic sewage and industrial effluents including presticides causing the ground-water pollution.

No other method will be efficient to decrease the solid content except the membrane separation technique. The groundwater may be recharged with the rainwater with the help of percolation wells.

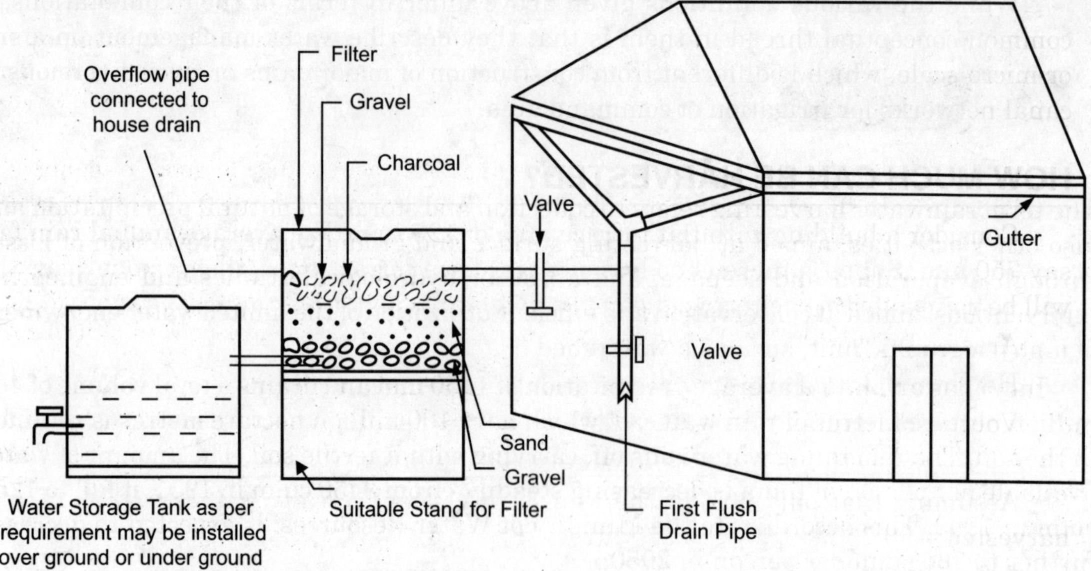

Fig.4.1: Roof Rainwater can be Stored in Tanks for Drinking and Cooking.

The high-level contents of the parameters observed may be minimised if the groundwater is recharged with the available storm water in the rainy season. This not only dilutes the constituents of the groundwater but also raises the groundwater level that depletes due to large-scale exploitation.

The various options, which can be covered under water harvesting are the following:

- Construction of storage structures for "in situ" collection of rain water.
- Farm ponds for collection of runoff, either for supplemental irrigation or for augmentation of groundwater.
- Check dams for storing runoff in first order (small) streams.
- Percolation tanks at appropriate sites for augmentation of groundwater.
- Construction of ponds and reclamation/revitalization of traditional water arresting structures.
- Artificial recharge through wells.
- Reduction of evaporation from surface water bodies and soils.
- Prevention of seepage losses in appropriate situations.
- Enhancement of runoff through mechanical and chemical treatment in catchment areas.
- Sub-surface dams to arrest base flow of groundwater.
- Soil and water conservation practices comprising contour and terrace bunding.

- Control of seawater incursion in coastal aquifers/prevention of sub-surface discharge of groundwater to sea.
- Control of transpiration without affecting normal plant growth.

While the various definitions given above differ in terms of their connotations, the common conceptual thread in them is that they describe water management on a small or micro-scale, which is different from construction of major dams or river diversions and canal networks for irrigation of command area.

HOW MUCH CAN BE HARVESTED?

Consider a building with flat terrace area of 125 sq.m The average annual rain fall is say 750 mm. Suppose there is no loss of water from terrace floor, then, in one year, there will be rainwater on the terrace floor to a height of 750 mm.

Area of Terrace	=	125 sq.m
Height of Rainfall	=	750 mm-0.75 m
Volume of Rainfall	=	125 × 0.75
	=	93.75 cu.m or 93.750 litres.

Assuming that only 60% of the total rainfall is effectively harvested Vol. of Water harvested = 93.750 × 0.60 = 56.250 litres.

A family of six needs about 21.900 litres of water per year.

Thus harvested water of 56.250 litres is more than the double required by the family of six.

Rain Water Harvesting in the Past

"In ancient times, water was revered as a valuable resource with such an intense ardour that it was almost sacred, the source of life - Amrit Varshini.

It was immortalised and its outlets became places of pilgrimage"—Tirth.

All civilisations developed near rivers, and as the need grew people found new ways of conserving and distributing water. Thus evolved various ways of catching and procuring water, a system of distribution to settlements—Jala Kosha. Water harvesting meant sustainable method of collecting water locally to ensure its availability to every village, town and city—Jala Sanchaya. The water in these macro catchments replenished the ground water table, and made it available at the surface for irrigation, in individual tanks and wells.

The earliest, historical evidence is available from 550 AD in Junagadh - a stepwell was cut into rocks—the Adi Kadi Vav. The central idea of a Kund evolved from this, an excavation lined with stone which allowed the water level inside the cavity to fluctuate freely with the surrounding water table. Soon the Kund developed as a place of worship and ceremonial ablutions, for Hindus and Muslims alike, also becoming pleasure pavilions (baradaris) with the advent of the Sultans in 13th century AD.

The Vav consists of three major constituent elements. There is a vertical well (kupa), with an arrangement for hauling up water by buckets (ghat yantra); a stepped corridor leading down several storeys beginning at the entrance pavilion and ending at water level wells; and lastly numerous intermediate tower like pavilions (kuta) built as open halls (mandapa) in the stepped corridor.

Lessons from the Past

- Evidence shows that during the Harappan period there was a very good system of water management, as can be seen in the excavation at Dholavira in Kachch.
- During the pre-independence days, people used to manage water resources considering it as part of the nature that is essential for their survival. This could be seen from the rainwater harvesting structures in the low rainfall areas of Rajasthan, harvesting springs in hilly areas and mountainous regions and percolation ponds and tanks in Southern India.
- In the ancient times in Tamil Nadu, people used to store rainwater in public places separately for drinking and bathing purposes and called them as Ooranies. They also formed percolation tanks or ponds for the purpose of recharging irrigation.
- In the villages of the Thar Desert, an ingenious system of rainwater harvesting known as kund or kundis - the local name given to a covered underground water tank - exists.
- Similarly, in the Spiti valley of Himachal Pradesh, kul (diversion channels) irrigation is utilised to carry water from the glaciers to the village.
- Jodhpur and Jaisalmer have havelis that channel every drop into a container usually, under the courtyard - to store rainwater.
- The Vice regal Lodge at Shimla, built in the 1880s, channels every drop of water from the roof of the building into two large tanks, one under the main front garden and the second next to the Council Chamber.

The "Sudarshana Lake" had been built for several centuries near Junagarh in Saurashtra area in Gujarat state. According to stone inscriptions, this irrigation dam was constructed in the 4th century BC and was occasionally repaired by many generations.

The *qanats* are underground channels constructed to collect water seeping into the colluvium (coarse grained sediment deposit) occurring at the foot of hills and transferred to the habitat through an elaborate system of channels having vents for air escape at the surface. They are found in India from 300 BC and are in working condition even today in the cities like Aurangabad in Maharashtra, where a fountain in a water reservoir and a grain-grinding mill actuated by the course of the running water, are still functional.

An elaborate system of tanks diverting the Ganges water for domestic use was discovered in excavations at Sringaverapura near Allahabad. The floodwaters were diverted through a 1 km. long, 11 m wide and 5 m deep channel to a series of tanks. The first tank was a settling chamber for depositing silt; the overflow from this tank was led back to the Ganges on the downstream side. According to Lal (1992), this system flourished in the 12th century BC.

Raja Bhoja constructed a huge lake at Bhopal in the 11th century AD. This lake, covering an area of 65,000 hectares at that time, comprised not only a dam across one river (Betwa) but also a diversion weir on another stream, which changed its course through ninety degrees.

More and more region-specific water harvesting structures evolved over the coming centuries. These can be broadly classified into a few categories. The basic principle of water harvesting in each category is the same. However, the nomenclature differs from region to region and dialect to dialect.

TRADITIONAL MANAGEMENT OF WATER

In desert areas and drought prove areas of Rajasthan and other parts of the country the people had devised traditional water harvesting techniques which enabled the cities to survive until the piped water supply came into being. Rajasthan has very few surface water channels and most of them are ephemeral in character. There are hardly any perennial river and the people have to depend greatly on groundwater reserves which are also brackish in nature. The wells are used for dual purpose-drinking and irrigation. Wells for irrigation were known as "Chahi" and "Nahri" and the Canals from large tanks were used to irrigate lands. These Nahri's had earthen embankments on both sides which were used to collect 1) rainwater 2) and to protect the water from animals.

After a good shower, the saturated land was usually used for cultivation. In the villages, *kund* (covered pits to collect rainwater), *sar* (simple excavation where the ground was hard) were constructed for storage of water and small ponds provided sufficient water to drink for 7 to 8 months in good season. Tanks were considered to be a very reliable system of water harvesting and were found both in cities and villages. Almost every household has underground tanks to collect rainwater from the roof. These underground tanks are usually circular holes on the ground and are internally lined with fine polished line. The use of water in Rajasthan is closely intervened with the social fabric of the area. The use of lakes, tables, excavated by kings, jagirdars were largely confined to them only. The different water bodies had different social dimensions. *Agor* (catchment) and *bund* (dam) were of people's concern. No houses were allowed to be constructed on *par* (areas of recharge or where water accumulates and seeps into the ground). Villagers were able to identify a par by the kind of foliage on it. These pars were also the source of sweet water. Normally in Rajasthan *polar water* (rainwater) and *water* (groundwater) were used for drinking. *wier water* (river water) and *bir* water (dam water) were used for irrigation and other purposes. Other wells like *Sur ka Kua,* (which were underground channels opening into wells) were also used by the community. wier water, was water collected in depressions during the rains were used for Rabi crops and Bir water spreads across the land like a huge sheet with a depth of 30-40 m and were used for Kharif crops.

In addition to all this the society of arid Rajasthan had several norms to ensure the optimal use of water. For example, ordinary guests were given butter milk and only important guests were given water. Moreover, in Rajasthan, water is still not served in glasses but in tumblers which cannot be touched by mouth. This saves wastage as the water remaining in the tumbler could be used by someone else. People had adopted innovative techniques to save this scarce commodity. In Jaisalmer district even today people bathe on a wooden platform under which a vessel is kept and water that drains into the vessel is used for animal consumption. Similarly, in Pokhran and Phalodi, people bathe on a stone block outside the house from which water drains into an animal watering tank.

Thus, water harvesting, distribution and use was very much regulated by the community. A lot of care was usually taken by the community for keeping tanks and lakes clean. Before the onset of monsoon the *Agor* or *Patyam* (catchment) of the tanks would be methodically cleared of all rubbish and wastes. Labour for this was contributed by the entire community. Thus with the help of these strategies the people could optimally use water. However, an important issue which has emerged over the years in these arid areas is the constraint in availability of water. Since the independence, most of the cities in the arid lands of Rajasthan have switched to piped water supply and this has led to the total negligence of the traditional water harvesting techniques. Today most of the urban settlements are surviving on the supplies from the Indira Gandhi Canal (the supply of which was meant for irrigation of agricultural lands).

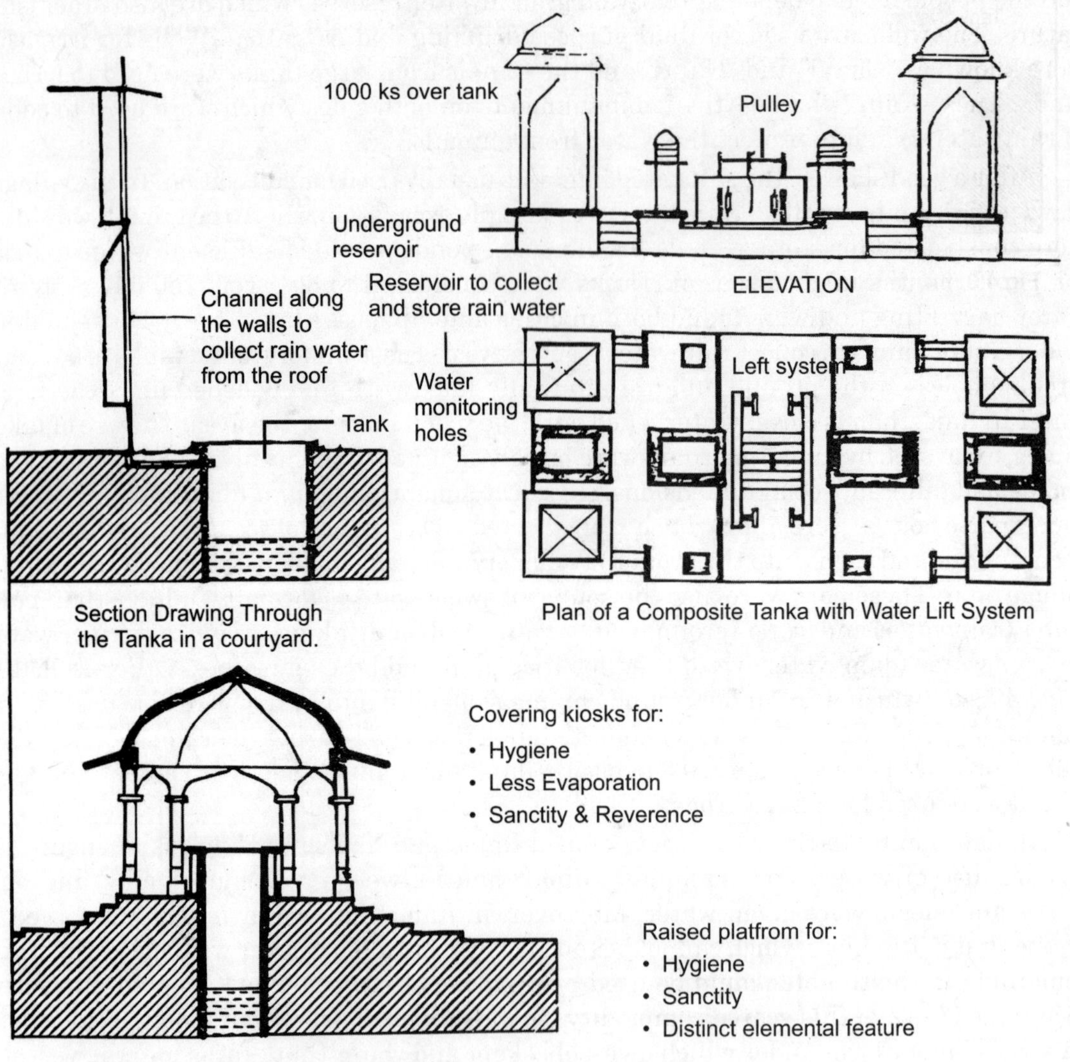

Fig.4.2: Water Harvesting System in Arid Areas

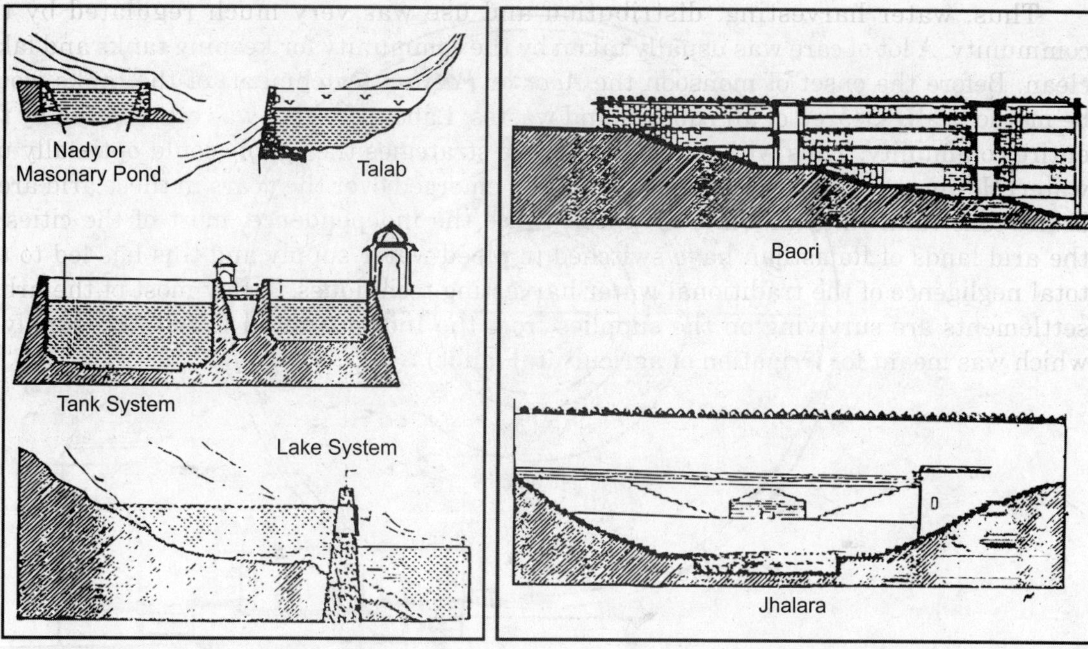

Fig.4.3: Indigenous Techniques of Storing Water in Arid Areas, Surface Water & Ground Water Bodies.

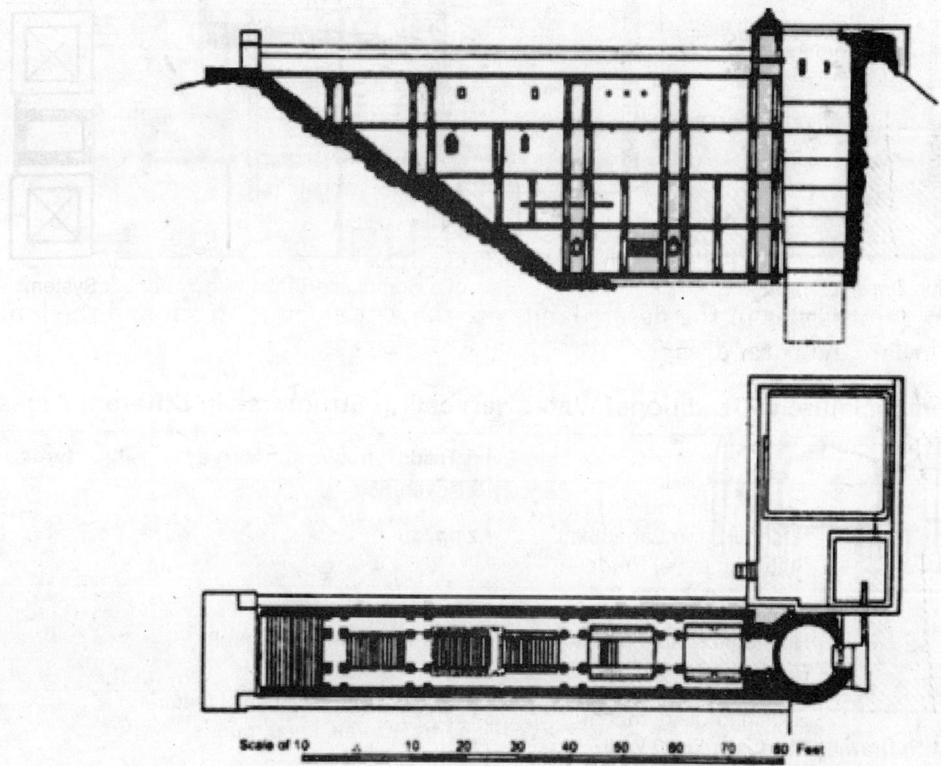

Fig.4.4: Stepwell for Water Harvesting at Vayad, Gujarat.

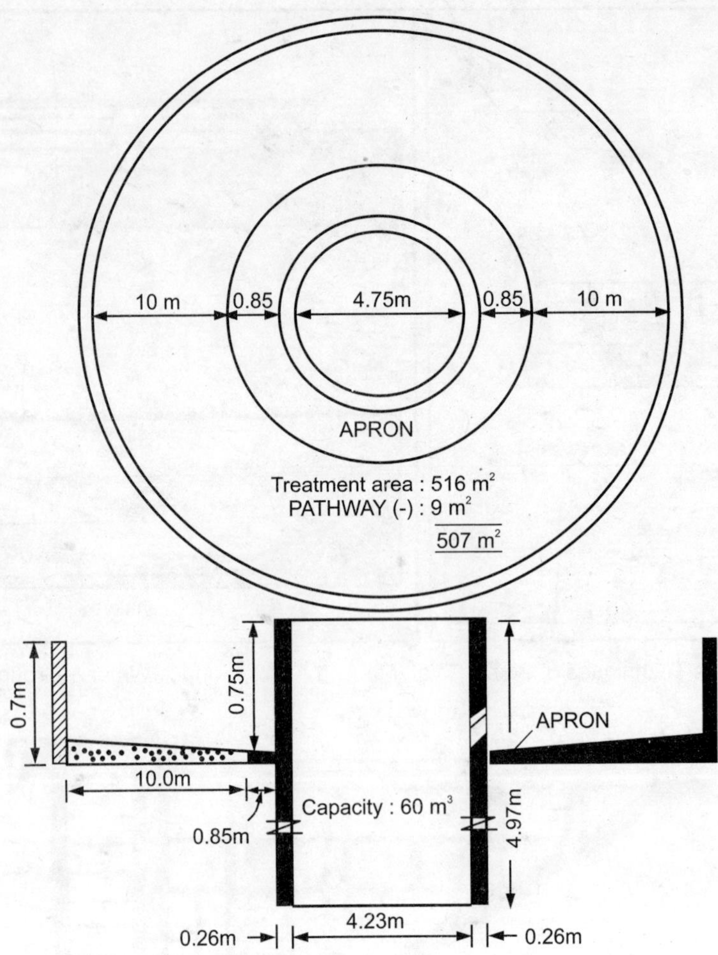

Fig.4.5: Plan and Section of a Tank.

Other important issues in the desert lands are the settlement pattern and the built form, both in urban and rural areas.

Table 4.1: Nomenclature of Traditional Water Harvesting Structures in Different Areas

Ecological Zones	Areas	Traditional Water Harvesting Structures
Trans-Himalayan Region	Ladakh (Jammu and Kashmir) and Lahaul and Spiti (Himachal Pradesh)	zing, kul
Western Himalaya	Jammu and Kashmir, Himachal Pradesh and Uttar Pradesh	kundal, kuhls, guhls, naula
Eastern Himalaya	Sikkim and Arunachal Pradesh	jhoras, kholas, khup, bunds

Northeastern Hill Ranges	Nagaland, Manipur,	bunds, zabo, bamboo drip
	Mizoram, Tripura, Meghalaya	irrigation, tuikhur
Brahmaputra Valley	Assam valley	jampois, dongs
Indo-Gangetic Plains	Punjab, Haryana,	jhalars, chuhi, bunds, nadi,
	Delhi, Uttar Pradesh,	kunds, baolis, dighis, pokhars,
	Bihar, West Bengal	johads, ahars and pynes, kana nadi, pukurs
Thar Desert	Rajasthan, Gujarat	tankas, nadi, talab, johad, bandha, sagar, samand, sarovar, kohar, kua, baolis, par, berls, tank, bavdis, baoris, jhalaras, toba, kundis, khadins, stepwells, virdas
Central Highlands	Rajasthan, Uttar Pradesh,	bunds, bavdi, talab, kunds, baori, lakes
	Madhya Pradesh	and tanks, haveli, bhandaras, jali karanj, pats
Eastern Highlands	Bihar, Orissa	ahars, pynes, katas, mundas, bandhas
Deccan Plateau	Maharashtra, Karnataka,	bhandaras, phad, talparges, kere, kattes,
	Andhra Pradesh	arakere, volakere, devikere, kunte, kula, anicut, kuntas, bunds
Western Ghats	Maharashtra, Karnataka, Kerala	surungam, qantas
Western Coastal Plains	Gujarat, Maharashtra,	virdas, kund, pats, keris,
	Karnataka, Kerala	kettes, anicuts
Eastern Ghats	Orissa, Andhra Pradesh, Tamil Nadu	bandhas, ayacuts, bunds
Eastern Coastal Plains	Orissa, Andhra Pradesh,	kunta, go kuta, dona, revu, oddu, anicuts,
	Tamil Nadu	kasam, eris, ryots, kulams
The Islands	Andaman and Nicobar,	bunds, split bamboo water harvesting
	Lakshadweep Islands	system, step wells, dug wells

User communities mostly managed the traditional water supply structures. There was little government interference. However, after the advent of the British colonial system of governance and establishment of minor irrigation departments, the role and the rights of the community in constructing and maintaining the water harvesting structures went into decline. The most dramatic effect of this can be seen in South India where several thousands of tanks got silted and went into disuse.

Table 4.2: Classification of Traditional Water Harvesting

Systems According to Rainfall Zones

Rainfall Zone	Approx. Geographic Area	Types of Water Harvesting Structures
1. High rainfall (Greater than 1500 mm/year)	Himalayan foot hills, northeast region, Western Ghats	a. Diversion of natural spring water using local material such as bamboo at some places. b. Diversion of part of mountain stream discharge to habitat or farmland through channels and return of excess water from further downstream.
2. Medium rainfall (1000 to 1500 mm/year)	Gangetic West Bengal, Odisha, Bihar plateau, Himachal Pradesh, Eastern Madhya Pradesh	Various types of ponds, diversion of flood flow from rivers, dug wells, water impoundment in farms.
3. Low rainfall (400 to 1000 mm/year)	Western Uttar Pradesh, Haryana, Delhi, Punjab, Jammu and Kashmir, Eastern Rajasthan, Western Madhya Pradesh, Andhra Pradesh, Maharashtra, Karnataka and Tamil Nadu, rain shadow portion of Kerala	Various types of ponds, Cascade of irrigation reservoirs, rapats (percolation tanks), dug wells, step wells, roof water harvesting systems for domestic use.

Table 4.3: Rain Water Harvesting Structures in Different Agro-climatic Zones

Agro-climatic Zone	State of the Art	Recommendations
1. Humid North-Western Himalayans	i. Roof water harvesting ii. Diversion of perennial springs & streams in storage structures iii. Village Pond iv. Collection from hill slope	Improvement in roof structures and use of proper material such as corrugated sheets for generating higher run-off and with arrangement of foul slush diversion system and proper storage structure for checking water from contamination.
2. Himalayan Fool Hills	i. Collection from hill slope ii. Village ponds	Improvement in roof top harvesting system as above.
3. Humid High Rainfall North Eastern Zone	i. Roof Water Harvesting ii. Diversion of perennial springs & streams in a storage structure (tank)	Improvement in roof top harvesting system as above.
4. Humid Assam Bengal Plains	i. Tank ii. Anicut/check dam iii. Gully plugging iv. Contour bunding	Improved design of tank for minimising evaporation and seepage losses, control of sediment load and water pollution.
5. Sub-humid & Humid Satluj Ganga Alluvial zone	i. Pond ii. Check dam iii. Gully plugging iv. Contour bunding	Improved design of tank for minimising evaporation and seepage losses, control of sediment load and water pollution.

6. North-Western Semi-arid & Arid Zone	i. Nadi ii. Tanka iii. Khadin iv. Percolation tank v. Anicut vi. Gully plugging vii. Contour bunding	i. Adoption of improved design of Nadi and Tanka. ii. Sand filled reservoir iii. Sub-surface barrier iv. Flat batter tank.
7. Central Semi-arid Vindhyan Zone	i. Pond ii. Check dam iii. Contour bunding iv. Gully plugging v. Sub-surface dykes	i. Adoption of improved design of Nadi and Tanka. ii. Sand filled reservoir iii. Sub-surface barrier iv. Flat batter tank.
8. High rain fall high run-off Chhotanagpur Plateau	i. Tank ii. Anicut/check dam iii. Gully plugging iv. Contour bunding	Improved design of tank for minimising evaporation and seepage losses, control of sediment load and water pollution.
9. Assured Rainfall Deep Black soil Malwa Plateau & Narmada Basin	i. Ponds ii. Check dams iii. Sub-surface dams	Improvement of Existing system.
10. Variable Rainfall South Central Deccan Plateau Zone	i. Pond ii. Check dam iii. Percolation tank iv. Bandhara v. Gully plugging vi. Sub-surface dam vii. Contour bunding	i. Flat batter tank ii. Selection of suitable site and improvement of existing system, better water management.
11. Chhatisgarh Plateau Zone	i. Pond ii. Check dam iii. Percolation tank iv. Bandhara v. Gully plugging vi. Sub-surface dam vii. Contour bunding	i. Flat batter tank ii. Selection of suitable site and improvement of existing system, better water management.
12. South Eastern Brown/Red Soil Zone	i. Ponds/Tanks ii. Percolation tanks iii. Sub-surface dams	Improvement of existing systems.
13. Southern Variable rainfall, Mixed Soil Zone	i. Pond/Tank/Kunta ii. Nadi iii. Check Dam iv. Percolation Tank v. Sub-surface Dam vi. Gully Plugging	i. Flat batter tank ii. Selection of suitable site and improvement of existing system for better water management.

14. Southern bimodal Rainfall Zone	i. Ponds/Tanks ii. Percolation Tanks iii. Gully Plugging iv. Contour Bunding v. Check Dams	Improvement of existing system.
15. Eastern Coromandal	i. Ponds/Tanks/Kunta ii. Nadi iii. Check Dam iv. Percolation Tank v. Sub-surface Dam vi. Gully Plugging	i. Adoption of improved design of Nadi and tank. ii. Selection of suitable sites and improvement of existing system for better water management.
16. Western Malbar	i. Pond/Tank/Kunta ii. Check Dam iii. Percolation Tank iv. Contour Bunding v. Bandhara vi. K.T. Weirs vii. Sub-surface Dam	i. Improvement in existing system, better water management. ii. Construction of structures at suitable sites.

Source: Dr. S.K. Sharma, Central Ground Water Board.

TRADITIONAL SYSTEMS OF RAINWATER HARVESTING

Apatani: This system harvests both ground and surface water for irrigation. It is practised by the Apatani tribes of Ziro in the lower Subansiri district of Arunachal Pradesh. Under this system, valleys are terraced into plots separated by 0.6 metre-high earthen dams supported by bambo frames. All plots have inlets and outlets on opposite sides. The terraced plot can be flooded or drained off with water by opening or blocking the inlets and outlets as and when required. The stream water is tapped by constructing a wall of 2-4 m high and 1 m thick near forested hill slopes. This is conveyed to agricultural fields through a channel network.

Virdas: These are shallow wells dug in low depressions called *jheels* (tanks). They are found all over the *Banni* grasslands, a part of the Great Rann of Kutch in Gujarat.

Katas/Mundas/Bandhas: These were the main irrigation sources in the ancient tribal kingdom of the Gonds (now in Orissa and Madhya Pradesh). Most of these *katas* were built by the village headmen known as *gountias,* who in turn, received the land from the Gond kings. A *kata* was constructed north to south, or east to west, of a village. Under this system, a strong earthen embankment, curved at either end, is built across a drainage line to hold up an irregularly-shaped sheet of water.

Surangam: Kasaragod district in the northern Malabar region of Kerala is an area whose people cannot depend directly on surface water. They therefore depend on groundwater, and on a special water harvesting structure called *surangam* or a tunnel, about 0.45-0.70 m wide and about 1.8-2.0 m high. *Surangams* are similar to the *qanats* found in Mesopotamia and Babylon around 700 BC. 1,2 By 714 BC. Today, the Kunnikannan Nairs keep this tradition alive.

Korambus: This is a temporary dam stretching across the mouth of channels, made of brushwood, mud and grass. It is constructed by fixing a strong horizontal wooden beam touching either bank of the canal. A series of vertical wooden beams of an appropriate height is erected with their lower ends resting firmly on the ground and the other ends tied to the horizontal beam. Closely knitted or matted coconut thatch is tied to this frame. These are built twice a year especially before the onset of the monsoon season in order to supply water during the winter and summer. In the Kasargod and Thrissur districts of Kerala, *Korambus* is known as *chira.*

Jackwells: The Shompen tribal of the Great Nicobar Island have made full use of the topography to harvest water. In lower parts of the undulating terrain, bunds were made using logs of hard bullet wood, and water would collect in the pits so formed. The Shompens make extensive use of split bamboos in their water harvesting systems. A full length of bamboo is cut longitudinally and placed along a gentle slope with the lower end leading into a shallow pit. These serve as conduits for rainwater, which is collected, drop by drop in pits called Jackwells. Often, these split bamboos are placed under trees to harvest the through falls (of rain) through the leaves. A series of increasingly bigger Jackwells is built, connected by split bamboos so that overflows from one lead to the other, ultimately leading to the biggest Jackwell, with an approximate diameter of 6 m and depth of 7 m so that overflows from one lead to the other.

Johads: Are the water harvesting structures traditionally built in North-Eastern Rajasthan. Recently 300 Johads have been constructed with the help of Tarun Bharat Sangh, an NGO. These structures have been able to raise the groundwater table significantly, turning vast stretches of barren land into fertile agricultural land. People have found employment locally and migration to urban areas has considerably been reduced. This has benefitted the flora and fauna of the area. Women now spend much less time in fetching water for domestic and animal use.

Another important impact of Johad has been to bring the village society together. They have learnt to plan, supervise, execute and maintain works connected with Johads. Their voluntary contribution for building these structures has gone up from 25 per cent to 50 per cent.

Kunds/Kundis: These look like upturned cups nestling in saucers. These structures harvest rainwater for drinking, and dot the sandier tracts of the Thar Desert in western Rajasthan and some areas in Gujarat.

Kuis/Beris: Found in western Rajasthan, these are 10-12 m deep pits dug near tanks to collect the seepage. *Kuis* can also be used to harvest rainwater in areas with meagre rainfall. The mouth of the pit is usually made narrow to prevent evaporation. The openings of these entirely *kuchcha* (earthen) structures are generally covered with planks of wood, or put under lock and key. The water is used sparingly, as a last resource in crisis situations.

Baoris/Bers: These are community wells, found in Rajasthan used mainly for drinking water. Most of them are very old and were built by Banjaras (mobile trading communities) for their drinking water needs.

Jhalaras: Artificial tanks, found in Rajasthan and Gujarat, these were often rectangular in design with steps on three or four sides. The oldest *jhalara* is the mahamandir jhalara dating back to 1660 AD.

Nadis: Found near Jodhpur in Rajasthan, these ponds are used for storing water from an adjoining natural catchment during the rainly season.

Tobas: This name is given to a ground depression with a natural catchment area. A hard plot of land with low porosity, consisting of a depression and a natural catchment area is selected for the construction of a *toba*.

Tankas: Are underground tanks, found traditionally in most houses in Bikaner. They were buolt in the main house or in the courtyard as circular holes dug in the ground, lined with fine polished lime, in which rainwater was collected. *Tankas* were often beautifully decorated with tiles, which helped to keep the water cool.

Khadin: A *Khadin,* also called a *dhora,* is an ingenious construction designed to harvest surface runoff water for agriculture. Its main feature is a very long (100-300 m) earthen embankment built across the lower hill slopes lying below gravel uplands. Sluices and spillways allow the excess water to drain off.

Vav/Vavdi/Baoli/Bavadi: Traditional stepwells are called *vav* or *vavadi* in Gujarat, and *baolis* or *bavadis* in Rajasthan and in northern India. Built by the nobility for strategic and/ or philanthropical reasons, they were secular structures from which everyone could draw water.

Paar system: *Paar* is a common place where the rainwater flows from the *agar* (catchment) and in the process percolates into the sandy soil, representing the most predominant form of rainwater harvesting in the region.

Source: Newsletter, Centre for Science and Environment, July 15, 2004.

TECHNIQUES OF RAINWATER HARVESTING

Rain water falling on roof is collected and injected into groundwater reservoir through following structures:

1. Abandoned Dug Well
   - A dry/ unused dug well can be used as a recharge structure.
   - The recharge water is guided through a pipe to the bottom of well or below the water level to avoid scouring of bottom and entrapment of air bubbles in the aquifer.
   - Before using the dug well as recharge structure, its bottom should be cleaned and all the fine deposits should be removed.
   - Recharge water should be silt free.
   - It should be cleaned regularly.
   - It is suitable for large buildings having the roof area more than 100 sq.m.
   - Periodic chlorination should be done for controlling the bacteriological contamination.
2. Abandoned/Running Hand Pump
   - An abandoned/running hand pump can be used for recharge.

- The structures are suitable for the small buildings having the roof area upto 150 sq.m.
- Water is diverted from the roof top to the hand pump through pipe of 50 to 100 mm dia.
- For running hand pump a closing valve is fitted in conveyance system near hand pump to avoid entry of air in suction pipe.
- Recharge water should be silt free.
- During recharging period, the water extracted from hand pump should be utilised after proper chlorination.

3. Recharge Pit
   - Recharge pits are constructed for recharging the shallow aquifer.
   - These are constructed generally 1 to 2 metre wide and 2 to 3 metre deep.
   - After excavation, the pits are refilled with pebbles and boulders.
   - Water to be recharged, should be silt free.
   - Cleaning of the pit should be done periodically.
   - It is suitable for small buildings having the roof top area up to 100 sq.m.
   - Recharge pit may be of any shape i.e. circular, square or rectangular.
   - If the pit is of trapezoidal shape, the side slopes should be steep enough to avoid silt deposition.
4. Recharge Trench
   - It is constructed when permeable strata of adequate thickness is available at shallow depth.
   - It is a trench of shallow depth filled with pebbles and boulders.
   - These are constructed across the land slope.
   - The trench may be 0.5 to 1 metre wide, 1 to 1.5 metre deep and 10 to 20 metre long depending upon the availability of land and roof top area.
   - It is suitable for the buildings having the roof area of 200 to 300 sq.m.
   - Cleaning of trench should be done periodically.
5. Gravity Head Recharge Well
   - Bore wells/tube wells can be used as recharge structure.
   - This technique is suitable where:
     - Land availability is limited.
     - When aquifer is deep and covered by impermeable strata (clay).
   - The roof top rain water is channelized to the well and recharges under gravity flow condition.
   - Recharge water should be silt free.
   - The well can also be used for pumping.

- Most suitable for the areas where ground water levels are deep.
- The number of recharging structures can be determined in limited areas around the buildings depending upon roof top area and aquifer characteristic.

6. Recharge Shaft
   - A recharge shaft is dug manually or drilled by the reverse/direct rotary method.
   - Diameter of recharge shaft varies from 0.5 to 3 metre depending upon the availability of water to be recharged.
   - It is constructed where the shallow aquifer is located below clayey surface.
   - Recharge shaft is back filled with boulders, gravels and coarse sand.
   - It should end in more permeable strata (sand).
   - Depth of recharge shaft varies from 10-15 metre below ground level.
   - Recharge shaft should be constructed 10 to 15 metre away from buildings for the safety of building.
   - It should be cleaned regularly by scraping the top layer of sand and refilling it periodically.

Typical Structure: Designing a rooftop rainwater harvesting system would mean taking the following steps:

- Sloping the roof appropriately preferably towards the direction of storage and recharge.
- Designing gutters and/or down-pipes depending on site rainfall characteristics and roof characteristics.
- Putting in place a rain separator to divert and manage the first 2.5 mm of rain.
- Filtering the water to remove solids and organic material. Storing the filtered water in appropriate storage units like sumps, tanks etc.
- Recharging the groundwater through open wells, bore wells or percolation pits.

METHODS OF HARVESTING

The places where the rains are throughout the year, rain water can be stored in tanks. However, at places where rains are for 2-3 months, huge volumes of storage containers would have to be provided. To such places, it would be feasible to use rain water to recharge groundwater aquifers rather than for storage.

As shown in fig. suitable size gutters, pipes, filter and storage tank should be provided. First 10-20 minutes of rain water is flushed off through first flush drain pipe then the valve closed and filter pipe valve opened. The water passing through filter is stored in suitable size storage tank. The first flush drain pipe and overflow pipe should be connected with house drain. The underground storage tank may be of masonry or re-enforced cement concrete, plastic/ high density polyethylene or made of Ferro-cement. The size of the tank depends upon requirement, catchment and rainfall. The down pipe should be of at least 100 mm diameter with 850 micron nylon wire screen at the inlet to prevent dry leaves and debris from entering in it.

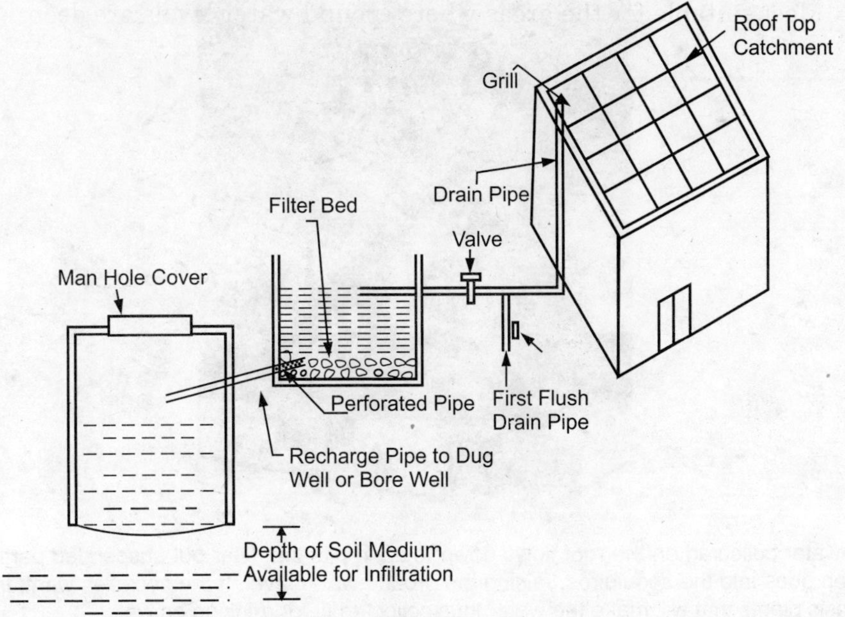

Fig.4.6: Recharge Assembly for Dug Well with Roof Top Runoff.

The underground storage tank should preferably have a suitable pump installed for withdrawal of water or pumping it to an overhead storage tank. Their top should remain at least 300 mm above the ground.

Before the tank is put into use, it should be thoroughly cleaned and disinfected with a suitable disinfectant such as chlorine, bleaching powder, potassium permanganate etc. Since the water shall remain stored for quite a long time, periodical disinfections of stored water is essential to prevent growth of pathogenic bacteria.

When the tank required to be cleaned or stored water is required to be disposed, it should be drained to the nearest nullah or any natural drains or disposed through a properly designed outlet system.

Rainwater that is collected on the rooftop of the building in diverted by drainpipes to filtration tank - for bore well, through settlement tank, from which is flows into the recharge well.

Dug well should have opening - weep-hole at regular intervals to allow seepage of water through the sides. Dug wells should be covered to prevent mosquito breeding and entry of leaves and debris. The bottom of recharged dug wells should be desilted annually to maintain the intake capacity. Any old well which has become defunct can be used for recharging, since the depth of such wells is above water table.

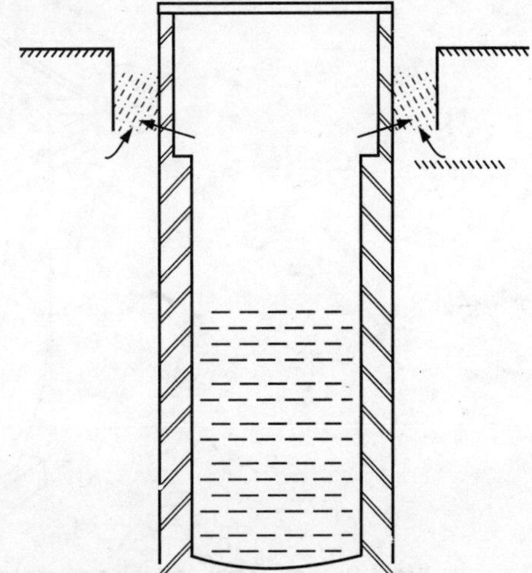

Fig.4.7: Recharge Assembly for Dug Well with Runoff from Ground Areas.

Fig.4.8: Rainwater collected on the roof flows down to sand pits that filter out suspended particulate matter. The water then goes into the acquihires, raising the ground water level. If the societies want, they can install reverse osmosis plants that will make the water thus collected fit for drinking as well.

Rainwater harvesting can be harvested from the following surfaces:

Rooftops: If buildings with impervious roofs are already in place, the catchment area is effectively available free of charge and they provide a supply at the point of consumption.

Paved and Unpaved Areas i.e., landscapes, open fields, parks, storm water drains, roads and pavements and other open areas can be effectively used to harvest the runoff. The main advantage in using ground as collecting surface is that water can be collected from a larger area. This is particularly advantageous in areas of low rainfall.

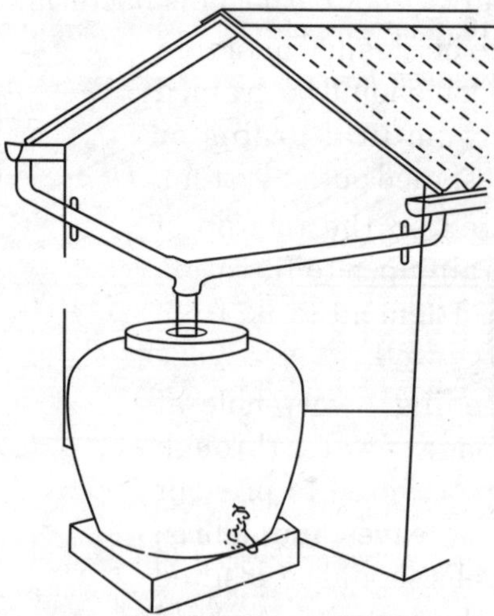

Fig.4.9: 2 m^3 'Thai Jar' wire reinforced cement roof catchment tank, popular in Thailand and other parts of South-east Asia. (Source: Waterlines)

Source: John Gould and E.N. Petersen, 1999, Rainwater Catchment System for Domestic Supply, IT Pub, London

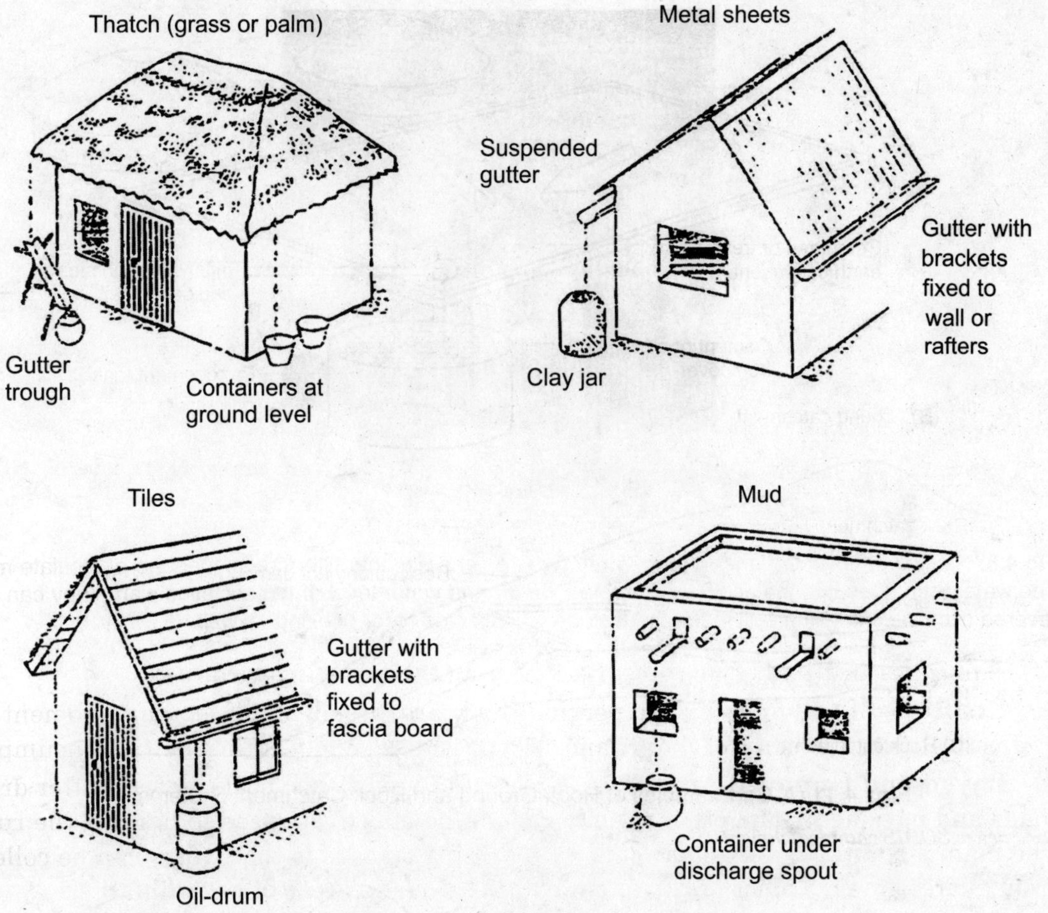

Thatch (grass or palm)
Gutter trough
Containers at ground level
Metal sheets
Suspended gutter
Clay jar
Gutter with brackets fixed to wall or rafters
Tiles
Gutter with brackets fixed to fascia board
Oil-drum
Mud
Container under discharge spout

Fig.4.10: Examples of Rainwater Harvesting Systems

Source: John Gould and E.N. Petersen, 1999, Rainwater Catchment System for Domestic Supply, IT Pub, London

Catchment area
Covered storage reservoir
Outlet tap
(A) Roof catchment

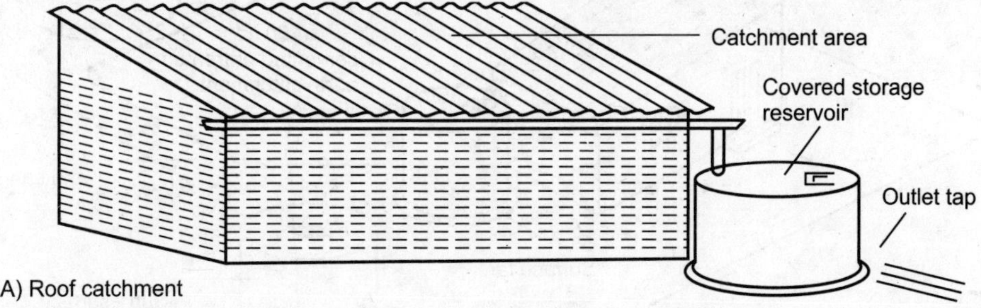

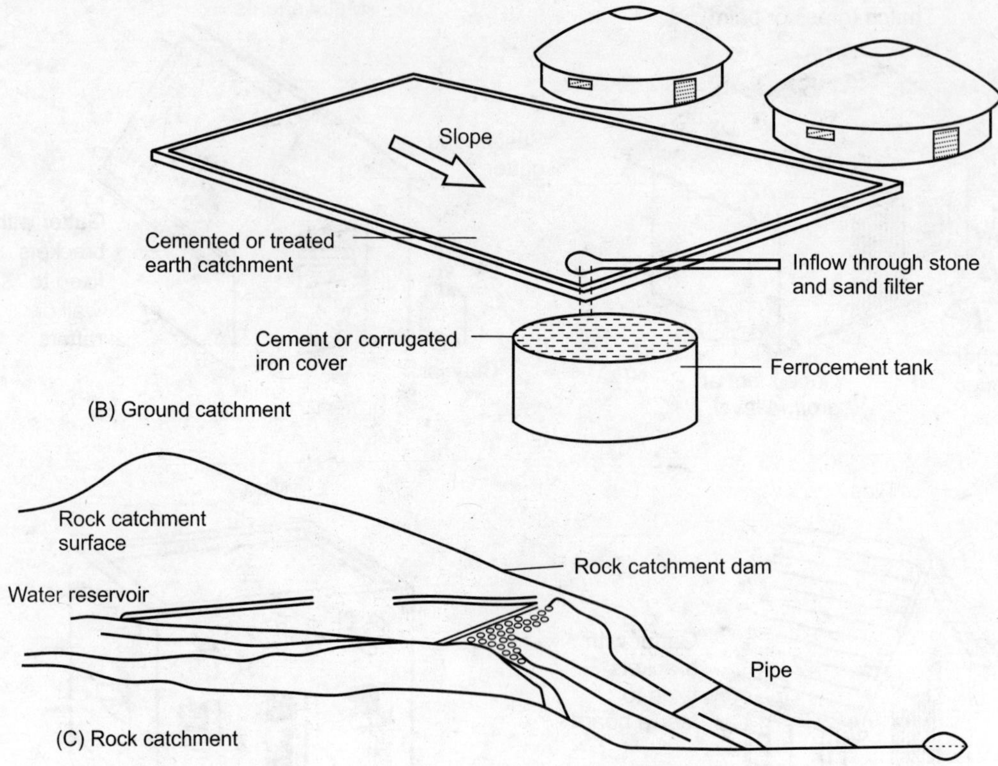

Figs.4.11 (A-C): Examples of Roof, Ground and Rock Catchment Systems.

Source: www.SODIS.chandcatchment

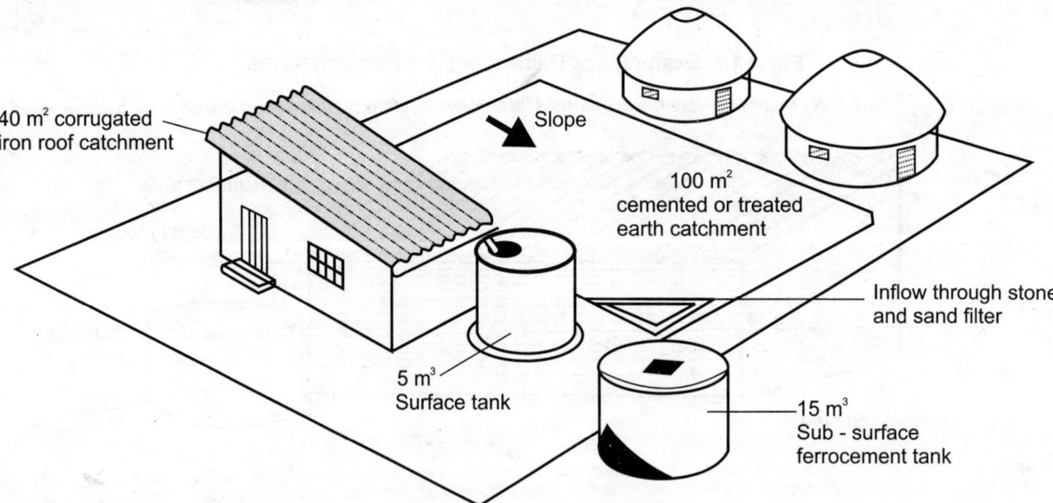

Fig.4.12: A rural household with roof and ground catchments and surface and sub-surface tanks in southern Africa. This system designed for a rural household in East Botswana (mean rainfall 400-500 mm/year) can meet essential household water needs in most years.

Source: Adapted from a drawing A. Gleske, 1995 and Pentair Water Treatment Pvt. Ltd., New Delhi

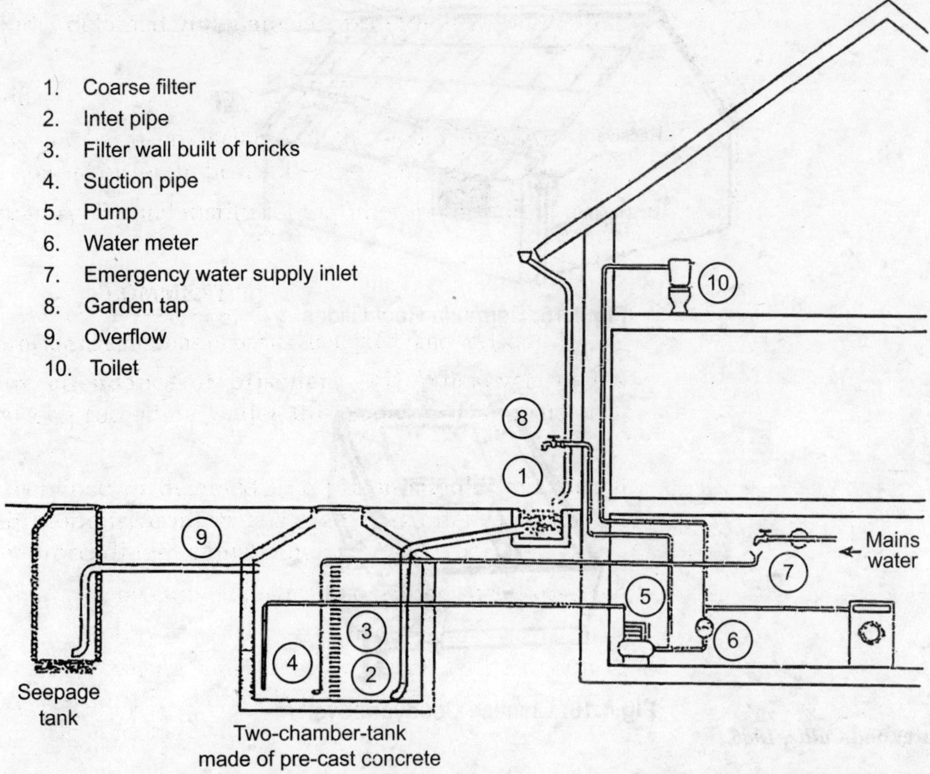

Fig.4.13: Double-chamber Sub-surface Rainwater Tank Non-potable Household Supply, and Seepage Well for Recharging Groundwater, Osnabr ck, Germany.

Source: John Gould and E.N. Petersen, 1999, Rainwater Catchment System for Domestic Supply, IT Pub, London

Fig.4.14: Rooftop Water Harvesting for a Village House

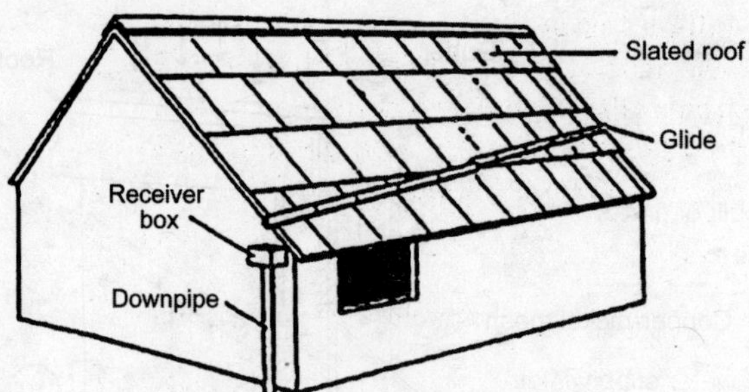

Fig.4.15: Bermuda Roof Glides

Fig.4.16: Chinese Courtyard System

Source: Pacey and Cullis, 1986,

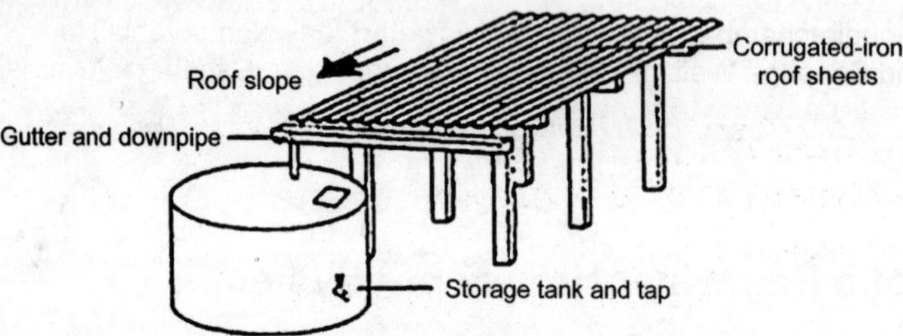

Fig.4.17: Purpose-built 40 m² corrugated-iron 'Roof' Catchment from Botswana.

Source: Pacey and Cullis, 1986

Fig.4.18: Stretched - woven Hassian Sacks Used for a 6 m² 'Sarisa' Catchment in Kenya.

Source Percey and Cullis, 1986

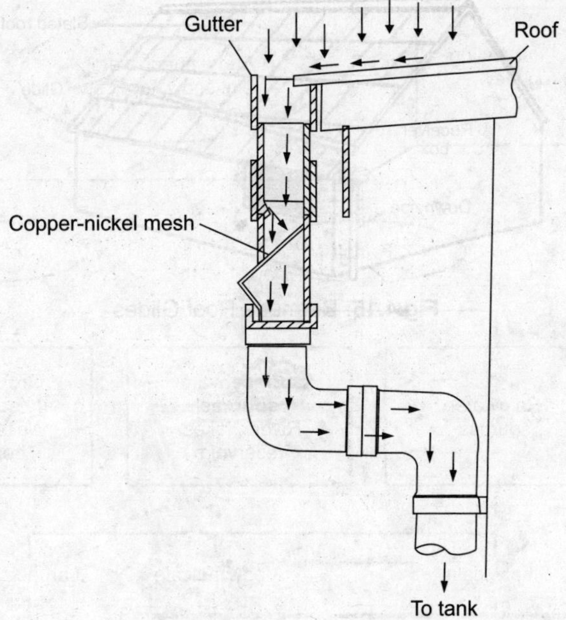

Fig.4.19: A Self-cleaning, Coarse Rainwater Filter after Finch, H., 1994; TIRUC, 1994, p. 336

Source: John Gould and E.N. Petersen, 1999, Rainwater Catchment System for Domestic Supply, IT Pub, London

Waterbodies: The potential of lakes, tanks and ponds to store rainwater is immense. The harvested rainwater can not only be used to meet water requirements of the city, it also recharges groundwater aquifers.

Storm water Drains: Most of the residential colonies have proper network of storm water drains. If maintained neatly, these offer a simple and cost effective means for harvesting rainwater.

Components of a Rainwater Harvesting System

A rainwater harvesting system comprises components of various stages - transporting rainwater through pipes or drains, filtration, and storage in tanks for reuse or recharge. The common components of a rainwater harvesting system involved in these stages are illustrated in Fig. 4.20.

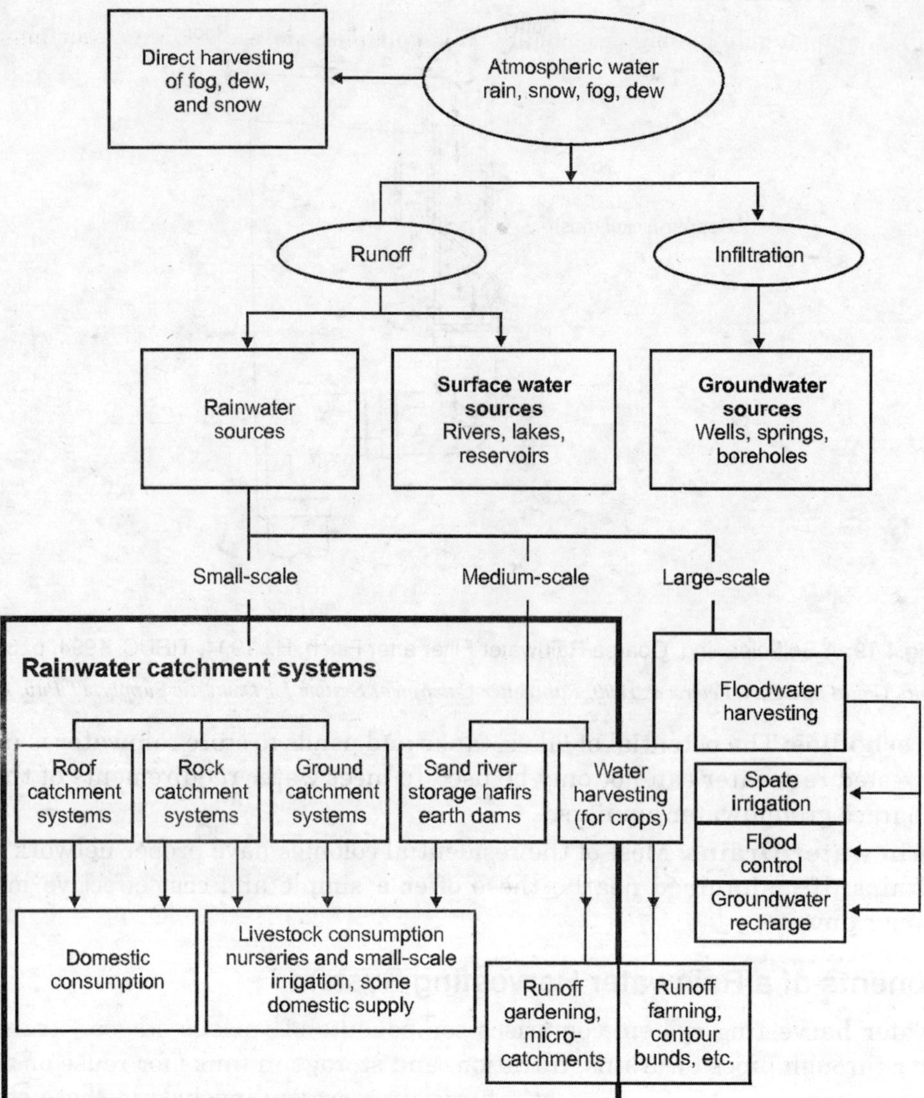

Fig.4.20: Sources of Rainwater Catchment.

WATERSHED MANAGEMENT

The terms watershed refers to a topographically distinctive area, which is drained by a stream. A watershed can be of a few hundred to few thousand hectares in size. If the topography is undulating, the area of the watershed is small and it increases as the undulation in a terrain is reduced.

(a) **On-channel Storage and Recharge of Storm Water Channels:** Several seasonal streams become activated during the monsoons providing the outlet to flood waters as well as local area runoff. The channels may run through urban areas or their hinterlands. With careful planning these waters can be stored on the channel itself. The waters can be detained through construction of appropriate located regulators. The storage capacity of channels can be enhanced by deepening and widening them wherever possible. Storm water channels can be enlarged to form on-channel lakes at locations conductive to recharge. The stored waters can be used from the surface or extracted from the ground while simultaneously they serve to recharge aquifers. These reservoirs also offer the possibility of creating rich urban bio-diversity habitats.

(d) **Off-channel Storage for Floodwaters:** As all storm water cannot be stored on channel, possibilities exist for the creation of off-channel reservoirs linked to the main channel. The creation of off-channel reservoirs, especially wherever a suitable recharge area exists in the neighbourhood of a channel, is a useful device for storing water for the lean season. These reservoirs also offer the possibility of creating rich urban bio-diversity habitats.

(c) **Storage in Lakes and Depressions:** In urban areas, lakes have often been misused for dumping of solid waste and sewage disposal. The storage capacity of these can be increased by desilting and widening. The water quality would need to be improved by replicating the principles of wetlands by the systematic induction of aquatic plants and fish. This is a very low cost treatment and subsequently only clean water would enter the underlying aquifer. The groundwater table is the most economical form of storage in that there is no expense in construction of reservoirs and water can thereafter be extracted from the aquifer by battery of shallow tube wells for a long duration extending into the lean season. The yet - untapped, potentials of the wetland habitats comprising of oxbow lakes, marshes, low lying lands on the floodplains, lakes, estuarine lands, abandoned quarries, artificial lakes over good recharge zones for the sustenance of not only the ecology but for other activities such as recreation, flood prevention, water storage, water extraction and natural water treatment.

(d) **Floodplain Reservoirs for Conjunctive Extraction:** Very often the most copious fresh water aquifers underline the floodplains. In urban areas, there is no place to store the monsoon discharge of the river due to lack of topographical relief. To overcome this difficulty simple reservoirs on the floodplain can be created by scooping out earth at appropriate locations and letting these reservoirs to be filled up by the expanded monsoon season flow of the river.

It should be ensured that watershed development should enhance the localised augmentation of supply and enhance and enrich the groundwater.

In order to adopt appropriate measures for augmenting ground recharge, it is necessary to spread water harvesting techniques by construction of check dams and percolation tanks, besides direct injection methods utilising the abandoned structures available in large numbers in the hard rock region. Before undertaking direct recharge into such abandoned structures, suitability of the structure for receiving recharge and also the quality of source water should be thoroughly examined to confirm the technical feasibility. Rain water harvesting and groundwater recharge can be suitably dovetailed with watershed development programmes aimed at soil and water conservation by arresting the run-off. This involves several actions as outlined below:

- Arresting groundwater decline and improving its accessibility
- Benefiting water quality in aquifers
- Preventing surface water run-off during monsoons
- Enhancing availability of surface water at a specific place and time
- Reducing power consumption
- Conserving urban waste water.

Key Action Areas

- Organising mass awareness campaigns through hand-outs, posters and community level involvement.
- Practising and implementing rain water harvesting and ground recharging through local bye laws.
- Motivating communities to participate in shramdan camps for cleaning ponds, tanks, wells and borewells; and reutilising them for groundwater storage.
- Financial support for construction of check dams and creation of catchment ponds.
- Recycling waste water subject to quality and dual piping system.
- Recharging industrial effluents after treatment.

Guidelines for Water Conservation

- Redesigning cisterns/tanks/reservoirs for conservation of water, e.g. traditional 3.5 gallon

  toilet cistern can be replaced by a 5 to 7 litres one with equal effect.
- Using urban waste water effluents after suitable treatment for horticulture etc.
- Recycling of waste water.
- Conserving their monsoon water in Johads, check dams, percolation tanks, village ponds etc. and regular desilting.
- Using existing or defunct wells and tube wells for purpose of recharging and rainwater harvesting.

Concepts and Measures

In view of the fast growing population and urbanisation, coupled with diminishing water resources, it is necessary to conserve available water, and harvest rain water to the maximum extent possible through various measures as given below:

- Leak detection and water supply network maintenance and rectification.
- Rain water harvesting, through in-situ tanks check dams etc.
- Controlling indiscriminate extraction of groundwater
- Recharging of aquifers.

- Meters to be installed along trunk mains and for bulk consumers and tariff review.
- Percolation structures, wherever suited, be encouraged in open grounds and road side will help recharge bore-wells and provide water for gardening, car washing etc.
- Mandatory system of roof-top rain water harvesting.
- Recycling and reuse of treated waste water
- Soft parking, landscape and pavements filled with pebbles and coarse sand to allow percolation.

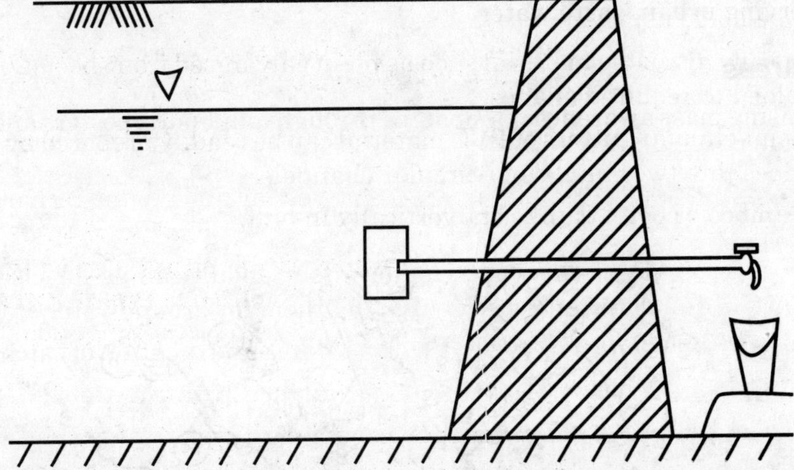

Fig.4.21: The sand storage dam is constructed above ground. Sand and soil particles transported during periods of high flow are allowed to deposit behind the dam„ and water is stored in these soil deposits.

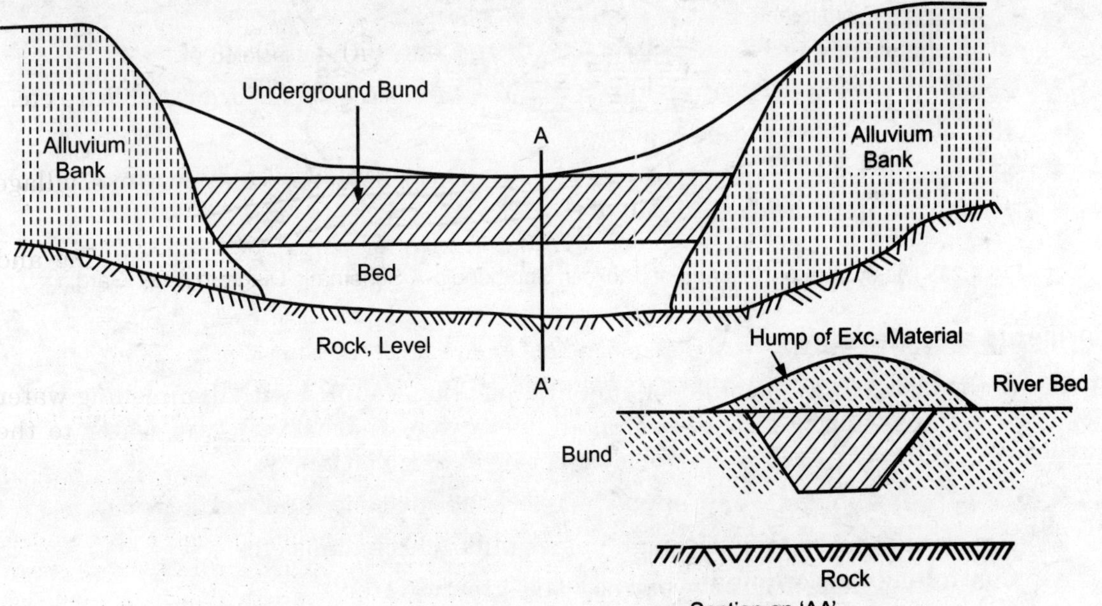

Fig.4.22: Schematic Diagram of Sub-surface Dyke

1. **Catchment:** The catchment of a water harvesting system is the surface which directly receives the rainfall and provides water to the system. It can be a paved area like a terrace or courtyard of a building, or an unpaved area like a lawn or open ground. A roof made of reinforced cement concrete (RCC), galvanised iron or corrugated sheets can also be used for water harvesting.
2. **Coarse mesh** at the roof to prevent the passage of debris.
3. **Gutters:** Channels all around the edge of a sloping roof to collect and transport rainwater to the storage tank. Gutters can be semi-circular or rectangular and could be made using:
   - Locally available material such as plain galvanised iron sheet (20 to 22 gauge), folded to required shapes.
   - Semi-circular gutters of PVC material can be readily prepared by cutting those pipes into two equal semi-circular channels.
   - Bamboo or betel trunks cut vertically in half.

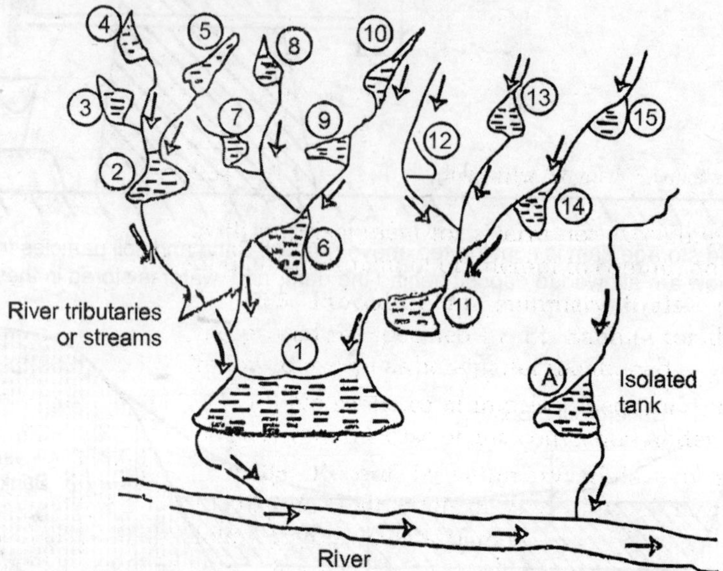

Fig.4.23: The system of Cascade of Tanks in South India-A Schematic Diagram (After Garg,

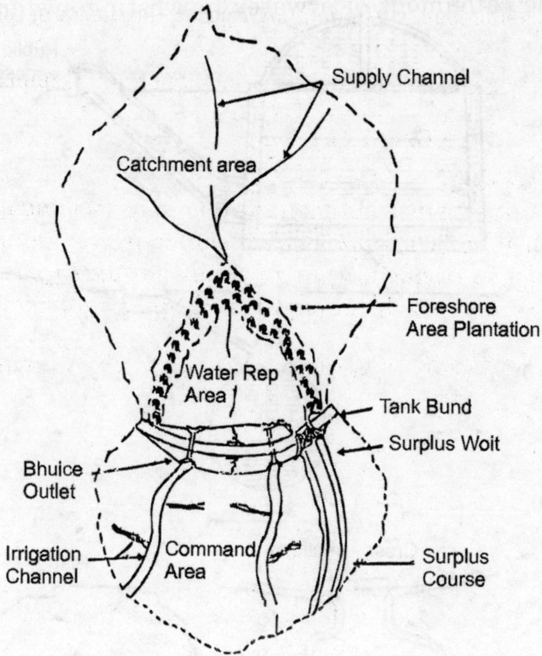

Fig.4.24: A Typical Irrigation Tank from Tamil Nadu (after Sreenivasan, 1999).

The size of the gutter should be according to the flow during the highest intensity rain. It is advisable to make them 10 to 15 per cent oversize.

Gutters need to be supported so they do not sag or fall off when loaded with water. The way in which gutters are fixed depends on the construction of the house; it is possible to fix iron or timber brackets into the walls, but for houses having wider eaves, some method of attachment to the rafter is necessary.

4. **Conduits:** Conduits are pipelines or drains that carry rainwater from the catchment or rooftop area to the harvesting system. Conduits can be of any material like polyvinyl chloride (PVC) or galvanized iron (GI), materials that are commonly available.

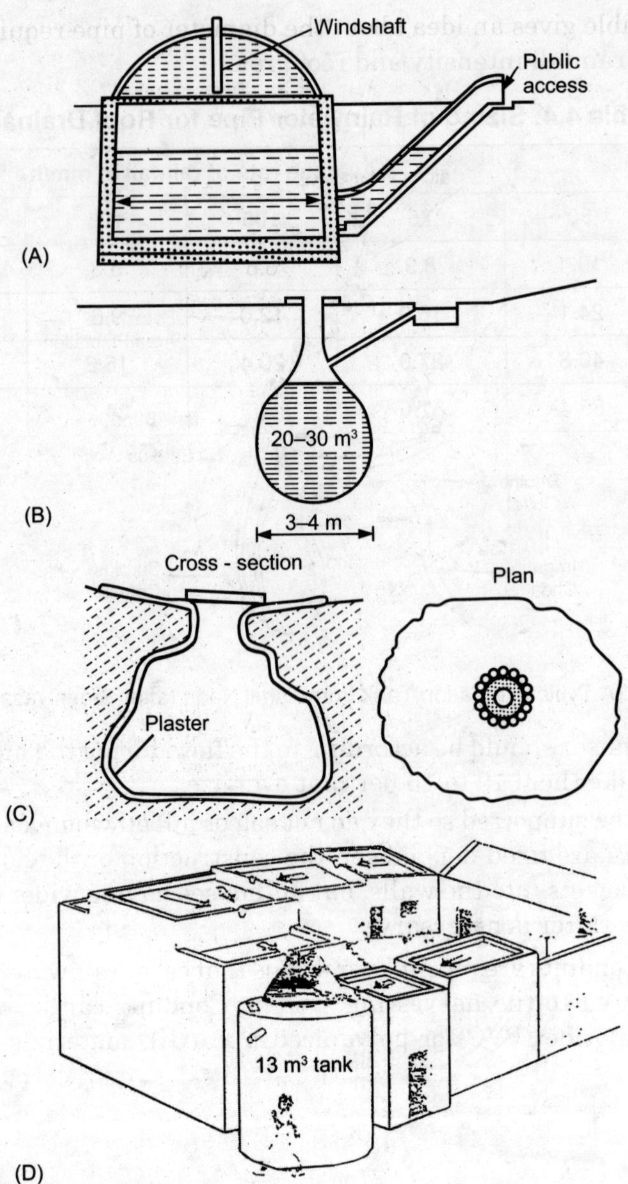

- Banbar, Additional sub-surface manual cistern, Iran.
  Source: Hassanizadeh, 1984.
- Shuijiao, Excavated clay-lined water cellar, loess plateau, Central China
  Source: UNEP, 1979.
- Chultun Mayan stucco cement/lime-plastered underground tank and ground catchment system in limestone region. Yucatan Peninsula, Mexico.
  (Source: Gordillo, 1982.)
- Adobe Granary Bins, Lined with ferrocement to convert to water tank, Mali.

Figs.4.25 (A-D): Some examples of Traditional Rainwater Catchment System

Source: www,SODIS.ch and catchwater

The following table gives an idea about the diameter of pipe required for draining out rainwater based on rainfall intensity and roof area.

Table 4.4: Sizing of Rainwater Pipe for Roof Drainage

Diameter of pipe (mm)	Average rate of rainfall in mm/h					
	50	75	100	125	150	200
50	13.4	8.9	6.6	5.3	4.4	3.3
65	24.1	16.0	12.0	9.6	8.0	6.0
75	40.8	27.0	20.4	16.3	13.6	10.2
100	85.4	57.0	42.7	34.2	28.5	21.3
125	—	—	80.5	64.3	53.5	40.0
150	—	—	—	—	83.6	62.7

mm/h-millimeters per hour; m-meters.
Source: National Building Code.

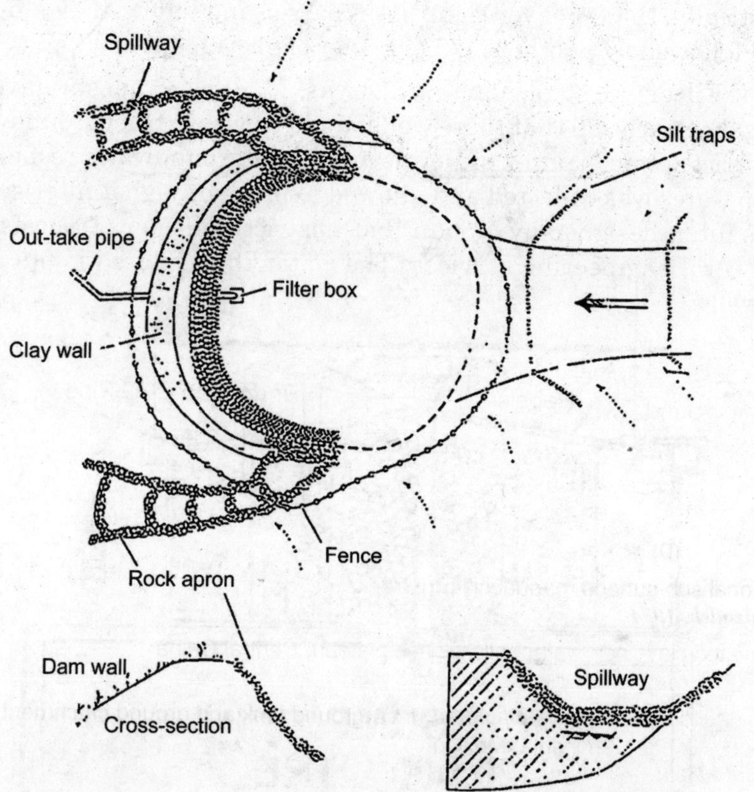

Fig.4.26: General Design for an Earth Dam with Curved Walls

(Source: Adapted from Lee and Visscher, 1992)

5. **First-flushing:** A first flush device is a valve that ensures that runoff from the first spell of rain is flushed out and does not enter the system. This needs to be done since the first spell of rain carries a relatively larger amount of pollutants from the air and catchment surface.

Filters

The filter is used to remove suspended pollutants from rainwater collected over roof. A filter unit is a chamber filled with filtering media such as fibre, coarse sand and gravel layers to remove debris and dirt from water before it enters the storage tank or recharge structure. Charcoal can be added for additional filtration.

- **Charcoal water filter:** A simple charcoal filter can be made in a drum or an earthen pot. The filter is made of gravel, sand and charcoal, all of which are easily available.
- **Sand filter:** Sand filters have commonly available sand as filter media. Sand filters are easy and inexpensive to construct. These filters can be employed from treatment of water to effectively remove turbidity (suspended particles like silt and clay), colour and microorganisms. In a simple sand filter that can be constructed domestically, the top layer comprises coarse sand followed by a 5-10 mm layer of gravel followed by another 5-25 cm layer of gravel and boulders.
- **Dewas Filter:** Most residents in Dewas, Madhya Pradesh, have wells in their houses. Formerly, all that those wells would do was extract groundwater. But then, the district administration of Dewas initiated a groundwater recharge scheme. The rooftop water was collected and allowed to pass through a filter system called the Dewas filter, designed by Mohan Rao, District Collector of Dewas, and engineers of the rural engineering services. The water thus filtered is put into the service tube well.

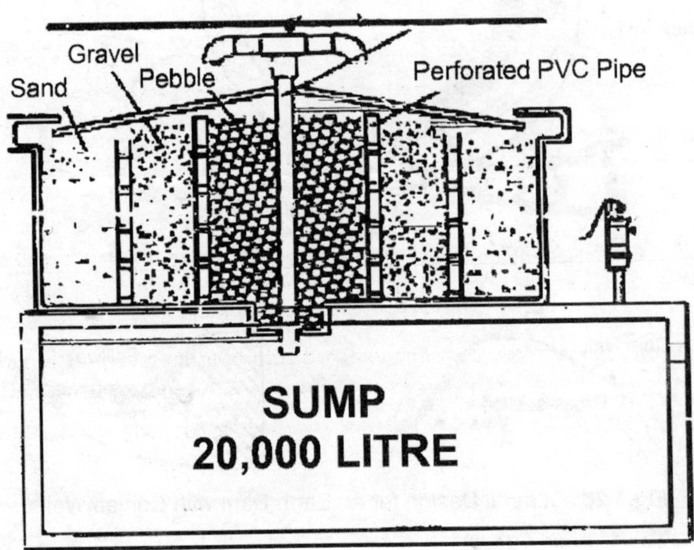

Fig.4.27: A Filter for Rainwater

The filter consists of a polyvinyl chloride (PVC) pipe 140 mm in diameter and 1.2 m long. There are three chambers. The first purification chamber has pebbles varying between 2-6 mm, the second chamber has slightly larger pebbles, between 6 and 12 mm and the third chamber has the largest - 12-20 mm pebbles. There is a mesh at the outflow side through which clean water flows out after passing through the three chambers. The cost of this filter unit is Rs. 600.

- **Filter for large rooftops:** When rainwater is harvested in a large rooftop area, the filtering system should accommodate the excess flow. A system is designed with three concentric circular chambers in which the outer chamber is filled with sand, the middle one with coarse aggregate and the inner-most layer with pebbles.

  This way the area of filtration is increased for sand, in relation to coarse aggregate and pebbles. Rainwater reaches the center core and is collected in the sump where it is treated with few tablets of chlorine and is made ready for consumption.
- **Horizontal roughing filter and slow sand filter:** The introduction of horizontal roughing filter and slow sand filter (HRF/ SSF) to treat surface water has made safe drinking water available in coastal pockets of Odisha. The major components of this filter are described below.
- **Filter channel:** One square meter in cross-section and eight metre in length, laid across the tank embankment, the filter channel consists of three uniform compartments, the first packed with broken bricks, the second with coarse sand, followed by fine sand in the third compartment. The HRF usually consists of filter material like gravel and coarse sand that successively decreases in size from 25 mm to 4 mm. The bulk of solids in the incoming water is separated by this coarse filter media or HRF. At every outlet and inlet point of the channel, fine graded mesh is implanted to prevent entry of finer materials into the sump. The length of a channel varies according to the nature of the site selected for the sump.

  Sump is a storage provision to collect filtered water from the tank through the filtered channel for storage and collection. While HRF acts as a physical filter and is applied to retain solid matter, SSF is primarily a biological filter, used to kill microbes in the water. Both filter types are generally stable, making full use of the natural purification process of harvested surface water and do not require any chemicals.
- **Rain PC:** AcquaSure, a consortium of three specialist Netherlands-based companies, has developed a system for the conversion of rainwater to drinking water in the form of a Rainwater Purification Centre (Rain PC).

  Rain PC is developed by scaling down the multi-staged water treatment method (MST), which involves screening, flocculation sedimentation and filtration and incorporating existing technologies like upward flow fine filtration, absorption and ion exchange. Coming in a small compact 26 kg unit, the Rain PC offers an affordable solution by converting rainwater into drinking water.

  Rain PC is made of ultraviolet resistant poly-ethylene housing and cover, stainless steel rods and bolts, a nickel-brass valve and as adapter for maintaining constant volume. Xenotex-A and activated carbon cartridge's along with ultra-membrane filtration or micro-membrane filtration modules incorporated in the RainPC has the capacity to deal with E.coli and the potential of meeting the Dutch as well as World Health Organisation (WHO) water regulation standards. The components can also be transported individually to be assembled at the site. Three product types are available based on their microbial contaminant removal capacity. This technology is ideally suited for virtually any situation and is a blessing particularly for those who have little or no access to regular safe drinking water.

The salient features of Rain PC are:

- Simple straight-forward installation.
- Easy to operate and maintain.
- Needs no power and operates at low gravity pressure (0.1 bar upward).
- The system is capable of providing a constant flow of about 40 litres of rainwater per hour, enough for a family of five for drinking, cooking and bathing purposes.
- Maintains nearly constant volume irrespective of water pressure.
- The Xenotex-A and activated carbon cartridge processes up to 20,000 litres and can be regenerated up to 10 times.

- **Rainwater harvester:** EA Water Pvt. Ltd. has launched a unique Rainwater Harvester, which filters runoff water from roads, which generally contains oil and grease. This system has been installed in the Gymkhana club, Sector-15, Faridabad, Haryana, which has a sand filter, which filters silt from runoff from roof, lawns and parking area.

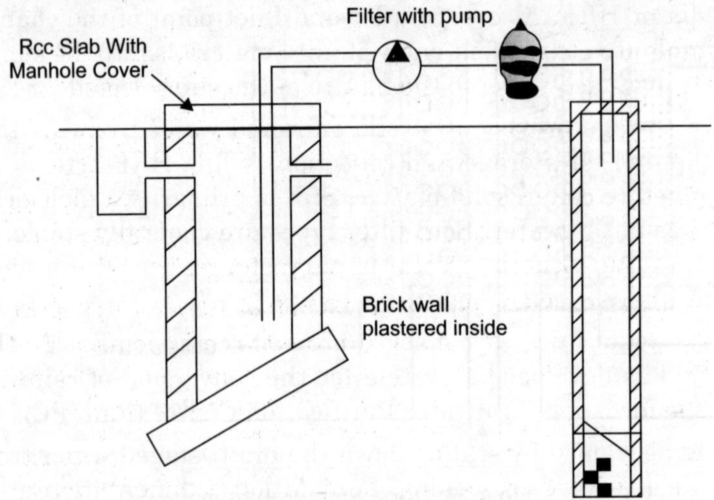

Fig.4.28: Diagram of Pressure Sand Filter Based Recharge System.

Source : Pentair Water Treatment Pvt. Ltd, www.pentair.com

- **Filters available in the German Market:** According to Wessels (1994), concerns over the possible negative health effects of rainwater utilisation led to some opposition. The Federal Office of Health, for example, initially objected to its use for washing clothes, personal hygiene and even for toilet flushing, due to possible risks of infection and allergic reactions. Long-term investigations by the health offices in Hamburg and Bremen, however, have yielded positive results with respect to the use of water for washing purposes and have confirmed that rainwater sources do not present a health risk.

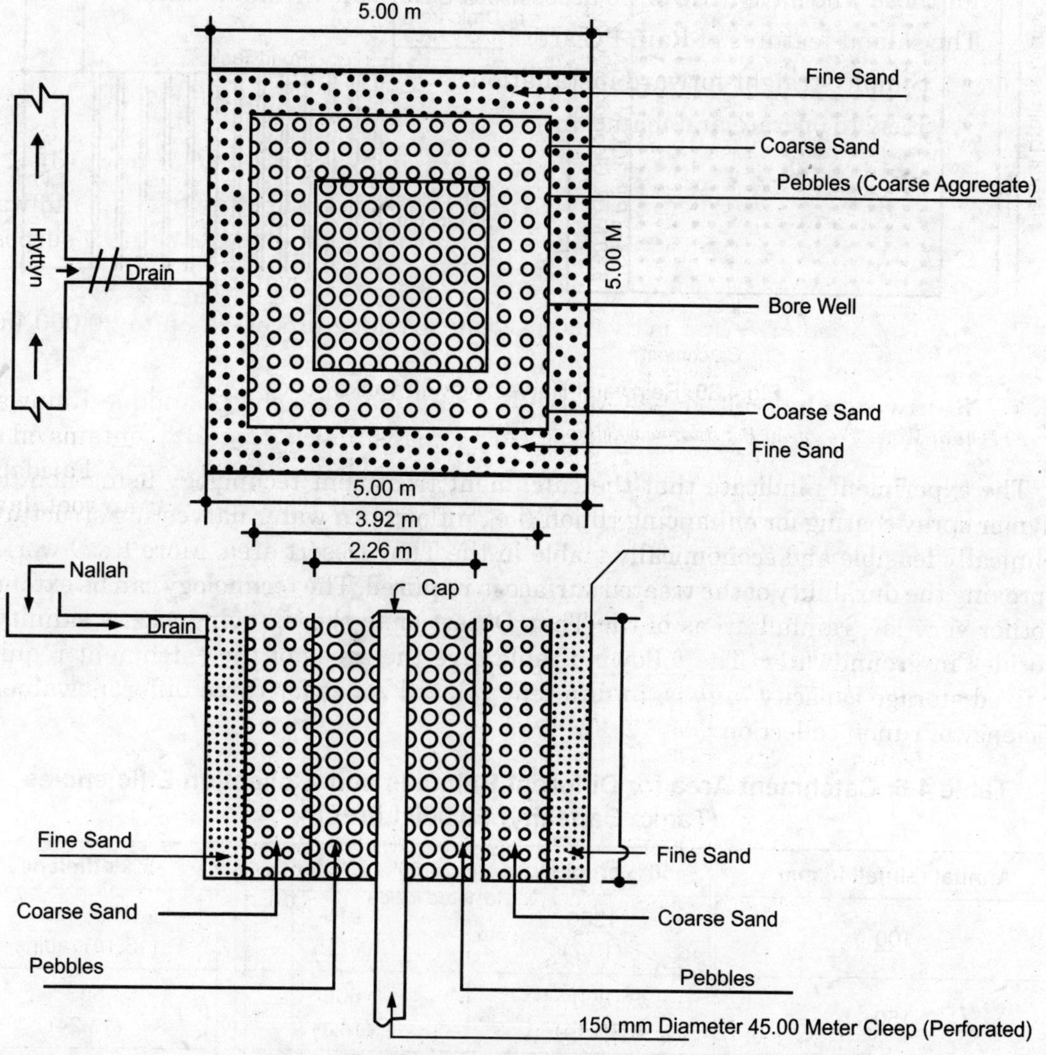

Fig.4.29: Drain Water Harvesting.

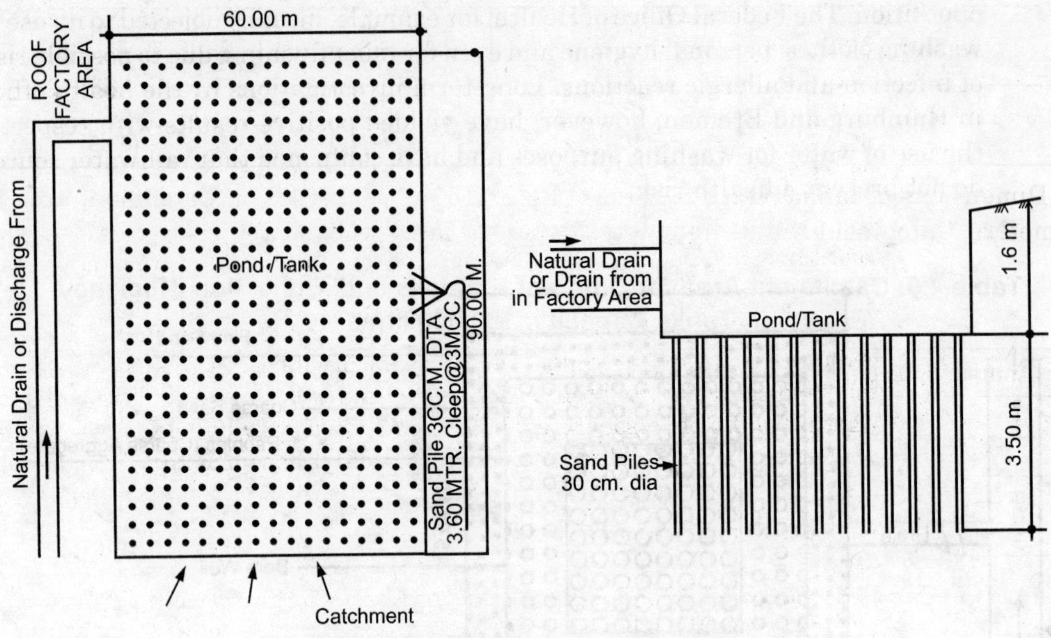

Fig.4.30: Rainwater Harvesting with Tanks/Ponds

Source : Pentair Water Treatment Pvt. Ltd, new delhi

The experiments indicate that the catchment treatment technology using non-toxic polymer spray coating for enhancing runoff (i.e., inflow) to a water harvesting structure is technically feasible and economically viable in the Thar Desert area. More R&D work on improving the durability of the treated surface is required. The technology can be extended to other very low rainfall areas of the Thar Desert or to those having excess salinity or fluorides in groundwater. The following table give the areas of the catchment required for fixed storage capacity *tankas,* in different rainfall zones for three different values of efficiency of runoff collection.

Table 4.5: Catchment Area for Different Rainfalls and Collection Efficiencies (Tanka Capacity: 60,000 Litres)

Annual rainfall in mm	50% efficiency	66% efficiency	75% efficiency
100	1200 (19.7)	910 (17.2)	800 area (16.14) radius
150	800 (16.14)	606 (14.1)	533 (13.24)
200	600 (14.0)	454 (12.26)	400 (11.53)
250	480 (12.6)	363 (11.0)	320 (10.4)

300	400 (11.5)	303 (10.1)	266 (9.5)
350	343 (10.72)	260 (9.4)	228 (8.85)

Dimensions of tanka: O.R.: 2.375 m, I.R.: 2.115 m, Depth: 4.3 m; Catchment area in sq. meters; Catchment radius in meters.

Table 4.6: Catchment Area for Different Rainfalls and Collection Efficiency Tanka Capacity: 100,000 litres

Annual rainfall in mm	50% efficiency	66% efficiency	75% efficiency
100	2000 (25.4)	1515 (22.2)	1333 area (20.8) radius
150	1333 (20.8)	1010 (18.2)	888 (17.1)
200	1000 (18.1)	757 (15.8)	666 (14.8)
250	800 (16.2)	606 (14.2)	533 (13.4)
300	666 (14.9)	505 (13.0)	444 (12.3)
350	571 (13.8)	432 (12.1)	380 (11.4)

Dimensions of kund: O.R.: 3.0 m, I.R.: 2.75 m, Depth: 4.22m Catchment area in sq. meters Catchment radius in meters

Rain Water Recycling and Purification

Rainwater is generally considered to be cleaner than grey water and to entail less risk of infection in the event of systems not operating properly. Its quality is generally good but it is contaminated to some extent by the gases, dust and living organisms in the atmosphere and by pollutants on the collection surfaces. Nitrogen and oxygen are the main gaseous impurities in rainwater; in and near urban areas the emissions of flue gases with CO_2 and sulphur dioxide can result in water of increased acidity and corrosiveness.

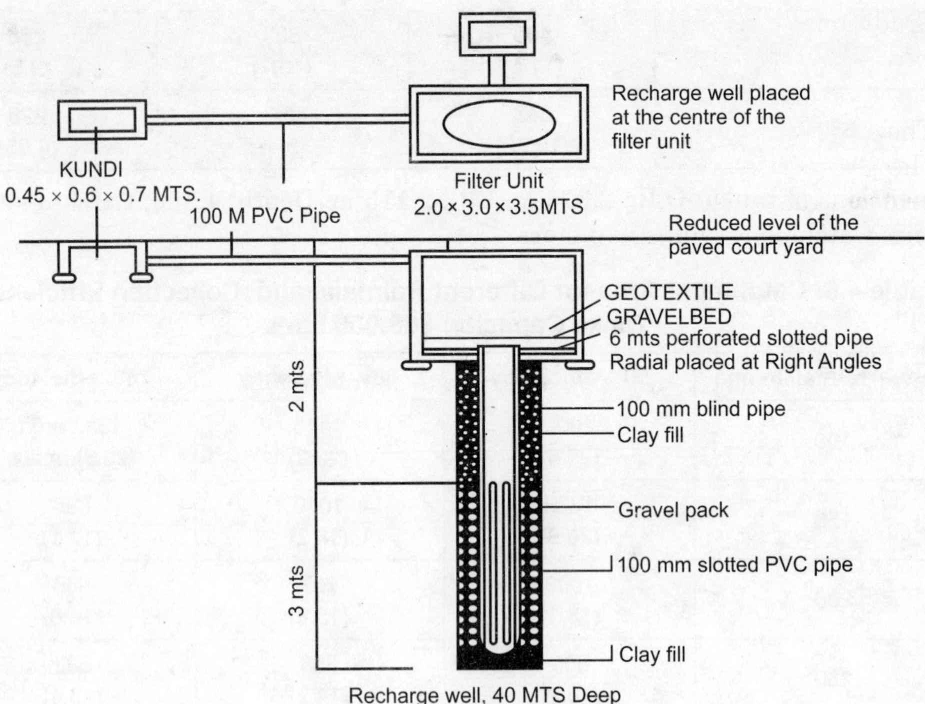

Fig.4.31: Recharge Well for Groundwater Recharge in Open Space.

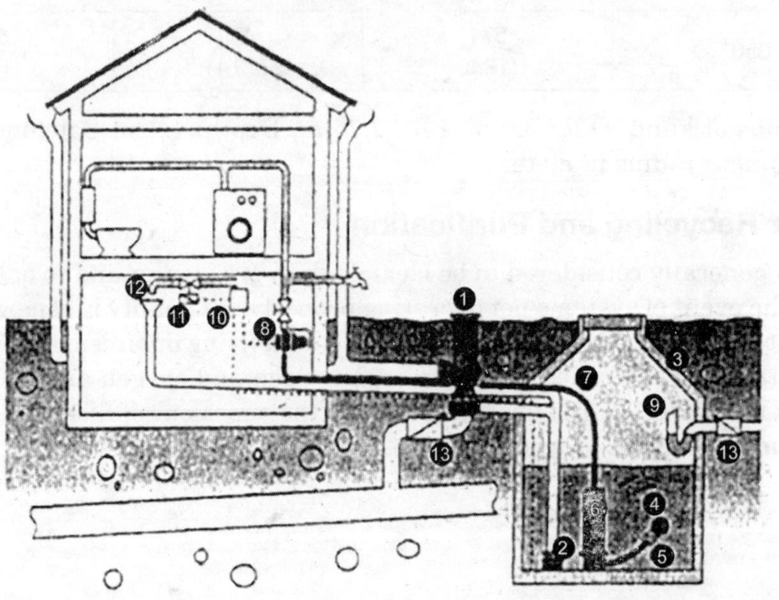

Fig.4.32: Rainwater Collection and Recycling System (1) Vortex underground filter;(2) Inflow smoothing filter; (3) Tank; (4) Floating fine suction filter; (5) Suction hose; (6) Multigo pressure pump; (7) Pressure hose; (8) Automatic pump control; (9) Trapped overflow; (10) Automatic mains back up; (11) Solenoid valve; (12) Type 'A' air gap; (13) Anti-surcharge.

Traditionally, rainwater has been stored in a butt and used for watering the garden. The next likely step is to collect the rainwater, filter it and store it in, say, a basement or underground storage tank of concrete or plastic. Such systems are commercially available.

The size of the tank will vary with the demand, the available area of collection surface and the percentage of demand that is to be met from the tank. (Obviously, a huge tank will protect against severe droughts but will have a cost penalty and other disadvantages.)

One of the primary requirements of a water harvesting system is that of containers to store the harvested water in a hygienic condition.

The Structural Engineering Research Centre (SERC), Ghaziabad, has done research and development on a large number of low-cost structures and implements like water/ grain storage containers, irrigation channels, biogas digesters and septic tanks, primarily using ferrocement. Ferrocement consists of a thin sheet of cement mortar which is reinforced with a cage made of wire mesh and steel bars.

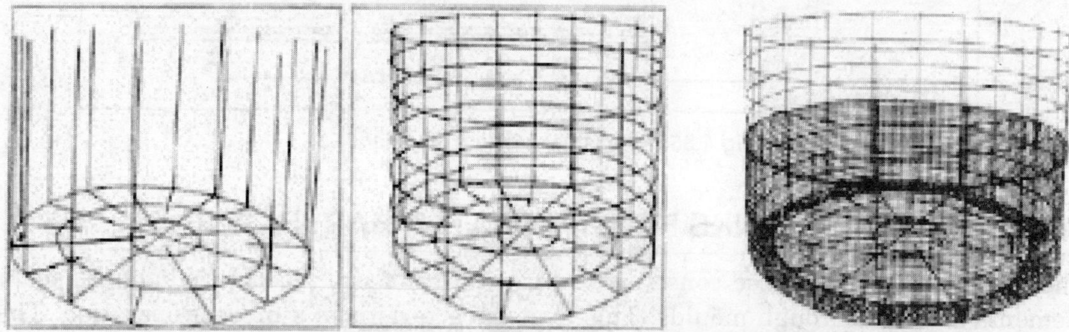

Fig.4.33: Reinforcement Cage for Ferro cement Tanks to Capture Rainwater.

Source; Action for Food Production and United Nation Children's Fund, Rooftop Rainwater Harvesting System

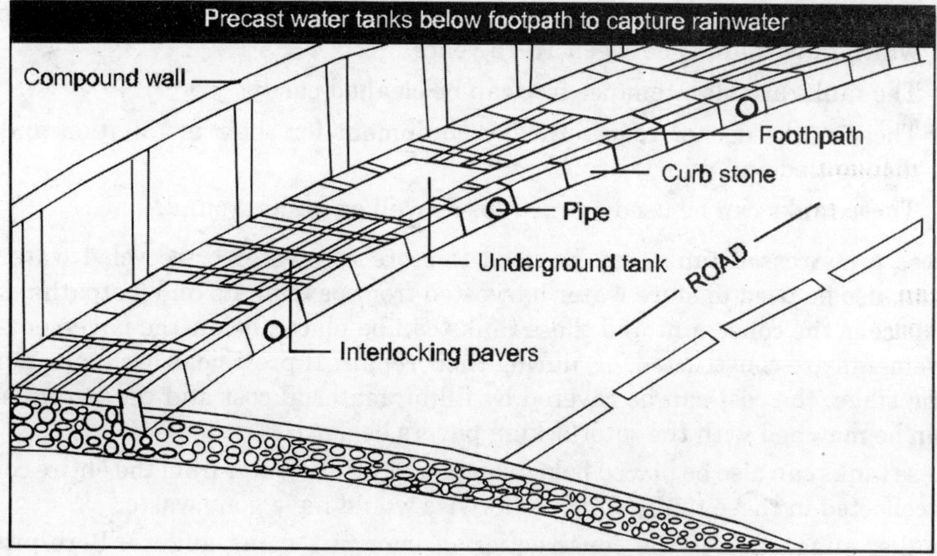

Fig.4.34: Precast concrete water tank

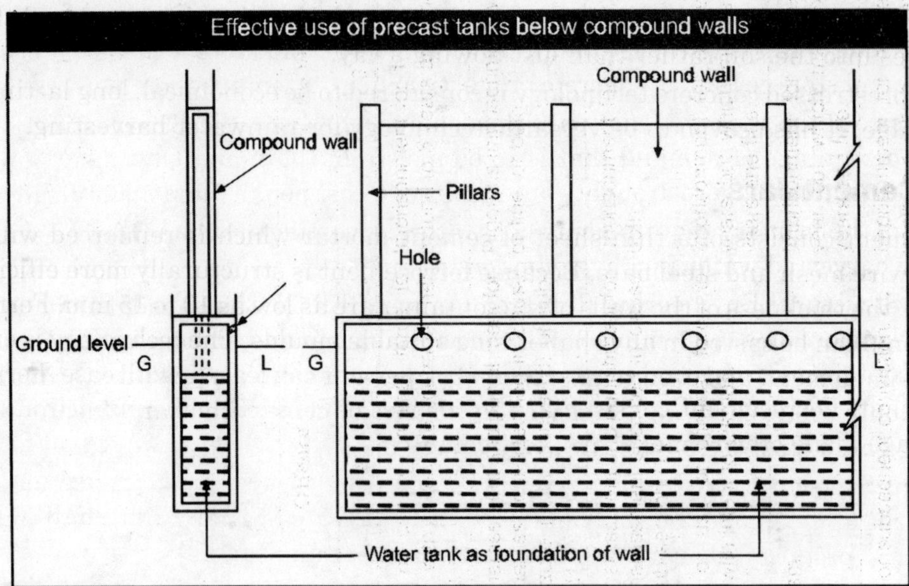

Fig.4.35: Precast tank below boundary wall

PRECAST WATER TANKS FOR RAINWATER HARVESTING

The process of prefabricated construction of water tank is a mechanised one. Repetitive elements are cast through moulds. This is a characteristic of a precast operation. These moulds are generally made of steel or wood and can be machined for low tolerance. High quality can therefore be ensured.

Prefabricated tanks for water storage offer several advantages:

(a) The life span of the tank is about 100 years as compared to ferro cement tanks which have a life of between 10-15 years.

(b) The tanks are maintenance free can be cleaned easily.

(c) These tanks do not require heavy equipment for their installation and can be dismantled and reinstalled.

(d) These tanks can be used on rooftops as well as underground.

These pre-stressed tanks can be used to store rainwater or recycled water. These tanks can also be used to store water harvested from pavements and footpaths. In urban areas, space is the constraint and these tanks can be placed below the pavements, before the pavements are constructed, or, during their repair. If provisions are made during the planning stage, the cost can be covered by filling material cost and the top finish of the tank can be matched with the interlocking pavers design.

These tanks can also be placed below compound walls. Water from the entire compound area is collected in these tanks which otherwise would have gone waste.

Another suggestion is the construction of bore pits using augurs. Bore pits of one metre depth can be dug using augurs on barren lands, roadsides, near play fields, forests, pathways and on bunds adjacent to fields before the monsoon. The purpose of these pits is to store water for longer periods of time so that percolation time increases and the water penetrates into the soil rather than just flowing away.

The prestressed concrete technology is considered to be economical, long lasting, strong and durable. It has proven to be versatile technology for rainwater harvesting.

Ferro Cement Jars

Ferro-cement consists of a thin sheet of cement mortar which is reinforced with a cage made of wire mesh and steel bars. Because ferrocement is structurally more efficient than masonry, the thickness of the walls of the container are as low as 10 to 15 mm. Ferrocement components can be casted in any shape using suitable moulds. The technology is extremely simple to implement, and even semi-skilled work persons can learn it with ease. Ferrocement requires only a few easily available materials - cement, sand, galvanized iron (GI) wire mesh, and mild steel (MS) bars - in small amounts compared to masonry and RCC.

A Pot Shaped Container: The process of construction of a pot shaped ferro cement container is quite simple. The only materials required are hessian cloth, chaff (waste from agricultural produce), GI wire mesh, MS bars, cement and sand.

Preparation of mould: The hessian cloth is first stitched into a sack resembling the shape of a container. It is then filled with chaff that is compacted in layers. Dry leaves or dry grass can also be used in place of chaff. Once the sack is filled with the filter material, it is beaten into the required shape by a wooden bat.

Laying of reinforcement: A GI wire mesh (22-26 gauge-see table) is tied around the mould leaving sockets at suitable locations for inlet, over flow and cleaning pipes. Typing 6 mm diameter MS bars at wide intervals both horizontally and vertically strengthens the reinforcement cage.

Preparation of cement mortar for plastering: Cement mortar of suitable proportion (Table 4.7) is prepared, having water content equal to 0.45 times the volume of cement.

Table 4.7: Specifications for Ferrocement Water Tanks

Capacity of container litres	Thickness of the walls (cm)	Ratio Cement: sand	Thickness of GI wire (gauge) (mm)
400	10	1:3	26
600	10	1:3	24
900	12	1:2:5	24
1500	15	1:2:5	22

Plastering: The mortar is plastered in two layers along the wall thickness, the second layer being applied 24 hours after the first. The ferro cement wall normally has a thickness of 10 to 15 cm, depending on the volume of the container. The cement mortar is applied ensuring a minimum clearance (cover) 3 mm between the reinforcement mesh and the outer surfaces of the wall.

Removal of mould: The mould of the container is removed 24 hours after casting of the wall is completed, by removing the filler material. The container can be brought into use after 10 days of wet curing.

Ferrocement Tank using Skeletal Cage:

Phases of construction

i. Selection of site
ii. Marking for circular foundation:
iii. Excavation for foundation
iv. Compacting the excavated pit
v. Placing cement concrete in foundation:
Prepare Plain Cement Concrete of 1:4:8 mix (1 cement: 4 sand: 8 stone aggregate of 40 mm size)
vi. Erection of mould/Preparation of elements of skeletal cage
vii. a. Preparation of Elements of Skeletal Cage

Table 4.8: Choose the diameter of foundation (Df) for required storage capacity from the table

Capacity of storage tank (litres)	5,000 and 6,000	7,000 and 8,000	9,000 and 10,000
Df	2.40 m	2.70 m	3.00 m

Skeletal cage is an assembly of 4 types of elements (of different shapes) made from mild steel rods. They are:

- 'U' shaped elements
- 'L' shaped elements
- 'ë' shaped elements
- 'O' shaped elements

Table 4.9: Dimensions of elements for tank capacities 5,000 litres to 10,000 litres.

Element	No.	Dimensions	Capacity of Storage Tank (in litres)					
			5,000	6,000	7,000	8,000	9,000	10,000
u	2	H	1.8	2.1	1.9	2.1	1.9	2.1
		W1	2.05	2.05	2.35	2.35	2.65	2.65
	4	H	1.8	2.1	1.9	2.1	1.9	2.1
L		W2	0.82	0.82	0.95	0.95	1.05	1.05
	8	H	1.8	2.1	1.9	2.1	1.9	2.1

L		W3	0.5	0.5	0.6	0.6	0.65	0.65
		D1	9Nos	11Nos	10Nos	11Nos	10Nos	11Nos
			2.05	2.05	2.35	2.35	2.65	2.65
	1	D2	1.25	1.25	1.41	1.41	1.60	1.60
	1	D3	0.62	0.62	0.71	0.71	0.84	0.80

Source: Action for Food Production and United Nations Children's Fund, Rooftop Rainwater Harvesting Systems

Notes:

- Refer to above table for dimensions and number of each of these elements for 5,000, 6,000, 7,000, 8,000, 9,000 and 10,000 litres capacity system.
- Use 6 mm diameter rods for preparing 'U', 'L' and 'ë' shaped elements. Use 4 mm diameter GI wire for "O" shaped elements (Circular rings).
- Straighten, cut and bend the mild steel rods to form these elements.
- Adopt an overlap length of 10 cm in forming the circular elements.

Assembling the elements

- Place the two 'U' shaped rods vertically over the foundation, perpendicular to each other.
- Place the outer, middle and inner rings over the two 'U' shaped rods, coinciding with the circular marking and tie the intersections with binding wires.
- Place and tie 4 'L' shaped elements on the center marking of each quarter, each rod extending upto the inner most ring.
- Place and tie 8 'ë' shaped elements on the remaining markings, each element extending to the middle ring.
- Place and tie all the rings of diameter 'D1' over the vertical reinforcement at a uniform spacing of 20 cm.
- For providing cylindrical shape to the skeletal cage, fix cross bars at the top of skeletal cage and *i.e.* with ropes, 3-4 vertical rods to wooden pegs pegged to the ground.

Typing of mesh over skeletal cage

Select the reinforcement mesh that suits the capacity of the tank from the table below:

Table 4.10: Capacity of Tank

Capacity of Tank (Lt)	5,000 & 6,000	7,000 & 8,000	9,000 & 10,000
	22 gauge	20 gauge	20 gauge
Specification Chicken wire mesh	12mm (1/2") Opening	25 mm (1") opening	25 mm (1") opening

Source: Action for Food Production and United Nations Children's Fund, Rooftop Rainwater Harvesting Systems.

Note: Woven wire mesh of rectangular opening, of same specifications mentioned above, can be used if chicken wire mesh is not available.

Wrap two layers of selected mesh, one layer on the outer side and one layer on the inner side of the skeletal cage. Tie the mesh with binding wire to the skeletal cage at all intersections of elements.

Provide a tucking length of 30 cm (1 foot) at the base.

Project the mesh 10 cm above the top of the skeletal cage.

Cut the skeletal cage and insert pipe fixtures such as overflow pipe, drain pipe and tap at appropriate places as given in Table 4.11.

Table 4.11: Position of Pipe Fixture

Over flow pipe	10 cm below the top of cage
Drain pipe	5 cm above the foundation
Tap	10 cm above the foundation

viii. Plastering the tank's outside wall

- Prepare cement slurry (cement mixed with water) and add anti-rust agent (chrometrioxide tablets).
- Apply one coat of cement slurry (mix of cement and water) over the mesh using a painting brush.
- Prepare cement mortar of depending on capacity of tank.
- Apply the first coat of cement mortar on the outer surface at a thickness of 1 cm.

Care has to be taken to fill the space between the two layers completely. This could be done by using a GI sheet, slightly curved in shape to be held close to the skeletal cage from inside by a person, while cement mortar is applied by another from outside.

- Leave 10 cm of mesh projected above the cage unplastered in order to join the skeletal dome to the tank.
- After two hours, apply a second coat of mortar of a thickness of 1 cm.

ix. Plastering the tank's inside wall

- After two hours of outside plastering, apply cement slurry to the inner surface of the tank wall.
- Prepare cement mortar of 1:3 mix and add waterproof compound in liquid form
- Apply first coat of cement mortar of 1 cm thickness on the inner surface, starting from bottom of the tank moving laterally and progressing towards the top.
- After two hours, apply second coat of mortar to attain a total wall thickness of 2 cm.
- Apply cement slurry as final coat on outer an dinner surfaces of tank and smoothen using coir brush.

x. *Removal of mould.*

xi. Casting of tank floor

- Sprinkle cement slurry over the foundation concrete.
- Prepare plain cement concrete of 1:2:4 mix (1 cement: 2 sand: 4 stone aggregate of 12 mm size), pour it over the base and compact to a thickness of 50 mm (2 inch).
- Finish the floor base using cement mortar keeping the slope towards the drain pipe.
- Finish the wall and base joints (inner and outer) with cement mortar.
- Twelve hours after setting the tank floor, add waterproof compound (liquid form) with cement slurry and apply it over inside surface of the tank and smoothen with coir brush.

xii. Curing the tank

- Cure the tank for 14 days by pouring water thrice a day or covering the tank with wet gunny bags.
- In coastal areas, after curing for 14 days, apply rust proof paint over the outer surface of tank wall.

xiii. Construction of roof for the tank

- An assembly of mild steel elements is prepared as a skeletal frame for the roof. Chicken wire mesh is tied over it and plastered in cement mortar.
- The roof is provided with two openings. One is an opening of diameter 35 cm for accommodating the filter container. Another is a manhole with a 60 cm opening. The opening for the filter will be on one side of the roof. The manhole is provided at the centre of dome.

QUALITY OF STORED WATER

To prevent leaves and debris from entering the system, mesh filters should be provided at the mouth of the drainpipe. Further, a first-flush device should be provided in the conduit before it connects to the storage container.

If the stored water is to be used for drinking purposes, a sand filter should also be provided. Methods to protect rainwater quality include appropriate system designs, sound operation and maintenance and use of first flush devices and treatment. Treatment is mainly appropriate as a remedial action if contamination is expected. First flush devices can be effective in reducing levels of contamination if properly maintained. Good system design, operation and maintenance are generally the simplest and most effective means of protecting water quality.

(a) System design: The best initial step to protecting water quality is to ensure good system design. Water quality will generally improve during storage provided sunlight and living organisms are excluded from the tank and fresh inflows do not stir up any sediment. The design should include:

- Clean impervious roof made from smooth, clean non-toxic material. Over hanging branches above the catchment surface should be removed.

- Taps or draw-off pipes on tanks should be at least five centimetres above the tank floor (more if debris accumulation rates are high). A tank floor sloping towards the sump can greatly aid tank cleaning as will a well-fitting access manhole.
- Wire or nylon mesh should cover all inlets to prevent any insects and other creatures from entering the tank. The tank must be covered and all light excluded to prevent growth of algae and other organisms. The grill at the terrace outlet for rainwater arrests most of the debris carried by the water from the rooftop like leaves, plastic bags and paper pieces.
- A coarse filter and/ or foul flush device should be fitted to intercept water before it enters the tank for removing leaves and other debris.

(b) Operation and maintenance: Proper operation and maintenance of rainwater harvesting systems helps to protect water quality in several ways. Regular inspection and cleaning of catchment, gutters, filters and tanks reduce the likelihood of contamination. Water from other sources should not be mixed with that in the tank.

(c) Treatment: Treatment of stored rainwater only makes sense if it is done properly and if hygienic collection and use of the water will ensure it does not suffer from recontamination. There are several types of treatment possible, the most common being chlorination, boiling, filtration and exposure to ultraviolet or natural sunlight.

i. ***Chlorination:*** Chlorination is most appropriately used to treat rainwater if contamination is suspected due to the rainwater being coloured or smelling bad. It should only be done if the rainwater is the sole source of supply and the tank should first be thoroughly inspected to try to ascertain the cause of any contamination. Chlorination is done with stabilised bleaching powder (calcium hypochlorite - $CaOCl_2$) which is a mixture of chlorine and lime. Chlorination can kill all types of bacteria and make water safe for drinking purposes. About 1 gm (approximately 1/ 4 tea spoon) of bleaching powder is sufficient to treat 200 litres of water.

ii. ***Chlorine tablets:*** Chlorine tablets are easily available in the market. One tablet of 0.5 g is enough to disinfect 20 litres (a bucketful) of water.

iii. ***Boiling:*** Boiling is a very effective method of purification and very simple to carry out. Boiling water for 10 to 20 minutes is enough to remove all biological contaminants.

iv. ***Direct sunlight:*** This can also be used to kill many of the harmful bacteria in water by exposing it in clear glass or plastic bottles for several hours. Although feasible in some circumstances, the water must be clear, the weather fine and the water cooled overnight before consumption.

SODIS: Solar disinfection method uses sun's ultra-violet (UV) radiation to improve the microbiological quality of drinking water. It has been proven that synergies induced by radiation and thermal treatment have a significant effect on the die-off rate of microorganisms. The processes involved are indicated in the illustration. (Fig. 4.36)

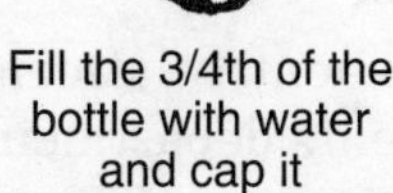

Clean the bottle	Fill the 3/4th of the bottle with water and cap it	Shake the bottle well	Keep the bottle on black well. Iron sheet for minimum 6 hours in sunlight before consumption

Fig.4.36: SODIS – Solar Water Disinfectant Method

Functioning of SODIS Depends on the Following Factors: (1) *Weather and climate:* SODIS requires sun's radiation and temperature to purify drinking water. The container needs to be exposed to direct sunlight for about six hours. If the temperature raises above 50 degrees Celsius, the disinfection process is three times faster. If weather conditions are not optimal the efficiency of the disinfection process can be increased by using halfblackened plastic bottles, which achieve approximately 5 degrees C higher water temperatures than fully transparent, non-painted bottles. Placing the plastic bottles on a black corrugated zinc sheet would also help in achieving disinfection.

- Water Turbidity: Suspended particles in the water reduce the penetration of solar radiation into the water and protect microorganisms from being irradiated. SODIS requires relatively clearer water with a turbidity of less than 30 N TU (Nephelometric Turbidity Unit- nephelometric is a modern commercial instrument used to measure turbidity).
- Material: Various types of transparent plastic materials are good transmitters of light in the UV and visible range of the solar spectrum. Plastic bottles made of PET (Poly Ethylene Terephthalate) are preferred because they contain less UV stabilizers than PVC (Poly Vinyl Chloride) bottles.
- Shape of Containers: More the area of bottle exposed to sunlight, more would be efficiency in achieving disaffection.
- Oxygen: SODIS is more efficient in water containing high levels of oxygen. Shaking a water container which is about 3/4th full for about 20 seconds before they are filled completely could increase oxygen levels. On reacting with this water, sunlight produces highly reactive forms of oxygen (oxygen free radicals and hydrogen peroxides). These reactive forms of oxygen kill microorganisms.

Although SODIS can neither treat turbid waters nor change the chemical quality of water, this method is ideal to disinfect small quantities of water used for consumption.

Polymer Spray Coating

The criteria laid down for the selection of a polymer liquid for spray coating of the catchment surfaces of tanks should be the following:

1. The material should be non-toxic.
2. It should work as soil (sand) stabilizer or binder and also the treated surface should be hydrophobic (water repellent).
3. The treated surface should last for a few years in the inhospitable weather conditions of the Thar Desert.
4. The costs should be comparable with those for providing water through tanker/ desalination plant/ long distance pipelines.
5. The finally evolved methodology of catchment treatment should be simple and robust for adoption by NGOs or Governments departments or villagers.

The research programme lasting over seven years comprised three phases:

1. Laboratory testing of the polymer using Thar Desert soil.
2. Field testing in Hyderabad on small plots under natural rainfall or with a rainfall stimulator on small, open to sky plots using Thar Desert sandy soil.

The results of full-scale field experiments, indicate that average runoff at treated sites varies from 40% to 82% of the monsoon rainfall, as compared to 1 to 3% for an untreated "control" plot in the same locality. It is to be noted here that only some of the rainfall events generate runoff and runoff quantum is not directly proportional to the total amount of rainfall. The cost of the polymer treatment was 25% of the cost of the traditional method of preparing the catchment surface using lime, *surakhi,* bricks and cement. The technology is socially acceptable, as the villagers had no hesitation in drinking the water flowing in the *tanka* from a 'chemical' coated surface.

The experiments indicate that the runoff generation efficiency of the treated surface decreased during the second year. This was due to the observed damage to the treated surfaces mainly due to vandalism:

Table 4.12: Design Criteria for Rectangular Plain Sedimentation Tanks

Design parameter	Guide range of values
Detention time, hours	0.5 to 3
Surface loading, m^3/m^2/day	20 to 60
Depth of tank, m	1.5 to 2.5
Length:width ratio	4:1 to 6:1
Length: depth ratio	5:1 to 20:1

Adapted from: Schulz and Okun, 1991.

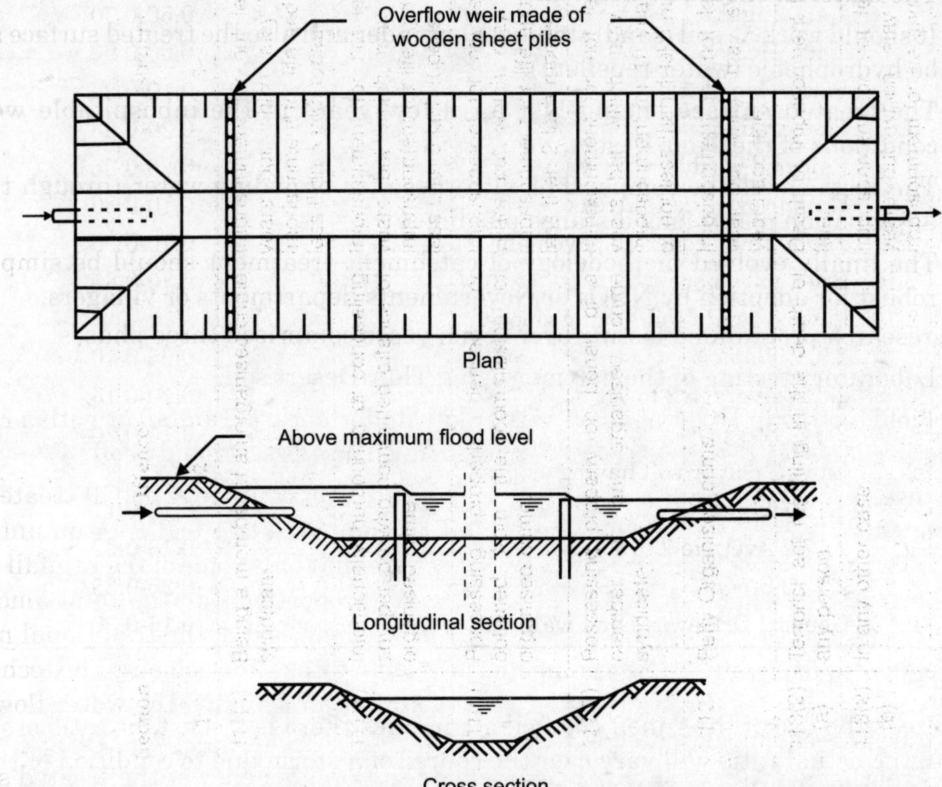

Fig.4.37: Excavated Sedimentation Tank.

Table 4.13: Runoff Co-Efficient of Various Surfaces

		Co-efficient
1.	**Roof Catchment**	**Co-efficient**
1.1	Tiles	0.8-0.9
1.2	Corrugated metal sheets	0.7-0.9
2.	**Ground Surface Covering**	
2.1	Untreated Ground Catchments	
2.1.1	Soil on slope less than 10%	0.0-0.3
2.1.2	Rocky material catchment	0.2-0.5
2.1.3	Business Area	
2.1.3.1	Down town	0.70-0.95
2.1.3.2.	Neighbourhood	0.50-0.70
2.2.	Residential Complexes in Urban Areas	
2.2.1	Single family	0.30-0.50
2.2.2	Multiunits, detached	0.40-0.60
2.2.3	Multiunits, attached	0.60-0.75
2.3	Residential Complexes in Suburban Areas Apartments	0.50-0.70
2.4	Industrial	

	2.4.1 Light	0.50-0.70
	2.4.2 Heavy	0.60-0.90
2.5	Parks, cemeteries	0.10-0.25
2.6	Play grounds	0.20-0.35
2.7	Railroad yard	0.20-0.35
2.8	Unimproved Land Areas	0.10-0.30
2.9	Aspheltic or concrete pavement	0.70-0.95
2.10	Brick pavement	0.70-0.85
2.11	Lawns, sandy soil having slopes	
	2.3.1 Flat 2%	0.05-0.10
	2.3.2 Average 2 to 7%	0.10-0.15
	2.3.3 Steep 7%	0.15-0.20
2.12	Lawns, clayey soil having slopes	
	2.12.1 Flat 2%	0.13-0.17
	2.12.2 Average 2 to 7%	0.18-0.22
	2.12.3 Steep 7%	0.25-0.35
2.13	General Driveways and walls	0.15-0.30

Source: ASCE and WPCF, 1969.

Whereas the use of the runoff coefficients implies there is a constant ratio of rainfall to runoff, the actual ratio will vary over the course of a storm due to condition of the area and the variability of the rainfall pattern. A common practice is to use average coefficients for various types of areas and assumed that the coefficients will be constant throughout the duration of the storm.

DEBRIEFING QUESTIONS

1. What is the significance of rainwater harvesting?
2. What are the various methods of rainwater harvesting?
3. Write short notes on the following:
   (a) Bund
   (b) Agor Tank
   (c) Stepwell
   (d) Kund
   (e) Baoli
   (f) Johad
   (g) Tanka
   (h) Sub-surface Dyke
4. Explain briefly the following:
   (a) Dug well

(b) Hand Pump
(c) Recharge Pit
(d) Recharge Trench
(e) Recharge Shaft
(f) Thai Jar
(g) Ferro-cement Jars/tanks

5. What are the simple, local methods of primary treatment of rainwater?
6. What are the measures for water consumption at family level?
7. Explain the various kinds and methods of filters used for water purification.

Section	Offence	Punishment	Cogn[illegible]
1	2	3	[illegible]
106(1)	Causing death by negligence.	Imprisonment for 5 years and fine.	Cogn[illegible]
	Causing death by negligence by registered medical practitioner.	Imprisonment for 2 years and fine.	Cogn[illegible]
106(2)	Causing death by rash and negligent driving of vehicle and escaping.	Imprisonment for 10 years and fine.	Cogn[illegible]
107	Abetment of suicide of child or person of unsound mind, etc.	Death, or imprisonment for life, or imprisonment for 10 years and fine.	Cogn[illegible]
108	Abetment of suicide.	Imprisonment for 10 years and fine.	Cogn[illegible]
109(1)	Attempt to murder.	Imprisonment for 10 years and fine.	Cogn[illegible]
	If such act causes hurt to any person.	Imprisonment for life or as above.	Cogn[illegible]
109(2)	Attempt by life-convict to murder, if hurt is caused.	[illegible] imprisonment for life which shall mean [illegible] of that person's natural life.	Cogn[illegible]
110	Attempt to commit culpable homicide.	Imprisonment for 3 years, or fine, or both.	Cogn[illegible]
	If such act causes hurt to any person.	Imprisonment for 7 years, or fine, or both.	Cogn[illegible]
111(2)(a)	Organised crime resulting in death of any person.	Death or imprisonment for life and fine [illegible] lakh rupees.	Cogn[illegible]
111(2)(b)	In any other case.	[illegible] not less than 5 [illegible] imprisonment for life and fine not less than [illegible] lakh rupees.	Cogn[illegible]
111(3)	Abetting, attempting, conspiring or knowingly facilitating the commission of organised crime.	Imprisonment for not less than 5 years but which may extend to imprisonment for life and fine not less than 5 lakh rupees.	Cogn[illegible]

5 Sanitation and Sewerage

Sanitation and sewerage are the core elements of physical infrastructure that determine the environmental hygiene and public health of a settlement. Development of appropriate and efficient sewage treatment is vital to facilitate healthy and ecological development. The existing capacity of sanitation and sewerage system in Indian cities is grossly inadequate, as only about 55 per cent of the population is covered under organised sewerage system and about 15 per cent under onsite sanitation systems. Rest of the population does not have proper access to sanitation facilities. The increasing pollution in the rivers and water bodies is largely due to lack of sewage treatment facilities.

The insanitation, open defecation and urination, lack of individual toilets make the environment unhygienic and unhealthy together with pollution of surface and underground sources of water and germination of mosquitoes, flies and harmful vermin. In the words of Nobel laurate V.S. Naipaul, *"Indians defecate everywhere. They defecate mostly beside the railway track. But they also defecate on the beaches; they defecate on the hills; they defecate on the river banks; they defecate on the streets; they never look for cover. Muslims, with their tradition of purdah, can at times be secretive. But this is a religious act of self-denial, for it is said that the peasant, Muslim or Hindu, suffers from claustrophobia if he has to use an enclosed latrine."*

Around 130 million households in India do not have access to toilets. Over 50 percent of India's population attend to nature's call in fields, along railway tracks, behind bushes, or by roadsides. Of the one billion people in the world who have no toilet, India accounts for nearly 600 million. The situation in Indian villages is worse as more than 68 percent villagers defecate in open.

Around 42 percent schools in India do not have toilets for girls. Unavailability of toilets tor girls and women has various other social consequences. Many teenage girls leave school when they start menstruating because they have no privacy. Sexual violence is another threat that women face in day to day life when they venture out to attend nature's call.

Open defecation is not merely a sanitation issue, but is linked with the health. A United Nations (UN) report on sanitation say's: "Open defecation perpetuates the vicious cycle of disease and poverty and is an affront to personal dignity." Open defecation becomes disastrous when practiced by groups in close contact with each other. As India's population is huge, growing rapidly and densely settled, it is impossible even in rural areas to keep away human faeces from crops, wells, food and children's hands. Ingested bacteria and worms spread diseases, especially of the intestine. They cause enteropathy, a chronic illness that prevents the body from absorbing calories and nutrients. That helps to explain why in spite of rising incomes and better diets, rates of child malnourishment in India do not improve faster."

According to the Unicef, nearby 50 percent Indian children remain malnourished and open defecation is one cf the major causes. It is linked to a numerous health problems such as stunting in children, diarrhoea, cholera, typhoid, hepatitis, polio, worm infestation, reduced physical growth, impaired cognitive function, and lower IQs. It is to be noted that diarrhoea is still the biggest cause of child deaths around the world and 25 percent of deaths occur in India.

Table 5.1: Open Defecation in Various Countries

Country	People defecate in open (in millions)
India	626
Indonesia	63
Pakistan	40
Ethiopia	38
Nigeria	34
Sudan	19
Nepal	15
China	14
Nigeria	12
Burkina Faso	9.7

Source: UNICEF

MANUAL SCAVENGERS

Manual scavengers constitute one of the most invisible groups in the country, performing the inhuman occupation of removing human waste or excreta from dry or insanitary latrines, which is in complete violation of the right to a dignified life. As per the 2011 Census, about 23 lakh insanitary latrines exist in the country and estimates indicate 60 million people are engaged in manual scavenging in India (PRS Legislative Research, 2013). It is particularly noteworthy that 95 per cent of manual scavengers are women (UNDP and UN Solution Exchange, 2012). This shows that even within caste, there is a gender dimension to poverty. Further, as the Baseline Assessment Report of the UN Women Fund for Gender Equality project reveals, women manual scavengers have limited access to basic income and social protection: 99 per cent women from the community earned less than Rs. 2000 a month; 82 per cent women had never been to school; 77 percent had no access to public health services; only 17 percent women had access to wage employment under MGNREGA; 4 per cent women had access to housing under the Indira Awas Yojana and 13 per cent women had managed to access the rehabilitation scheme for manual scavengers and dependents (Jan Sahas Social Development Society, 2014).

Manual scavengers have faced historic, systematic and inter-generational stigma, discrimination and exclusion from mainstream society, due to their caste and gender identities. Manual scavenging is a deep and complex phenomenon based on entrenched societal attitudes and rooted in the caste system. The conditions of employment of a manual scavenger are appalling - a scavenger usually receives Rs. 15-20 per month from each household where he/she cleans latrines, besides rotis given once a day. Manual scavengers are forbidden from entering temples and other places of public significance. They often face threats, verbal and physical abuse and are at the receiving end of violent atrocities. Although legislation and schemes have existed outlawing manual scavenging as an illegal occupation, little progress has been made on discontinuing this practice on the ground. (UN Women, 2014 Engendering Rural Livelihoods: Supporting Gender Responsive Implementation of the NRLM, New Delhi).

As per the Prohibition of Employment as Manual Scavengers and their Rehabilitation Act, 2013, (...) "A manual scavenger means a person engaged or employed (...) for manually cleaning, carrying, disposing of, or otherwise handling in any manner, human excreta in an insanitary latrine or in an open drain or pit into which the human excreta from the insanitary latrines is disposed of, or on a railway track or in such other spaces or premises (...)".

The Supreme Court in 2014 in *Safai Karmachari Andolan v. Union of India* directed that all people working as manual scavengers be rehabilitated *"based on the principles of justice and transformation"* and reiterates that states have a duty to implement the 2013 Act. Shortly before taking office as prime minister, Narendra Modi said: *"My identity is of a Hindutvawadi, but I say build toilets before you build temples."*

According to a World Health Organization (WHO) report, an average of Rs. 6,500 per person is lost in India due to lack of cleanliness and hygiene. The report also suggests that a nine times return for every rupee spent on sanitation is possible. These returns are in the form of improved health, lowered mortality rates, higher school attendance rates, and disease control. Ending open defecation will bring immense benefits by helping India in improving health standards, and in turn show affirmative progress on health indicators such as Infant Mortality Rate (IMR), malnutrition among children and women among others.

Bill Gates, Founder of Microsoft, during his visit to India said *"Making toilets either at the community or household-level and having them affordable and attractive-which in the case of toilets implies that they don't smell badly-that's a tough challenge. If you were to go to every city in India and get piped clean water for every toilet in every home and a pipeline that treat the water, you're talking about literally hundreds of billions of dollars."* Building toilets does not mean that people will use them and there seems to be a host of cultural, social and behavioural aspects. Open defecation is not only linked with non-existence of toilets in every household but it is a social-cultural habit. People can be seen urinating even outside the public toilets. The reasons could be a behavioural issue or poor cleanliness of the public toilets. Alongside building toilets, the government needs to make a mechanism to maintain those toilets and promote their use.

The government initiated Total Sanitation Campaign (TSC) in 1999, which was later named Nirmal Bharat Abhiyan. Under the scheme, the government spends over Rs. 7000 crore every year to subsidise rural toilet construction in the country. However, the scheme did not pick up and it was difficult to see that the poor construct and maintain the toilets.

The government in the year 2012 created a five-year 'Sanitation and Hygiene Advocacy and Communication Strategy Framework' to advise states on how to counter the culture of open defecation, including setting up local education committees. However, the initiative did not take off. More than half of the country's sanitation education budget since 1999 was not spent, according to the Ministry of Drinking Water and Sanitation.

The backlog of toilet is linked with the sanitation and sewage facilities. In most cities, sewage lines do not cover all households, which usually exist only in the so-called planned colonies. Today the cities have expanded into surrounding peri-urban and rural areas, where underground drainage and sewerage infrastructure is absent. Putting up sewage lines in already-built, crowded, congested and haphazard urban areas is a difficult task. This challenge is compounded by the fact that even where sewage lines exist, they are broken, choked and is a state of disrepair. Vast areas of city are without toilets, posing the problems of sanitation, hygiene, public health and ecological degradation. This poses a profound environmental, social and cultural challenge and implications. Defecation is a private human activity and habits may largely differ between regions and cultures. In many cultures handling of excreta is subject of strong taboos and may be related to aspects of human dignity. Knowledge and perception of the reuse of excreta largely varies between cultures and regions and sanitation practices differ.

National Urban Sanitation Policy

The aim of the National Urban Sanitation Policy (NUSP), 2008 is to transform Urban India into community-driven, totally sanitized, healthy, and liveable cities and towns. Basic features laid down in NUSP given below should be adhered for planning of the cities

- Cities must be open defecation free
- Must eliminate the practice of manual scavenging and provide adequate personnel protection equipment that addresses the safety of sanitation workers
- Municipal sewage and storm water drainage must be safely managed
- Recycle and of treated sewage for non?potable applications should be implemented wherever possible
- Solid waste collected and disposed?of fully and safely
- Services to the poor and systems for sustaining results
- Improved public health outcomes and environmental standards.

The objective of public waste water collection and disposal system is to ensure that sewage or excreta and sullage discharged from community is properly discharged, collected, transported, treated to the required level of degree and finally disposed off without causing any health or environmental problems 161.

As per CPHEEO manual, 80% of water supply may be expected to reach the sewers, however it recommends designing the system by considering minimum wastewater flow of 100 litres per capita per day.

Table 5.2.: Recommended Design Period for Sewerage System Components

S. No.	Component	Recommended Design Period (Years)	Clarification
1.	Collection system *i.e.* Sewer Network	30	The system should be designed for the prospective population of 30 years, as its replacement is not possible during its use.
2.	Pumping Stations (Civil Works)	30	Duplicating machinery within Pumping station would be easier/ cost of civil works will be economical for full design period.
3.	Pumping Machinery	15	Life of pumping machinery generally 15 years
4.	Sewerage Treatment Plant	30	The construction may be in a phased manner as initially the flows may not reach the designed levels and it will be uneconomical to build the full capacity plant initially.
5.	Effluent Disposal and Utilisation	30	Provision of design capacities in the initial stage itself is economical.

Source: Manual on Sewerage and Sewage Treatment, CPHEEO.

On-site Sanitation Systems

Conventional on-site sanitation systems have been used for centuries to provide excreta disposal at the household level. There are two basic types, dry systems (pit toilets) and water based systems (flush toilets). The former have been used in the "North" until the mid 19th century, when water-borne sewerage started its march across industrialized nations and became the accepted standard for excreta disposal. Given the limited financial means of most developing countries, the pit toilet have remained their principal means of excreta disposal, except for the well to Issues and do classes, who are often provided with government subsidized sewerage systems. On-site pour flush toilets are today popular and widely used in India.

Neither pit toilets nor pour flush toilets included resource recovery as part of their function. Indeed, the health impact of underground waste disposal, particularly its impact on groundwater, was rarely considered. Fortunately, pathogenic organisms as a rule, did not survive in the ground, and in any event did not travel very far. However, the increasing density of population often has led to situations where the required minimum distance between drinking water well and pit toilets cannot be respected.

In many densely populated areas, the use of pit toilets has also led to nitrate concentrations in groundwater, which exceed the maximum level recommended by the WHO for drinking water and which have been linked to serious health problems, particularly for babies. This became a concern when low income societies adopted the western custom of bottle feeding their babies, thus exposing them to the danger of methemoglobinemia (blue baby syndrome).

Shallow groundwater is still a major source supply in rural and peri-urban areas, especially for the poor. The design of the conventional "drop and store" pit-toilet (and of most other on-plot systems) pollutes this precious groundwater as it deliberately aims to retain only solid matter in the pit and infiltrates as much of the liquids as possible into the subsoil. As these liquids contain all the soluble elements of the excreta as well as viruses and pathogens, this type of sanitation, depending on the hydro-geological situation, can be a highway to groundwater contamination.

Septic Tanks

In most of the Indian settlements, where municipal sewerage does not exist, local on site sanitation systems have been adopted. The most popular 'on–site 'system is the age-old septic tank and its versions. The septic tank comprises an underground concrete or brick tank separated by a suspended baffle wall. The system is basically a horizontal continuous flow sedimentation tank in which sewage moves very slowly so to be retained for a period sufficient to permit most of the suspended matter to settle in the form of sludge at the bottom of the tank. Some of the lighter solids including grease and fat rise to the surface of the sewage to form a floating scum. The scum and the sludge so formed are retained within the sewage tank for several months during which they are decomposed anaerobically to form gases and liquids through the process of sludge digestion. This also reduces the volume of the sludge. The septic tank effluent is generally disposed of in a soak/seepage pit, normally filled with brick or stone aggregate. Methane produced in the tank is removed through a vent pipe which is high enough to avoid bad smell. The septic tank suffers from a number of problems. Periodically the tanks have to be desludged. Although the cost of construction is generally low, but because of spatial and financial constraints the soakage pit is usually not constructed and the highly septic effluent is allowed to flow into storm water drains or in open land. To function properly sufficient water supply is necessary.

A serious flaw is that if any crack occurs in the floor or wall of the tank then the hydraulic pressure inside leaks the liquid contents to the surrounding soil creating a no–flow situation in the tank which stops its functioning. Another risk is that when septic tanks are cleaned manually, they expose the cleaner to grave danger. The disposal of tank contents in insanitary manner contaminates subsoil water.

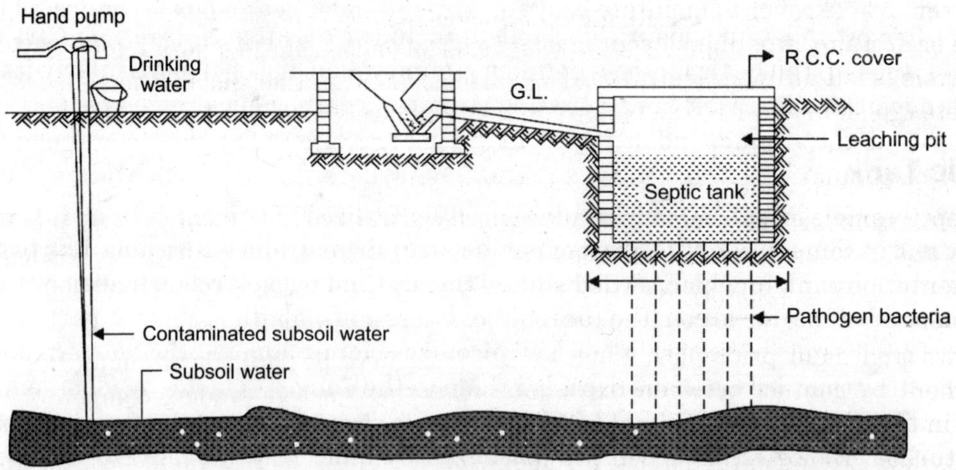

Fig.5.1: Contamination of Sub-soil Water due to a Leaking Septic Tank/Soak pit

Single Pit Latrine

Single pit latrine consists of three main parts: the dug pit, the superstructure and the squatting pan. The pit can be offset or the latrine may be directly sited over the pit. The pit is mostly unlined and of varying dimensions though, generally, not exceeding 1 m × 1 m in area (or 1 m diameter) with a depth of about 1.5 m. The superstructure is made from local materials like bamboo, straw, jute matting or wood and sometimes includes a corrugated tin/asbestos sheet as roofing material. Generally, there is no formal ceramic or fibre glass squatting pan with water seal. Instead, this is improvised from sheet metal set into a bamboo or wooden floor and formed into a chute which in inclined towards the open pit.

When the septic tank pit is filled, it is either excavated or a new pit is dug. The existing superstructure is either moved over to the new pit or a new superstructure is constructed. In cases where the pit is to be reused, pit contents are disposed of in nearby low– lying areas. Otherwise the old pit is simply covered over with soil. The liquids reach into the soil by way of the unlined pit while the solids are retained. When the pit is filled, the top slab is moved to a newly dug pit and the old one is covered with soil. The main problem with all single pit latrines is that when the latrines are full there is dislocation of its use for some time which may extend to a number of days. The single pit contents endanger the health of the scavenger and final disposal of sludge is an environmental hazard.

Two Pit Latrine

The two-pit pour flush (TPPF) waterfall latrine provides an alternative and intermediate on-site sanitation system. It is comparatively easy to construct, operate and maintain, and is economical in quantity of water required. The system consists of a squatting pan connected to a junction box outside the latrine, further leading to two underground pits. The pan has a steep bottom slope of 30 degree to the horizontal, a narrow neck of 70 mm and a 20 mm waterfall trap set into the floor of the latrine. This makes it amenable to complete flushing with only two liters of water. The junction box has a Y–junction so that the flow can be directed into one pit at a time by blocking one branch of theY–junction. It has a removable cover to facilitate easy inspection. The junction box is connected through a pipe or drain to two–lined leaching pits with the bricks laid in a honeycomb pattern. The pit bottom is left unlined. The pit can be rectangular or circular and is fitted with reinforced concrete top slab.

Septic Tank

The septic tank is most common, small–scale decentralized treatment system. It is compact, robust and in comparison to the cost of its construction, extremely efficient. It is basically a sedimentation tank in which settled sludge is stabilised by anaerobic digestion. Dissolved and suspended matter leaves the tank more or less untreated.

Two treatment principles, a mechanical treatment by sedimentation and a biological treatment by contact between fresh wastewater and active sludge, compete with each other in the septic tank. Optimal sedimentation takes place when the flow is smooth and undisturbed. Biological treatment is optimized by quick and intensive contact between new inflow and old sludge, particularly when the flow is turbulent. The way the new fluent flows through, decides which treatment is more effective.

A septic tank consists a minimum of two compartments. The first compartment occupies about half of the total volume because most of the sludge and scum accumulates here. The following chamber(s) are provided to calm down the turbulent liquid. They are of equal size and occupy the other half of the total volume. Scum accumulates must be removed every two years. Scum does not harm the treatment process, but it does occupy tank volume.

Borehole Latrine

A borehole latrine consists of a circular hole of 30-40 centimetre diameter and vertically dug into the ground to a depth of four to eight metres, most commonly six meters. On the top of the hole, a reinforced cement concrete (RCC) slab with circular opening is placed. This technology promoted way back in 1930s is still in use in many parts of the country.

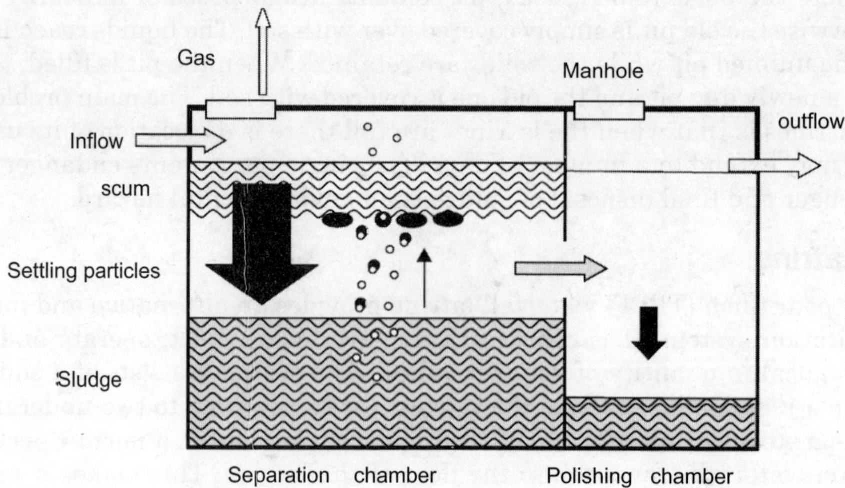

Fig.5.2: Cross-section of a Typical Septic Tank Showing Principle of Sanitation

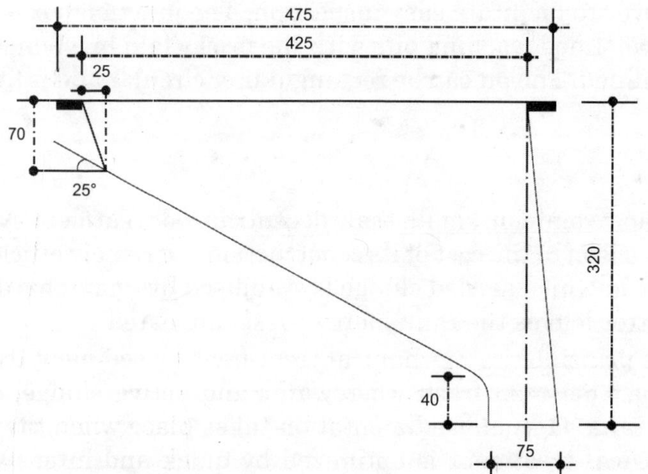

Fig.5.3: Squatting Pan and Trap for Pour Flush Toilet

Note: All measurements in millimeters

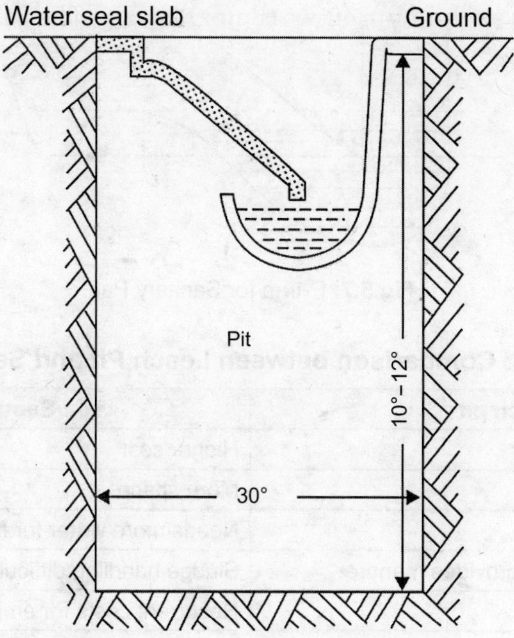

Fig.5.4: Hand Flush Water Seal Toilet

In 1949-50, an improved version of borehole latrine was developed by Rural Training and Health Centre in Singur, West Bengal, which was known as dug well latrine. This type of latrine had a circular pit of 0.75 meter diameter and 3 to 3.5 meter depth. If required, the pit was lined with pottery rings to prevent caving of the soil. A concrete squatting plate was placed on the top of the pit.

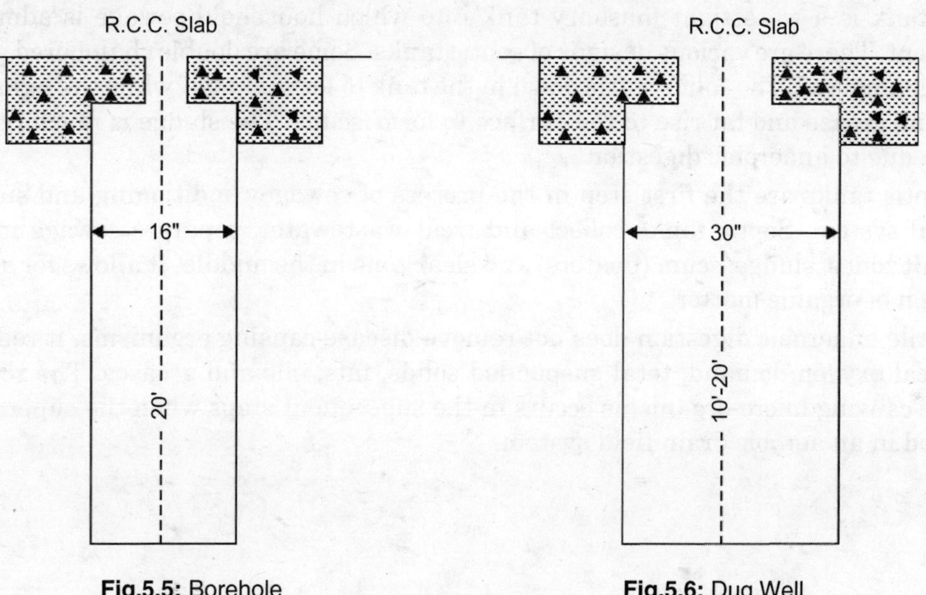

Fig.5.5: Borehole Latrine

Fig.5.6: Dug Well Latrine

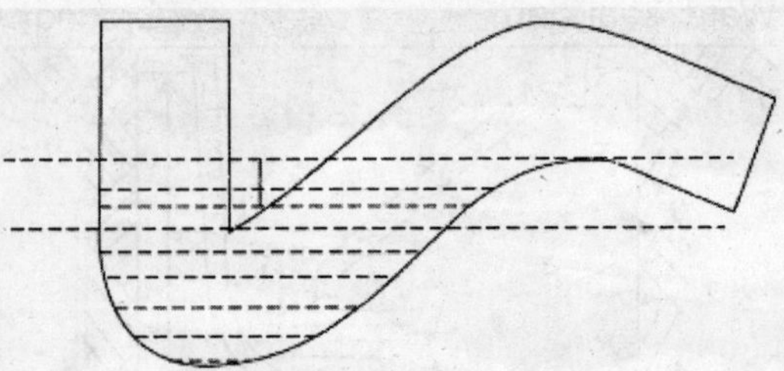

Fig.5.7: P-trap for Sanitary Pan

Table 5.3: Comparison between Leach Pit and Septic Tank

Leach pit	Septic tank
Low cost	Higher cost
Less space	More space
Needs little water	Needs more water for flushing
Sludge handling easy, provides manure	Sludge handling difficult
No recurring cost	Recurrent costs for emptying
Pit emptying easy	Safe disposal of effluents-pollution
No mosquitoes	Mosquito menace

Source: UNICEF, 2007

SEPTIC TANK SANITARY SYSTEMS

Septic tank sanitary systems are widely used in urban areas and also in rural areas. The septic tank is a watertight masonry tank into which household sewage is admitted for treatment. There are various designs of septic tanks. Some are double chambered and some single chambered. The solids settle down in the tank to form 'sludge' while the lighter solids including grease and fat rise to the surface to form 'scum'. The sludge is much reduced in volume due to anaerobic digestion.

Septic tanks are the first step in the process of sewage conditioning and subsurface disposal system. Septic tanks collect and treat wastewater, separate sewage into three different zones: sludge, scum (floaters) and clear zone in the middle. It allows for anaerobic digestion of organic matter.

While anaerobic digestion does not remove disease-causing organisms, it reduces the biological oxygen demand, total suspended solids, fats, oils and greases. The removal of disease causing micro-organisms occurs in the subsequent steps when the supernatant is disposed in an aerobic drain field system.

Fig.5.8: A Septic Tank Being Installed

As a final stage of disposal, the treated effluent from the septic tank is discharged to the leach field where it percolates through suitable 'septic zones' and finally into the subsoil for further purification. The functioning of septic tank is correlated to the design, usage, its correct size for occupancy and long term storage of the sludge. Settling of solids and flotation of scum requires a calm glow to promote growth of bacteria. Design and functionalities are critical in improving the effluent quality leaving the tank.

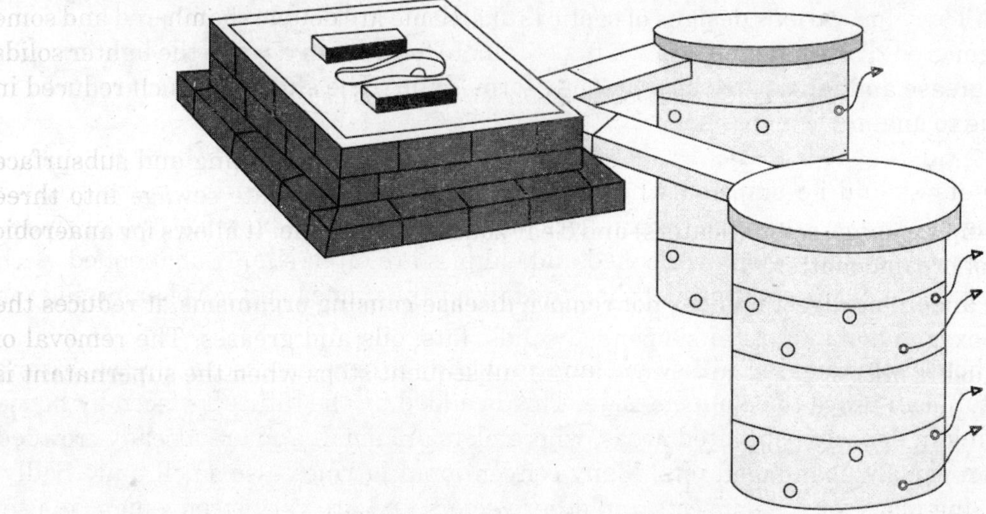

Fig.5.9: Off Site Double Pits with RCC Rings

Source: UNICEF, 2007

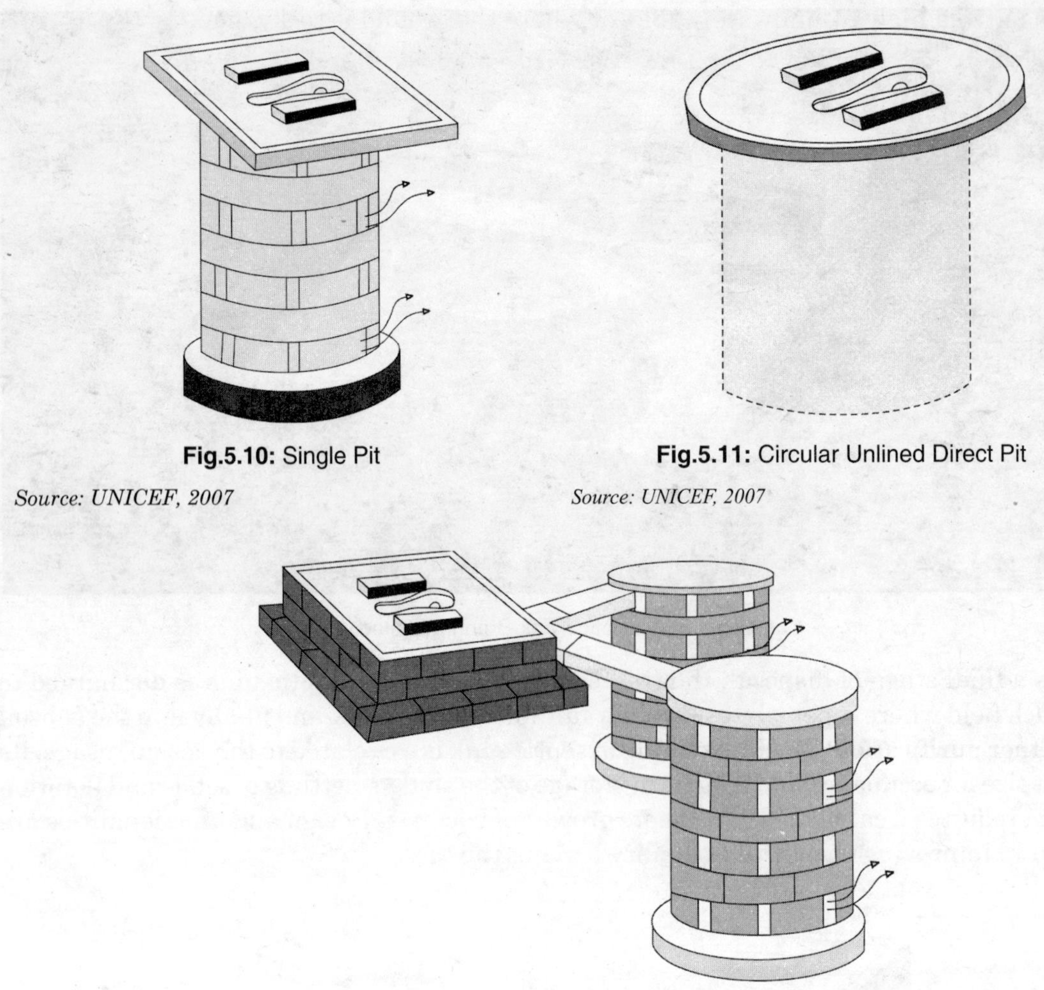

Fig.5.10: Single Pit

Source: UNICEF, 2007

Fig.5.11: Circular Unlined Direct Pit

Source: UNICEF, 2007

Fig.5.12: Off-site Double Pit-Brick Honeycomb Latrine

Source: UNICEF, 2007

Pit toilets should be emptied when they are full, with the content being treated before being put to any other use (e.g. in agriculture). In practice, however, faecal sludge management aspects are often overlooked and old pits are often simply abandoned, with users preferring to dig a new pit toilet than to attempt the unpleasant job of emptying the old one.

For a household, digging a new pit, and providing a new superstructure each time the old is full, can prove to be a expensive. This is added by the difficulty faced by house owners living in densely populated areas, where plots are small and are already crowded with old, previously abandoned pits. Many conventional latrines also smell quite badly, are a breeding place for flies, insects and other vectors and are very inconvenient to use, especially for children, women and girls, as they have to be built at a distance from the house, making night-time visits undersirable. Added to this are the problems of pit construction in areas with a high groundwater table, or where the ground is rocky, making the use of a pit extremely impractical. Pits are also susceptible to flooding during heavy rainfall and subsequent collapse.

SEWERAGE SYSTEMS

Conventional sanitation systems, based on water or waterborne sewerage, are the accepted manner for removal of human waste. However, they have been unable to make a significant impact on the backlog of nearly half of the population. Even if a sufficient investment could be made, so that conventional sanitation systems could be provided to people who lack access to adequate sanitation, the resulting sanitation systems would not be sustainable.

In many places, sewered sanitation results in polluted ground and surface waters. It can therefore lead to a whole new series of problems. In India, the idea of every family having access to a flush toilet may sharply increase drinking water consumption for flushing and lead to increased water pollution and health hazards. Problems to be addressed for sustainable sanitation systems include:

- Household motivation to invest in sanitation;
- Household expectation from the system; and
- Lack of awareness of hygienic risks and practices, which may rise or may pose servere hygienic risks.

Barriers to Total Sanitation

Studies identify the following 10 barriers in achieving total sanitation;

- Lack of political will;
- Low prestige, priority and recognition;
- Poor policy at all levels;
- Weak institutional framework and unclear distribution of responsibilities;
- Inadequate and poorly used resources;
- Inappropriate approaches;
- Failure to recognize defects of current excreta management systems;
- Neglect of consumer preferences;
- Ineffective promotion and low public awareness; and
- Women and children do not get preference.

It is important to consider how the toilet is connected to its transportation system. The flush toilet within the house may lead to an open drain and this, in turn, can create even more unhygienic conditions and disease. Therefore, a complete sanitation solution is the only answer *i.e.*, a toilet that works and a sewage system that can convey and treat sewage for safe disposal.

When the latrine is flushed, the excreta passes to the pit. The liquids percolate out into the surrounding soil through the honeycomb openings and down through the open pit bottom while the gases produced are absorbed into the soil. Only the solids accumulate in the pit. Each pit is designed to last for about three years before it gets filled. When this happens the connection in the junction box is switched so that excreta can pass to the second pit while the first one is taken out of use. The filled pit is left untouched for 14-18 months during which time all the pathogens die out and the contents trun into a rich organic humus which is safe to handle. The pit can then open during any dry period and the dry contents can be easily removed. With its established high fertilizing value, it can be used as manure. The empty pit is then ready to be back into use.

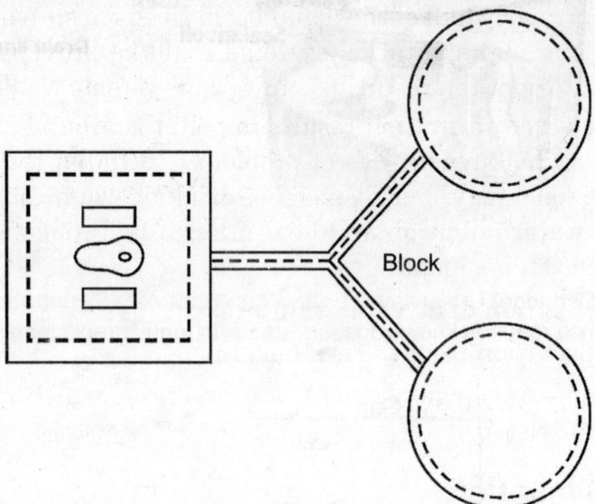

Fig.5.13: A Typical two pit pour flush latrine

New Systems

Today, we have a whole new range of environment friendly technologies. The adoption of ecological sanitation is essential for improving the living condition as well as for environmental sustainability. The development of simple, small eco-technologies, which facilitate community management and recycling, provide environmentally sustainable alternatives to the present sanitation and pollution caused by it.

Bindeshwar Pathak, a Gandhian social scientist, has developed cost-effective, environment-friendly sanitation system which aims to make scavenging obsolete. In 1970, Pathak established the Sulabh International, which introduced low cost yet hygienic latrines and sanitation technology. The two-pit pour flush toilet uses only two litres of water to flush (compared to 12.5 liters used by a modern flush toilet) and generates bio-gas energy. One pit is used while the other is a standby. When one gets filled up, the waste flow is directed to the second. After 18 months, the waste from the first pit gets incorporated into the ground and may be used for manure. The public latrine complexes developed by Sulabh International have also become centres of bio-gas production which is used to light streets, heat water and to cook.

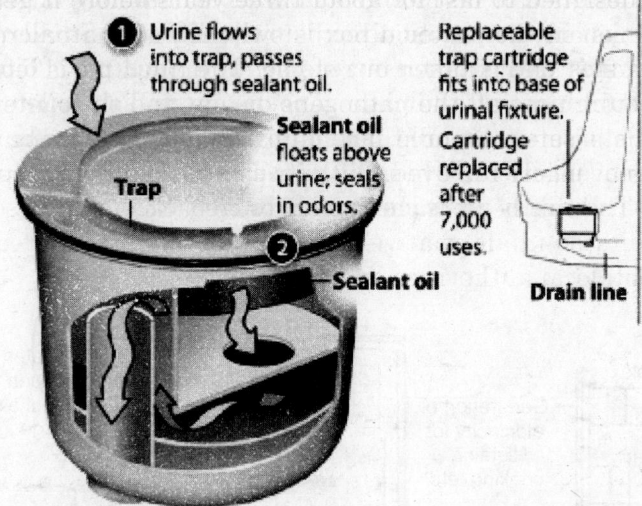

Fig.5.14: Waterless (Cartridge) Urinal saves up to 40,000 gallons (152,000 litres) of water yearly. However, maintenance requires knowledgeable staff with regular replacement of costly parts.

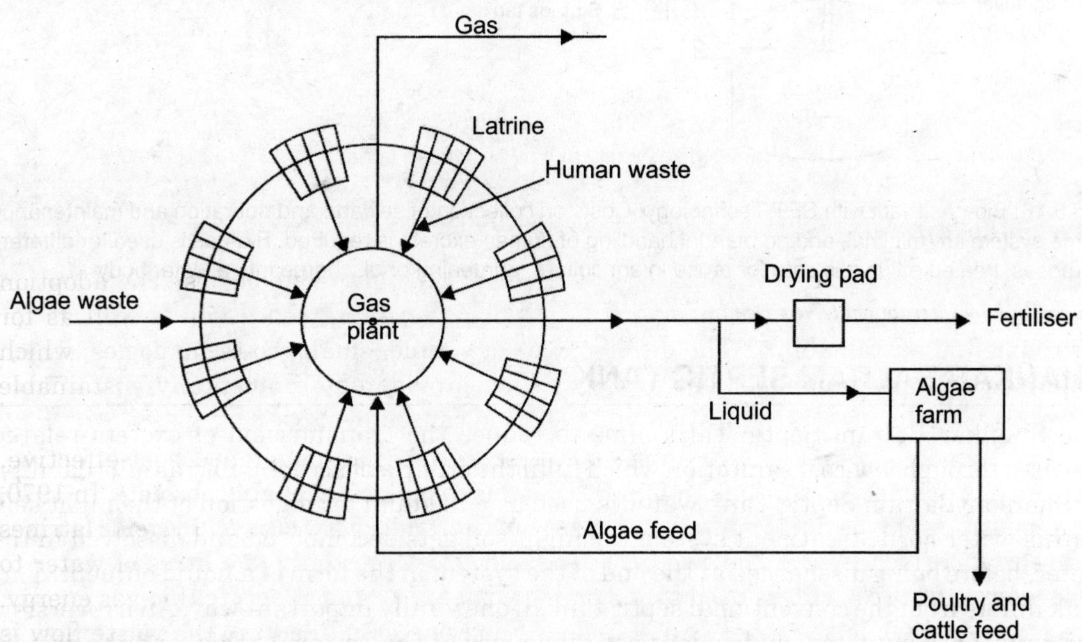

Fig.5.15: A Centralized Composite Bio-gas Plant

Bio-gas plant is a composite unit of a digester and a gas holder. The gas holder floats on the top of digester, wherein gas is collected and delivered at a constant pressure to the gas appliance through the pipe-line. The underground digester made of brick, is covered by a reinforced masonry dome. While the bio-gas plants working only on the night soil are not very successful, mixing of night soil with cow dung producing bio-gas has been encouraging. Such plants can be intergrated with community latrines where WCs are without flushing tanks and connected directly to the digester. Small quantity of cow dung and other organic wastes are mixed for increasing the production of biogas, which can meet most of the fuel requirement of a kitchen About, one million toilets and 4000 "pay and use" community toilets, serving more than 11 million people daily, have been installed all over India. 40,000 scavengers have been liberated from the menial task of clearing bucket latrine and 240 towns have been made scavenger-free. Meanwhile, scavengers are being trained and rehabilitated in alternative vocations. Sulabh has forged close partnerships with the government, local authorities and communities involved.

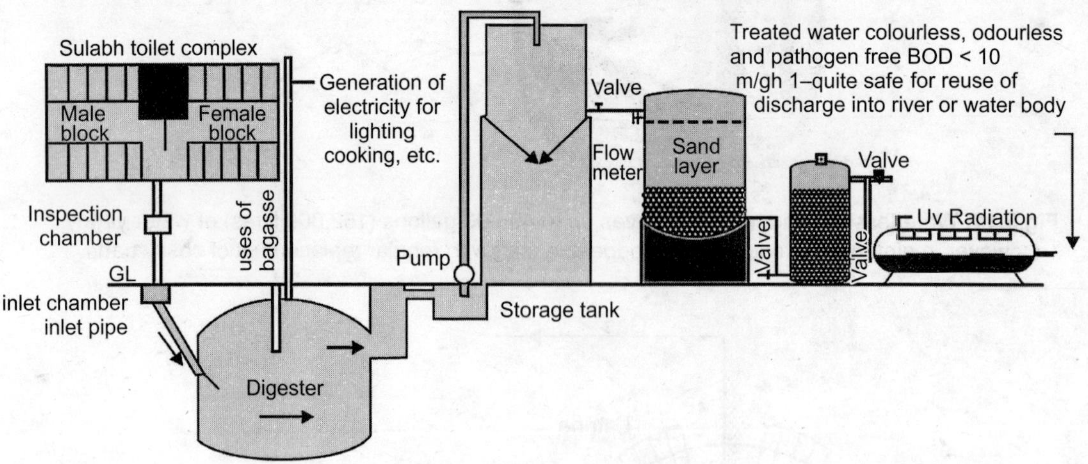

Fig.5.16: Bio-gas Plant with SET Technology: Costs on collection of sewage and operation and maintenance of the system are minimal, and no manual handling of human excreta is required. Bio-gas is used for different purposes, treated effluent is safe for reuse in agriculture, gardening or discharge into a water body.

Source: Sulabh International, New Delhi.

SHANKAR BALRAM SEPTIC TANK

The Shankar Balram Septic Tank aims to reduce the transmission of excreta-related diseases through low-cost sanitation which fulfil the criteria of acceptability as affordability. In Shankar Balram Septic Tank water is used as a medium for digestion of the night soil. In this water borne treatment system the night soil gets decomposed and dissolved in the water, before being discharged at the end of the system in the form of a liquid effluent. This tank differs from the conventional septic tank in one vitally important way. An irreversible build up of accumulated sludge does not take place at the bottom of this, and the system does not clog, requires no periodic cleaning and remains maintenance free for decades. This tank consists of two cement concrete industrial hume pipes, the bigger one serves the purpose of main tank while the small one as its subsidiary. The size of bigger pipe varies in its capacity depending upon the number of users, while the size of subsidiary one remains constant. Both the pipes are connected together by means of a connecting pipe. The top of two hume pipes are at the same level. The exact height of the connection between the two pipes is vital for the successful working of the tank. It is so designed and fixed that the outlet pipe does not allow the flow of raw night soil. The bottom of both hume pipes is concreted with a rich mixture to ensure the water tightness of the tanks. A smaller connecting pipe is installed in the 12' buffer above the water level, so that only one outlet ventilation pipe is needed for both the tanks. Concrete covers are fabricated at the site for installation, after which the entire rim and all joints are sealed with a rich cement mixture to ensure watertightness of the entire system. A vent projects from the cover of the main tank to a suitable height above the building level. An open drain is usually built to carry the effluent from the tank, and the household waste water from the bath and kitchen also gets mixed with the effluent from the septic tank.

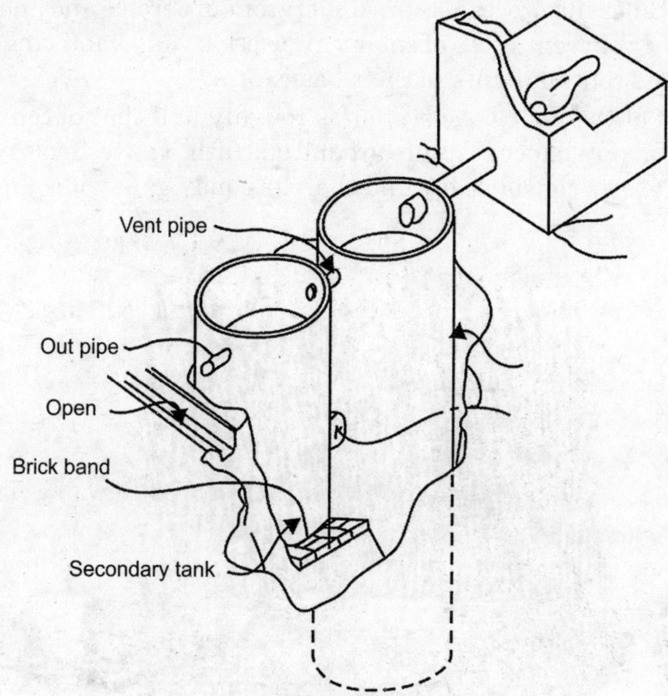

Fig.5.17: Shankar Balram Septic Tank

The RCC pan is so designed and fixed that it serves the purpose of automatic sealing without the need of separate water seal. The septic tank latrine must be used after both tanks are filled with the water.

ROOT ZONE SYSTEM

Developed in the Germany, the Root Zone process has been emerging as an economical and efficient method to treat industrial and domestic effluents. Bio-diversity is the key to the Root Zone process. More than 2000 types of bacteria and thousands of fungi exist in the reed bed. These microbial organisms oxidize impurities in the waste water. Since the process occurs underground, aerobic and anaerobic zones and process induce different types of chemical reactions and balancing bacterial growth, thus decomposing the contaminants to their basic form.

Three integrated components are essential to the system— reeds, reed bed and microbial organisms. The system involves running contaminated water through the root zone of the specially designed reed beds. The reeds and the reed bed together provide efficient effluent treatment, and the released water is environmentally acceptable (Figs. 5.18). Phosphates, sulphur compounds and nitrogenous materials reduce their elemental forms, while BOD and COD are drastically brought down.

Bio-diversity is the key to the Root Zone process. More than 2000 types of bacteria and tens of thousands of fungi exist in the reed bed. These microbial organisms oxidise impurities in the waste water. Since the process occurs underground, aerobic and anaerobic zones exist side by side, inducing different types of chemical reactions and balancing bacterial growth, thus decomposing the contaminants to their basic form.

The capital cost of the Root Zone system is roughly half that of conventional systems while operation and maintenance costs are around one fifth. The technology has the potential to reduce treatment costs considerably and therefore may gain wide application.

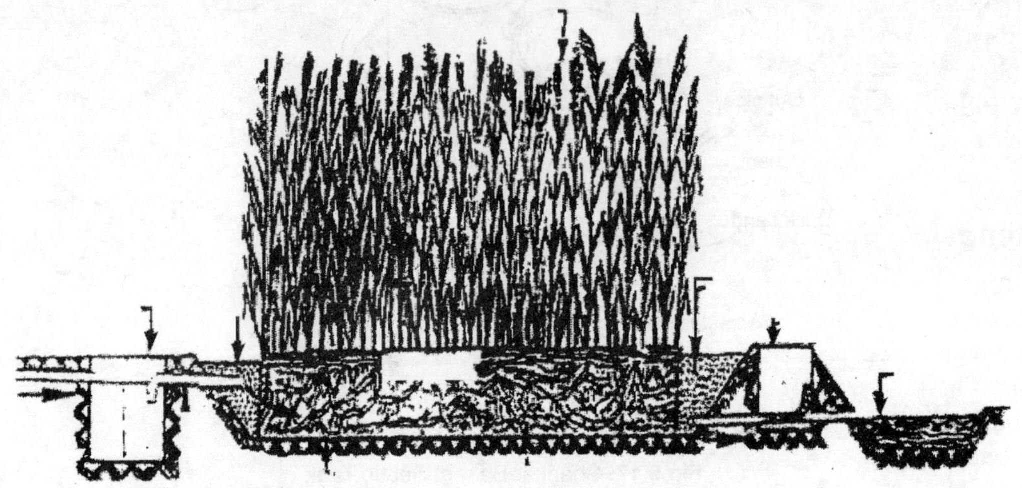

Fig.5.18: Root Zone System

In the two-tier waste-water recycling system, while one pipe collects kitchen and sink water, toilet waste flows through the other. The sewage is discharged and the remaining water is treated before being reused for non-potable purposes.

Decentralized Wastewater Treatment System

The Decentralized Wastewater Treatment System (DEWATS) is an innovative technological option for the treatment of sewage in areas where there are no treatment systems or underground sewerage. DEWATS is an on-site wastewater treatment system that is environment friendly, and has low maintenance cost (it works on the principle of gravity flow thus there are no electricity costs).

BORDA, a Germany based organization, has partnered with a non-governmental organization, Exnora International, for setting up the first DEWATS-linked community toilet in East Devadanam. Tiruchirapalli. The DEWATS module includes a biogas digester and an aerobic baffle reactor, which acts as the primary treatment system, followed by an aerobic planted gravel filter and collection tank, which acts as the secondary system. The wastewater from the toilet flows into the biogas settler through underground pipelines. The primary treatment system is treated with cow dung slurry for micro-organism during the commissioning of the plant. The gas production takes place after a digestion period of 30 to 90 days. The biogas is used in the community canteen. The digested slurry along with the sludge flows into the anaerobic baffle reactor where the sludge settles down due to gravitational flow in a series of chambers. The clarified wastewater is further treated in the planted gravel filter, where plants such as Cana Indica and common reeds absorb the nutrients present in the wastewater and the aerobic bacterial inoculums present in the planted gravel filter bed further digest the sludge, if any. The wastewater produced is clear with no odour and can be let out into any water body. The treated water has a biochemical oxygen demand value of less than 30 mg/litre, which is the standard set by the Bureau of Indian Standards for safe discharge of wastewater into any water body.

In East Devadanam, the DEWATS is designed to treat nine cubic metres of wastewater per day. The gas produced from the biogas digester is utilized for the lighting of toilet and for heating purposes. The treated wastewater is used for irrigation in the nearby organic vegetable garden maintained by self-help groups that is a source of income for them. The entire operation and maintenance is carried out by the Sanitation and Hygiene Education Team.

Extended Aeration Technique

An innovative system of sewage treatment is extended aeration technique. The processing unit can be designed in the form of a compact, modular and a self-contained system. It can be installed independent of the overall city system and is suitable for sewerage needs of remote areas. The process is based upon the phenomenon of self-oxidation by the bacteria. When the bacterial cells reach their ageing stage, the consumption of organic matter is maximised. The bacterial cells draw their own oxygen for additional energy which they require. The process of self-oxidation leads to stabilization of sludge and purification of the effluents. A sequence of the treatment process, viz, screening, aeration, setting and thickening takes place within the package unit. Thus, open land can be saved from spill over of sullage and waste. The unit can be installed underground and the site can be landscaped from above. It is catholically protected against electrical currents and corrosion. The unit can also be constructed in the masonry/ concrete over the ground.

For scattered villages, townships, unauthorised colonies and suburbs which often lack the availability of city sewerage system, the concept of package unit has been developed. The package unit contains the compartments required for the treatment process in one metallic parallel piped shaped casing:

- Screening unit
- Aeration zone

- Settling zone
- Thickening zone

The package unit is attached with an adjacent motor to lift the sewage water to the inlet of the unit.

Generally, the unit is buried and does not affect the appearance of the site where it is installed. Earthwork is usually started a few days before delivery of the package unit so as to limit any risk of flooding.

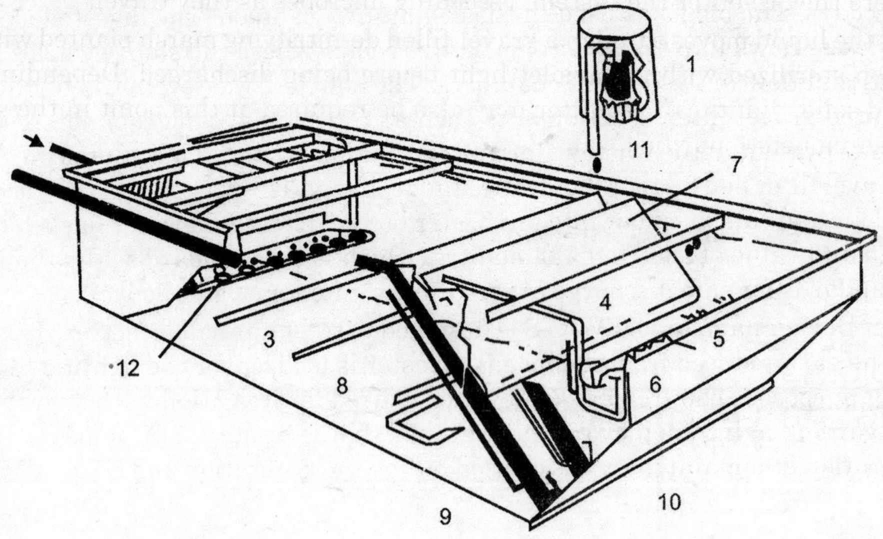

1. Electric Blower Set
2. Raw Water Inlet
3. Extended Aeration
4. Calcification
5. Treated Water Collection
6. Scum Collection
7. Baffle
8. Communication Port
9. Sludge Recirculation
10. Scum Pumping (Optional)
11. Air Feed
12. Air Diffuser

Fig.5.19: Extended Aeration Technique

WASTEWATER RENOVATION AND CONSERVATION CYCLE

Land application of waste water, after secondary treatment, is an advanced method. This approach (the 'living filter') considers effluent and nutrients as resources rather than as a product for treatment and disposal. Treatment is provided by natural biological and chemical processes as the water moves through the living filter provided by soil, plants, micro-organisms and related ecosystems. The renovated water then percolates to recharge the ground water reservoir.

Solar Aquatics Septage Treatment System

As a typical septage treatment plant (Fig. 5.20), effluent discharged into a headworks consisting of a receiving station and DE gritter. From there it flows by gravity into in-ground tanks where the effluent is blended to compensate for variations in loads and is preconditioned. Following equalization and preconditioning, effluent is pumped to a clarifier. From here sludge is pumped to the sludge stabilization tank where it is aerobically stabilized. It then flows to the reed bed for passive composting. Underflow from the reed bed is recycled by pumping it back into the equalization tank for further treatment.

Liquid from the clarifier enters a series of solar silos that contain the plants and worms involved in the clean-up operation. After moving through the silos by gravity flow, the liquid enters a second clarifier. Solids from this clarifier are recycled back to the equalization tank, where they re-enter the system, reseeding microbes as they travel.

Next, the liquid moves through a gravel-filled de-nitrifying marsh planted with grasses and is then sterilized with ultraviolet light before being discharged. Depending on total suspended solids limits, a sand filter may also be required at this point in the system for final polishing before purified water is discharged.

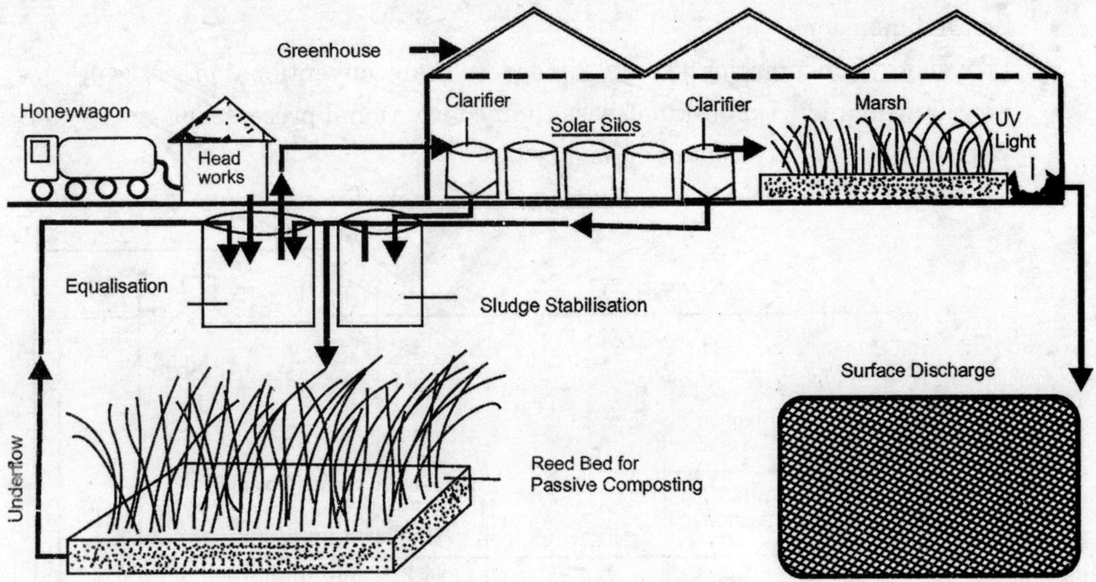

Fig.5.20: Solar Aquatic Septage Treatment System

Source: Ecological Engineering Associates, 13 Marcom Lane, Marion, MA, USA

BIOPAQ-UASB TECHNOLOGY

The heart of the BIOPAQ system is the anaerobic reactor based on PAOUES Upflow Anaerobic Sludge. Blanket (PAQ-USAB) process (Fig.5.21). The reactor consists of a large corrosion resistant tank which incorporates a unique 3-phase settler to separate the sludge, biogas and effluent. The settler is located at the top of the reactor and is designed for specific COD reactor loadings and hydraulic throughout.

A flow distribution network is located at the base of the reactor. This network is designed to distribute the flow evenly throughout the bottom of the reactor. This eliminates short-circuiting and promotes the proper formation of the granular sludge which is a critical factor in reactor operation. The distribution network is designed to facilitate easy cleaning, thereby eliminating potential plugging problems.

The anaerobic process comprises of a UASB reactor called BIOPAQ. The raw effluent is first taken to the buffer tank. The capacity of the buffer tank is calculated such that acid genesis phase is triggered immediately. The effluent is then pumped to Biopaq reactor through a series of distribution pipes. The multiple distribution ensures a uniform flow of liquid throughout the sludge blanket making maximum use of available high bacterial population. The liquid rises to the top of Biopaq reactor along with biogas generated and also some sludge particles.

Methanisation

Methanisation is a process of anaerobic biological treatment suitable for concentrated effluents. Methanisation is designed for treating effluents from industry and human settlements. The following are the advantages of methanisation :

- Small dimensions
- Low energy consumption (2 to 10 times less than conventional processes)
- Low sludge production (3 times less than conventional processes)
- High dod and cod treatment efficiency
- Production of a re-usable gas (heating, steam generation)

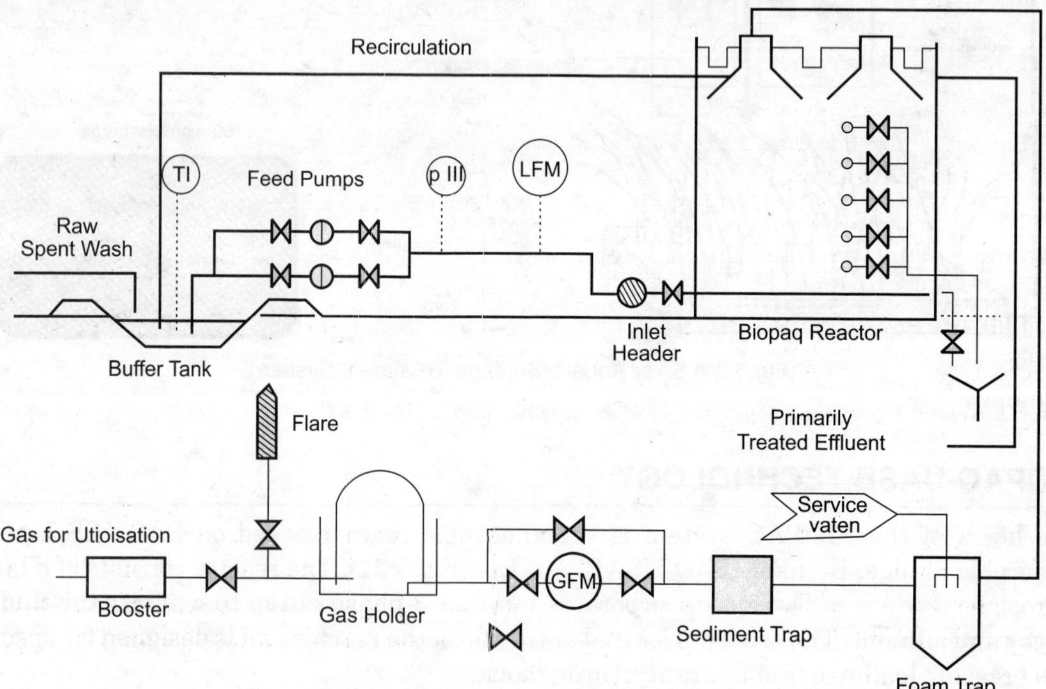

Fig.5.21: Biopaq Treatment (Schematic)

Source: Degrcmont India Ltd.

Methanisation involves two main stages each of which uses specific bacteria, depending upon the nature of the effluent to be treated.

- Suspended growth
  - Contact
  - Sludge blanket
- Attached growth
  - Fixed bed
  - Fluidized bed

Contact Suspended Growth: The contents of the digester thoroughly mixed by stirring with effluent to be treated recirculated settled sludge.

Sludge Blanket Suspended Growth: The effluent to be treated flows through a sludge blanket maintained at the bottom of the digester. The bacterial mass is attached to a solid support medium to obtain a higher concentration of active matter. The result is a more compact treatment unit. The volume obtained in relation to complete mixing can range from two times less with fixed bed and up to 10 times less with a fluidised bed.

Fixed Bed Attached Growth: The effluent to be treated flows up through a filter medium on which the treatment bacteria develop.

Fluidised Bed Attached Growth: The effluent to be treated flows up through a mobile granular support medium kept in suspension and on which the treatment bacteria develops.

FFR Technology

Sewage Sludge Treatment: Based on NEERI's FFR technology- a demonstration project at Bhubaneshwar-presents an innovative system of Sewage Treatment-cum- Energy Recovery (Fig. 5.22). As against the conventional treatment of sewage which is generally a dilute stream, separation of sewage solids in the form of a sludge and treating the same has several advantages. This technique has been practised in developed countries for several years and package plants are available for community sizes ranging from 10,000 population to over 250,000. Considering the importance of the process in providing cost effective alternative to the energy intensive conventional aeration process, this is a promising option for the Indian conditions.

The patented gas-mixed system ensures:

- Turbulent mixing of fresh substrate
- Distribution of end-products of bacterial degradation
- Supply of fresh substrate to the bacteria
- Control and stirring up of settled sludge
- Control, mixing and destruction of skimming

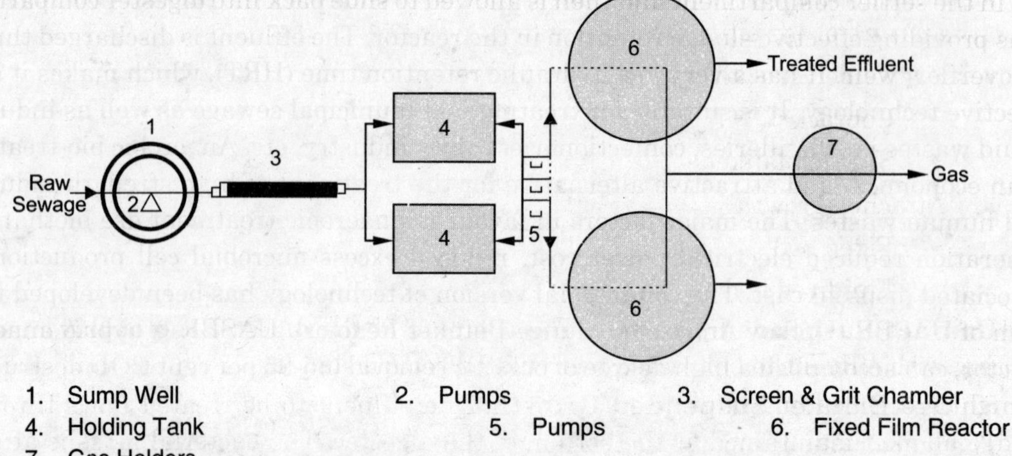

Fig.5.22: Treatment and Biogas Recovery from Sewage, RRL Bhubaneshwar

Automation: BIMA digester can be completely automated and can be controlled with a computer. Two operation modes are possible. These are pressure controlled and time controlled modes.

Dry Type Gas Holder: In place of conventional gas holder (storage tank with either fixed for floating roof), a dry type gas holder. The dry type gas holder consists of a gas balloon enclosed inside a protection silo. A mounting device on top of the silo is used for mounting the balloon. The gas balloon is of reinforced, high quality foil. The thickness of the foil is controlled and custom designed to store the gas. The gas membranes are available in sizes from a minimum of 110 cu.m. to a maximum of 6,000 cu.m.

Manpower: Since the entire operating sequence can be automated, no manpower is required to operate the digester. Moreover, as BIMA has no moving parts, there is no need for maintenance personnel.

Energy Generation: The biogas generated as a result of anaerobic degradation can be coupled to a gas engine to generate electricity. The electrical conversion efficiency gas engine is about 36 per cent and approximately 2.2 kwh of electricity can be generated from about 1 cu.m of biogas. Besides power generation, gas engines could be coupled to waste heat recovery to recover valuable heat from the off gas of the engine. The gas engine and the waste heat recovery system can be completely automated to generate power as and when the need arises and to flare the gas the rest of the time.

Upflow Anaerobic Sludge Blanket (UASB) Reactor

The UASB reactor (a Dutch technology) has a dense blanket of granular or flocculated nonaerobic sludge. The sufficient contact between sludge and waste water is maintained with the production of biogas, because the rising gas bubbles provide a gentle mixing. After passing through the sludge bed, a mixture of biogas, sludge and water enters a three-phase separator. The biogas is separated in a gas collector, whilst the sludge aggregates, settle out in the settler compartment and then is allowed to slide back into digester compartment thus providing effective sludge retention in the reactor. The effluent is discharged through an overflow weir. It has a very low hydraulic retention time (HRT), which makes it a cost effective technology. It is suitable for treatment of municipal sewage as well as industrial liquid wastes i.e. distilleries, confectioneries, dairy industry, etc. Anaerobic bio-treatment is an economical and attractive alternative for the treatment of high strength industrial and human wastes. The major factors in favour of anaerobic treatment are methane gas generation reduced electrical power cost, reduced excess microbial cell production and associated disposal cast. The commercial version of technology has been developed in the form of UASBR (Upflaw Anaerobic Sludge Blanket Reactor). UASBR, a hybrid anaerobic reactor, can accomplish a high degree of organic removal (85-95 per cent COD destruction) at high organic loading rates.

The reactor can be constructed as either circular or rectangular, open or closed tank to suit the local site conditions. The reactor consists of inlet distribution, sludge blanket and gas solid separators. The unique distribution allows for intense mixing of the effluent with the sludge. In addition, the gas rising through the sludge bed (generated during anaerobic degradation) also assists in mixing process. The liquid, gas and sludge are separated at the top of the reactor with gas-liquid-solid separators. These separators not only collect the gas but also facilitate solid-liquid separation and prevent carry-over the sludge.

In order to monitor the performance of the plant and for process control, the pH, temperature and oxidation-reduction potential (ORP) prevailing in the reactor, and the quantity as well as the composition of the gas generated, control systems arc built in to adjust the flow and/or add nutrient/ buffer, depending on the conditions in the reactor.

BIMA Digesters

A commonly available version of UASB Reactor is BIMA digester. The digester is divided into different functional compartments, the lower main lei mental ion chamber and the upper chamber which are connected by special halls. There is also a gas connecting pipe between the two centibars close to the roof. The gas produced in the digester is discharged into the gas holder from the upper chamber by closing an automatic valve in the gas pipe between the two chambers the gas produced in the main chamber is collected, which in turn displaces an equal amount of digested substrate into the upper chamber, building up a level difference and thus a gas pressure in the main chamber. When the required level difference is achieved (mixing pressure), the gas pressure is suddenly released by opening the automatic valve in the gas connecting pipe.

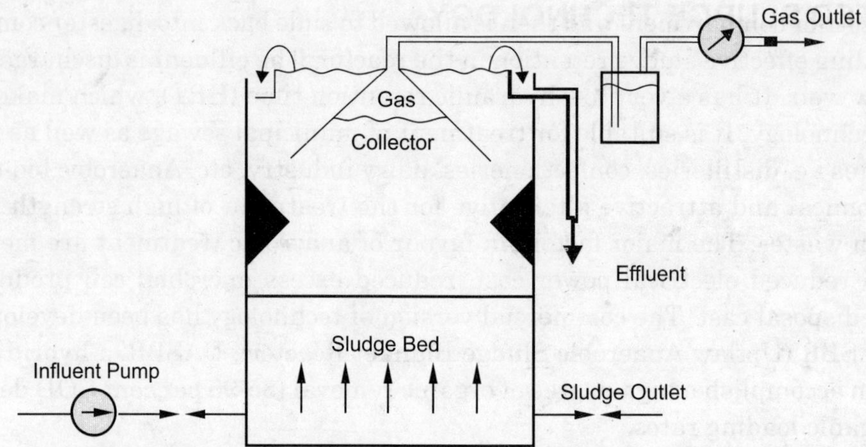

Fig.5.23: Upflow Anaerobic Sludge Blanket (UASB) Reactor System

Source: Enkem Engineers, Chennai

Thus, the substrate displaced into the upper chamber flows back to the main chamber with high velocity. On account of this reflux action as well as the special hydraulic design of the digester, fresh substrate, scum and sediments are perfectly remixed with the contents of the main chamber. The whole digester is completed automated and can be operated in two modes, pressure controlled modes are applicable for substrates with average or high specific gas production Decentralised rates, time controlled modes are used in the case of substrates with low specific gas production rates. In the former case, the cycles are determined by the velocity of gas production and pressure rising, while in the latter a small gas blower supplements the microbial gas production to ensure a certain number of mixing cycles per day.

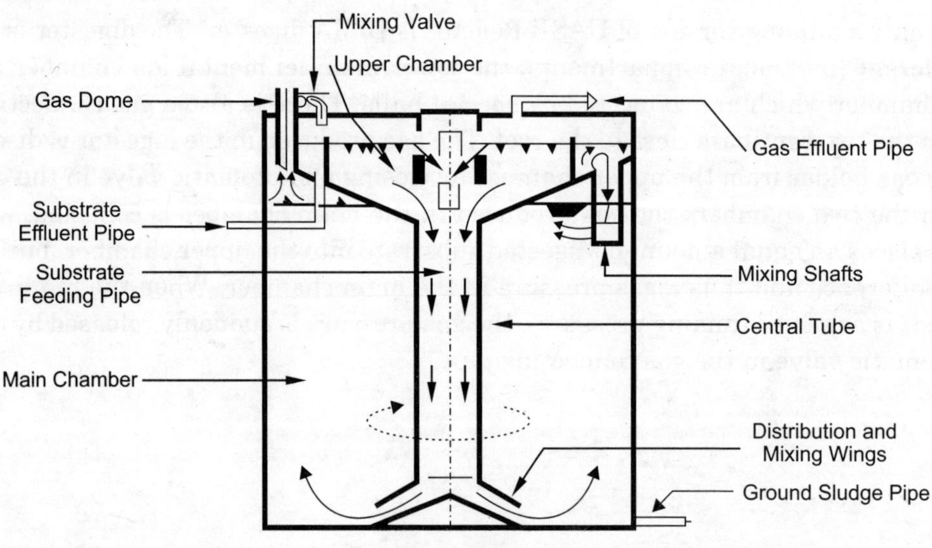

Fig.5.24: Bima Digester

Source: Enkem Engineers, Chennai

ACTIVATED SLUDGE TECHNOLOGY

It involves production of an activated mass of micro-organisms which is kept under suspension by mechanical mixing. This does not contain any filter media for bacterial growth. It is commonly used for treatment of municipal sewage and other industrial liquid wastes containing organic matters. The stabilisation of organic matter takes place in suspension form.

Ecological Sanitation (Ecosan)

"Ecological sanitation" or Ecosan works on the principle of separating the urine and faeces at source and putting both the urine and faeces back into the local nutrient cycle. Designing a toilet pan where the urine and faeces go in different directions ensures this. The faeces drop straight down in a small storage chamber made of concrete or other impervious material. The urine goes to a tank. The faeces is stored and allowed to decompose by a process of aerobic digestion. Ash, or other organic absorbing material like sawdust, is used to cover the faeces to aid in the drying out process. Time allows heat, given of by decomposition and normal soil bacteria to kill all the disease carrying organisms. Within six months the faeces is reduced to humus.

The process to sanitise faeces can be split up into two types; with urine it is a composting, without it dehydrating process. In both, the action of time and soil microbes destroy pathogens. To reduce health risks from handling urine and excreta in agriculture the protocols are to be followed for physio-chemical and biological factors that affect the survival organisms in the environment: temperature, pH, ammonia, moisture, solar radiation, UV light, nutrients, and the presence of other micro-organisms. For the comparatively harmless urine in small family based systems, the urine may be used directly or after short periods of storage if the crops are intended for family use. The likelihood of transmission is larger between family members than through urine fertilized crops.

The human excreta contain pathogens–enterovirus, salmonella, shigella, faecal coliform and others. These bacteria and virus make human excreta deadly, which need to be killed and the nutrients recycled. The bacteria is biodegradable, which needs the sun to burn it and the soil to degrade it.

The re-engineered toilets are built on the principle of ecological sanitation. In this approach urine, which is pathogen–free, is diverted so that it can be used as fertilizer, while the faecal matter is composted to destroy the bacteria and pathogens and returned to the land. Alternately human waste is dosed with natural but cultivated bacteria–mega–bacillus and other–designed to hasten decomposition of the pathogens. The bacteria-dosed waste is then 'cleaned' using plant filters, oxidation and by bio remediation using microbes.

Ecosan systems have the overriding aim of improving public health and hygiene. However, the basic premise of ecosan of closing the nutrient loop between sanitation and agriculture, means that designers of ecosan systems must consider a much wider range of factors than those of conventional systems. This also leads to the consideration of soil conservation, increased long-term food security, and the sustainable use of resources in the form of nutrients, organics, water and energy. As ecosan aims to solve sanitation problems and ensure reuse at the lowest possible level, promoters of these systems also have to consider the active engagement of the local private sector, contributing to job creation and poverty reduction.

Modern ecosan toilets are most useful where laying sewage lines is a problem. In Sweden composting toilets were first introduced more than 50 years ago. Though a wide variety of models are being used, the' Clivus' Multrum Composting Toilet is one of the most popular model. The Clivus Multrum is a single vault–composting toilet where urine, faeces and organic household wastes are combined and processed together. The model is available as a unit and consists of three main components—a composting vault with a slanting floor, air conduits and a liquid storage tank.

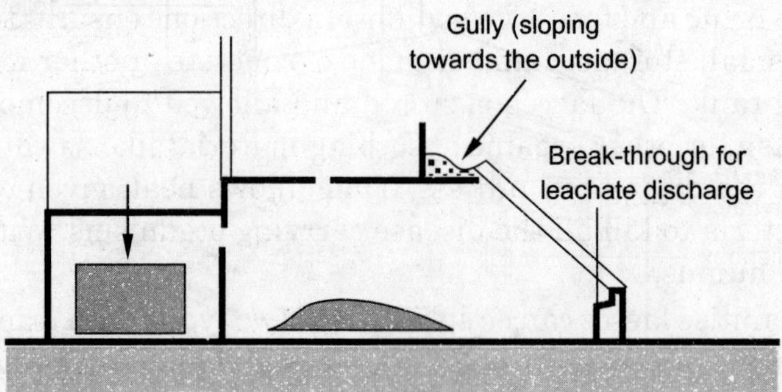

Fig.5.25: Urine-separation vermi composting toilet

Source: GTZ, (2006)Concepts for Ecological Sustainable Sanitation

Faeces, urine and toilet paper along with all kinds of kitchen and organic household wastes go into the Multrum. The contents slide down slowly along the Multrum sloping floor with the fresh deposits at the upper end down to the storage part of the vault. The heap decomposes, reducing to less than 10 per cent of its original volume and gradually forms humus. The humus produced has similar bacterial as that of soil and directly used as a fertilizer and soil conditioner. The humus produced in this storage space at the lower end. Besides these, a tube connects the toilet seat riser with the receptacle and a sloping channel for the kitchen waste. process is taken out after five years for the first time and later once a year. In Sweden, this model is used in houses, weekend houses, institutions and for public toilets. One problem is that since there is no diversion of urine the slanting floor poses a risk of liquid accumulation at the lower end of the composting vault. To stop this a container for liquid storage below the composting vault has been provided in a newer version.

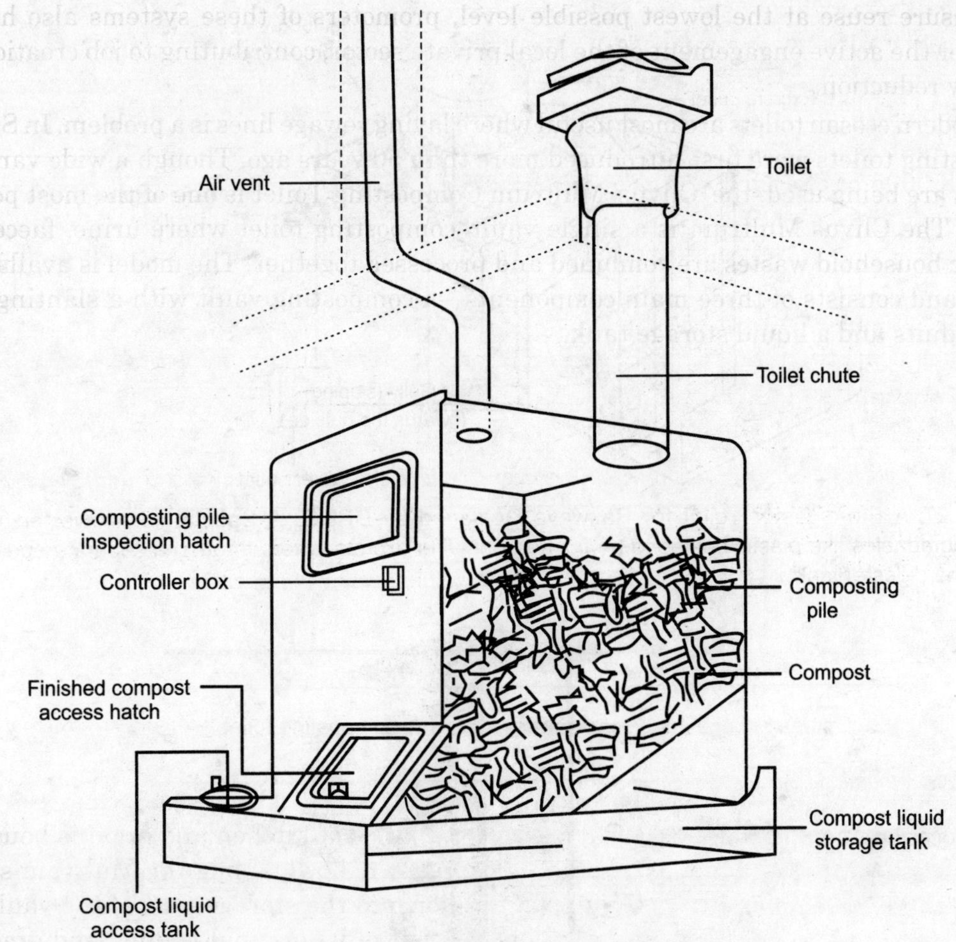

Fig.5.26: Clivus Multrum Composting Toilet

Source: New Age Approach, Down to Earth, February 28, 2002.

Dr. Vinod Tare of IIT, Kanpur has developed an innovative toilet design for the segregation and recycling of human excreta and urine. When these toilets are flushed, the vortex movement of water cleans the pan surface and pushes the solid waste downwards into a tank at the centre. While the centrifugal force— acting outwards from a centre of rotation—presses water to the surface of the pan, the geometric design of the surface moves it through a circular path downwards towards the separator. At the separator, the water is guided into pipes in the sides that take it to another tank. The separation is close to 99 per cent. The water is re–used for flushing the toilet, rather than discharging it along with the excreta. Indian railways are going to experiment with this type of toilet, besides working on few more technologies using micro–organism to decompose excreta and treating the liquid before it is thrown on the track. The eco–sanitation principles can be adopted for designing toilets for railways.

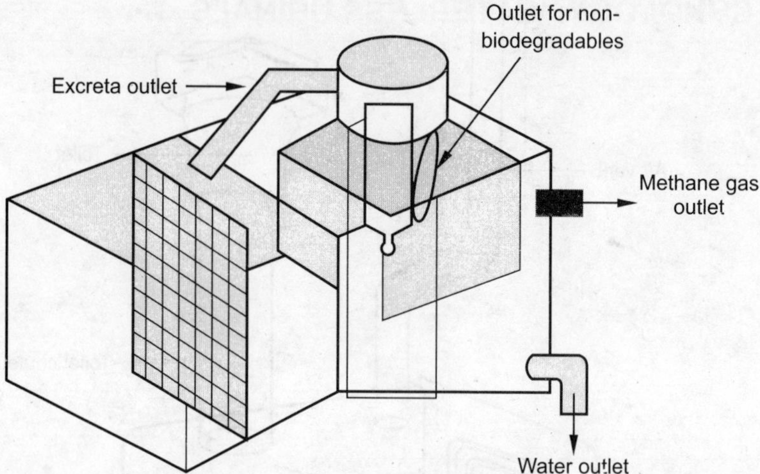

Fig.5.27: A Bio-digester Toilet for Railways Designed by DRDO. A Special lid allows for non-biodegradables like plastic bottles to be flushed out. The excreta enters through a separate opening into the digester tank

Fig.5.28: An E-Toilet

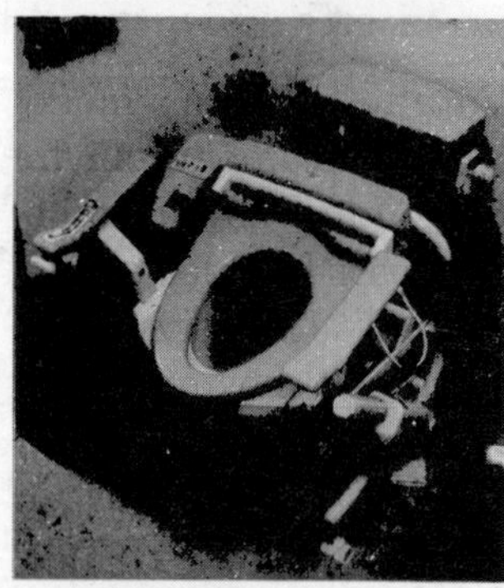

Fig.5.29: Super Hi-Tech "Throne" with soft seats and back, telephone, PC, etc.

Source : *The Hindustan Times, New Delhi, 3 October, 2005.*

VACUUM TECHNOLOGY, WATERLESS URINALS

Toilets originally designed for ships and airplanes are now being adapted for houses also. The vacuum toilets also reduce the amount of water. Electric incinerator toilets fall into this category where the faeces and urine mixture is dried by electric fans and then burnt. Today vacuum technology, waterless urinals diverting toilets are available as innovative eco–sanitation systems.

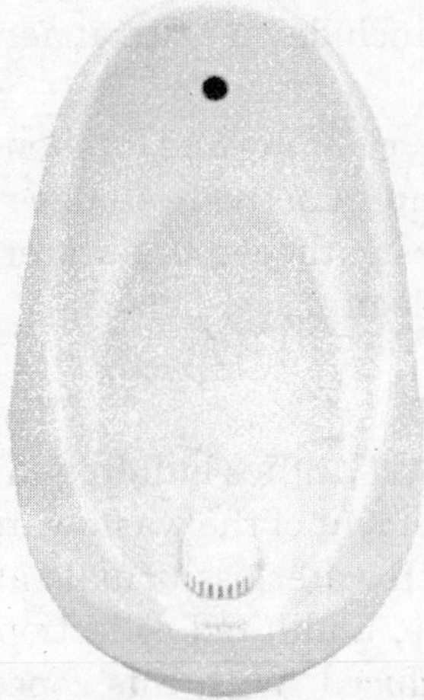

Fig.5.30: A waterless urinal achieves zero water consumption as it relies on the simple vertical trap principle where the urine flows into the drain insert passing through a floating layer of a patented liquid which acts as a barrier that prevents sewer vapours from escaping to the restroom atmosphere. The urine accumulated under the barrier overflows into the central tube and runs down to conventional drain line.

Examples of vacuum based ecosan project include the headquarters of the KFW-Bank in Frankfurt, Germany, or the Building Initiative "Wohmen & Arbeiten", Freiburg, Germany. Waterless urinals are becoming increasingly common. A recent extension of the headquarters of the KFW Bankkengrouppe, in Frankfurt, Germany, offers space for around 300 workplaces and has 13 apartments on its two floors. In the building an innovative water and wastewater management system has been installed that includes vacuum technology and meets economic and ecological requirements.

PEEPOO – A SANITARY REVOLUTION

The PEEPOO story starts with Anders Wilhelmson, a professor of architecture at the Royal Institute of Technology at Stockholm, who was visiting a number of burgeoning megacity slums, and discovered that 40 percent of world's population lack access to toilet. Several questions began to occupy his thoughts- shouldn't it be possible to produce cheap toilet so simple that even a young child could use it, and which could also wipe out pathogenic bacteria and other microorganism? The result of his research was Peepoo, a bio-degradable, a single use toilet in the form of a thin bag with an ammonia based powder at the bottom. The producer, Peepoople, is now distributing it via schools and web retailers, 'micro entrepreneurs', in slum areas, but also via aid organisations to disaster areas and refugee camps. Peepoole has realised something important. While decision makers and media must talk about these basic hygienic needs, it is also vital to respect people's privacy, particularly that of women. The development of small germ killing toilet has been about understanding the situation of women and helping them solve an embarrassing problem.

Office areas and apartments are equipped with vacuum toilets and vacuum urinals which are connected to the central vacuum pumping station in the basement by a vacuum pipe system. Blackwater from toilets and urinals is discharged to the communal sewerage system. However, it is planned to include on– site anaerobic treatment of the blackwater in the next phase of the project.

Grey water from hand washing, cleaning and kitchens is collected in a separate gravity pipe system. It is treated using compact activate sludge reactors, membrane filtration and UV light. The treatment guarantees pathogen free water for reuse. The treated greywater is used for toilet flushing and for cleaning.

Urine Separation and Reuse

During the renovation of the main GTZ office building in Eschborn, a modern, ecologically sustainable concept for the management of the wastewater from toilets has been installed. The main building is equipped with waterless urinals and water flushed urine diversion (UD) toilets. Through the separate, undiluted collection of urine, the water demand for flushing toilets is significantly reduced. With this concept, the GTZ not only saves 900 million gallons of water per year, but also the load of substances and nutrients from the urine on the water treatment facilities is reduced.

After treatment, the urine can be used as agricultural fertiliser. The research will help to improve agricultural production with fertilizer originating from urine. The system can be a model for countries where water is scarce and fertilizer is needed for agriculture. For the treatment and reuse of the brown water originating from the UD-toilets an activated sludge reactor, followed by membrane filtration, is being explored as a technological option.

Vacuum Sewerage Systems

Vacuum sewerage systems, developed in India by Subhash Deshpande, are wastewater collection systems. An air stream generated by differential air pressure drives the wastewater towards a vacuum station. In contrast to conventional gravity systems, the pressure within the system is maintained below atmospheric pressure (negative pressure). Vacuum sewerage systems have three elements:

- Collection chambers in each house/ building;
- Vacuum sewer lines; and
- Central vacuum station.

Vacuum technology is based on differential air pressure. Rotary vacuum pumps generate an operating pressure of -0.4 to -0.6 bars at the vacuum station. Vacuum pump is the only element of the vacuum sewerage system, which needs electricity. Interface valves installed inside the collection chamber work pneumatically. Sewage flows by gravity into collection sump at each house. After a certain level is reached, the interface valve controlled by a controller opens. The resulting differential pressure (1.4 to 1.6 bar) between atmosphere (one bar) and vacuum (-0.4 to-0.6 bar) becomes the driving force, and transports the wastewater towards the vacuum station using air (instead of water) as carrier. Therefore, no inspection manholes are required in vacuum systems.

The vacuum sewers are filled with air at a pressure of -0.4 to -0.6 bar. Air-to-liquid ratio is the key for a reliable operation. When a system is well designed, the sewers contain only very small amounts of sewage. Sewers should never be totally filled with sewage since the hydraulic friction loss would be too high. This is ensured by using the interface valves. The proper selection of equipment ensures the automatic optimization of the air-to-liquid ratio. It also offers highest operational security, optimization of hydraulic transporting conditions, and thus reduction in energy consumption of a plant.

Considering that the vacuum is the means of transport, sewers can be laid in flat terrain and up to certain limits may also be counter–sloped. The saw-tooth profile keeps sewer lines shallow. Expensive trenching, as in case for gravity sewers having 1:100 slope, is avoided. Once sewage arrives in the vacuum storage tank at the vacuum station, it is pumped to the discharge point, which could be a gravity sewer or the treatment station directly. Discharge of effluent in a Vacuum System takes place via a Collection Sump with an Interface Valve.

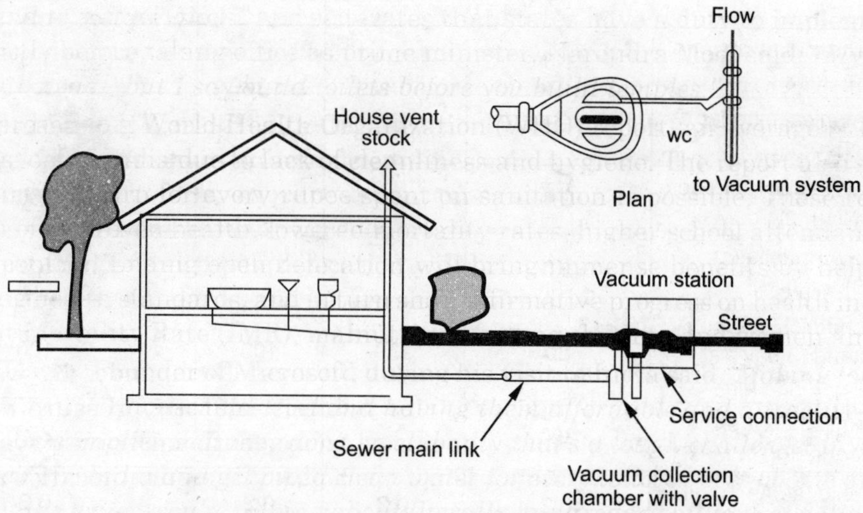

Fig.5.31: Schematic Vacuum Sewerage Technology

Source: Subhash Deshpande, Vacuum Sewerage Systems, IPT Journal (p.32-45) January–February 2006

EFFLUENT STANDARDS FOR SEPTAGE MANAGEMENT

As per MoEF, the general effluent standards for discharging the waste water are given in Table 5.4.

Table 5.4: General Effluent Standards for Discharge

S.No.	Parameter	Standards			
		Inland surface water	Pubic Sewers	Land for Irrigation	Marine Coastal areas
1.	Colour and Odour	All efforts should be made to remove colour and unpleasant odour as far as practicable			
2.	Suspended solids (mg/g. Max)	100	600	200	For process waste water 100 For cooling water effluent 10 per cent above total suspended matter of influent.
3.	Particulate size of suspended solids	Shall pass 850 micron IS Sieve	–	–	(a) Floatable solids, max 3 mm (b) Settleable solids, max 850 microns
4	pH Value	5.5 to 9.0	5.5 to 9.0	5.5 to 9.0	5.5 to 9.0
5	Temperature	Shall not exceed 50C above the receiving water temperature			Shall not exceed 50C above the receiving water temperature.
6	Oil and grease mg/L. Max.	10	20	10	20
7	Total residual chlorine mg/m. Max	1.0	–	–	1.0
8	Ammonical nitrogen (as N). mg/L. Max.	50	50	–	50
9	Total Kjeldahl Nitrogen (as NH_3), mg/L. Max	100			100
10	Free Ammonia (as NH3), mg/L. Max.	5.0	–	–	5.0
11	Biochemical Oxygen demand (3 day's at 270C), mg/L. Max.	30	350	100	100
12	Chemical Oxygen Demand, mg/L. Max.	250	–	–	250
13	Arsenic (as As). mg/L. Max.	0.2	0.2	0.2	0.2
14	Mercury (as Hg), mg/L. Max.	0.01	0.01	–	0.01
15	Lead (as Pb). mg/L. Max	0.1	1.0	–	2.0
16	Cadmium (as Cd). mg/L. Max	2.0	10	–	2.0
17	Hexavalent Chromium (as Cr^{+6}). mg/L. Max.	0.1	2.0	–	1.0

18	Total Chromium (as Cr.), mg/L. Max.	2.0	2.0	–	2.0
19	Copper (as Cu), mg/L., Max.	3.0	3.0	–	3.0
20	Zinc (as Zn.), mg/L. Max	5.0	15	–	15
21	Selenium (as Se) mg/L. Max	0.05	0.05	–	0.05
22	Nickel (as Ni), mg/L., Max	3.0	3.0	–	5.0
23	Cyanide (as CN), mg/L. Max	0/2	2.0	0.2	0.2
24	Fluoride (as F), mg/L. Max	2.0	15	–	15
25	Dissolved Phosphates (as P), mg/L. Max	5.0	–	–	–
26	Sulphide (as S). mg/L. Max	2.0	–	–	5.0
27	Phenolic compounds (as CSHS OH, mg/L. Max	1.0	5.0		5.0
28	Radioactive materials:				
a	Alpha emitter micro curie/mL	10–7	10–7	10–8	10–7
b	Beta emitter micro curie/mL	10–6	10–6	10–7	10–6
29	Bio-assay test	90% survival of fish after 96 hours in 100% effluent	90% survival of fish after 96 hours in 100% effluent	90% survival of fish after 96 hours in 100% effluent	90% survival of fish after 96 hours in 100% effluent
30	Manganese (as Mn)	2 mg/L.	2 mg/L.	–	2 mg/L.
31	Iron (as Fe)	3 mg/L.	3 mg/L.	–	3 mg/L.
32	Vanadium (as V)	0.2 mg/L.	0.2 mg/L.	–	0.2 mg/L.
33	Nitrate Nitrogen	10 mg/L.	–	–	20 mg/L.

Source: Manual on Sewerage and Sewage Treatment Systems, CPHEEO 2013.

Recommended norms for Sewage

The recommended treated sewage quality is proposed to be achieved for the stated reuse as given in the table 5.5.

Table 5.4: Recommended Upper Limits of Treated Sewage Quality for Specified Activities at Point of Use

S.No.	Parameter	Toilet flushing	Fire Protection	Vehicle Exterior washing	Non-contact impoundments	Landscaping, Horticulture & Agriculture Crops which are eaten			
						Horticulture Golf Course	Non edible crops	Raw	Cooked
1	Turbidity (NTU]	<2	<2	<2	<2	<2	AA	<2	AA
2	SS	Nil	Nil	Nil	Nil	Nil	Nil	Nil	30
3	TDS	2100							
4	pH	6.5 to 8.3							
5	Temperature °C	Ambient							
6	Oil & Grease	10	Nil	Nil	Nil	10	10	Nil	Nil
7	Minimum Residual Chlorine	1	1	1	0.5	1	Nil	Nil	Nil
8	Total Kjeldahl Nitrogen as N	10	10	10	10	10	10	10	10
9	BOD	10	10	10	10	10	20	10	20
10	COD	AA	AA	AA	AA	AA	AA	AA	30
11	Dissolved Phosphorous as P	1	1	1	1	2	5	2	5
12	Nitrate Nitrogen as N	10	10	10	5	10	10	10	10
13	Faecal Coliform in 100 mL	Nil	Nil	Nil	Nil	Nil	230	Nil	230
14	Helminthic Eggs/ litre	AA	AA	AA	AA	AA	<1	<1	<1
15	Colour	Colourless	Colourless	Colourless	Colourless	Colourless	AA	Colourless	Colourless
16	Odour	Aseptic which means not septic and no foul odour							

Source: Manual on Sewerage and Sewage Treatment Systems, CPHEEO 2013.

In order to achieve desired water quality, excess chlorination, granular activated carbon adsorption/ Ozonation and/ or various kind of filtration including membrane are recommended. For recreational impoundments for non-human contact, residual chlorine is not required so as to protect aquatic species of flora and fauna. As per the Manual of Sewerage and Sewage Treatment, the following provisions are to be followed for usage of treated sewage, sludge and biogas utilization:

(a) Reuse of treated sewage should be taken up after discussions between ULB, water boards, PHEDs / Jal Nigams and the public, as the case may be. Various possible reuse methods such as farm forestry, greenbelt development and lawns in road medians.

(b) Utilization of sludge in public areas is not possible due to issues of public acceptance and hence it is best to focus on farm forestry.

(c) Utilization of alternative energy, like in plant energy to be harnessed from bio-methanation and to evaluate the ambient temperature suitability or hearing of sludge vs economics.

(d) Reuse of treated sewage to a minimum extent of 20% by volume shall be mandatorily explored and the proposed use for achieving this 20% target shall mandatorily form part of the CSP.

(e) Utilization of sludge as a construction material (as porous pavement, bricks etc.)

Table 5.6: Fact Sheet for Various Treatment Processes

S. No.	Recycling Technique	Applicability	Land Requirement (Per MLD in Hectares)	Capital Cost (per MLD in INR)	O&M Cost (Million/ Year/MLD)
1	Activated Sludge Process (ASP)	• The most widely used option for treatment of domestic wastewater for medium to large towns where land is scarce.	0.15-0.25	2-4	0.3-0.5
2	Waste Stabilisation Pond Systems (WSPS)	• In warm conditions • Easy land availability • Where power supply is expensive, low or unreliable. • Where social preference for aquaculture	0.8-2.3	1.5-4.5	0.06-0.1
3	Upflow Anaerobic Sludge Blanket Process (UASB)	• The suitability of this technology may be doubtful as a standalone secondary treatment option	0.2-0.3	2.5 - 3.6	0.08-0.17
4	Duckweed Pond System (DPS)	• Low strength domestic wastewater after sedimentation with influent BOD < 80 mg/L • In combination with existing WSP • Easy land availability	2-6	1.5 - 4.5	0.18

		• As a polishing pond for an existing activated sludge plant or other technology based STPs.			
5	Biological Filtration and Oxygenated reactor (BIOFOR) Technology	–	0.04	6.5 - 8.1	0.86
6	High rate Activated Sludge Biofor-F Technology	–	0.08	5.2	0.18
7	Fluidized Aerated Bed (FAB)	• Small to medium flows in congested locations • Sensitive locations • Decentralised approach • Reliving existing overloaded STPs.	0.06	3-5	0.6 - 0.75
8	Submerged Aeration Fixed Film (SAFF) Technology	• Small to medium flows in congested locations • Sensitive locations • Decentralised approach • Reliving existing overloaded trickling filters	0.05	7	1.14
9	Cyclic Activated Sludge Process (CASP)	• Small to medium flows in congested locations • Sensitive locations • Decentralised approach • Reliving existing overloaded trickling filters	0.1-0.15	Not Available	Expected to be higher than ASP

Source: Compendium of Sewage Treatment Technologies, National River Conservation Directorate, MOEF, 2009.

Land availability plays an important role in providing such facilities. Land requirement may vary based on the technology adopted for sewage treatment. However, it is evident to mark the required land on development plan. It is recommended to decide the most suitable technology in advance based on the various parameters such as cost, environmental aspects, social acceptability, ease of maintenance, etc.

For effective septage management plan, robust data on septage arrangements, their quantity and locations of its generation etc. are required. The ULBs would need to make arrangements to collect baseline data, like type of latrine disposal, effluent disposal arrangement, size, age, when it was last cleaned, access to the on-site system, arrangement for disposal of effluent, if any of existing installations, to plan for workable de-sludging schedules. It is advisable to divide the city into different sanitary zones (if not already done) and carry out the baseline survey in one or a few of these zones, pilot de-sludging area wise schedules to learn operational issues and devise solutions, before up-scaling to the entire ULB can be taken up. The selection of zone could be based on availability of septage disposal sites – existing STPs could be potential septage disposal/application sites or trenches provided in solid waste landfill sites or suitable urban forestry sites where the septage trenches would serve to fertilize the plants. In order to be economical and financially competitive, it is suggested that households in demarcated septage management zone should be within 20 to 30 km travel distance from the identified treatment and disposal sites. Further a two-step process is to be followed for selection of the treatment system:

1. To determine the appropriate treatment option on the basis of size of town, land availability, proximity/availability of sewage treatment plants and proximity to residential areas; and
2. To conduct a techno-economic feasibility to choose the most appropriate technology on the basis of capital, operations and maintenance costs.

The steps in planning and implementing septage management are given below:

- Collect data on the households and other properties with on-site arrangements in the city.
- List out the municipal, private and other septic tank/pit cleaning services active in the city
- Identify catchment-wise land for septage treatment facility such as use existing STP where available; or acquire land if not available for construction of septage treatment facility
- Formulate draft regulations for septage management
- Choose technology for septage treatment: prepare design of Septage Treatment and Disposal Facility (STDF) along with operations and maintenance costs
- Conduct techno-economic feasibility of the STDF
- Implement construction of septage management and disposal facility
- Purchase vehicles and vacuum trucks etc.
- Launch awareness campaign
- Initiate training and capacity building
- Provide cleaning services incrementally in areas completing surveys of tanks and pits.

Further, the guidelines for selection of septage management disposal system are summarized in Table 5.7.

Table 5.7: Guidelines for the Selection of Septage Disposal System

Town/ category	Conditions	Recommended Technologies	Capital cost	O&M cost	Management

Unsewered Class-III, IV and V towns and rural communities	Remote land area available with suitable site and soil condition	Sludge drying beds and waste stabilization pond	Low	Low. User fees to recover O&M costs	Municipality or private (if implemented by private sector through a management contract)
	Land available but close to settlements	Lime stabilization, Sludge drying beds and waste stabilization pond	Low to medium	Low to medium. User fees to recover O&M costs	Municipality or private (if implemented by private sector through a management contract)
	Inadequate land area with unsuitable site and soil condition, but available STP capacity within 20-30 km distance	Disposal at STP	Low to medium	Low to medium. User fees to recover O&M costs	Municipality
Partially sewered medium size (Class-II towns)	Land area available with suitable site and soil condition but close to settlements	Lime stabilization, sludge drying beds and waste stabilization pond	Low to medium	Low to medium. User fees to recover O&M costs	Municipality or private (if implemented by private sector through a management contract)
	Inadequate land area, but available STP capacity	Disposal at STP	Medium	Medium. User fees to recover O&M costs	Municipality or private (if implemented by private sector through a management contract)
	Inadequate land area; no available STP capacity	Disposal at independent mechanical treatment facility	High	High. User fees to recover O&M costs	Municipality or private (if implemented by private sector through a management contract)
Class-I and metro cities	Available STP capacity	Disposal at STP	Medium	Medium. User fees to recover O&M costs	Municipality or private (if implemented by private sector through a management contract)
	No available STP capacity	Disposal at independent mechanical treatment facility	High	High. User fees to recover O&M costs	Municipality or private (if implemented by private sector through a management contract)

Source: Advisory Note – Septage Management in Urban India, MoUD 2013.

RECOMMENDED NORMS FOR PUBLIC TOILETS IN PUBLIC AREA

The general standard/guidelines for public toilets in public area and modified norms for public toilets in public places and roads recommended in the draft CPHEEO manual are given below.

Table 5.8: Norms for Public Toilets in Public Area

Type	Norms for Toilets
Public Toilet	On roads and for open areas @ every 1 km, including in parks, plaza, open air theatre, swimming area, car parks, and fuel stations. Toilet shall be disabled- friendly and in 50-50 ratio (M/F). Provision may be made as for public Rooms.
Signage	Signboards on main streets shall give directions and mention the distance to reach the nearest public convenience of visitors. Helpline number shall be pasted on all toilets for complaints/ queries.
Modes	Pay and use or free. In pay and use toilets entry is allowed on payment to the attendant or by inserting coin and user gets 150-20 minutes.
Maintenance/ Cleaning	The toilet should have both men and women attendants. Alternatively automatic cleaning cycle covering flush toilet bowl, seat, hand wash basin, disinfection of floor and complete drying after each use can be adopted, which takes 40 seconds. Public toilets shall be open 24 hours.

Source: Manual on Sewerage and Sewage Treatment Systems, CPHEEO 2013.

Critical Factors in Implementing Sanitation Systems

The Table presents some of the factors that may either encourage or discourage different stakeholders to opt for ecosan solutions. The table does not present an exhaustive list of all the motivating and constraining factors all the stakeholders may have, but should serve to provoke thought on what these factors may be. In many cases the motivating factors represent expectations of the stakeholders involved, while the constraints represent their fears and risk which need to be dealt with by the promoters or local body.

Table 5.9: Critical Factors and Constraints in Implementing Sanitation Programmes

Principal stakeholders	Examples of motivating factors	Examples of constraints
Users of sanitation facilities (households, neighbourhoods, tourists, employees, etc.	• Hygiene improvement • Structural stability • Local physical factors (high groundwater table, rocky ground, etc.) • Reduced costs • Increased comfort • Improvement of quality of life • Greater security • Interest in recyclables • Prestige • Ecological reasons • Water scarcity • Unreliable water supply	• Culture, habits, taboos • Hygiene concerns • Unfamiliarity • Fear of loss of comfort • Unavailability of structural elements • Legislative restriction • Economic factors

User of recyclables	• Economic reasons • Local and reliable availability of agricultural inputs (water, nutrients, organics) • Increase of crop yields for either the market or for family needs • Improvement of self sufficiency • Ecological reasons	• Culture, habits, taboos • Lack of logistics • Fear of negative consumer perfection • Fear of negative long term effects on soil
CBOs and self-help groups	• Failure of conventional / existing sanitation system • Local improvement of quality of life • Interest in recyclables • Reduced costs • Local physical factors (high groundwater table, rocky ground)	• Culture, habits, taboos • Lack of information • Insufficient financing • Inappropriate legislation • Influence of interest groups • Hygienic concerns
NGOs • Failure of conventional / existing sanitation systems	• Economic reason, • Ecological reasons, • Agricultural reuse of recyclables • Improve quality of life,	• Culture, habits, taboos, • Lack of information, • Insufficient financing, • Inappropriate legislation, • Influence of interest groups, • Hygienic concerns,
Local authorities governmental Institutions	• Political reasons • Economic reason • Ecological reasons • Failure of conventional/existing sanitation system • Possibility of financial support • Sustainability of system • Support regional self-sufficiency • Promotion of (urban) agriculture • Job (and income) creation • long-term security of social service (water supply, etc.)	• Culture, habits, taboos • Lack of information • Lack of start-up funds / insufficient financing • Monitoring of treatment / handling etc. more difficult for decentralized system, • Distrust of alternative systems • Not recognized as state of the art technology • Reluctance to change status quo • Contradiction of existing legal framework / long term plans • Powerful lobby from conventional centralized sanitation industry • Corruption

Service providers Planners / consultants: constructors (builders): Maintenance service providers; producers of equipment; providers of collection, treatment, transport and marking of the recyclables	• Increase profit, • Opening up of a potentially huge new market, • Request / need for particular product, • Further develop their own know how, • Ethical / ecological reasons	• Absence of technical knowledge • Absence of products • Inappropriate legislation • Lack of suitable tools • Economic interest of (waste) water monopolies • Fear of failure (economic risk) • Not yet recognized as state of the art • Reluctance to make the necessary increase in effort • Lack of experience in decentralized planning / participation • Lack of start-up funds, • Fear of reduced profit margins in smaller/ decentralized projects • Regulatory obstacles
Developers & investors	• Increase attractiveness of developments (eco-label), • Safe and secure "disposal" (especially in tourist areas), • User satisfaction, • Economic reasons, • Legal requirements,	• Absence of service ogistic • Culture, habits, taboos, • Lack of information • Lack of start-up funds, • Monitoring of treatment / handling etc. more difficult for decentralized system, distrust of alternative systems • Not recognized as state of the art technology • Reluctance to change status quo, • Contradiction of existing legal framework / long term plans • Powerful lobby from conventional centralized sanitation industry • Corruption, • Less commission for ecosan projects,
Financial Institutions	• Economic reasons • Failure of existing / conventional systems • Improving sustainability • Guarantee repayment of credit,	• Absence of specific financing instruments • Not recognized as state of the art technology • Need for research and development
Research Institutions	• Need for research and development, • Availability of research funds, • Ecological reasons,	• Absence of research funds, • Lack of Prestige,

Source: Adapted from GTZ, 2006, Concepts for Ecological Sanitation in Formal and Continuing Education

SUSTAINABILITY ASSESSMENT

The criteria for the assessment of the sustainability of sanitation systems can be grouped into five main categories: health, environment, economy, socio–culture, and the technical. The protection of human health is the main aim of providing sanitation. To stress this fact, health has been taken as a category in itself and not as an element of environmental or social consideration. The technical function of the system is also an important criteria for sustainable operation. The criteria can provide a checklist and assist planners, decision makers and other stakeholders to identify specific sustainability indicators for a suitable sanitation system.

Health

The sanitary system should minimize risks and safeguard public health. This covers the use of the sanitary installation, collection, transport, treatment and destination of the treated products. For instance, the risk of being infected through ground water or surface waters, contaminated from leaking sewers, pit toilets, or wastewater disposal should also be considered, as well as the risk of being infected when consuming vegetables from fields irrigated with insufficiently treated wastewater and bathing in lakes or the sea near overflows or waste water treatment plant outlets.

The following are critical condition for health

- Exposure to pathogens and risk of infection related to all system elements including collection, treatment reuse and final destination of products/wastes
- Risk of exposure to hazardous substance: heavy metals, medical residues, organic compounds
- Health benefits due to improved hygiene, food
- production, nutrition status, livelihoods

Environmental

The sanitation systems aim to protect the environment against possible detrimental effects of the discharge. When considering the environmental sustainability, both emissions to different recipients (water, soil, and air), as well as resources used by different sanitation systems during the construction and operation phase must be accounted for. Moreover, it is important to consider the suitability of the treated product for possible reuse in agriculture or aquaculture. The following indicators can be considered under environmental sanitation:

- Use of natural resources, construction
- Land
- Energy
- Construction material
- Chemicals
- Fresh water
- Discharge to water bodies

- BOD/COD
- Impact on eutrophication
- Hazardous substances: heavy metals, persistent organic compounds, antibiotics/ medical residues, hormones
- Air emissions
- Contribution to global warming
- Odour
- Nutrients, Energy, Organic material and Water
- Quality of recycled product (released to soil)
- Hazardous substances: heavy metals, persistent organic compounds, antibiotics\ medical residues, hormones

Economy

The ability to pay for sanitation is clearly a decisive factor in choosing the system. Among the users, the ability to pay is an important criterion for sustainability. However in the end it may be their willingness to pay that will define within what range the costs (both for construction and O&M) can vary and services be sustained by the population. Some experts argue that the other criteria may eventually be reducible to purely economic considerations. Considerations such as the impact on local development (e.g. local private sector), health benefits, etc. might be expressible in monetary terms. However, this reduction of criteria to their estimated economic value is not sufficient to allow a thorough consideration of all of the aspects. The following are critical condition for economy

- Annual costs, including capital and maintenance costs
- Ability to pay – user (% of available income)
- Environmental and health costs
- Benefits from reuse
- Local development, business and income generation effects

Socio-cultural Aspects

The objectives of sanitation are to protect human health and the environment. Sustainability in sanitation is based on these objectives. There is an obvious need to include social criteria as they are crucial to the sustainability of the use the services provided by the system. It is possible to distinguish at least three very different types of important criteria within this category, namely cultural acceptance, institutional requirements, and perceptions of sanitation. Society is more dynamic and flexible than criteria such as human health and the environment and therefore, the socio-cultural criteria are subject to a more dynamic change with time than human health and the environmental criteria. Although improving human health and protecting the environment are the main objectives of sanitation this might not be enough to sell the sanitation concept to future users. It is also important to recognize that the prime driver for sanitation might be security and status rather than health and environment some of the recent studies have shown. Another sanitation driver could be the possibility of increased food security, if the sanitation solution can provide hygienically safe fertilisers. The socio-cultural aspects include the following:

- Acceptance by the users and willingness to contribute through work and or money for sanitation services
- Convenience (comfort, personal security, smell, noise, attractiveness, adaptability to different age, gender, and income groups)
- Current legal acceptability and institutional compatibility
- Appropriateness to current local cultural context (acceptable to use and maintain)
- System perception (complexity, compatibility, observability–including aspects of reuse)
- Ability to address awareness and information needs

Women and Sanitation

Women suffer disproportionately from inadequate sanitation facilities. Women are often forced to defecate in the open and due to the stigma associated with this they have to wait until dark, when they also face the risk of physical abuse. Public lavatories, are few and far apart, besides being dirty and dark. Instead of using filthy lavatories, women go along the drains after ten at night or early in the morning at four or five o'clock. The health and other risks associated with poor sanitation facilities are considerable for women in low-income areas, even where there are public toilets. These obstacles can be grouped under six action areas as follows:

1. To give the poor a voice. The voice of the poor too often is not heard and misperceptions about them persist.
2. Neutralize vested interests. Water vendors, organized crime, corrupt and dishonest utility staff may have a vested interest in preventing better services for the poor.
3. Eliminate administrative and legal barriers. Land ownership and tenure issues often create a barrier to the provision of service to the poor. The poor may be unaware of administrative and legal requirements, or find it difficult to understand them and comply.
4. Strengthen capacity, autonomy, and accountability and provide incentives to serve the poor. Public service providers sometimes lack the autonomy, as well as financial and human resources and incentives required to provide service to the urban poor. Municipalities and utilities are not held accountable for the provision of satisfactory water supply and sanitation services. The services provided by small private service providers (SPSPs) are not recognized, encouraged, and regulated.
5. Adopt appropriate financial system keeping in view the reality that the tariffs do not cover the full cost of services, and poor household find it difficult to pay connection fees upfront, and also monthly bills. Small-scale service providers need help as they lack adequate finance to extend networks into peri-urban informal settlements.

6. Overcome physical and technical barriers. The over exploitation and degradation of water resources affects the poor disproportionately. Physical and technical challenges make extending formal sewerage networks into informal and unplanned settlements more difficult. Technical guidelines for design, construction, operation and maintenance for the LCS programme have to vary according to the different geological conditions, safeguarding against soil and ground water pollution. Uniform financing terms and conditions cannot work throughout the country, but a combination of beneficiary contribution, loan and subsidy with easier terms for lower income groups needs to be adopted. Various options need to be viewed from the perspective of (a) sustainability, (b) management, and (c) planning norms and standards for public health and hygiene, design, construction and maintenance.

Technical Performance

The technical reliability and performance of the system is the key to its success and ultimately its sustainability. One of the more important sub-criteria under this category is robustness and flexibility, both within the system (to be able to receive varying loads) and externally (to be able to withstand varying extreme environmental conditions as well as user abuse). The technical functioning of the system is seen as perhaps the most flexible group of criteria. In this regard, it is important to adapt technologies to the needs of the situation rather than trying to make a situation comply with the demands of a favoured technology. The technological considerations include the following:

- System robustness: risk of failure, effect or failure, structural stability,
- Robustness against extreme conditions
- Robustness of use of system: shock loads, abuse of system
- Possibility to use local competence for construction and O&M
- Ease of system monitoring
- Durability/Lifetime
- Complexity of construction and O&M
- Compatibility with existing systems Flexibility/adaptability (to user needs and existing environmental conditions – high groundwater level, geology, etc.)

The requirements of sustainability are dictated by context and can change with time. The concept of sustainability criteria should be the basis of the planning processes, which should address the local issues in the given context. A multiple sustainability criteria need to be an extremely complicated process, but it should be simple enough to facilitate working with grass root level groups.

Maintenance and Management

General cleaning and maintenance of a In India the maintenance of public urinals and toilets often takes a backseat. Even if a public toilet is provided, its poor maintenance nullifies the facility and wastes the entire resource and investment. Even in the capital of the country, it is common to find public toilets squatted upon, broken, without doors, water and electricity. The posh toilets built during the Commonwealth Games, 2010 in New Delhi remain locked or unused because of maintenance and management deficit.

Generally cleaning and maintenance of public toilet should be carried out daily. It should follow a systematic sequence to prevent the toilets areas from becoming dirty, stinking, wet and soiled. The general cleaning should be divided into spot and thorough cleaning. Spot cleaning refers to the process whereby only specific areas are cleaned (i.e., those that are soiled). Thorough cleaning refers to the cleaning of the entire restroom and is usually carried out once a day. The sequence of cleaning should follow this checklist:

- Replace all expendable supplies;
- Pick up litter and sweep floor;
- Clean and sanitize commodes and urinals;
- Clean and sanitize basins;
- Clean mirrors and polish all bright work;
- Spot-clean walls, ledges, vents and partitions;
- Wet-mop floors; and
- Inspect work and correct the error.

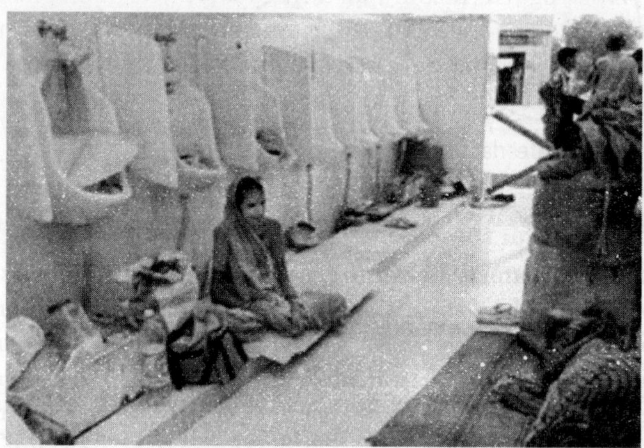

Fig.5.32: Squatters in a public toilet opposite AIIMS, New Delhi. The picture speaks volumes in respect of neglected maintenance of toilet block, its improper location and lack of low cost staying arrangement for the poor.

Source: The Hindustan Times, New Delhi, August 3, 2011.

An inspection card should be used in supervising and monitoring of the daily maintenance of the toilet. This card should be placed at the back of the entrance door of the toilet.

Schedule of Cleaning: Scheduled cleaning should be carried out on daily weekly, fortnightly or monthly basis (different surface, wares and fitting require different cleaning periods to maintain their cleanliness). Scheduled cleaning should be carried out during off-peak hours to avoid inconveniencing the users.

The timing and frequency of cleaning should be determined by the crowd flow. Cleaning should be done more often during peak hours and less during off-peak hours. Frequency of cleaning is usually determined by expectation and the standard of maintenance required by the management of the property and also the budget available for the maintenance of toilets. Toilets in shopping centres will require more frequent cleaning than condominiums.

Basic Equipment and Supplies: Different equipment for different joints and corners, as well as different disinfectants, should be used in the cleaning of different sanitary wares and fittings. Cleaners of public toilets should be trained in the proper usage of specific cleaning agents and equipment for different types of materials and finishes in the toilets, e.g. tiles, mirrors, stainless steel, etc.

Mechanical Ventilation System: Mechanical ventilation system should be properly maintained to ensure maximum efficiency and optimal operating conditions. It should be checked and serviced on a monthly basis. Cleaning of the systems should also be done weekly by wiping or dusting.

Training and education: Toilet cleaners should be properly trained and certified to perform the task well. Supervisors should also be trained with the right knowledge and skills to effectively supervise the cleaners. Toilet operators who engage in cleaning contractors for toilet cleaning should specify in their contract a performance-based outcome rather than head count-based outcome. The performance-based contract should also stipulate a requirement for the trained cleaner.

Having public education messages in the toilet can help persuade users to do their part in keeping toilets clean. In order to be effective in persuading people to do their part, message has to be:

- Attended to,
- Assimilated, and
- Remembered (for future action).

People readily attend to visuals. This makes the use of visuals an important part of the design of the message. Generally, visuals should be:

i. Simple and uncluttered,
ii. Attractive, and
iii. Eye-catching.

The language of public education has to be kept simple. This helps ensure that the message reaches all users regardless of their educational level. It ensures that the message is attended to, understood and remembered for future action. Slogans can be very effective because they are short, catchy and easy to remember. Subtle humour through cartoon characters can be used to deal with the personal and sensitive issues surrounding toilet use.

The usual means of message placement in public toilets are posters and stickers. To maximize the effectiveness of the right medium and manner of display should be selected. Generally, stickers should be used, if the main purpose of user education is to address specific behavioural concerns such as littering, careless aiming or the flinging of water everywhere. For display, stickers should be made of vinyl material, rather than paper.

These can be placed strategically at the spot where the problem behavior occurs, such as, on the wall above the urinal to encourage better aiming, at the wash basin area to discourage flinging of water everywhere. Poster can be used to convey generic messages such as "Help Keep This Toilet Clean and Nice". Posters should be displayed in a way that makes them repellent to water, e.g. laminated on both sides or protected by acrylic sheets. These can be mounted with non-marking adhesives such as scotch tape. Double-sided tape may damage certain types of wall surface and should therefore be avoided.

The toilets must be clean and hygienic. Due to the Lack of hygiene, bacteria can enter the body through unclean toilets. Special attention needs to be given to maintenance of plumbing, light and ventilation. The circulation of air will help in getting rid of the foul smell. To make a public toilet successful, regular maintenance is very important. Preventive maintenance of a public toilet depends upon the specification of materials, fittings finishing, plumbing and also the quality of construction. These contribute in lesser overheads and problems of cleaning and maintenance.

DEBRIEFING QUESTIONS

1. Explain the salient features of National Urban Sanitation Policy.
2. Explain the salient features of:
   (a) Total Sanitation Campaign
   (b) Nirmal Bharat Abhiyan
   (c) Swachch Bharat Mission
3. What are the types of on-site Sanitation Systems?
4. Explain how a septic tank works and its advantages and disadvantages?
5. Write short notes on:
   (a) Two Pit Latrine
   (b) Borehole Latrine
   (c) Direct Pit Latrine
6. What are the barriers in achieving Total Sanitation?
7. Write short notes on
   (a) Waterless (cartridge) Urinal
   (b) Composite Bio-gas Plant
   (c) Clivus Multrum Composting Toilet
8. Explain the purpose and process of Ecological Sanitation (Eco-san).
9. Write short notes on:
   (a) Bima Digester
   (b) USAB Reactor
   (c) FFR Technology
   (d) Methanisation

(e) Solar Aquatic Septage Treatment System
(f) Extended Aeration Technique
(g) Root Zone System
(h) Anaerobic Sludge Blanket
(i) Vacuum Sewerage System

10. Give the norms for public toilet.
11. Explain the indicators to be considered for Environmental Sanitation.

6 Drainage and Storm Water Management

Frequent and large scale flooding in Indian cities disrupt transport, water supply and sanitation systems and result in diseases and epidemics. These have been witnessed in the floods that have occurred in Chennai (Nov-Dec, 2015), Assam (2012), Uttarakhand (2013), Jammu and Kashmir (2014) and Gujarat and Maharashtra (2005). In poorly drained areas with inadequate sanitation, urban run-off mixes with excreta, spreading pathogens around communities and increasing risk to health from various water-borne diseases. Infiltration of polluted water into low-pressure water supply systems can contaminate drinking water and is frequently a source of gastro-intestinal disorders. Drainage plays an important role in traffic and transportation. It is common that with monsoon, the roads get sub-merged and pot-holed mainly due to poor drainage. In many cities, buildings, industrial and residential colonies come up in flood zone or low lying areas, the drains are not joined with the main storm water drains, and the gradient of the drains is faulty which result in water logging.

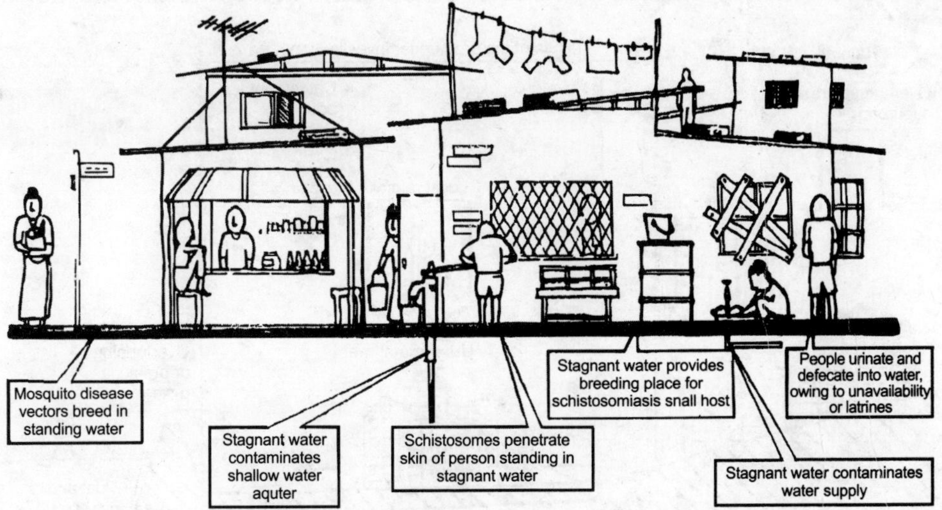

Fig.6.1: Stagnant Water and Disease Transmission–The Health Consequences of Poor Drainage.

Source: Cairn Cross, S. and E.A.R. Quano (1991), Surface Water Drainage for Low-income Communities, WHO/ UNEP, World Health Organization, Geneva, Switzerland.

A large number of natural drainage channels wind their way across settlements on their way to the river channels. Many of them become the corridors of encroachments and the carriers of polluted water, wastes and discharge from open lavatories. Open drainage channels are potential sources of infection and disease, especially to children who play in them, and polluted water from drains is often used for agriculture, where water resources are scare. Flooded septic tanks and leach pits, and blocked drains provide breeding sites for mosquitoes, which transmit malaria and filariasis, that can lead to elephantiasis and swelling of the legs.

URBAN DRAINAGE PLANNING

Urban drainage involves collecting, reusing and disposal of waste/ storm/surplus water. The global practices integrate systems of public health and water management with local hydrology, by which the waste and storm water is replenished and recycled. As is being increasingly realized that drainage is the foundation of sustainable urban development, it is necessary to factor in freak weather patterns, locational features of the city and the speed and magnitude of its growth while designing the drainage system.

Urban drainage system includes of the network of sewers, drainage channels, storm water and flood runoff. Storm water networks carry mainly surface runoff from streets, roofs, parking lots, and other surfaces towards the receiving water body. Sanitary sewers. Sanitary sewers carry household and commercial sewage and industrial waste waters towards a treatment plant. Combined sewers are used for collection and conveyance of both surface runoff and sanitary sewers in one conduit. Combined sewer invariably suffers from the disadvantages of sluggish flow during most part of the year leading to the depositing of sewage solids and creating foul and offensive conditions. Further it is costly. In view of these reasons combined system is not usually recommended for Indian conditions.

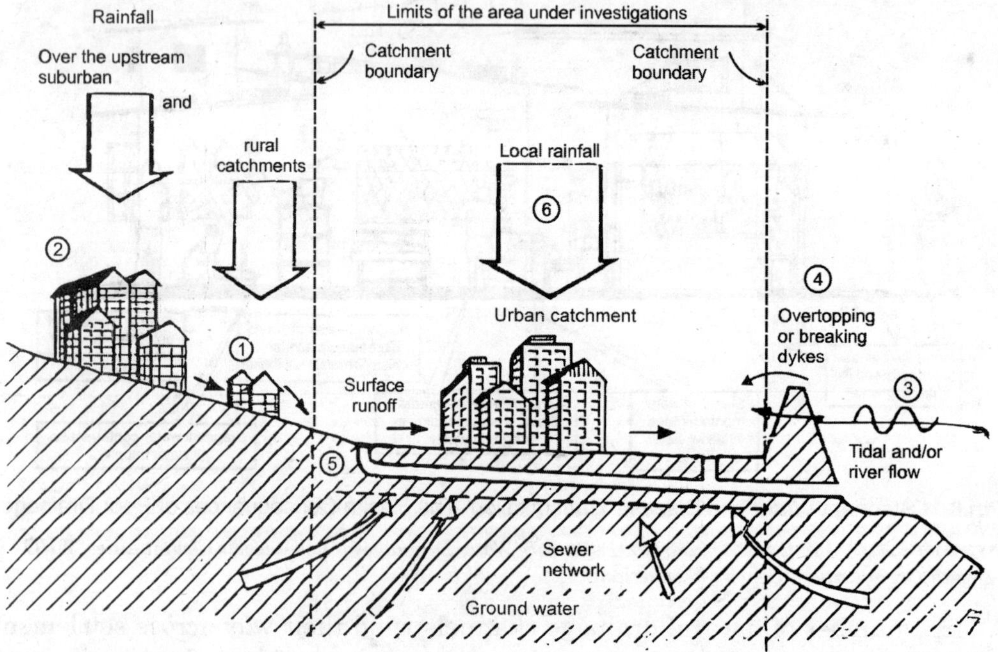

Fig.6.2: The Origin of Water in Urban Areas Relevant to Urban Storm Drainage.

Storm water in urban areas originates from different sources:

- Water originating from rural and suburban areas flowing through the urban zones.
- Water originating from the adjacent urban areas.
- Surface water from rivers and the other streams flowing under extremely high levels near the areas considered. Water can enter the city either through streets and open areas or through the underground sewer network.
- Water from overtopping or breaking of dykes.
- Water from rising ground water levels.
- Water from the rainfall.

The framework of drainage needs effective planning, appropriate technologies, accountable administration and community participation. It needs reducing the generation of waste and introducing recycling and primary treatment of city drains before discharging it in water bodies. The drainage system is integral part of land use planning and conservation of natural drainage. This requires understanding and characteristic of floods and strategies for their control.

Table 6.1: Typology of Flood Types, Characteristic and Effects

Flood type	Characteristics of flooding and effects
Type A	Localized flooding caused by inadequate drainage of storm water runoff, which can happen virtually every time it rains where the provision of drainage infrastructure is poor. The main effects of these events are related to a deterioration in environmental health conditions, especially water-borne diseases.
Type B	Flood events of this type occur less frequently than type A may include temporary disruption to transportation systems and inconvenience to city life. These events contribute to the propagation of water-borne diseases and can cause structural damage.
Type C	Large-scale inundations causing widespread disruption and damage affecting communities and businesses throughout cities. These events are infrequent and often reach a dramatic scale of the effects and structural damage; e.g. inundation of Chennai (2015), Assam (2012), etc.

Source: Adapted from Cairn Cross S. and E.A.R. Quano (1991), Surface water drainage for low income communities. WHO/UNEP.

Flood control strategies are developed to address the specific problems, as given in the following table:

Table 6.2: Flooding Problems and Strategies for Flood Control

Flooding problems	Strategies
Flood avoidance	Preventive response which aim to ensure that flood problems are avoided.
Flood mitigation	Immediate responses to flood warnings before a flood, and the course of action to take during a flood to reduce risks.
Flood recovery	Action to be taken after a flood event to enable communities to recover from the impact of a flood event.

Solid waste	Controls of solid waste to avoid blockages (such as plastic bags, packaging) and reduction in hydraulic capacity of the drainage system.
Pollution mitigation	Reduction in the discharge of pollutants into the storm-water drainage system.
Vector control	Improved practices to reduce vector transmission of diseases related to urban drainage and flooding.

Planning of urban drainage is based on the following performance requirements:

- Hydraulic
- Environmental
- Structural
- Operational

Accordingly, a drainage plan combines a number of options, both structural and non-structural, keeping in view various factors such as rainfall, catchment area, levels/ contours, land configuration, environmental parameters, traffic impact, utilities and services etc. It incorporates storm water quality enhancement and pollution reduction by various methods of sedimentation, filtering, infiltration, biological uptake and straining. Thus, appropriate options-grass buffer/swale, porous pavement, extended detention basin, wetland basin, retention pond, sand filter, wetland channel and tunnel are integrated with the proposed drainage system.

Calculation of Volume of Flow for Drainage System

During rainfall periods there can be a considerable amount of storm water that does not infiltrate into the ground surface and most of this becomes the excess overland flow or direct surface runoff. There are many contributing factors for analysing the quantity and temporal variations of this flow; these include geology of the land, topography, geography, rainfall intensity and pattern and the land use type.

Estimation of such runoff reaching the storm sewers, is dependent on the intensity and duration of precipitation, characteristics of the tributary area and the time required for such flow to reach the sewer. There are various methods available to calculate the runoff, the two of the below mentioned methods can be adopted to calculate runoff:

- Rational Method
- Kirpich Equation Method

Rational Method

It is based on the assumption that the entire precipitation over the drainage district does not reach the sewer. The characteristics of the drainage district, such as, imperviousness, topography including depressions and water pockets, shape of the drainage basin and duration of the precipitation determine the fraction of the total precipitation which will reach the sewer. This fraction known as the *coefficient of runoff* needs to be determined for each drainage district.

The National Disaster Management Authority (NDMA) guidelines on urban flood management has prescribed that all future storm water drainage systems for peak flow for any city should be designed after taking into consideration a runoff Coefficient of up to 0.95164 using the rational method. The runoff reaching the sewer is given by the expression,

$$Q = 10\ CiA$$

- Where Q is the runoff in m^3/hr;
- C is the coefficient of runoff
- i is the intensity of in mm/hr and
- A is the area drainage district in hectares.

It may be reiterated that Q represents only the maximum discharge caused by a particular storm.

The portion of rainfall, which finds its way to the sewer is dependent on the imperviousness and the shape of the drainage area apart from the duration of storm. The percentage of imperviousness of the drainage area can be obtained from the records of a particular district. In the absence of such data, the table below may serve as a guide.

Source: National Disaster Management Guidelines—Urban Flood management, 2010.

Table 6.3: Runoff Coefficients for Stated Surfaces

S.No.	Type of Area	Percentage of Imperviousness
1.	Commercial and Industrial Area	70-90
2.	Residential Area	
	– High Density	61-75
	– Low Density	35-60
3.	Parks and undeveloped areas	10-20

Source: NDMA.

When several different surface types or land use which comprise the drainage area, a composite or weighted average value of the imperviousness runoff coefficient can be computed, such as:

$$I = 1/A \times (A_1 \times I_1 + A_2 \times I_2 + A_3 \times I_3 + \dots\dots\dots + A_n \times I_n)$$

Where, the subscripts refer to respective sub-drainage area types, and 'A' is the total drainage area.

Kirpich Equation Method

The time of concentration is defined as the time it takes for a drop of water in the remotest point in a drainage basin to travel to the outlet. As calculation methods go, the Kirpich formula is one of the most widely used methods.

$$Tc = 0.01947 \times L0.77\ S–0.385$$

Where:

Tc = Time of concentration (hours)

L = Maximum length of water travel (m)

S = Surface slope, given by H/L (m/m)

H = Difference in elevation between the remotest point in the drainage basin and the outlet (m)

The Kirpich equation is normally used for natural basins with well-defined channels. If there are many undefined channels that are grassed or vegetated throughout, the Kirpich formula will likely underestimate the time of concentration, and so a factor of 1.3–1.5 should be added. If most of the drainage basins are non-natural (in urban system) with concrete or other smooth channels, the result should be decreased by about 40 – 60%.

Drainage system may carry sewage flow and storm water flow separately or in combination. When conduit carry only sanitary sewage without rainfall inputs, computation is simple. It starts with the projection of population and then go to the water supply level. Per capita sewage production is a function of standard of living and water use. Climate is yet another factor. In arid area 40% water used reaches the sewer. In intensively developed area it is 90% and generally 80%. Industries and commercial establishment may have their own source of water supply, but discharge their used flow into the public sewer. Often industrial waste may damage the sewer because of acidity. Tolerance limits of industrial effluent need to be checked before it is allowed to discharge in to the public sewer. The provision should be allowed for infiltration of ground water to the sewer. Based on materials of sewer, type of joints, position of ground water level; infiltration may range from 500 to 5000 litres a day per km of sewer per cm diameter of sewer and 250-500 litres/day per manhole. The above guidelines will help to determine the total daily-flow in the sewers.

Drainage Options

Various options as given below are available for urban drainage planning which can be adopted according to flood drainage catchment area delineation, flood regime, slope and broad land use pattern:

1. **Natural Catchments**
   - Natural channels
   - Streams
   - Water courses
2. **Partly Sewered**
   - Man-made ditches
   - Open channels
   - Sewers
3. **Combined Sewer System**
   - Open channels
   - Combined sewers
   - Overflows
   - Treatment facilities

4. **Fully separated Sewer System**
   - Separate sewers and trunk
   - Sewers
   - Treatment facilities
   - Outfalls
5. **Sustainable Urban Drainage Systems**
   - Swales, filter strips and soak ways
   - Retention basins/structure
   - Infiltration ponds and source control, storage basins, etc.

Structural measures for urban drainage require detailed surveys, analysis and planning. The drainage should, so far as possible, follow the line of natural drainage. Drains should be in straight lines, or in long easy curves. Sub-mains should also follow the line of natural drainage. Laterals should be laid in the line of greatest slope. Intercepting drains are an exception generally being placed across the slope. Three principal types of lateral sub-drainages system are the herringbone, gridiron and parallel systems. The parallel system is the most economical type because it involves the least duplication of drainage by laterals and mains. The preferable arrangement is short mains and low laterals rather than the reverse. The spacing of laterals depends on the physical composition and text of the soils. A minimum slope of 1 per cent (1 ft in 100 ft) is recommended. Steeper slopes are better.

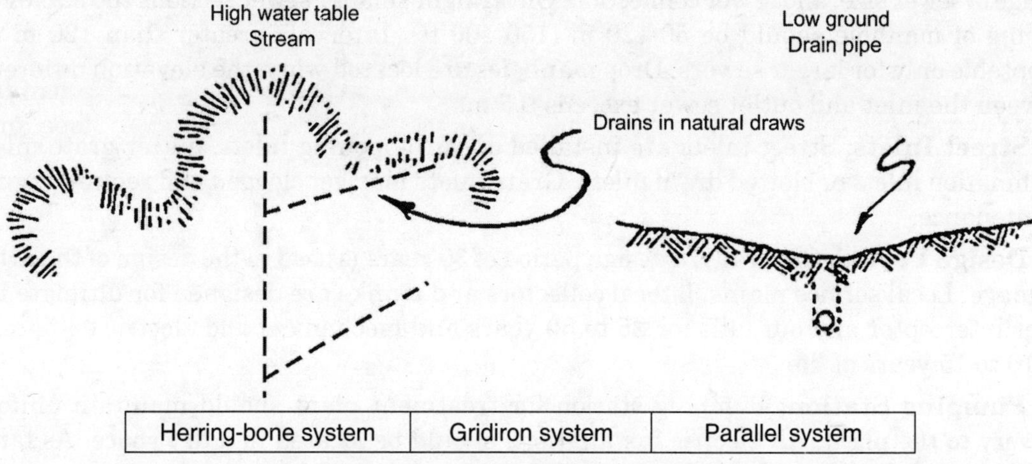

Fig.6.3: Drainage Nework Planning

Source: Handbook of Drainage and Construction Products. ARMCO Drainage and Metal Products, Inc., Middletown, Ohio.

Data requirement for the layout of urban drainage system include road maps, location of building and other utilities, floor elevation of building, soil, topography (contour map), undeveloped area and future expansion. Since gravity flow is used in drainage system a detailed map of site level may be highly useful. Having collected all data, plot on a base map, tentative location of sub-mains, interceptors and trunk mains. Interceptor and trunk mains shall pass through the lowest point in the drainage area and extend through the entire area of the drainage divide. Lateral and collecting sewers are connected to the interceptor sewer and proceed along slope. Collecting sewers should be located in all streets and right of way. Manholes shall be provided at all changes in grade, pipe size, direction of flow and quantity of flow. Maximum manhole spacing in a straight stretch cannot be more than 200 m. (600 ft) for pipes more than 600 mm and for 200 to 300 mm dia pipes it is 133 m (400 ft). For larger pipes no maximum is prescribed.

Drainage and sewerage systems comprise the following components:

Building Connection: Building connections connect the building plumbing to the nearest lateral or branch public sewer line. The minimum recommended size of building connection pipe is 100 mm (4 in) and a slope of 2%.

Lateral and Branch Sewer: Lateral and branch sewers are used to collect waste water from building connections and convey it to a main sewer. A lateral has no other sewer tributary except building connections, the minimum diameter being 200 mm (8 in).

Trunk Sewer: Main (trunk) sewer accepts flow from several branch or lateral sewers and convey it to the treatment plant or to an intercepting sewer.

Interceptors: Interceptors are large sewers that were originally designed to accept dry weather flow from a number of trunk sewers and carry the flow to a treatment facility. The layout of main sewers and interceptors should follow the natural slope and terrain.

Manholes: Manholes are installed at every change of slope, direction of sewer lines, change of sewer size, and sewer connection. On straight smaller sewer sections the maximum spacing of manhole should be 50-120 m (150-400 ft). Intervals greater than 120 m are acceptable only for larger sewers. Drop manholes are located when the elevation difference between the inlet and outlet sewer exceeds 0.5 m.

Street Inlets: Street inlets are installed as curb-opening inlets, gutter grate inlets, combination inlets or slotted drain inlets. Grate inlets may get clogged and require regular maintenance.

Design Period: Generally, a design period of 30 years is used in the design of the urban drainage. Local service mains, lateral collectors and trunks are designed for ultimate life; other interceptor and out falls for 25 to 50 years and mechanical and electric equipment for 10 to 15 years of life.

Pumping Station: Pumping station for treatment plant should maintain uniform delivery to the plant. All electric installations should be located in a dry space. As far as possible pumping is avoided in any well-designed sewerage.

Calculation of Volume of Flow for Drainage Planning

Drainage system may carry sewage flow and storm water flow separately or in combination. When conduit carry only sanitary sewage without rainfall inputs, computation is simple. It starts with the projection of population and then go to the water supply level. Per capita sewage production is a function of standard of living and water use. Climate is yet another factor. In arid area 40% water used reaches the sewer. In intensively developed area it is

90% and generally 80%. Industries and commercial establishment may have their own source of water supply, but discharge their used flow into the public sewer. Often industrial waste may damage the sewer because of acidity. Tolerance limits of industrial effluent need to be checked before it is allowed to discharge in to the public sewer. The provision should be allowed for infiltration of ground water to the sewer. Based on materials of sewer, type of joints, position of ground water level; infiltration may range from 500 to 5000 litres a day per km of sewer per cm diameter of sewer and 250-500 litres/ day per manhole. The above guidelines will help to determine the total daily-flow in the sewers.

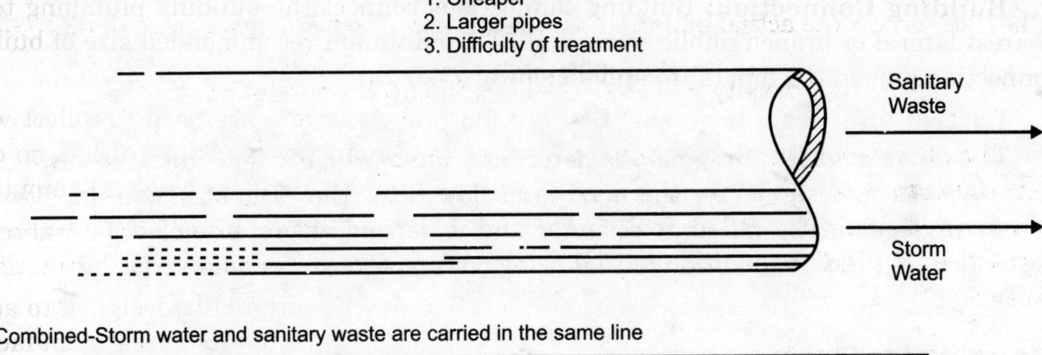

Fig.6.4: Combined storm water and sanitary waste are carried in the same line

Fig.6.5: Two separate lines are used for storm water drainage and sanitary waste disposal

Source: Water Supply and Sewerage, 3rd Edition, McGraw Hill Book Co., Inc.

Table 6.4: Urban Population, Density Profiles

Town Population	Density persons/ha
Up to 5,000	75-150
5,001 to 20,000	150-250
20,001 to 50,000	250-300
50,001 to 100,000	300-350
100,000 plus	350-1000

For planning purpose it may be more convenient to express sewage flow per unit area than per capita. If,for example, average sewage flow is 300 litres/ capita-day, the population density 100 persons/ ha and 100% increase for infiltration, the peak flow rate is 1/ 14 of average low, then allowing infiltration; the unit flow rate is 2*300 (litres/ cap day) *100 (persons/ ha)/ 14*60*60 –1.2 litre/ (s-ha).

Alternatively, if floor area per person and Floor Area Ratio (FAR) (which is the ratio of all floor area to plot area); or Floor Space Index (FSI) are known; density can be computed as follows:

For example if FSI=0.5 and floor area/ person = 9 sq m/ person, R is the residential area in hectares; then the actual built up floor area is equal to residential area, say R * 0.5 and

$$\text{density} = \frac{R \times 10000 \times 0.5}{9} = 555.56\ R\ \text{persons / ha}$$

The above guideline may help to determine the total daily flow in the sewer, but the sewer should be designed for the maximum flow rate. The ratio of maximum flow rate to average flow rate is called peak factor and is depend ant on size of the contributing population. For Indian condition the following peak factors are recommended for the design of sewer:

Storm Water Flows

Storm water runoff is the portion of precipitation which drains over the ground surface. Estimate of such runoff reaching the storm sewer depend s upon; intensity and duration of precipitation, characteristics of the tributary area and the time required for such flow to reach the sewer. This may be estimated by rational method, hydrograph method, rainfall-runoff correlation studies and digital computer models, inlet method or empirical formula. Here the rational method will be discussed.

The runoff reaching the sewer is given by the expression: Q = 10 CiA

where "Q" is runoff in cubic metre/ hr, "C" is the coefficient of runoff, "i" is the intensity of rain fall in mm/ hr and "A" is the area of drainage district in hectare.

Coefficient of run off "C" is the percent of rainfall that becomes runoff. The losses from rainfall to runoff from rainfall, infiltration and less from evaporation, transpiration and retention in vegetation and depression.

Table 6.5: Peak Flow Factor for Urban Population

Population	Peak Flow Factor
Upto 20,000	3.50
20,0001 to 50,000	2.50
50,001 to 7,50,000	2.25
Above 7,50,000	3.00

Table 6.6: Runoff Coeficient for Rational Formula

Description of Area	Runoff Coefficient
Business	
Downtown	0.7 to 0.95
Neighbourhood	0.5 to 0.7
Residential	
Single family	0.3 to 0.5
Multi units detached	0.4 to 0.6
Multi units attached	0.6 to 0.75
Residential	0.6 to 0.75
Suburban	0.25 to 0.4
Apartment	0.5 to 0.7
Industrial	
Light	0.5 to 0.8
Heavy	0.6 to 0.9
Parks, cemeteries	0.1 to 0.25
Playground	0.2 to 0.35
Unimproved	0.1 to 0.3
Parks and garden	0.0 to 0.1

Table 6.7: Runoff Coefficients Related to Types of Surface

Surface Character	Runoff Coefficient
Pavement	
Asphalt and concrete	0.70 to 0.85
Brick	0.75 to 0.85
Roofs	0.75 to 0.95
Lawns Sandy Soil	
Flat 2% or less	0.05 to 0.10
Average 2-7%	0.10 to 0.15
Steep greater than 7%	0.15 to 0.20
Lawns Tight Soils	
Flat 2% or less	0.13 to 0.17
Average 2 to 7%	0.18 to 0.22
Steep greater than 7%	0.25 to 0.35

For finding C value, measure surface by type and find percentage area and then multiply the percentage with the respective C. If the drainage area is not homogeneous, the area must be first divided into homogeneous segments (roofs, streets, lawns) with partial areas. The average runoff coefficient with partial areas *A1*. can then be computed by using the coefficient for the partial surface areas as follows:

Analysis of past records over a period of years is required to arrive at the estimate of the intensity "i" –duration for given frequencies. The longer the period available more dependable is the forecast. The rainfall intensity of the rational formula is obtained from rainfall intensity-duration-frequency curves established from the past meteorological observations. The intensity of rainfall decreases with duration. These curve represents the relationship between the intensity of rainfall in mm/hr (in/hr) or litre/s-ha, its duration and the statistical probability of storm occurrence expressed as a recurrence interval, in years or storm frequency.

The relationship may be expressed by a suitable mathematical formula. The following two equations are commonly used:

where i = intensity of rainfall (mm/hr),

t = duration of storm (minutes) and a, b and n are constants.

The value of "i" can be found for any time of concentration. Time of concentration is the time required for the rain water to flow over the ground surface from the extreme point of the drainage basin and reach the point under consideration. It may generally vary from 3 to 5 minutes.

The frequency of storm for which the sewers are to be designed depends upon the importance of the drainage area. Commercial and industrial land use should be subjected to least flooding than residential area.

Hydraulic Design of Sanitary Sewer

- First draw a line to represent the proposed sewer in each street. It can pass through a side of the street or to the centre, but shall be away from water supply main and preferably at a lower elevation.
- The arrow is drawn near the line to indicate the direction of flow of sewage based on the slope of the ground.
- Locate manhole and number it.
- Then sketch the limits of the service area for each lateral.
- Measure the several service area.

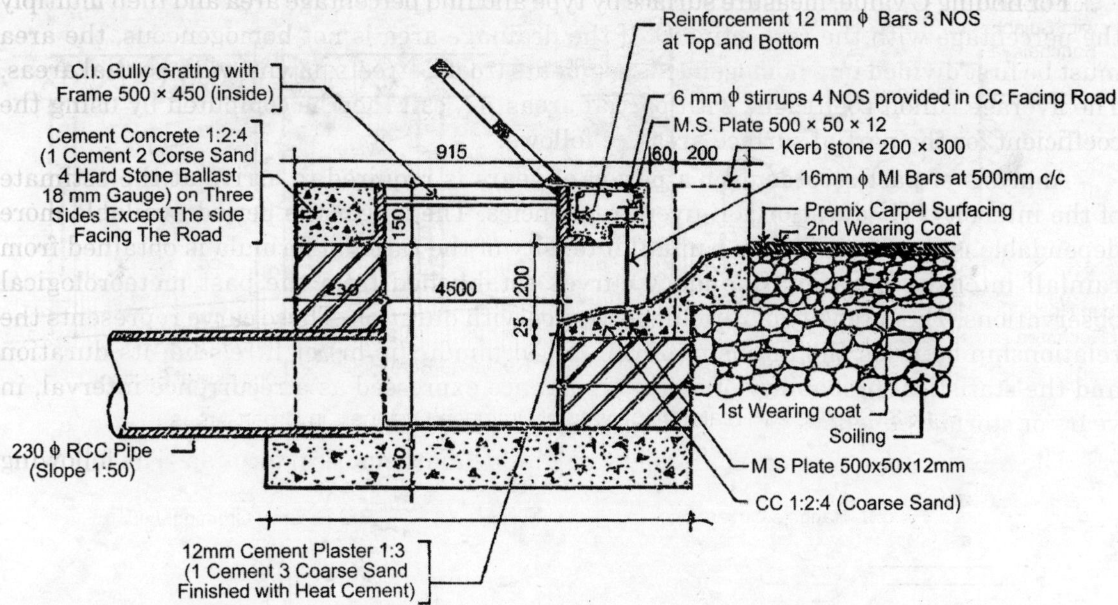

Fig.6.6: Typical Section showing Position of Gully Chamber. At 15 m c/c Both sides of the Road is staggered Position Connected to the Roadside Drain with 230 mm ϕ RCC Pipe

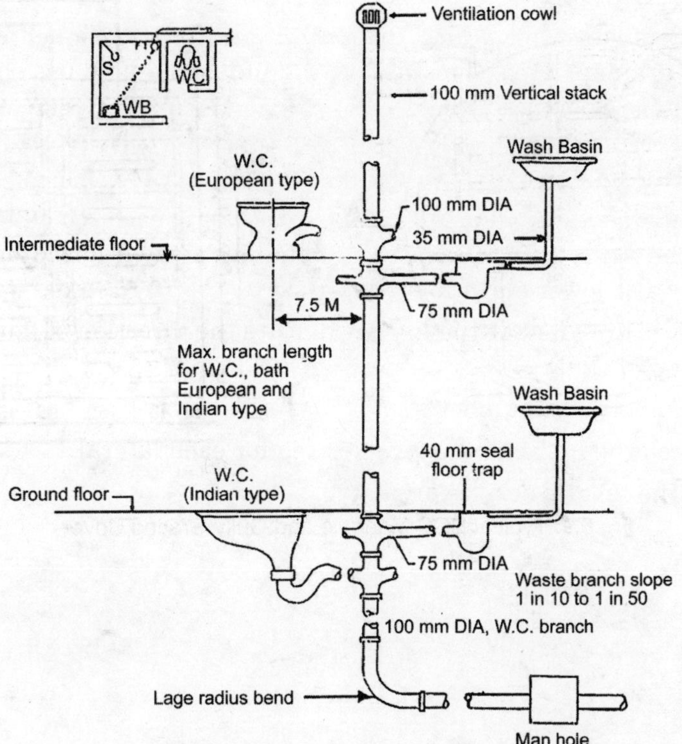

Fig.6.7: Single Stack System of Building Drainage

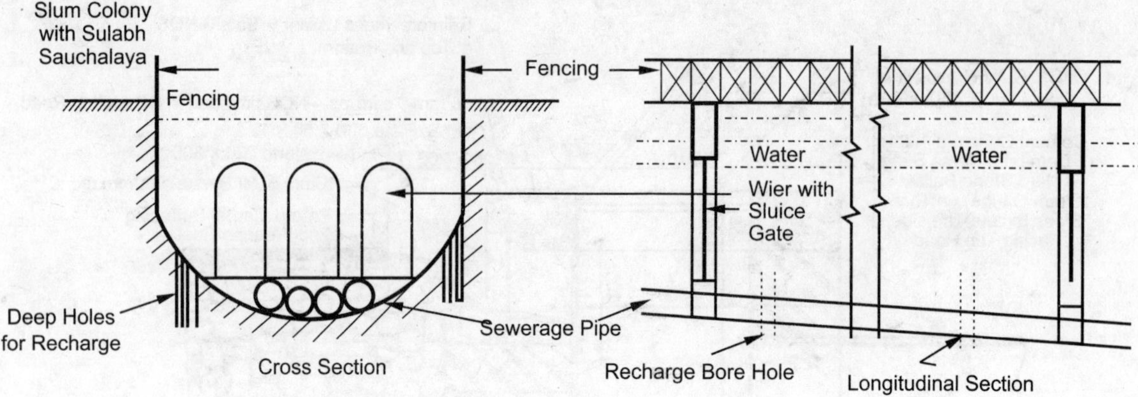

Fig.6.8: Suggested Treatment of Natural Drainage Channels in Urban Areas

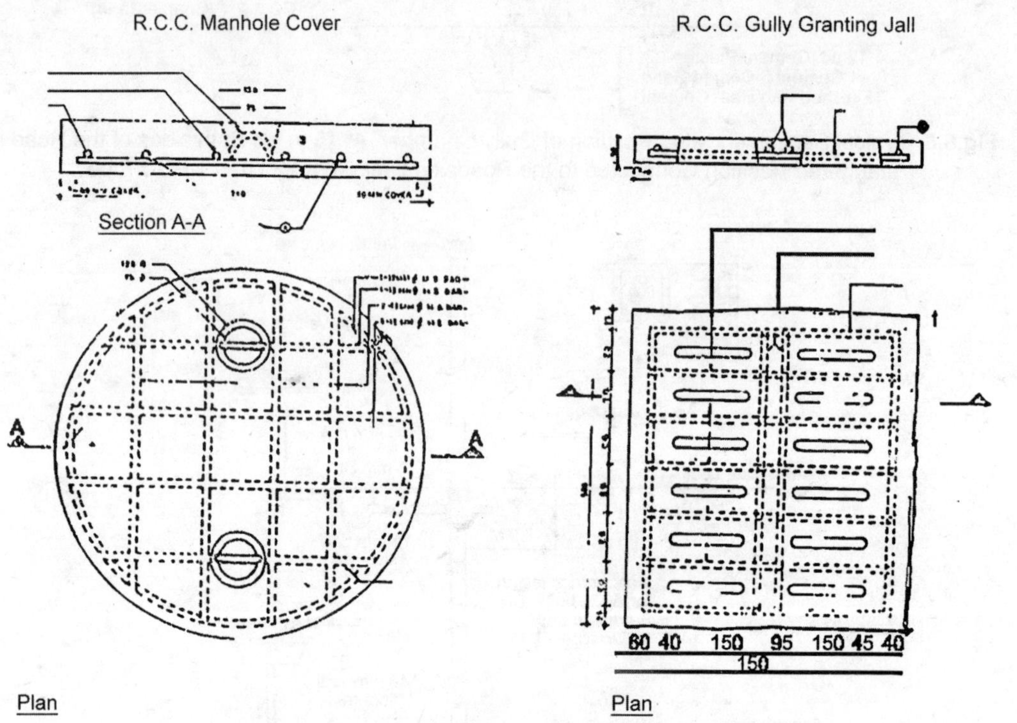

Fig.6.9: Typical RCC Manhole and Gully Grating Cover

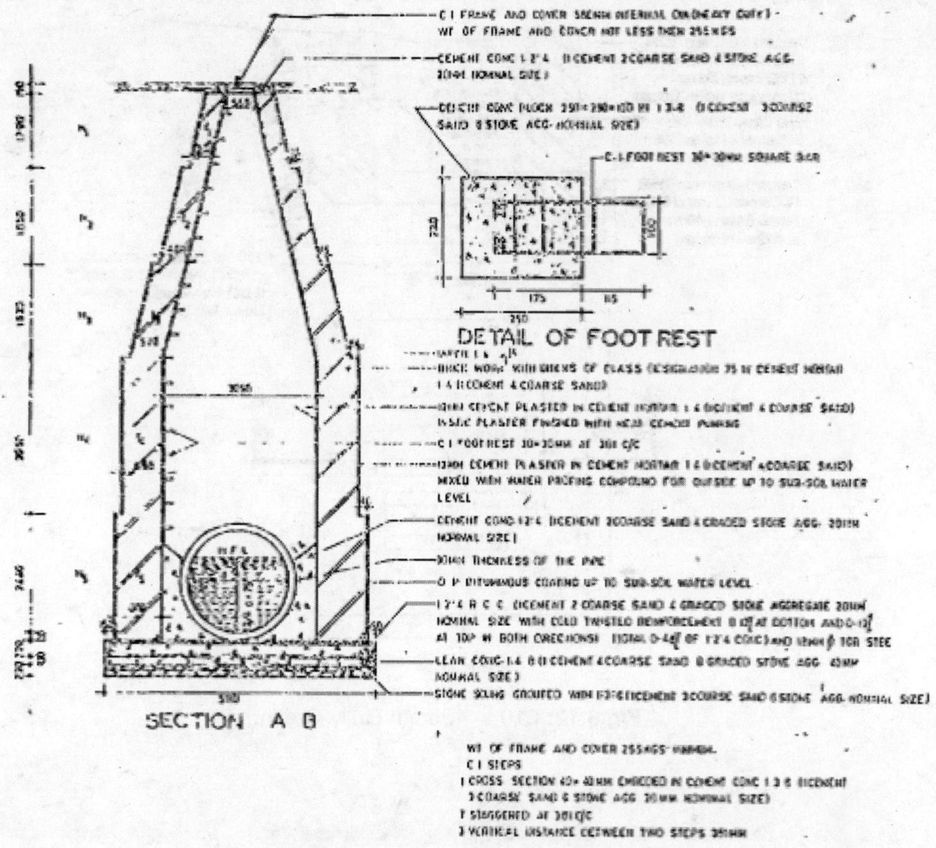

Fig.6.10: RCC Manhole Detail

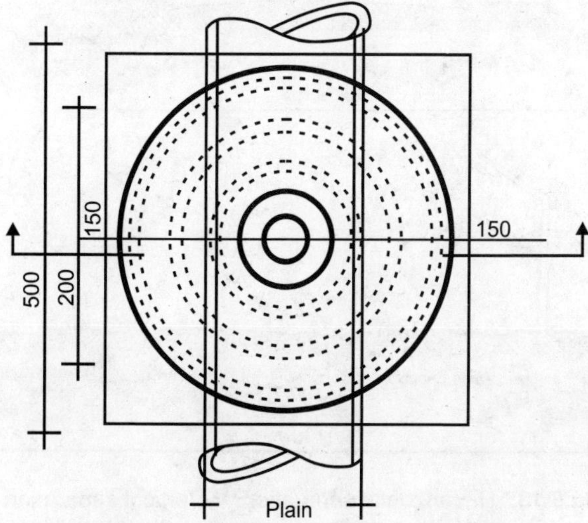

Fig.6.11: 3050 dia Manhole for 2400 Barrel

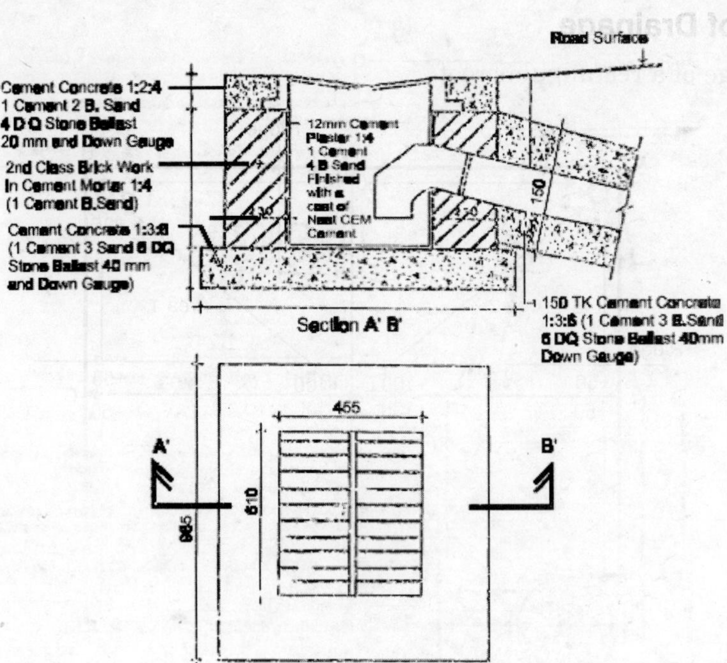

Fig.6.12: 610 × 455mm Gully Grating

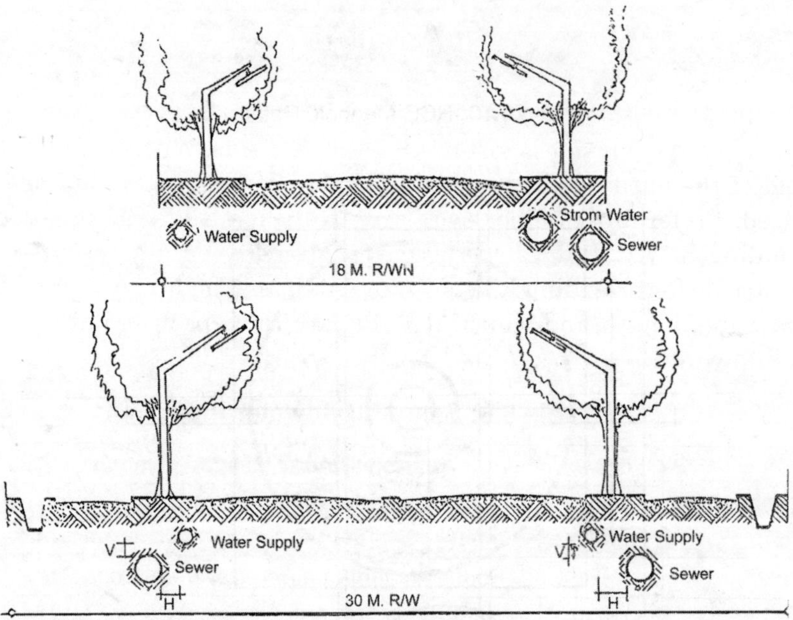

Fig.6.13: H-Horizontal separation, V-Vertical separation

Note:- Horizontal separation of 3 metres & vertical separation of 0.3 metre to be maintained.

Building Roof Drainage

Case 1: Drainage of a rectanguiar roof

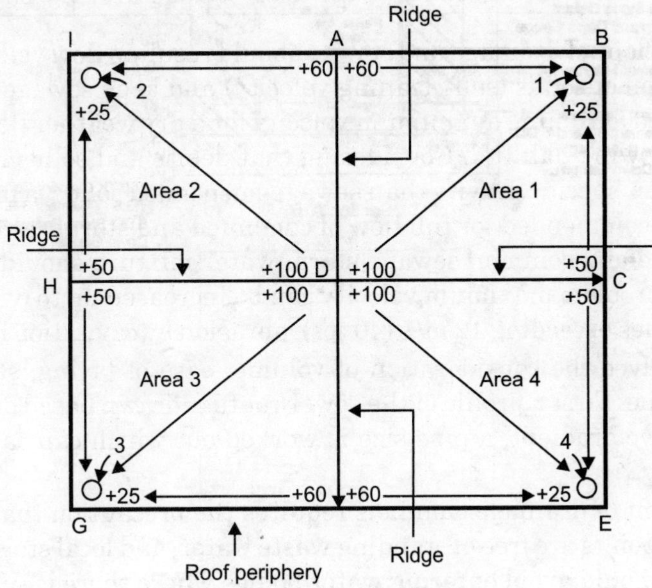

Fig.6.14: Roof Drainage Plan

Notes: +25, +50 etc. shows the thickness of screen over structural slab for providing slope for water draining purpose.
Portion ABCD (Area 1) will drain the water in RWP No. 1
Portion ADHI (Area 2) will drain the water in RWP No. 2
Portion CDFE (Area 4) will drain the water in Pipe No. 4
Portion FDHG (Area 3) will drain the water in Pipe No. 3

The following procedure is adopted for drainage of a building.

Mark centre point of all sides. Extend them to that side. Wherever all these lines meet, is the centroid of the entire drainage area. Number of areas now obtained are the no. of R.W.P. required. Preferably number each area and each R.W.P so that each area drains into corresponding R.W.P. Ridges are dividing line between adjacent areas. Please note that this thumb rule for providing R.W.P. is for small and medium size roofs only. For roofs having considerable length and width, R.W.Ps. are to be provided at certain c/c spacing along length and width.

Table 6.8: Size of Rainwater Pipe

Dia of pipes mm	Average Rate of Rainfall in mm/hr					
	50 mm/hr	75 mm/hr	100 mm/hr	125 mm/hr	150 mm/hr	200 mm/hr
	Roof drainage Area (m^2) Covered by one Pipe					
50	13.4	8.9	6.6	5.3	4.4	3.3
65	24.1	16.0	12.0	9.6	8.0	6.0

75	40.8	27.0	20.4	16.3	13.6	10.2
100	85.4	57.0	42.7	34.2	28.5	21.3
125	–	–	80.5	64.3	53.5	40.0

The minimum slope of drainage and sewers should result in flow velocities large enough to prevent deposition of solids (self-cleaning velocity) and keep sewage afresh. Commonly slopes are calculated assuming a minimum velocity for a dry weather flow of 0.6 m/ s (2 fps) when sewers are flowing half full. Considering that debris and solids in storm water will normally have higher specific gravity than sewage-suspended solids, a minimum velocity of 0.9 m/ s (3 fps) is recommended for full flow of combined and storm sewers. The maximum velocity for concrete and common sewage-storm water mixture should be 3 m/ s (10 fps). When grit is not a problem maximum velocity can be increased up to 6 m/ s (20 fps) or 8 m/ s (27 fps). At velocities exceeding 12 m/ s (40 fps), abrasion by cavitation becomes a problem. Pipe selection involves the consideration of volume, ease of laying, structural strength, economy and hygiene. First a profile of the sewer route is drawn based on the ground slope. Then using tentative gradient, a pipe size is worked out which can carry the flow under full condition.

The development of drainage channels requires the precaution that the areas around shelters and water points are free of standing waste water, and local storm water drains are kept clear. People should avoid entering water bodies where there is schistosomiasis risk. The most effective way to avoid drainage problems is in the choice of site and lay out of the emergency settlement. It may not be practicable to address the drainage problems of some sites, or of nearby water bodies. To avoid stagnating household wastewater, it may be appropriate to involve the population concerned. Technical support and tools may be needed, and it may be necessary to provide information and alternatives. Special care is needed to ensure that latrines and sewers are protected from flooding in order to avoid structural damage and leakage.

Non-structural Storm Water Management Strategies

Non-structural storm water management strategies for mitigation of flood focus upon preventive action and rely predominantly on behavioural changes in order to be effective. These are particularly relevant in low-income communities in tropical climates, where flooding is inevitable and resources are scarce. A well-coordinated and balanced combination of both structural and non-structural measures is required as part of the long-term flood mitigation strategies. Non-structural storm water management strategies include sustainable urban drainage systems, filter strips, swales, infiltration basins, bio-drainage and zero run-off, retention systems, etc.

Sustainable Urban Drainage

Storm Water Management in an urban area is done by the following strategies:

- Surface water moves through swales and filter trenches that remove entrained pollutants.

- The peak river discharge is delayed and reduced with incorporation of retention basins, enabling storage of water for re-use.
- Storage in retention/detention pond manages flood as well as enables ground water recharge
- Infiltration of water to ground through infiltration basins and soak ways improves the quality of water in river and decrease peak river discharge, as well as enables ground water recharge.

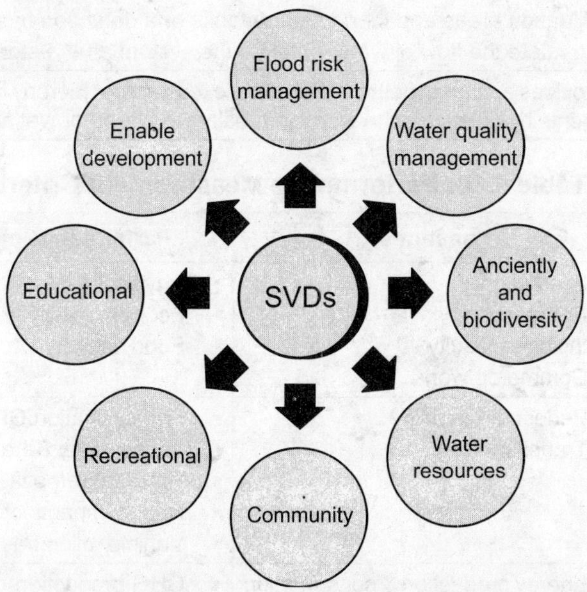

Fig.6.15: Functions of Sustainable Urban Drainage System

Sustainable Urban Drainage System (SUDS) is defined as "Collective group of approaches which are used to manage surface water that take into account the water quantity (flooding), water quality (pollution) and amenity issues". SUDS is regarded a sequence of management practices, control structures and strategies designed to efficiently and sustainably drain surface water, while minimising pollution and managing the impact on water quality of local water bodies. It mimics the nature and typically manages rainfall close to where it falls. SUDS can be designed to slow water down (attenuate) before it enters streams, rivers and other watercourses, they provide areas to store water in natural contours and can be used to allow water to soak (infiltrate) into the ground or evaporate from surface water and lost or transpire from vegetation (known as evapo-transpiration) techniques. These include green roofs, permeable surfaces, infiltration trenches, filter drains and filter strips, swales-shallow drainage channels, detention basins, purpose built ponds and wetlands etc. SUDS works on the principles and its performance is measured as given in the following tables.

Table 6.9: Principles of SUDS

Prevention	• Good site design and effective housekeeping measures • Includes within the site management plan
Source control	• Achieved with features such as green roods, soakways, rain gardens and permeable pavements • Harvested water can be reused for non-potable uses
Site control	• Planned management of water in a local area or site • Approached include routing water from building roofs and car parks to soakway or infiltration areas and then to storm tanks or a detention basin • Attenuate the flow of water entering the system after a storm event
Regional control	• Provides a natural method of handling excess water thereby reducing the risk of flooding events by collecting the water in a balancing pond or wetland

Table 6.10: Performance Measurement Criteria

	Functions	Performance measurement criteria
Social system	• Interaction • Recreation • Inclusion/equity • Commerce/work	• Physical & psychological well-being • Economic activity • Food security
Circulation system	• Pedestrian mobility • Transit mobility	• Transportation GHG reductions • Accessibility & travel costs • Pollutant removal • Health impacts of fuels • Vehicle miles travelled
Energy system	• Energy production & consumption • Material production • Recycling & reuse	• GHG production • Air quality • Recycling rates • Local food availability
Biological system	• Climate moderation • Ecosystem restoration • Wildlife habitat	• Biodiversity • Health of indicator species • Habitat connectivity
Hydrologic system	• Water retention • Groundwater recharge • Water quality • Water reuse	• Quantity and quality evaluation • Hydrologic connectivity • Pollution abatement & removal • Amount of water retained / detained
Geologic system	• Support for all the other systems • Soil and surface drainage • Natural hazard mitigation	• Slope and soil type • Seismic activity

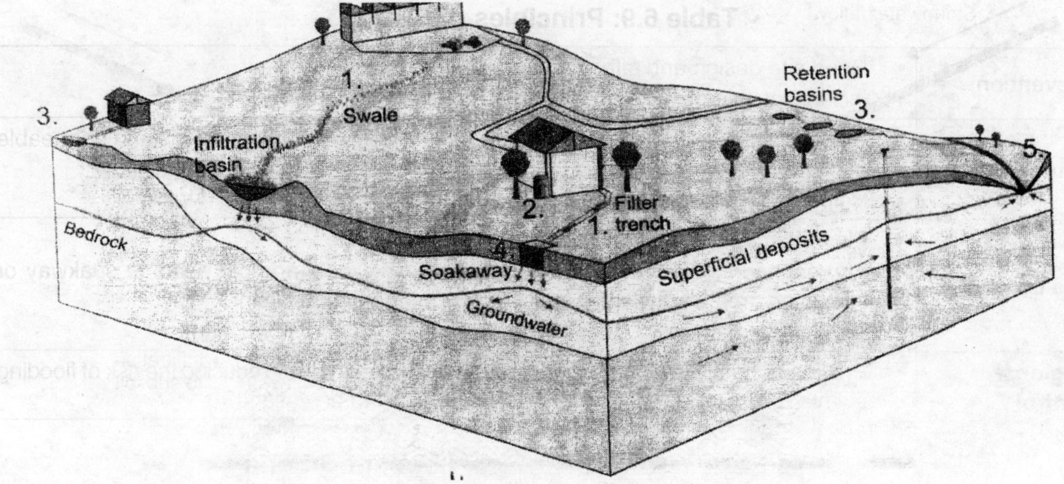

Fig.6.16: SUDS Model

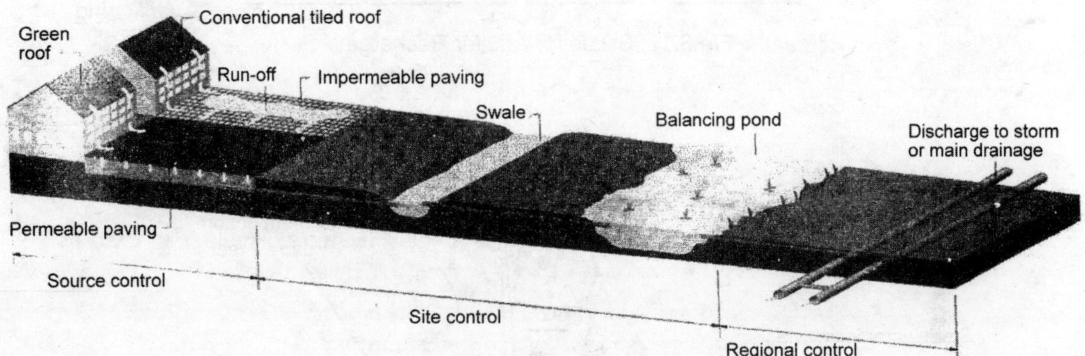

Fig.6.17: SUDS Management Train

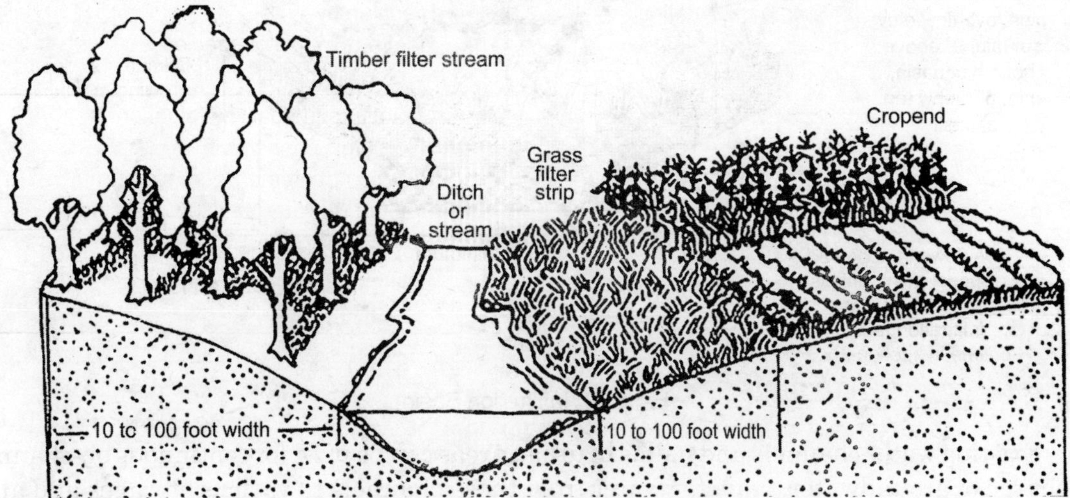

Fig.6.18: Landscape Filter Strips

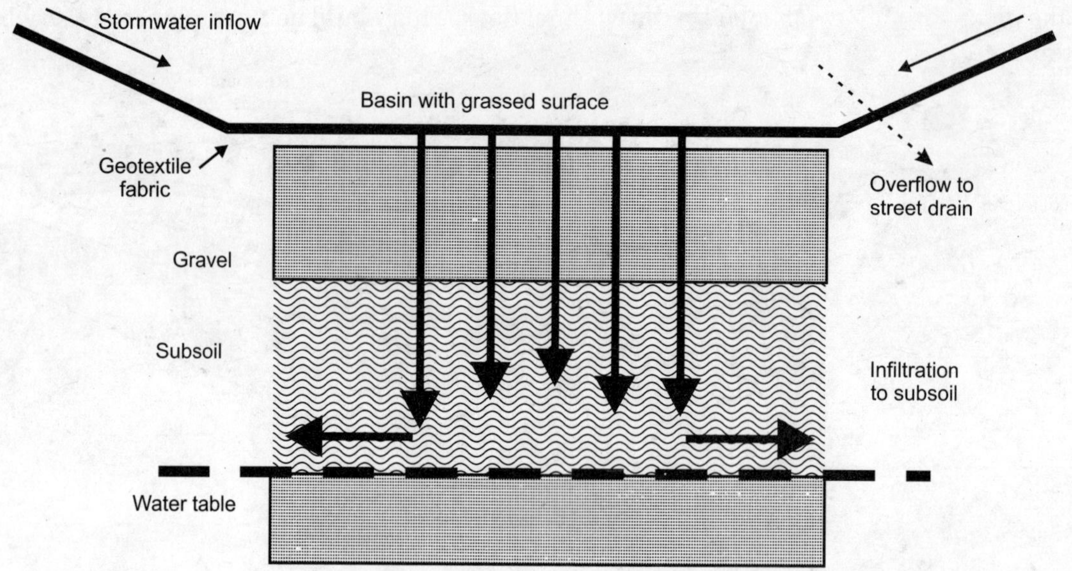

Fig.6.19: Swale for Aquifer Recharge

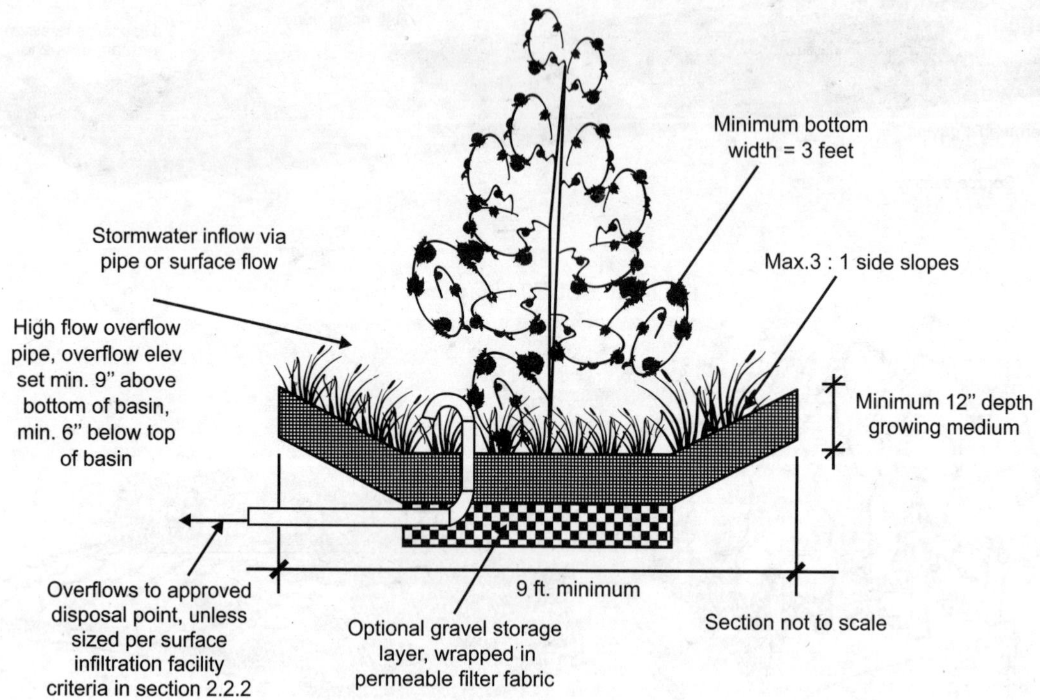

Fig.6.20: Infiltration Basin

Existing water channels and nallas in urban areas can be used for continuous on-channel recharge of groundwater, transfer water from these channels to various storages and also store water from the channels with the help of sluice gates. Continuous recharge would take place through deep holes on the channel floor. This would be a complex exercise which would require computer modelling of channel capacities, transfer mechanisms, storage capacities, flood prevention and system operations

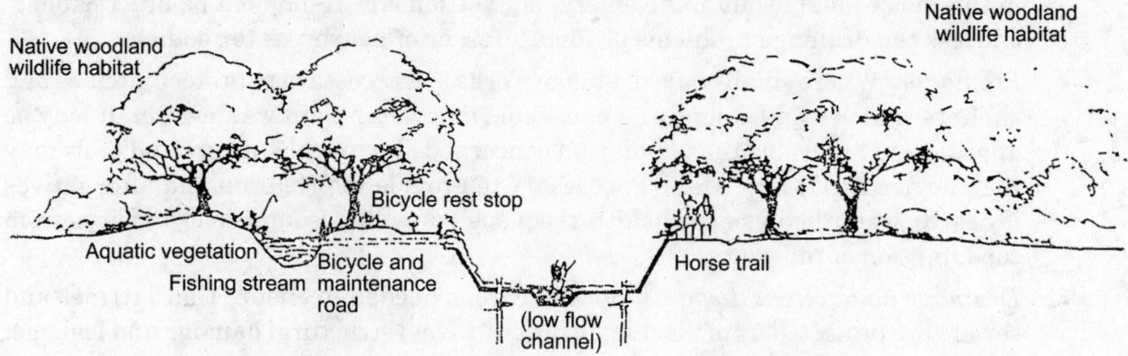

Fig.6.21: Proposed Landscape of an existing Nalla (Najafgarh Drain, Dwarka, Delhi)

The channel once filled with water thus could constitute urban waterways. Their cleanliness would be maintained with the use of root zone system, mosquito weed, guppy fishes to consume mosquito larvae, and aeration. The channel would be shaped, landscaped, deepened to increase the water storing capacity wherever possible. Their banks would be the green lungs of city. While serving the primary function of on-channel groundwater recharge they would also modify the micro-climate. The conjunctive use of groundwater along with on-channel recharge would also be possible. Their use for urban water sports, recreation, flotels, floating offices, waterborne transport (with Panama canal type sluice gate barriers) would be examined and integrated in the system as revenue generating components.

Fig.6.22: Urban Drain as Green Corridor.

The development of drainage channels requires the precaution that the areas around shelters and water points akm re free of standing waste water, and local storm water drains are kept clear. People should avoid entering water bodies where there is schistosomiasis risk.

- Areas around shelters and water points are free of standing wastewater, and local storm water drains are kept clear.

- There is a demand for tools for drainage works.
- People avoid entering water bodies where there is a schistosomiasis risk.
- *Site selection and planning:* The most effective way to avoid drainage problems is in the choice and lay out of the emergency settlement. It may not be practicable to address the drainage problems of some sites, or of nearby water bodies.
- *Promotion:* Where small-scale drainage works are necessary to protect latrines and shelters, and to avoid stagnating household and water point wastewater, it may be appropriate to involve the population concerned. Technical support and tools may then be needed. It may also be necessary to provide information and alternatives if nearby water bodies pose health risks such as schistosomiasis or hazards from consumption of the water.
- *Drainage and excreta disposal:* Special care is needed to ensure that latrines and sewers are protected from flooding in order to avoid structural damage and leakage.

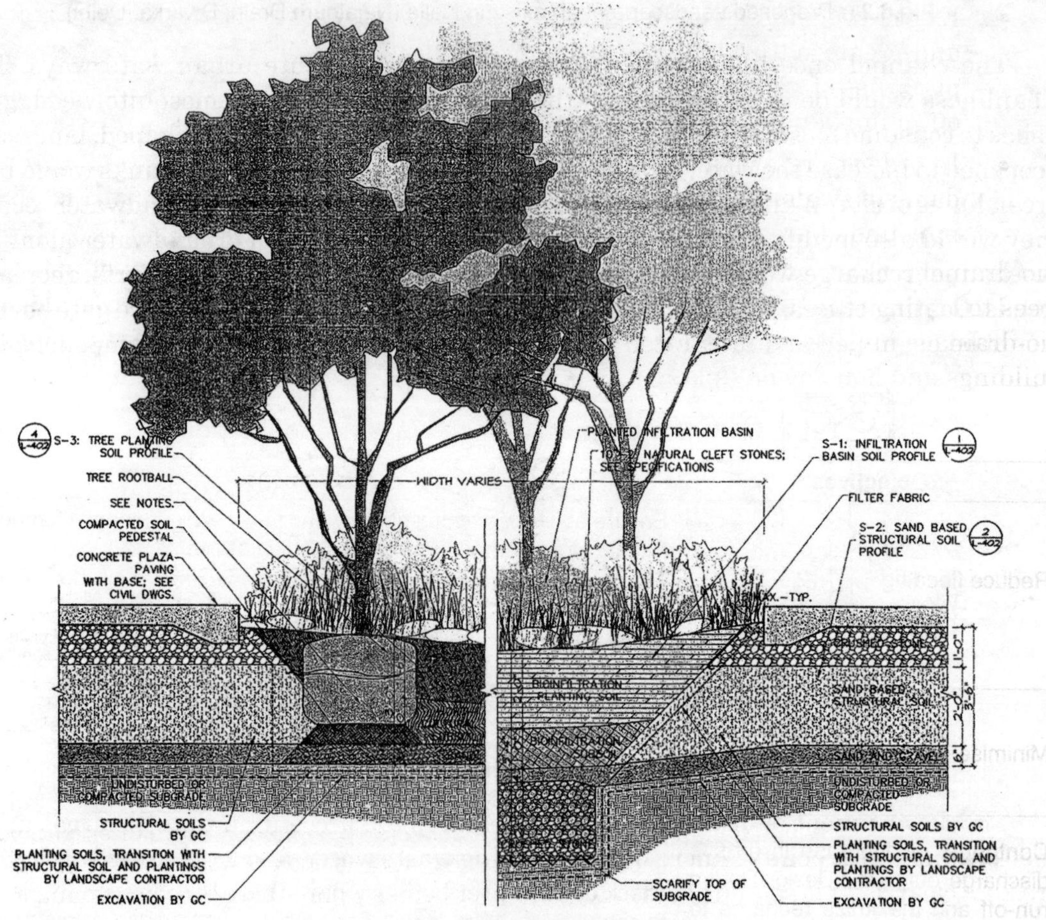

Fig.6.23: Infiltration Basin Detail

Source: HMWhite Site Architecture, Landscape Architecture Magazine, June 2014.

Concept of Bio-Drainage and Zero Run-off

Bio-drainage is an ecological option for controlling water logging, flooding and salinity, combined with recharging of ground water aquifer and plant physiology. Bio-drainage seeks adaptation of water channels for green corridors and siting of roads, buildings and housing with the following principles:

- To reduce flooding, and ensure ability of existing primary and secondary drainage channels to handle storm run-off by defining drainage easements.
- These drainage easements will be determined by the 25 year floodplain (an area, which had inundated anything during last 25 years). However, the vegetation easement of undisturbed forest and undergrowth must be respected.
- In order to minimize erosion and siltation, there must be a legal and enforceable prohibition on clearing of ground cover, shrub undergrowth or trees within drainage casements.
- Existing channels should be enhanced where necessary with berms (side barriers) and create natural swales (water holding depressions) by introducing layered plantings of native vegetation. It is necessary to provide adequate storage of run-off in impoundments or temporary water storage ponds.
- In order to retard run-off and to enhance recharge, it is proposed to build check dams in swales over permeable soils to enhance recharge. Install trickle tubes in impounded areas to permit even flow.

Bio-drainage as an environment-friendly option, is based on the property of plants and trees to transpire water and is used to remove excess water and salinity. In urban areas bio-drainage involves adaptation of water channels for green corridors and siting of roads, buildings and housing on uplands.

Table 6.11: Bio-drainage Objectives and Adaptations

Objectives	Adaptations
Reduce flooding	Ensure ability of existing primary and secondary drainage channels to handle storm run-off by defining drainage easements. These drainage easements will be determined by the 25 year floodplain, however, a minimum vegetation easement of undisturbed forest and understory must be respected: 90 m for primary drainage channels and 30 m for secondary drainage channels.
Minimise erosion and siltation	Prohibit clearing of ground cover, shrub understory, or trees within drainage easements. Enhance existing channels where necessary with berms and 'create' natural swales by introducing layered plantings of native vegetation.
Contribute no increase in off-site discharge during the Design Retard run-off and maximize recharge to even base flow of stream	Provide adequate storage of run-off generated by Design Storm in impoundments or the storm temporary water storage ponds. Use check dams in swales and on lots to slow flow over permeable soils to enhance recharge instail trickle tubes in impounded areas to permit even flow.

Source: Wallace, Roberts and Todd, 'Woodlands New Community, Guidelines for Site Planning Report', 1973 Philadelphia.

The factors that influence natural drainage including:

- the amount of sealed and unsealed ground to be drained;
- the type of soil with its rate of permeability; and
- the intensity and duration of typical rainfalls and severe storms.

In the early stages of design it is important to allow for enough land to be able to implement a natural drainage system that can absorb all rainwater falling on all sealed and unsealed ground on the site. This can occur on a small scale on individual building plots, or larger development sites where a sophisticated interconnected water system can provide a important amenity function as well.

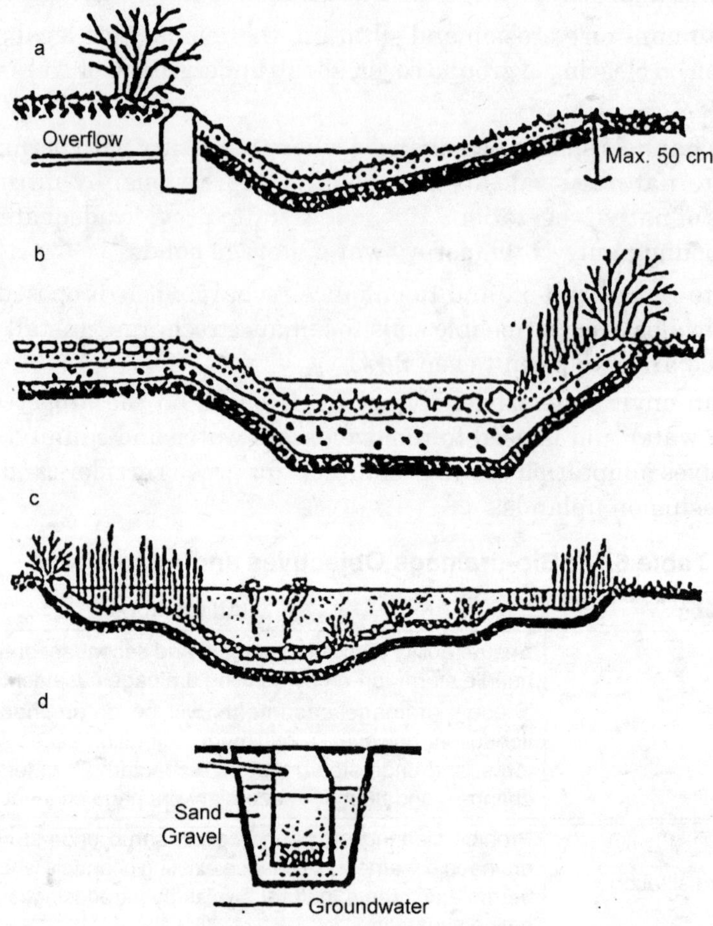

Fig.6.24: Different Retention System.

By segregation of rainwater from urban waste water disposal, it can be used for recharging of aquifers, irrigation, etc. Zero run-off drainage envisages complete retention of rainwater into the area by creating retention ponds, lakes and swales for flood/rainwater discharge. The present method of draining of storm water through masonry drains entails huge costs and effort, which also endangers the groundwater aquifer. The creation of water bodies in the low lying areas not only solves the drainage problem, but also provides a source of recreation, cattle rearing, fishing and improvement for the micro-climate.

The advantages of zero run-off and bio-drainage include the following:

- An efficient economical and effective drainage system which will increases ecological sustainability.
- Availability of self-sustaining greenery, recreational and water reserves for recreation, boating, fishing, etc.
- The enrichment of the aquifers, with continual replenishment.
- An economic and ecological process of development of the human settlements.
- A source of water for the lean periods, especially for horticulture and agriculture.

Retention Systems

The possible retention systems and their description are given below.

- Low area drain (soakaway, in landscaping terms): Water collects at the lowest part of the site. It is generally designed as an extension of the lawn where the soil can become highly saturated during rain. Visually the area under drain function like a lawn and can be used a green trail.
- Ditches and swales: There can also be grassed planted with water-loving plants and provide a habitat for animals. Ditches will form temporary water features after rainfall. They take up less space than a low area drain. They are permanent feature and cannot easily perform a different function when dry. Swales can form part of a surface water collection network that feeds into a pond.
- Permanent retention ponds: These are permanent water features that have a higher amenity values than the above option. Ponds require more space than ditches as water levels needs to fluctuate and more water is permanently retained on site. Retention ponds can be well supplemented with reed-bed filtration system prior to discharge into a local river. They score slightly higher than ditches and swales in environment terms as they combine aesthetic values with new habitat creation and filtration.
- Underground collection cistern: This is the most expensive option with the lowest environment rating and also limited capacity. Its strongest point is the minimal space required. Water is collection and slowly discharged into the groundwater through a sand and gravel pit. Filtration is therefore minimal water quality inferior to that of the other systems.

A comparative description of various retention systems is given below:

Table 6.12: Possible Run-off Retention Systems to Aid Natural Drainage

Retention	General description	Approx. land-take	Maximum Construction depth	Environmental	
				cost	rating
System Low area drain	Water collects at the lowest part of the site. It is generally designed as an extension of the lawn where the soil can become highly saturated during rain. After heavy rainfalls it can take up to three days for the water to drain away. Visually the area drain functions like a lawn and in summer can be used as such.	5-15%	0.5 m	Low	Medium High
Ditches and Swales	This can also just be grassed or planted with water-loving plants and laid out with rocks. The ecological value of ditches and swales is therefore higher since they provide a different habitat for animals. Ditches will form temporary water features after rainfall. They take up less space than a low area drain, as they can be a lot deeper - more than a metre. They are permanent features and cannot easily perform a different function when dry. Swales can form part of a surface water collection network that feeds into a pond.	2-5%	1.5 m	Low	High
Permanent retention ponds	These are permanent water features that have a higher amenity value than the above options. Ponds require more space than ditches as water levels need to fluctuate and more water is permanently retained on site. Retention ponds can be supplemented with reed-bed filtration systems prior to discharge into a local river. They score slightly higher than ditches and swales in environmental terms as they combine aesthetic values with new habitat creation and filtration.	3-5%	Varied	Medium	High
Underground collection cisterns	This is the most expensive option with the lowest environment rating and also limited capacity. Its strongest point is the minimal space required. Water is collected and slowly discharged into the groundwater through a sand and gravel pit. Filtration is therefore minimal and water quality inferior to that of the other systems.	>1%	N/A	High	Low

Source: Jain A.K. , 2007, Water- A Manual for Engineers, Architects, Planners and Managers, Daya Publishing House, New Delhi

Depending on the specific aims to be achieved in drainage and flood control, various retention systems offer different advantages. The land-take necessary (the site area required for natural drainage systems) depends on the depth of the drain, which in turn depends on the local soil. The approximate land-take is a percentage of the total site area to be drained and forms a drought guide. This enables the planning of natural drainage systems during the early stages of development. The exact land-take for a natural drainage system relates to the actual quantity of scaled land on site and the degree of impermeability of materials used.

The suitability and advantages of each system depends on the topography and the aims of the development in each case. The land-take necessary (the site area required for natural drainage systems) depends on the depth of the drain, which in turn depends on the local soil and slope. The approximate land-take is a percentage of the total site area to be drained and forms a rough guide, to enable the planning of natural drainage system. The exact land-take for a natural drainage system relates to the zoning of flood regimes, actual quantity of scaled land on site and the degree of impermeability, as given below:

Zone 1 : This is permanently submerged zone, which may be a water body, lake, river, flood zone, etc.

Zone II : In this zone the flow regime changes depending on both the downstream and upstream water levels. During periods of high water levels in receiving streams it is permanently drowned. During periods of low downstream levels water can flow with a free surface. Several intermediate conditions occur during storms;

Zone III : In this zone backwater effects exist only when high downstream water levels and high discharges coincide during heavy storms

Zone IV : This zone exists in the upstream parts of the network, such as at elevations high enough to avoid surcharging from downstream. The surcharging of pipes occurs in this zone only during peak flows.

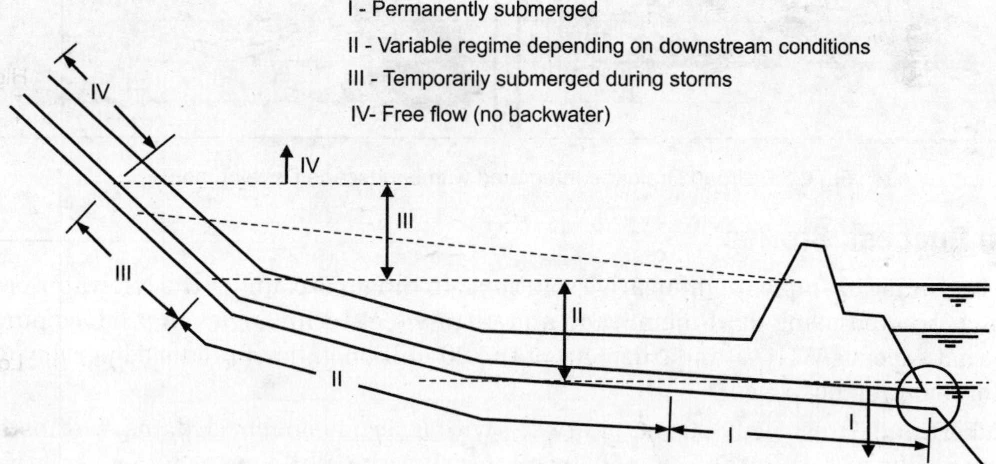

Fig.6.25: Zones in a Storm Sewer with the Different Flow Regimes.

Landscape Development of Drains

Retention ponds, swales, lakes, canals, water channels or nallas in urban areas provide continuous on-channel recharge of groundwater, and are also valuable landscape elements. Continuous recharge would take place through deep holes on the channel floor. This requires computer modelling of channel capacities, transfer mechanisms, storage capacities, flood prevention and system operation. However, storm water and recycled/filtered waste water can be supplement with the urban water needs, particularly for irrigation and horticulture, improve the micro-climate and landscape

The channel filled with water could constitute urban waterway. Their cleanliness should be maintained with the use of root zone system, green bridge, mosquito weed, guppy fishes to consume mosquito larvae, and aeration. The channel should be shaped, landscaped and deepened to increase the water storing capacity, wherever possible. Their banks can be the green lungs of city. While serving the function of on-channel groundwater recharge they also modify the micro-climate. The conjunctive use of groundwater along with on-channel recharge would also possible. Their use for water sports, recreation and waterborne transport could be possible. The existing dirty drains can be converted into attractive landscape trails and greenways, as have been developed in Pune (Oshodham Park Nallah) and at other places.

Fig.6.26: Urban Drainage Integrated with Landscape Development

Some Success Stories

There are ample examples of innovative solutions. In Singapore, under the NE water project sewage is treated using dual-membrane and ultra violet technologies to produce purified water that exceeds WHO standards. Out of the 20 million gallons produced per day 6% is used for drinking purpose.

In Portland, Oregon a 2330 feet bio-swale, with intermittent check dams, was installed to prevent silt and pollutants runoff from entering the Willamette River. As a result suspended solids including garbage entering the river system reduced by 50%. In Vienna, the 3-km long and 30m deep wastewater tunnel, Wiental-Kanal running under the river has an automatic distribution and pumping system that controls the discharge of waste-water, preventing pollution of the River Wine.

The UN Environment Programme's Integrated Wetland System uses examples from Kerala and West Bengal to prove that the biochemical reaction can utilize the sun's energy for recycling. Treated wastewater is routed into the wetlands for aquaculture and irrigation, generating food and employment.

The 'stream daylighting' project in Zurich revitalized 1.6 km of the city's streams, and 31% of the city's extra rainwater was diverted into them. The city has installed rot proof impact resistant polyester silt traps that capture increasing quantities of silt from surface runoffs and keep the drainage system clean. Ecologically engineered porous pavements that can suck in road runoff, drain blockers that seal manholes to prevent overflows, algorithms that predict surface run offs, computer simulations that calculate flooding patterns of cities and GIS technologies that map flood prone areas are being increasingly used to enhance urban drainage.

In Tokyo some 750 buildings are using RWH for fire-fighting and as drinking water. The 8,400 sq.m rooftop of the Ryogoku Kokugikan Sumo-wrestling Arena collects rainwater, drains it into a 1,000 cubic meter underground tank, which is then used for flushing and air conditioning. Tokyo has dug infiltration pits that make its runoff sink into the soil. Lyon in France has built gravel filled filter drains that prevent sediments from reaching recycling pumps. Ontario in Canada has constructed detention ponds near its rivers to reduce downstream flooding.

Mumbai has installed 30 automatic weather stations comprising state-of-the-art tipping bucket rain gauges at 26 locations to know exact rain patterns in real time. Bangalore Water Supply and sewage Board (BWSSB) is using a GIS application called BISON(Bangalore Info systems on Network) to manage and maintain the city's water drainage and wastewater networks. Kolkata is using a mathematical 3D flood simulation model to help map various flood situations. Calicut under MATURE (Mission for Application of Technology for Urban Renewal and Engineering) Programme, has undertaken an integrated development plan that aims to regenerate the city's ponds, and model its drainage and water facilities.

The Central Road Research Institute (CRRI) promotes the 'eco-phalt' road technology that uses a special ingredient called Drain Asphalt Modified Additive that keeps roads dry by sucking in the surface water. In 2006 a trail road paved with this technology has already been laid at Dhaula Kuan (National Highway 8).

Indian cities can learn from global best practices of integrated water and drainage management. UNESCO's Institute of Water Management cautions that the success of all such systems would hinge on the integrated planning and governance. The investments could be nullified by the environment costs of poor drainage, compounded by poor administrative and financial governance.

DEBRIEFING QUESTIONS

1. Explain the link between drainage and environment.
2. Explain the typology of floods and strategies for its control.
3. Explain sustainable urban drainage system.
4. Write short notes on:
   (a) Swales

(b) Soak-ways

(c) Retention Basins

(d) Infiltration Pond

5. Explain the process of drainage planning.
6. Explain the following:

(a) Single Stack System of Building Drainage

(b) Roof Drainage

(c) Separation of Storm Water and Waste Water

7. Explain the concept of Bio-drainage/Zero Run-off Drainage

7 Solid Waste Management

Solid waste Management, the collecting treating, and disposing of solid material that is discarded because it has served its purpose or is no longer useful. Improper disposal of Municipal solid waste (MSW) can create unsanitary conditions, and these in turn can lead to pollution of environment and to outbreaks of vector – borne diseases – that is diseases spread by rodents and insects.

The problem of solid waste management in Indian cities is assuming serious proportions due to increasing population, urbanization, changing lifestyle and consumption patterns. It is estimated that nearly half of the garbage is not collected which create hygienic and health problems. Out of this about one third is treated.

Fig.7.1: Types of urban wastes

The following table indicates waste generation per capita per day:

Table 7.1: Waste Generation Per Capita per Day

S.No.	Land use type	Estimated waste generation
1.	Residential refuse	0.3 to 0.6 kg/cap/day
2.	Commercial refuse	0.1 to 0.2 kg/cap/day
3.	Street sweepings	0.05 to 0.2kg/cap/day
4.	Institutional refuse	0.05 to 0.2 kg/cap/day

Source: Manual on Solid Waste Management, CPHEEO – (2000)

MUNICIPAL SOLID WASTE GENERATION IN INDIA

India is one of the mega underdeveloped countries of the world with respect to population status and dynamics. The growing rate of population is a direct factor responsible for enhancement of municipal solid waste generation.

According to the Press Information Bureau, India generates 62 million tonnes of waste (mixed waste containing both recyclable and non-recyclable waste) every year, with an annual growth rate of 4% (PIB 2016). 62 million tonnes annually averages out to about 450 grams of waste per person per day. However, there is a lot of variability in per capita waste (MSW) generation in India. Daily household municipal solid waste generation ranges from 170 grams per person per day in small towns to 620 grams per person per day in large cities.

Following is the list of top MSW generating cities in India 2016.

S.N.	City	Population in Millions	MSW Tonnes per day
1.	Delhi	19.1	9620
2.	Mumbai	20.0	8600
3.	Kolkata	14.7	6000
4.	Chennai	10.1	5000
5.	Bangluru	10.4	4200
6.	Hyderabad	9.1	4000
7.	Ahmedabad	7.5	2500
8.	Pune	5.8	2300
9.	Surat	5.8	1680
10.	Kampur	3.0	1500
11.	Lucknow	3.3	1200
12.	Nagpur	2.7	1000
13.	Jaipur	3.5	1000
14.	Ludhiana	1.7	850
15.	Indore	2.5	850
16.	Coimbatore	2.6	850
17.	Agra	2.0	790

* *Urban Agglomerations*

Sources : *State Pollution Control Boards, Municipal Corporations, U.N. Population estimates. ICRIER, New Delhi, 4th January 2019*

Waste Categorization

Waste Categorization is the process by which the composition of different waste sources are analyzed. Waste categorization plays an important part in any treatment of waste which is proposed.

Solid waste can be classified in the following categories:

(1) Municipal Solid Waste (MSW)

Municipal solid waste consists of household waste, construction and demolition debris, sanitation residue and commercial units waste. The term garbage is generally used to putrescible or bio-degradable food waste. Rubbish is used to denote non-putrescible or non-biodegradable materials – Both these (Garbage and Rubbish) together have kitchen food wastes, paper, cloth, plastic, glass, metals, masonry, packing material etc. It may also include ash, cinder, dead animals, abandoned vehicles, pipes, sludge etc. Latest addition to it being e-waste.

As per Ministry of Environmental and Forest (MoEF)," Municipal solid waste (MSW) includes commercial and residential waste generated in municipal or notified areas in either solid or semi-solid form excluding industrial hazardous wastes but including treated bio-medical wastes."

(2) Industrial Solid Waste (ISW)

In majority of cases it is termed as' hazardous waste' as they may contain toxic substances. Hazardous industrial waste may be ignitable, corrosive, reactive, or radioactive. All industrial waste is not hazardous.

(3) Bio-Medical Waste or Hospital Waste

It is any kind of waste containing infectious (or potentially infectious) materials. This may include Sharps, soiled waste, disposables, anatomical waste, cultures, discarded medicines, chemical wastes etc., usually in the form of disposable syringes, swaps, bandages, human excreta etc. Obviously, these can be serious threat to human health, if not properly treated before disposal.

(4) Agricultural Waste

The waste generated during various processes of crop production and livestock rearing is called agricultural waste. This includes Plant Remains like husk, straw, cotton and tobacco waste, coconut waste, nutshells; Animal waste includes animal excreta (Gobar) and Processing waste as food crops are processed to extract grain. In this category huge amount of waste is generated particularly in villages.

(5) Nuclear Waste

Radioactive waste from the nuclear power and weapon industries has recently become a matter of great concern. The most important is the spent nuclear fuel, which remains active even for hundreds of years. Parts of cell phones, computers etc are also source of nuclear waste.

Sometimes radioactive leak and accidents in nuclear plants or willful explosion of nuclear devices cause greatest of hazards.

Critical Issues in SWM

- Storage of waste at source is lacking
- Segregation of recyclable waste is not done at source

- Primary collection of waste is not done at place of generation
- Design and location of municipal waste storage depots are inappropriate, resulting in littering of garbage
- Street sweeping is not done everyday
- Waste transportation is done in open vehicles
- Waste processing is partially practised in large number of cities
- Final disposal is done through crude dumping
- Inadequate arrangement of municipal bins/dump sites cause insanitary conditions

Management Issues

- Lack of land for landfills, SWM plants, waste recycling and procesing facilities
- Lack of planning for waste management
- Lack of proper institutional set-up for waste management, and its planning in urban local bodies
- Lack of technically trained manpower
- Lack of community involvement
- Lack of expertise and exposure to city waste management using modern techniques/ best practices
- Lack of awareness generation mechanism
- Lack of Management Information Systems
- Lack of funds and user charges

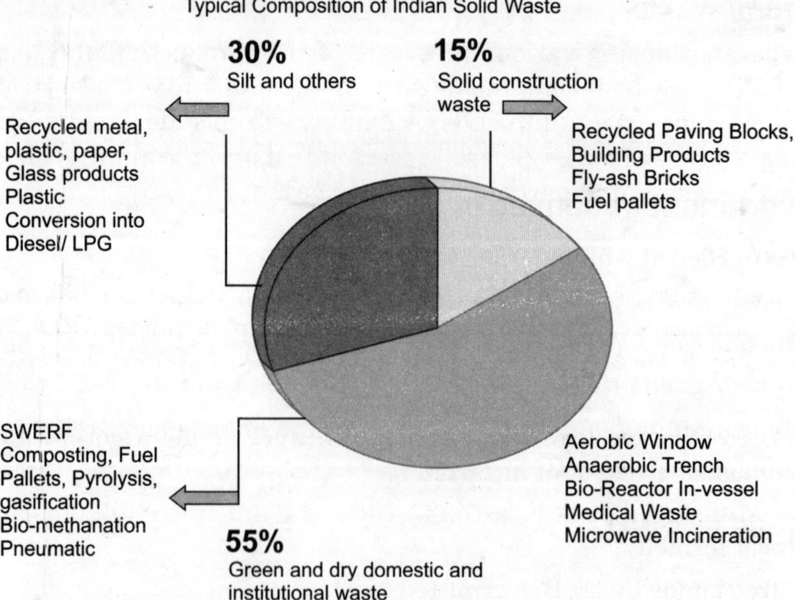

Fig.7.2: Municipal waste composition

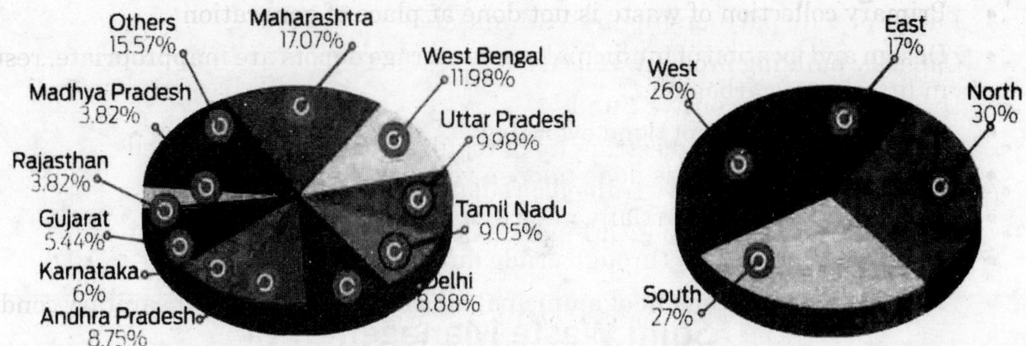

Fig.7.3: State-wise maharashtra generates 17 per cent of total solid waste volume, while west india generates nearly one-fourth of waste volume

Source: Annepu, Ranjith (2014), Efficient Handling for Cleaner Tomorrow. Urban Update, May

Fig.7.4: Solid waste land fill is an environment disaster and a harbour of pathogen and virus

Some Government Initiatives in SWM

- Swachh Bharat Abhiyan (2014-19)
- Bio-medical Waste Handling Rules, 1998 notified
- Municipal Solid Waste Management Rules, 2000 notified
- Reforms Agenda (fiscal, institutional, legal) cleared
- Technical Manual on Municipal Solid Waste Management released
- Technology Advisory Group on Municipal Solid Waste Management formed
- Inter-Ministerial Task Force on Integrated Plant Nutrient Management from city compost formed
- Tax-free bonds by ULBs permitted by government
- Income-tax relief granted to waste management agencies

- Public-private partnership in SWM initiated
- Capacity building brought in focus
- Urban Reforms Incentive Fund
- Guidelines for PSP and setting up of regulatiry authority finalised
- Model municipal by-laws framed/circulated for benefit of ULBs
- Financial Assistance by 12th Finance Commission

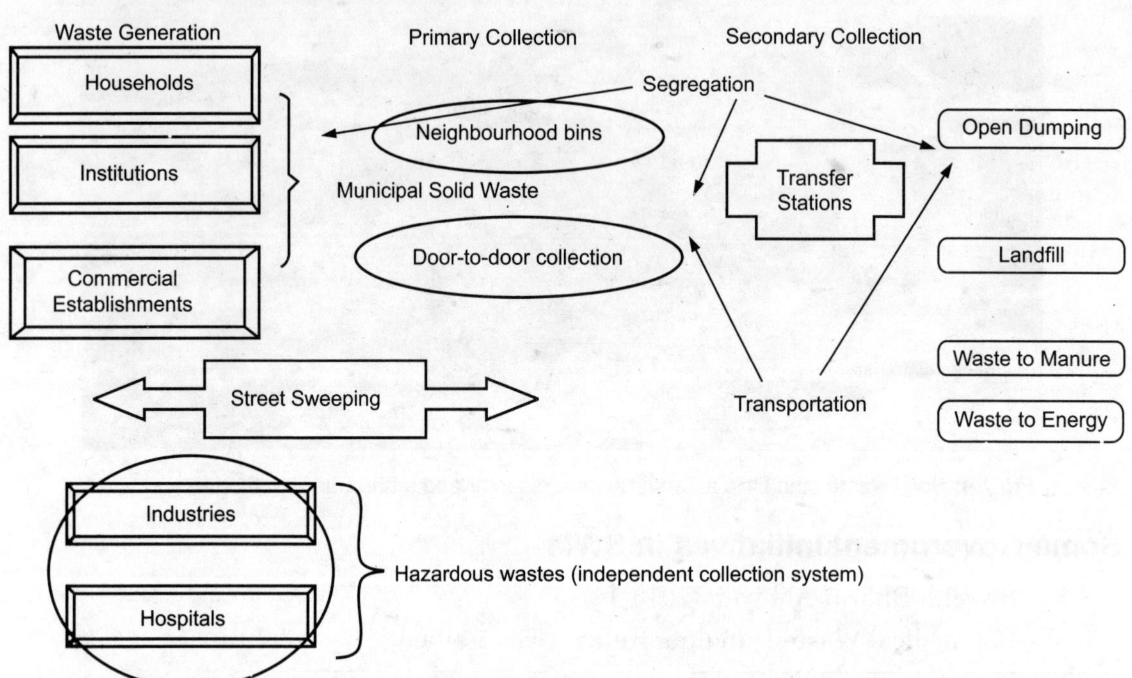

Fig.7.5: Municipal solid waste management: collection (60 per cent) and treatment (30 per cent)

Fig.7.6: The Swachh Bharat Abhiyaan (clean India campaign) was launched on 2nd October, 2014, on Gandhi Jayanti at Rajghat, New Delhi. The campaign includes 4041 statutory towns for the cleaning of the roads and infrastructure. The primary goal of the campaign is to achieve the vision of a 'Clean India' by october 2019 which will mark the 150th birth anniversary of Mahatma Gandhi. The urban development ministry has earmarked rs. 62,009 Crores and the ministry of drinking water and sanitation will spend Rs.1,34,000 Crores for the rural programme.

the Ministry of Environment and Forests notified the "Municipal Solid Waste (Management & Handling) Rules, 2000", that require the following compliances:

- Development and Improvement of properly engineered and constructed sanitary landfill (SLF), from which pollutants do not escape and having an impervious bottom layer for leachate collection and treatment
- Identification of landfill sites for future use and making site(s) ready for operation.
- Setting up of waste processing and disposal facilities.
- Monitoring the performance of waste processing and disposal facilities once in six months.
- Segregation of solid waste at point of generation, which includes organic, recycled, (plastic, glass, metal, paper, etc.).
- Specialised handling of construction and demolition waste, bio-medical waste, industrial/hazardous/toxic waste, and e-waste (electrical and electronic waste).

Inert material in the municipal solid waste has been steadily increasing (30-50%), a majority of which is generated from the demolition and construction activities. This needs a specialized storage, collection and transportation. Organic food wastes can be locally recycled and composed, which would reduce transportation cost and reduce landfill site demand. Bio-medical waste is generated in relatively smaller quantity in comparison to municipal solid waste but due to its infectious and toxic nature, it needs careful handling which is regulated by a Bio-medical Waste (Management and Handling) Rules, 1998.

Storage of waste at source is the first essential step of Solid Waste Management. Every household, shop and establishment generates solid waste on day?to?day basis. The waste should normally be stored at the source of waste generation till collected for its disposal. Biodegradable waste and non?bio?degradable waste should be collected in separate bins from the source.

- Green coloured bins – Waste bins for biodegradable waste
- White coloured bins – storage of recyclable wastes
- Black coloured bins – storage of other wastes

Primary collection of waste is the second essential step of Solid Waste Management activity. Primary collection system is necessary to ensure that waste stored at source is collected regularly and it is not disposed of on the streets, drains, water bodies, etc.
Local bodies should arrange for the primary collection of waste stored at various sources of waste generation by any of the following methods or combination of more than one method:

- Doorstep collection of waste through non-motorised and motorised vehicles with active community participation.
- Collection through community bins
- Doorstep or lane-wise collection of waste from authorised/unauthorised slums or collection from the community bins to be provided in the slums by local bodies

There is a strong opposition to location of new landfills, which are seen as an environmental disaster, causing health risk from flies and rodents; air pollution and water pollution through leaching. The Govt. of India recommends that:

"Open dumping of the solid waste should be discouraged, if necessary, by legislative action to be enforced in a phased manner starting with the metropolitan cities. Recycling and resource recovery should be encouraged. It should be made mandatory by the Governments that the organic waste should be used for resource recovery and it should not be disposed by the land fill method. Proper regulations should be framed to regulate the land fill site and its maintenance."

A study conducted by the World Health Organisation titled "Management of Solid Wastes in Developing Countries" concluded that to achieve the lowest cost, a policy of encouraging multiple small composting plants should be followed. Various studies conclude that recycling wastes is by far the cheapest alternative to landfills; and plants that convert garbage into other products are cheaper than those that burn garbage. As such, it may be more practical to build several small, decentralised waste facilities rather than large one.

The Role of the Rag Pickers

The informal rag pickers and kabaris help to keep the city clean. In Delhi, they recycle about 3000 mt of waste per day. However, they face serious problems of health and essential facilities. In Bagota incentives like free bus ticket, food coupons and cinema tickets are given to waste pickers for collecting and depositing waste in the designated municipal waste collection centres. Learning from Colombia and Brazil where the rag pickers are officially recognized, Pune Municipal Corporation has also brought them in municipal SWM network and provides them with remuneration, uniforms, waste bins and bags, gloves, pushcarts and health insurance. In US cities, like Chicago, property tax incentives are given for reduced waste disposal.

Transportation of the waste

Transportation of the waste stored at waste storage depots at regular intervals is essential to ensure that no garbage bins/containers overflow and waste does not litter on the streets. Hygienic conditions can be maintained in cities/towns only if regular clearance of waste from temporary waste storage depots is ensured. Transportation system has to be so designed that it is efficient, yet cost effective. The system should synchronize with the system of waste storage which is easily maintainable.

Solid Waste Treatment Technologies

There are various technologies available for treatment and processing of waste in an environmentally sound manner. However, a technology suitable for one may not be appropriate for others. The comparison of these technologies is shown in table below:

Table 7.2: Comparison of Different Solid Waste Treatment Technologies

Element	Compositing	Refuse derived fuel	Biomethanation	Gasification/ Pyrolysis	Incineration
Technically and economically feasible size of operation per day fresh waste	50 TPD and above	100 TPD and above	1 TPD at small scale and above 50 TPD at larger scales of pure organic waste	500 TPD and above. Due to high moisture in our waste, suitable only for segregated dry waste.	500 TPD and above due to high moisture in our waste. Suitable only for segregated waste. However, sizes as small as 10-50 TPD of waste are available for commercial sale but not advisable due to high miming costs.
Adopted Capacity for study	500 TPD	500 TPD	500 TPD	500 TPD Plant	500 TPD
Land required for adopted capacities	6 Ha	3 Ha	4 Ha	10 Ha	4 Ha
Waste Characteristics	Moisture Content >50% Organic Matter >40% C/N Ratio between 25-30	Moisture Content <45% Volatile Matter >40%	Moisture Content >50% Organic Matter >40% C/N Ratio between 25-30	Moisture content <45% Net Calorific Value >1200 kcal/Kg	Moisture Content <45% Net Calorific Value >1200 kcal/kg

Waste Suitability	Suitable for MSW Characteristics of India	Not suitable for MSW characteristics in India but workable with use of Auxiliary Fuel	Suitable for MSW characteristics of organic waste in India	Not suitable for MSW characteristic in India but workable with use of Auxiliary Fuel	Not suitable, due to high moisture in our waste.
Typical investment for assumed capacities (excluding cost of land)	INR 17-20 Cr. For a 500 TPD Plant	INR 17-20 Cr. For a 500 TPD Plant	Approximately INR 75-80 Cr for a 500 TPD Plant	INR 80-90 Cr. For S00 TPD Plant	NA
Recuning cost	INR 300 per ton of input waste	INR 290 per ton of input waste	INR 100 per ton input waste	NA	–
Recoverable	250 kgs of compost per ton of waste	200 kgs pellets perton of waste	80 cum of bio gas / ton of waste plus 200 kgs of manure / ton	NA	–
Volume reduction	45-55%	55-65%	55-65%	>80%	>80%
Environmental issue	Impurities in compose due to mixed waste, traces of heavy metals, leachate runoff	Problems in burning exhaust	Problems if mixed feed stock	Ash handling and Air Pollution	Ash handling and Air Pollution (emission of particular matter, chlorinated compounds dioxins / furans)
Technology Reliability	Running successfully in India	Running successfully in integrated facilities	Small scale organic treatment plant operational but mixed waste large scale plants failed in India	Insufficient operational experience for MSW	Only Plant in India failed due to mismatch in waste quality. MSW 2000 has recommended for incineration of waste only after doing a waste suitability analysis and proriding adequate flue gas management methods.

Limitation	Large Land Requirement, Non acceptance of compost as soil enrichener in some areas of the Country Process depends highly on factors such as waste quality & climatic conditions	Fluff / Pellets can be used a fuel in large industries, e.g. In cement kilns with necessary permissions from the PCBs and required pollution control measures.	The technology requires pre-segregated homogenous biodegradable waste as mixed waste retards efficiency of the process. Hence applicability is limited to highly organic and homogenous waste streams like market wastes.	Requires waste with high calorific value. Expensive flue gas remediation methods to attain achievable outputs.	Expensive technology, waste criteria must have low moisture content and high calorific value, which is not found in Indian Waste. Costly flue gas remediation methods to attain achievable outputs.

Source: Toolkit for Solid Waste Management Jawaharlal Nehru National Urban Renewal Mission, MoUD.

Waste after treatment must be disposed in a manner that does not create any instance of environmental pollution and public nuisance. The MSW Rule 2000 defines waste disposal as an activity, which involves "final disposal of municipal solid wastes in terms of the specified measures to prevent contamination of ground-water, surface water and ambient air quality".

The landfill design shall be aimed to minimize the following:

- The ingress of water into the landfill,
- The production of leachate, its subsequent outflow and uncontrolled dispersions into surrounding aquatic environment,
- The accumulation, migration and uncontrolled release of landfill gas into the atmosphere.

Solid Waste Energy and Recycling Facility (SWERF)

A sustainable and feasible solution for the management of municipal solid waste is setting up a Solid Waste Energy and Recycling Facility (SWERF). SWERF reduces the need for future acquisition of land for landfills by 90 per cent. Odours and health risks are reduced as the SWERF is totally enclosed. There is sterilized handling of recyclables and rejects as the municipal solid waste is first processed in an autoclave. Greenhouse gases are reduced by efficient conversion technology. The SWERF system has the following features:

- Waste Reception, that includes the rag-pickers and kabaris in the loop
- Autoclaving of waste
- Mechanical source separation of recyclables
- Organic pulp washing, drying and storage
- Gasification of organic pulp to create syngas
- Electricity generation from syngas

The existing technologies that are being used by SWERF include:

- Split bin collection for municipal wastes.
- Sorting and processing technologies utilized in existing materials recovery facilities.
- Value adding process for glass, plastics, metals and paper, in which rag pickers and kabaris have an important role.

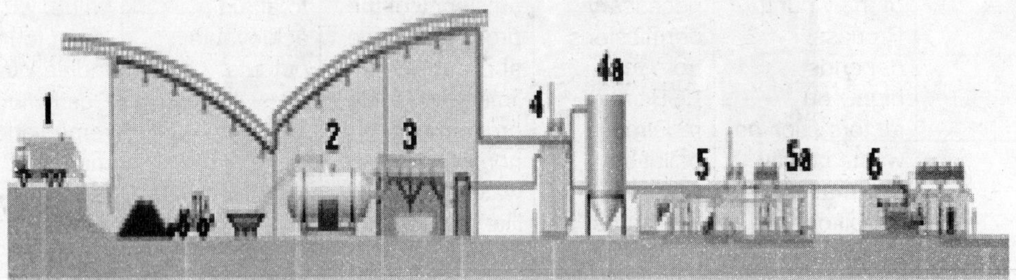

1. Household waste delivered
2. Waste sterilisedin autoclave
3. Waste separated into organic pulp and recyclables
4. Organic pulp washed and dried

4a. Pulp storage

5. Organic pulp converted to syngas

5a. Syngas cleaning

6. Electricity generated from syngas

Fig.7.7: Solid Waste and Energy Recycling Facility (SWERF) Technology

Source: http://www.sovereignty.org.uk/features/eco/swerf.html

Fig.7.8: Waste pre-treatment and separation (autoclave): waste pre-treatment involves the receival of the waste which is then sterilised with steam in an autoclave (with heat and pressure). This process is like cooking the waste in a pressure cooker

source: http://www.sovereignty.org.uk/features/eco/swerf.htm

Fig.7.9: Advanced Thermal Conversion (Gasification Module): In the SWERF® process, the washed and dried organic pulp is fed into a sealed metal tube and heated from the outside at a high temperature. This converts the dry pulp into energy-rich synthetic fuels and a solid residue. The solid residue is inert and can be landfilled, further processed or used in other applications. The synthetic fuels are cleaned in preparation for the use in the power generation equipment.

source: http://www.sovereignty.org.uk/features/eco/swerf.html

Fig.7.10: Electricity generation from waste (generator module)

Source: http://www.sovereignty.org.uk/features/eco/swerf.html

Fig.7.11: Biofertilizer (autoclaved green waste)

Source: http://www.sovereignty.org.uk/features/eco/swerf.html

Materials Recovery Facility

A Materials Recovery Facility at the solid waste dump site can take in source-separated recyclables, such as newspapers, metals and bottles and recycle these in the traditional methods. The benefit of integrating this is that any materials which cannot be recycled, like dirty paper are used to produce electricity instead of going to landfill.

Production of organic-based fertiliser and soil amendment products represent an exciting and incremental extension to the solid waste treatment technology. Historically, it has been cost prohibitive to produce large quantities of high-grade organic-based fertiliser and soil amendment products. This problem is addressed with integration and utilisation of the solid waste treatment infrastructure. The green waste is sterilised, separated, shredded, pelletised and dried. Ash and grit fractions from the treatment process can be added to the product to enhance mineral content and improve drainage. The process yields a premium mulch product with commercial applications and is free of pathogens, parasites and fugitive seeds.

Waste Recycling and Segregation

Recycling of waste can be effectively achieved by segregation at the source, reducing the waste by more than 50 per cent. The waste can be converted into a reusable commodity through advance processes at the recycling centers. Recycling services require segregation of waste materials into different types. The pyramid of most desirable option begins with reducing waste and then proceeds to reuse, recycle and recover. Disposal of waste in landfill is least desirable option.

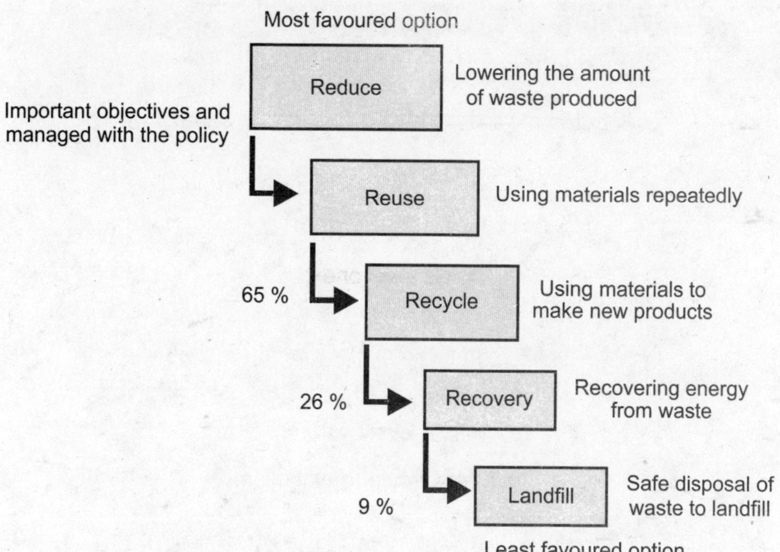

Fig.7.12: Danish waste hierarchy: 65 per cent waste is recycled, 26 per cent is incinerated and 9 per cent is landfill. Incineration does not count as recycling.

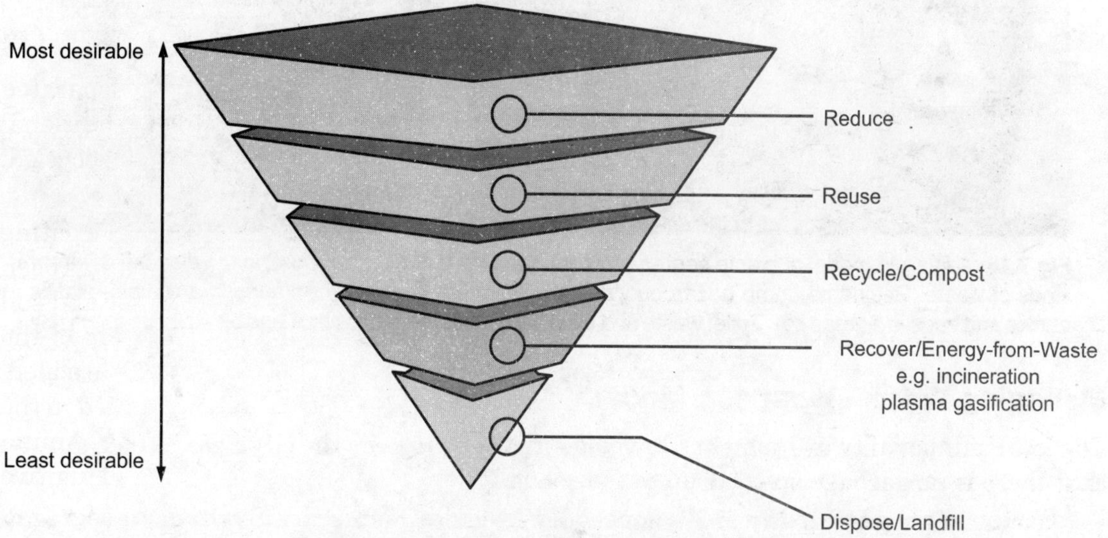

Fig.7.13: Inverted pyramid of waste management

Source: Ottawa.ca

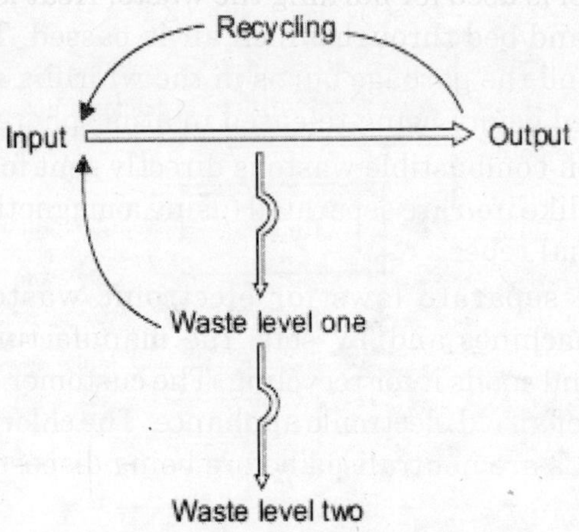

Fig.7.14: Circular metabolism

Fig.7.15: A five-bin model of waste segregation: the families use different dustbins to dispose different kinds of waste. Residents dump hazardous waste like batteries and razor blades in a red bin, plastic bottles and bags in a green bin. Toilet waste and wet kitchen waste go into black and blue bins respectively

Managing Waste-Japanese Model

The local municipality in Japan has invoked strict laws to keep the city clean, which ensures that there is no garbage spitting on to the roads.

Strict segregation laws: At household level, residents are required to segregate garbage into four types-combustible (kitchen waste), non-combustible (plastic), resources trash (newspapers, magazines) and large-sized trash (furniture, home appliances).

Garbage Collection: The municipal authorities have designated specific days in a week to collect different types of trash. For example, if the days specified for collecting non-combustible waste are Monday and Thursday, then residents would throw out only combustible waste on those days. To get rid of other trash, they would have to wait for the specified day.

Treating waste: Instead of dumping the waste directly into landfill sites, the waste is processed, treated and then dumped into a landfill.

Incineration: No fuel is used for burning the waste. Heat is generated by friction, the incineration unit has a sand bed through which air is passed. The air causes the sand to whirl like boiling water and the garbage burns in the whirling sand at 600° celsius. Toxic gases produced are treated before being released in atmosphere.

Recycling: While non-combustible waste is directly sent for recycling, in case of bulk waste, valuable resources like iron are separated (using a magnetic belt, etc) and compressed and then sent for industrial reuse.

e-waste: There are separate laws for electronic waste. In case of computers, refrigerators, washing machines and TV sets, the manufacturing company collects the appliance from the user and sends it for recycling. The customer has to shell out a recycling fee whenever he disposes of an old electronic appliance. The chlorofluorocarbon (CFC) gases from refrigerators and ACs are neutralised before being disposed of.

Source: *Hindustan Times, July 24, 2006*

Fig.7.16: Segregation bins for solid waste

Waste to Energy (W2E)

By adopting scientific methods, energy can be recovered from waste. As the organic matter decays it produces biogas, known as landfill gas (LFG), landfills have to be managed so that the LFG does not pose a threat to the surroundings. Rather than discharging it to the atmosphere, LFG can be recovered for providing a useful energy. Indian municipal waste has a potential to generate 150-250 cu. mtr. of biogas/LFG per ton depending upon the quality of wastes. By employing a set of generator and transformer, bio-gas can be converted in electrical energy.

The task force of the Planning Commission, headed by K. Kasturirangan proposed a target of setting up 215 Waste to Energy plants by 2031 to generate 1,075 MW of power and setting these through public-private partnerships (PPPs) with viability gap funding of up to 40 percent. The report stresses for an integrated approach towards municipal solid waste management, and the need for segregation of waste at source with private sector help. Since the urban local bodies (ULBs) lack the financial and institutional capacity necessary for integrated management of municipal solid waste, which requires investments, especially for 'Waste to Energy' projects, the report suggests to transfer the commercial risks to the private sector in order to ensure an efficient system for collection, transportation and processing of waste for generation of electricity. It is proposed that 'Waste to Energy' projects can be set up in cities with population above two million, generating more than 300 tonnes per day or more of combustible waste.

The report proposes exemption from corporate income-tax for the first five years, immediate refund of value-added tax and a feed-in tariff, which means higher price for waste management and energy, among other things.

As a pilot project, the Delhi Energy Development Authority since 1986 has been extracting gas from a sanitary landfill in Timarpur. Eight wells were bored into this 80 acre dump, 20 to 35 feet deep, have been yielding 33 cubic feet of gas a minute. This can be used either for cooking or to power a 30 KVA generator, using diesel or a mixture of diesel and gas as fuel. The gas, similar to bio-gas, is 55 to 60 percent methane, with a calorific value of 550 British Thermal Units and is supplied to 20 outlets in Balak Ram Hospital twice a day for six hours. The gas has not been used to generate electricity over a length of time though the provision is there and the necessary electrical connections to the hospital have been made. The generator consumes six litres of fuel an hour but when mixed with gas a saving of 4.5 litre of the fuel is affected. To extract the gas, a four inch diameter perforated pipe is sunk. Another perforated pipe, two inches in diameter, is lowered into it and the surrounding area is filled with bricks to prevent decaying matter from clogging the pores. In Delhi new W2E plants have invited strong opposition from the neighbouring localities due to pollution, soot, odour and garbage processing. This needs improvement in plant technology and waste transport, as well as locating such plants outside the residential zone.

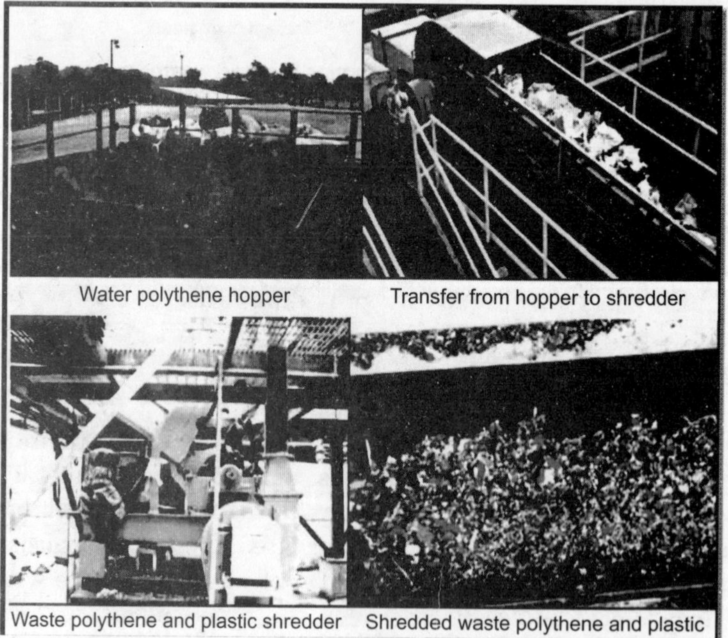
Water polythene hopper

Transfer from hopper to shredder

Waste polythene and plastic shredder

Shredded waste polythene and plastic

Fig.7.17: Converting polythene waste into gasoline and diesel

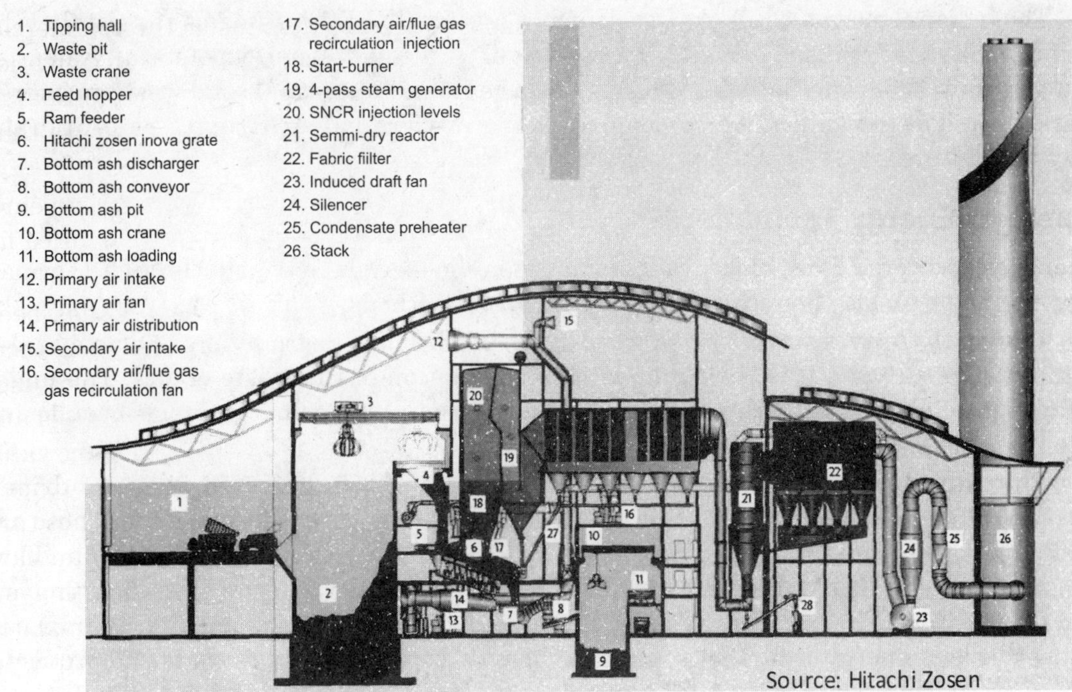

Fig.7.18: Waste to energy plant

Source: Hitachi Zosen

In Waste to Energy System, prior to the collection of solid waste, recyclables are sorted out and retrieved for processing to prolong their life-spans. The solid waste that remains is then collected and sent to the various waste-to-energy plants for incineration. Incineration reduces the volume of solid waste by about 90% and produces steam that runs turbine generators to generate electricity. The incinerated ash and other non-incinerable wastes are then transported to the Transfer Station and then to the landfill. Licensed collection vehicles deliver incinerable solid waste to the plants. To prevent odours from escaping into the environment, the air in the refuse bunker is kept below atmospheric pressure. High-capacity rotary crushers reduce the size of bulky solid waste to improve its burning efficiency. The solid waste is then fed into the incinerator by a grab crane. As the incinerator is heated to temperatures between 800 and 1,000 degrees Celsius, a lining of silicon carbide tiles protects the incinerator walls from the extreme heat and corrosion. Each load of solid waste is reduced to about 10% of its original volume in about five hours. Catalytic fabric filter systems and two-zone electrostatic precipitators remove pollutants from the flue gas before it is released from the plant while ash is collected and the ferrous material is removed to be sold and recycled.

Jindal Ecopolis, which is operating the new plant in Okhla, claims to be using better technology but the problem remains that waste is not segregated. The Okhla plant is in the middle of a residential area, which of inviting stiff resistance from the residents, who suspect that if waste is not segregated, the plant is burning everything, from plastic to chemicals. For 16MW of power, it is too high a price to pay.

The Indian Renewable Energy Development Agency (IREDA) estimates suggest that the country has so far realised only about 2 percent of its waste-to-energy potential. A market analysis from Frost and Sullivan predicts that the Indian municipal solid waste-to-energy market could be growing at a compound annual growth rate (CAGR) of 0.7 percent in the very near future.

Waste-to-Energy Technologies

Energy is recovered from biodegradable and non-biodegradable waste through thermal, thermo-chemical, bio-chemical and electrochemical methods.

Thermal Conversion: The process involves thermal degradation of waste under high temperature and it is then that complete oxidation of the waste occurs. The major technological option under this category is incineration. But incineration has been losing attention because of its emission characteristics.

Thermo-chemical conversion: This process entails high temperature-driven decomposition of organic matter to produce either heat energy or fuel oil or gas. These are useful for wastes containing high percentage of organic non-biodegradable matter and low moisture content. The main technological options under this category include Pyrolysis and Gasification. The products of these processes (producer gas, exhaust gases, etc.) can be used as heat energy or further processed chemically to produce a range of end products.

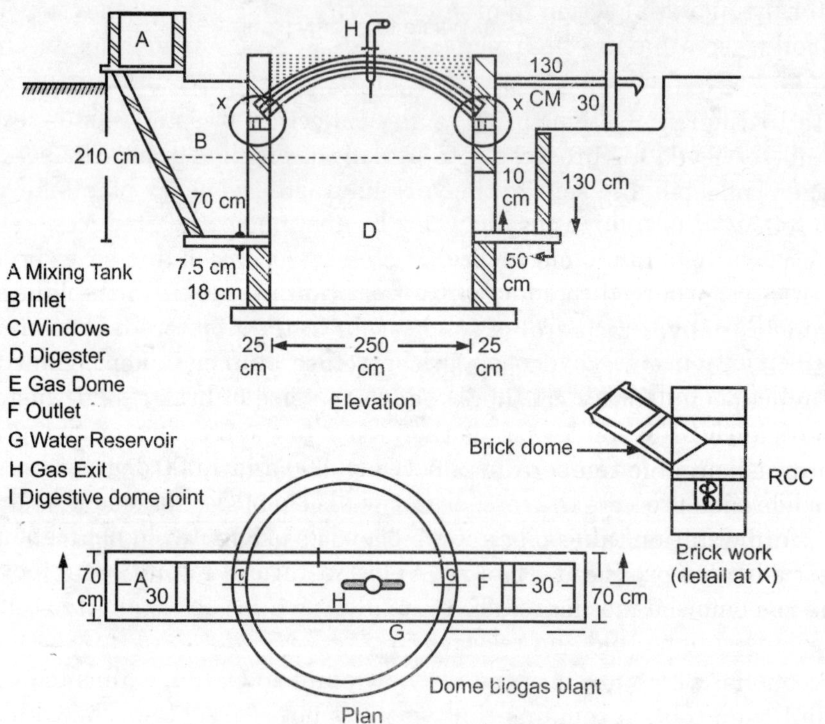

Fig.7.19: Waste producing gas

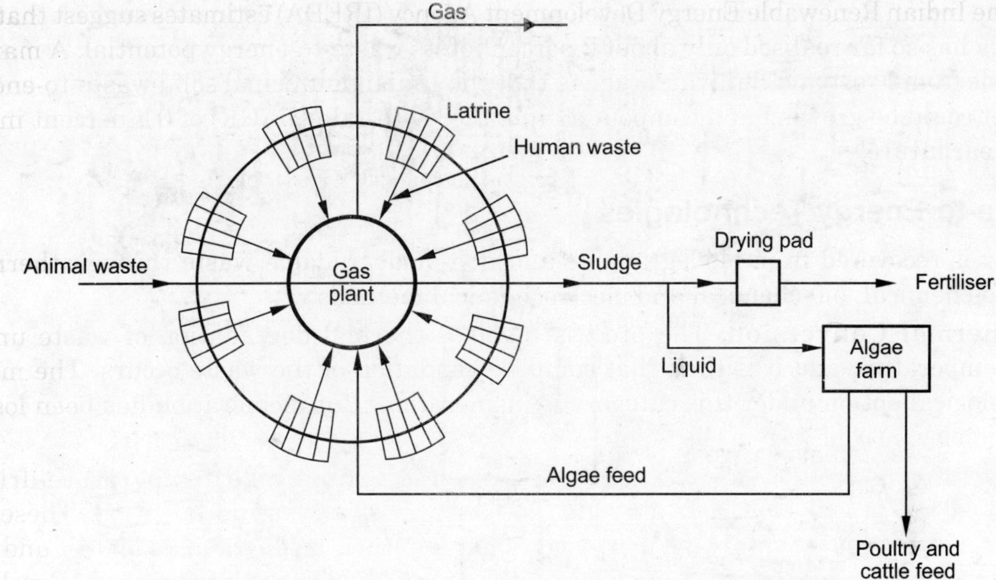

Fig.7.20: Bio-gas generation from human and animal waste

bio-chemical conversion: This process is based on enzymatic decomposition of organic matter by microbial action to produce methane gas, and alcohol, etc. This process is preferred for wastes having high percentage of organic, bio-degradable (putrescible) matter and high level of moisture/water content, which aids microbial activity. The major technological options under this category are anaerobic digestion (bio-methanation) and fermentation. Of the two, anaerobic digestion is the most frequently used method for waste to energy generation, while fermentation is still emerging.

Electrochemical conversion: Electrochemical conversion in the context of waste to energy refers typically to microbial fuel cells (MFC). These systems are developed to trap energy from wastes, where the reduction-oxidation machinery of immobilised microbial cells is catalytically exploited, for the accelerated transfer of electrons from organic wastes to generate electricity and bio-hydrogen gas. However, this methodology needs extensive evaluation studies on bulk scale liquid waste treatments and stands at a nascent level in India as well as worldwide.

In this process organic matter of solid waste is segregated and thereafter it is fed directly into a bioreactor, where in presence of methanogenic bacteria, and under an aerobic condition, the fermentation takes place and biogas is produced. In addition, a high quality organic manure is also produced. Energy can be recovered by digesting certain organic wastes and recovering the methane rich bio-gas which provides heat and electricity.

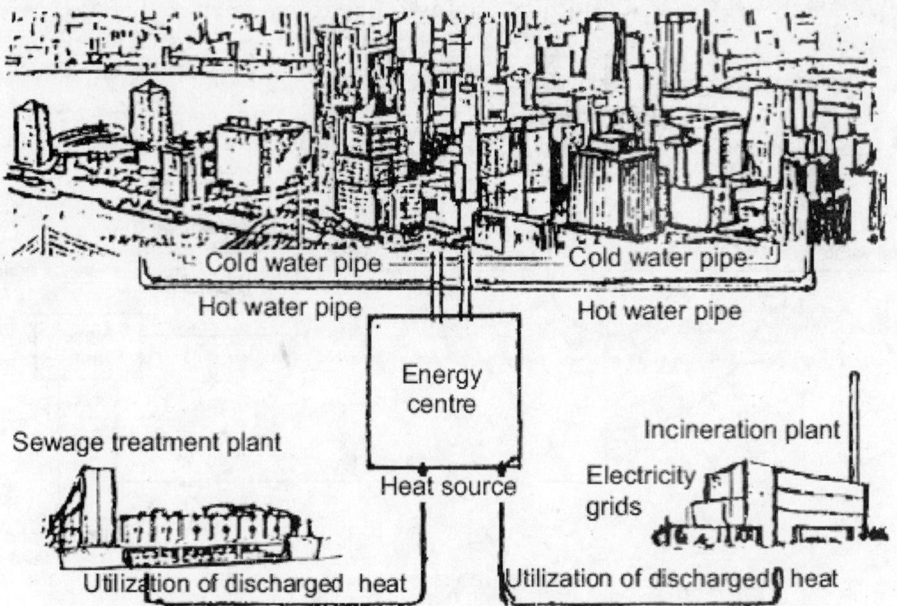

Fig.7.21: Energy centre as a link between waste treatment and the city

CONSUMER PLASTICS—USES AND FATE

Most packaging and products in the waste stream are made of a small group of commodity plastics, including polyethylene (PE), polypropylene (PP), polyethylene terephthalate (PET), polyvinyl chloride (PVC), polystyrene (PS) and polyamide (PA), better known as nylon. These plastics have different properties, reflecting their intended uses. Their different properties may afect their durability and fate in the water. For example, PE and PP are less dense than seawater and will tend to be buoyant, whereas PS, PA and PET are denser and will tend to sink. All of these plastics can be recovered and recycled if there is appropriate infrastructure and willingness on the part of the public. Collecting and recycling mixed types of plastic remains a challenge, although separation based on density difference can be effective. Consumer plastic objects often find their way to the drains, rivers and water bodies through a combination of poor waste management practices, inadequate policies and regulation, ineffective enforcement, and the attitude and behaviour of individuals.

The major drivers of plastic use are improved physical or chemical properties compared with alternative; low cost; mass production capability; and a reduction in the use of resources. Moreover, life-cycle analysis has shown that using plastic, rather than alternatives, often results in significant reductions in energy consumption and greenhouse gas emissions in applications ranging from food containers to vehicles and aircraft.

The applications of plastics in consumer products are many and varied. There are also significant regional differences in their use and disposal. Polyethylene bags are commonly used in West Africa to provide safe drinking water, but they often end up in water bodies due to a lack of waste disposal facilities. In Europe, approximately 30 per cent of plastics are used for disposable packaging. Quantitative data for many countries are difficult to obtain, particularly on the use and fate of single-use items such as bottles, carrier bags and food packaging.

Use of plastic materials reached approximately 100 kg per year per capita in North America and Western Europe in 2005 and is expected to increase to 140 kg by 2015. Rapidly developing Asian countries constitute the world's largest potential growth area, with current use of around 20 kg plastic per year per person estimated to increase to 36 kg by 2015. Rates of plastic recycling and re-use vary greatly, even within developed regions. For example, in 2009 more than 84 per cent of used plastics were recovered—that is, recyclied or reused for energy generation—in seven EU countries, as well as in Norway and Switzerland, Several European countries recovered only 25 per cent of less. Improving waste management operation is an often overlooked opportunity for innovation and job creation, especially in developing countries, where only a small percentage of the plastics produced are recovered.

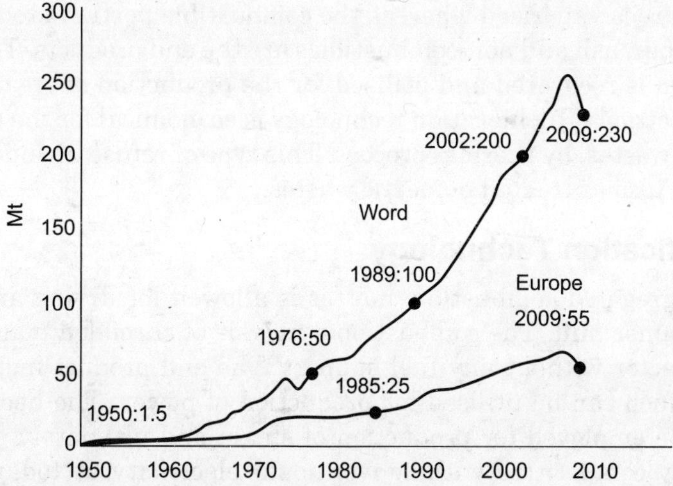

Fig.7.22: Growth in plastics production, 1950-2009. After five decades of continuous growth in world plastics production, there was a drop in production in 2008 due to the economic downturn. Close to 25 per cent of world production takes place in Europe. Plastics represented in the figure are thermoplastics, polyurethanes, thermosets, elastomers, adhesives, coatings and sealants, and polypropylene fibres. PET, PA and Polyacrylic Fibres are not included.

Source: *Plastics Europe (2010)*

Fuel-Pellets

Municipal Solid Waste (MSW) comprises various materials such as paper, plastics, glass, metals, vegetable matter, rags, rubber, etc. It is essential to segregate the organic materials from MSW in such a way that it is free of sand, moisture and other ferrous & non-ferrous materials. The segregated organic matter can be dried, ground and pelletised. This ground organic matter can be added by biomass to enrich calorific value of fuel pellets.

A pilot project of the Department of Science and Technology had been set up at Mumbai for producing about 80 tones per day of refuse derived fuel pellets as a coal substitute. The palletisation technology from has been developed as an indigenous technology with pallets having a Calorific Value of 3500 Kcal/Kg, as a substitute of coal.

In Gujarat, Vadodara Municipal Corporation has taken up the production of petro-coal pellets. It is produced from combustible garbage plastic, dry leaves, discarded clothes and vegetable waste supplemented by oil or petro-waste available from oil refineries. The process that turns combustible garbage into petro-coal is similar to the natural process that yields ordinary coal from the mines. After dehydrating and pulverising the garbage, it is mixed with petro-waste. Palletisation makes the material dense and hard, ensuring that it burns slowly and uniformly. The fuel is about half the price of coal and emits lower levels of sulphur dioxide and carbon monoxide. The entrepreneurs have also taken up conversion of biomass like, paddy husk, straw, groundnut shell and other waste materials of low calorific value into briquetted fuel which is an efficient source of energy.

Incineration Technology

Incineration is a process of controlled combustion of solid wastes and residue. During combustion moisture is vaporised whereas the combustible portion produces heat Carbon dioxide, water vapour, ash and non-combustibles are the end products. The heat generated during incineration is recovered and utilised for the production of steam, heating water, and generating electricity. Incineration technology is economical for the treatment of large quantities of solid wastes, by thermal process. This type of refuse includes not only metals and laminates but also untreated domestic wastes.

Pyrolysis/Gasification Technology

In this process segregated combustible matter is allowed for drying and thereafter it is shredded in a hammer mill. The combustion/pyrolysis of shredded matter takes place in a fluidised bed reactor without any fuel support. The end product includes combustible "producer" gas, which can be utilised for production of power. The heat produced in the process can also be employed for production of steam and ultimately generating power. Although not fully competitive with conventional electricity at today's energy prices, extracting energy from wastes offers environmental benefits, helping to reduce fossil fuel consumption and, amongst other things, the problem of methane emissions from landfill sites.

Composting Option

Incineration is an effective solution, but it is expensive for wastes having a high proportion of vegetable and putrescible matter. Economic constraints favour composting, as the city wastes contain a significant amount of nitrogen, phosphate and potash, as well as organic soil supplements.

A study by NEERI of 33 cities of India indicates that waste contains moisture varying from 10 to 40%, non-combustible materials or ash 50 to 80%, and volatile or combustible material only 10 to 25%. Modern day garbage also contains a large amount of various non-recyclables, viz. wax paper, silver printed packets, Tatra-packs, foamy plastic, cigarette foil, soft squeeze tubes, sachets, cellophane paper and thermocol. Thus, incineration is not always a viable and practical process to dispose and treat the waste. Another common problem is mixing of MSW with sewerage. In absence of sanitation and sewerage facilities in many areas, especially in inner city, village, squatter settlements and slums, night soil and sludge get discharged in the river, local drains, ponds, garbage dumps or streams. In Delhi, it is assessed that the BOD load of river Yamuna is as high as 1,50,000 Kg. per day. High quantities of sludge endanger the soil and sub-soil water by percolation of the pollution. This calls for adoption of simple and economical processes, like composting.

In spite of its simplicity and relative cost advantage, composting has remained a challenge. Commercial viability has so far been elusive except occasional reports. The MSW Rules of 2000 stipulate stringent quality requirement, especially in terms of minimum concentration of eight metals—As, Cd, Cr, Cu, Pb, Hg, Ni and Zn. However, for bio-degradable matter composting is one of the most popular system of garbage disposal. Apart from saving land, valuable products like compost, biogas, heat, electrical power, recycled paper, plastics, glass, metals, etc. are obtained. Composting can be done in different ways:

- Aerobic window composting
- Anaerobic trench composting
- Vermi-composting
- In-vessel aerobic composting

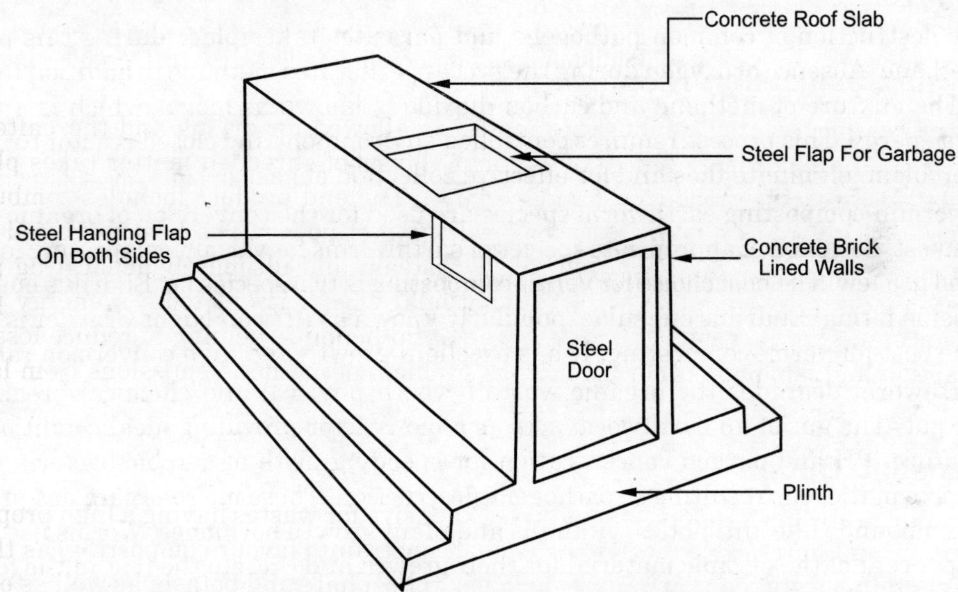

Fig.7.23: Bio reactor waste compositing

Effective Micro-organisms

Effective Micro-organisms (EM) is a combination of various beneficial, naturally-occurring micro-organisms mostly used for a found in foods. It contains beneficial organisms of 3 main kinds—phototrophic bacteria, lactic acid bacteria and yeast. These effective micro-organisms secrete beneficial substances such as vitamins, organic acids, chelated minerals and anti-oxidants when in contact with organic matter. They change soil micro-flora and fauna so that disease-inducing soil becomes disease-suppressing soil which in turn has the capability to develop into a zymogene soil. The anti-oxidation effects of these micro-organisms pass directly to the soil or indirectly to plants maintaining their NPK and CN ratio. This process increases the humus content of the soil and is capable of sustaining high-quality food production.

Effective micro-organisms were developed in liquid form by Prof. Dr. Teruo Higa, University of the Ryukyus, in 1982. At first, EM was considered an alternative for agricultural chemicals. But its use has now spread to applications in environmental, industrial and health fields. However, it must be stressed that EM is not a synthetic chemical nor a medicine.

Vermi-composting

Composting of biodegradable waste by earthworms, including human excreta, is gaining popularity. Organic waste is allowed to be decomposed by micro-organisms already present in the waste. The process can be accomplished either in presence or in absence of oxygen known as aerobic or decomposition respectively. During aerobic decomposition, organic compound gets oxidized to oxides of carbon and nitrogen and temperature of the mass rises to 70 °C.

The destruction of common pathogens and parasites takes place during this period. The methane. Absence of oxygen during the process generates methane in addition to other gases. The mixture of methane and carbon dioxide is known as biogas which is a useful source of energy. This process requires controlled environment and closed reactor to reduce odour problem, eliminate flies and for effective collection of gas.

In vermin-composting earthworm species are used for the conversion of organic waste into compost. Selection of appropriate species of earthworms for vermin-composting in India is limited to a few. The best choice for vermi-composting is two species i.e. Eudrilus eugeniae and Eisenia fetida. Eudrilus eugeniae, popularly known as African Night Crawler is found to be the best for vermi-composting. It has excellent growth and high conversion ratio.

Earthworm degrades the organic waste by both physical and chemical breakdown in their gut. The gut of an earthworm acts as a bio-reactor providing ideal conditions for temperature. PH and oxygen concentration for speedy growth of aerobic bacteria which outcompete pathogens resulting in pathogens destruction. These micro-organisms produce useful compounds like antibiotics, vitamins and plant growth hormones. Worms use about 5 to 10 per cent of the organic material for their growth and excrete the rest in the form of granular cast which is known as vermin-compost. The granular loose vermin-cast provides oxygen rich, nutrient rich media for aerobic microbes which further promote decomposition process. About 2,000 worms are required for a volume of I × 1 × 0.5 m. On an average, 5 kg of waste is partially digested by 1,000 worms in a day.

Community Vermi-culture and Vermi-Composting

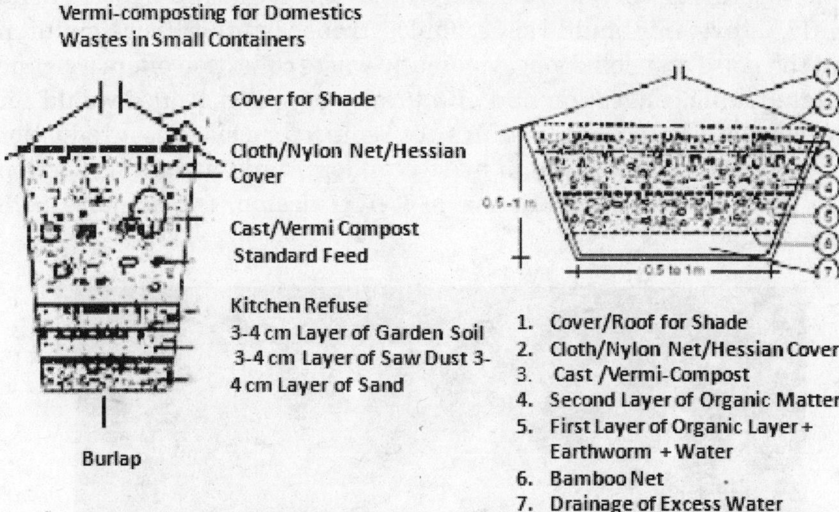

Fig.7.24: Vermi-compositing for domestic waste

Fig.7.25: Vermi-compositing for large volumes in small container

PNEUMATIC WASTE COLLECTION

Garbage transportation is the one of the largest overheads of local bodies in Delhi. To obviate this problem, at least in new areas, pneumatic waste collection can be adopted. It consists of an integrated collection station, piping system and discharge valves that are situated below vertical chutes outside the residential, commercial and other areas. Waste of one type is collected into the garbage chute, where it is stored above a garbage valve between the emptying cycles. There is a main pipe network under the valve that connects all valves and transports the waste to the collection station. These systems work best in a radius of about a kilometre. The new colonies and urban extension can be pre-planned with pneumatic waste ducts, gravity, chutes and

Fig.7.26: Pneumatic waste collection system

source: www.envoegroup.com

The current system which uses more than 500 waste-hauling trucks is very costly compared to the operating cost of mechanized system. Disadvantage of current system include unsightly, unhygienic landfill sites, dhalos, transportation, truck maintenance cost, labour cost and the noise and pollution. Pneumatic waste collection offers a cleaner solution that can alleviate traffic congestion and encourage recycling. It also would improve the overall city hygiene, aesthetic, and landfill sites required, besides cost reduction in waste collection. The processing of garbage is mostly underground at local collection stations. Fewer garbage trucks lead to a reduction in CO_2 emission, traffic noise pollution and reduced haulage costs.

Fig.7.27: Pneumatic solid waste management has been installed at Mecca

pneumatic Solid Waste Management has been implemented at several places, including Mecca, at and Holy city's Grand Mosque. Vacuum conveying technology specialist Mari-Matic has supplied a waste collection system to deal with the massive waste management challenge facing this city. Completed in 2013, Mari-Matic's Metro Taifun System in Mecca automatically conveys 900 tonnes of solid waste per day during peak season like the annual Hajj pilgrimage. Waste bags are fed into gravity chute intakes linked to a network of 30 km of pipelines running beneath the grand mosque complex. Transporting waste through airtight pipelines instead of using trucks minimizes problems of noise, smell and dust. Waste is conveyed through the pipelines to a collection terminal well away from the mosque. At the terminals the waste is compacted into containers for transportation by truck to a recycling and landfill site. The system uses much less energy than trucks or conventional automatic waste collection systems.

Bio-reactor Landfill—A New Concept

Disposal of waste in bioreactors is an innovative solution that represents a new step in the reduction of the impact on the environment in the critical area of waste management. Increasingly efficient techniques combine physical, chemical, biological and concentration processes to obtain purified leachates that can be disposed of in the natural environment.

After capture, bio-gases are normally burned in flares, which prevents the atmospheric emission of methane which has a greenhouse gas impact that is 21 times greater than that of CO_2. There is also an increasing number of projects that make use of biogas in either electricity or heating networks.

The bio-reactor landfill disposal involves control over degradation, with attempts to accelerate the process so as to rapidly stabilize the volume of waste. A bio-reactor is a chamber in which the waste degradation process is accelerated by re-injection of the leachates. This helps to control humidity, a critical parameter in degradation, and create optimal conditions for the development of bacteria.

Re-injection of leachates helps to bring about stabilization of the waste in less than ten years or so, whereas at least 20 to 30 years are needed in the case of disposal in conventional cells. The main advantages expected from bio-reactor processing include:

- The reduction of long-term environmental risks by more rapid stabilization of waste, since the performance of sealing systems can deteriorate over time;
- Reduced biogas leakage, thanks to a surface geo-membrane which ensures that disposal cells are leak tight. This results in a reduction in foul odors and greenhouse gas emissions;
- An increase in the number of sites capable of making use of energy conversion, as bio-reactors increase the volume of captured gases.

- The floor and walls of bio-reactor cells use the same sealing systems as conventional cells.
- They are covered with a hermetic geomembrane, which means no biogases are lost.
- Vertical pipes inject leachates that humidify the entire volume of waste.
- The re-injected leachates are then recovered at the bottom of the cell, then purified in the site's processing station.
- Suction piping then captures the biogas and routes it to the flare where it is disposed of.

Fig.7.28: Hierarchy of C&D waste treatment

source: *ADEME, 2006*

Ecological sustainability is ensured by geological and hydro-geological studies prior to the project, the distancing of the site from populated areas, very strict technical specifications such as, increased leak tightness of disposal cells using geomembranes on the bottom and sides of the disposal cells, capture and processing of leachates and biogases.

Management Reforms

Promoting alternative approaches in solid waste management need municipal reforms, change in attitude (particularly towards rag pickers), capacity development and a partnership approach involving the private and community sectors. Suitable mechanisms need to be evolved to overcome the barriers in promotion, development and dissemination of new approaches to waste management. This requires proper coordination and dovetailing of the resources of the government/local bodies, development authorities, NGOs, RWAs, industry, cooperatives and private sector. A clue can be taken from the British Local Authority Act, whereby the concept of 'Compulsory and Competitive Tendering', the monopoly of local bodies in the solid waste management and other services has been disbanded and they have to compete with the private sector. Indian cities can learn from such success stories and incorporate suitable legal, managerial and financing reforms that will bring investments, together with new, smart technology and operational efficiencies.

Integrated planning and management should include, maintenance and replacement of machinery, equipment and vehicles, manpower training and research, identification of appropriate sites for the dustbins, dalaos, transfer stations, SWERF and SLF sites, compost plants, incinerators, workshops. The operational planning involves planning and development of gravity chutes, pneumatic ducts, route optimisation for transportation of waste; marketing of bye-products, and effective involvement of rag pickers, community groups and resident welfare associations. Regular analysis of waste characteristics on seasonal basis both in terms of quantity and constituents may help in improved SW management. Preparation of Waste Management Plans in consultation with the inter-connected agencies would help in synergizing the efforts and improving overall performance.

The solid waste/management operation has the following broad stages: (*a*) waste generation and storage, (*b*) waste collection, (*c*) waste transfer/transportation, (*d*) treatment, recycle and recovery, and (*e*) disposal. There are successful examples of public-private partnerships in practice in different cities in India and abroad that could be examined for adoption at different stages of the entire opera-tion. Resident Welfare Associations, Non-governmental Organisations and Community-based Organisations have demonstrated substantial reduction in cost of waste management and improvement in services and a better environment. Certain indigenous techniques of waste treatment have also been developed that could be explored for adoption, depending on the local conditions.

Construction and Demolition Waste

It is estimated that waste generation during construction varies between 40 kg to 60 kg per m2 of floor area. Waste generation during renovation and repair work is estimated to be 40 kg to 50 kg per m^2. Demolition of buildings, on average generates waste between 300 kg to 500 kg per m^2 of floor area. CDW and other inert matters makes up almost one third of the total urban waste.

The total quantum of waste from the construction industry is estimated to be between

12 million to 14.7 million tonnes per annum, out of which seven to eight million tonnes are concrete and brick waste. In construction sector, 70% of the waste is not recycled. The Bureau of Indian Standards (BIS) and other codal provisions do not provide specifications for the use of recycled products in construction activities.

Fig.7.29: Hierarchy of C&D waste treatment

The reports state that C&D wastes in India amount to nearly one-third of the total MSW. On that basis, the amount of C&D wastes can be estimated to be nearly 70 million tonnes per year. The data indicates that the proportion of concrete in demolition wastes to vary from 23 to 35 percent. The proportion of soil varies between 31.5 to 43 percent. About 30 to 35% of urban solid waste comprises construction and demolition waste. The SWERF provides its segregation from other wastes and encourages its recycling in association with the builders, PWD, Municipal Corporations, etc. Construction and demolition waste is compulsorily segregated for reusable items and recycling. This can be used for land filling, road soling, pavement base material, production of pre-cast tiles and blocks and recycled concrete aggregate.

Table 7.3: Amount of C&D Waste in Various Countries

Country	Amount (MT per year)	Year
Germany	223	2003
Australia	19	2008-09
China	200	2005
Japan	77	2012
S. Korea	61.7	2013
India	14.7	2001

Table 7.4: Constituents of C&D Waste

Constituent	Quantity Generated in Million tons p.a.
Soil, Sand and Gravel	4.20 to 5.14
Bricks & Masonry	3.60 to 4.40
Concrete	3.60 to 4.40
Metala	0.6 to 0.73
Bitumen	0.25 to 0.30
Wood	0.25 to 0.30
Others	0.10 to 0.15

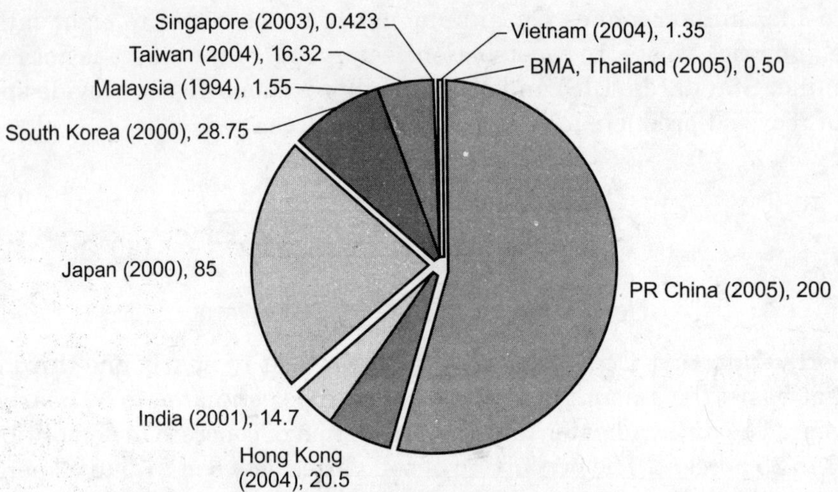

Fig.7.30: Estimates of C&D Wastes in Some Asian Countries

Environmentally sustainable management of C&DW has become a top priority keeping in view the following:

- Some C&DW are hazardous because of the materials, such as asbestos, lead, tars, paint and preservative residues, adhesives, bonding agents, sealants and plastics.
- Some materials become hazardous by environmental effects, e.g. a factory where surface reactions between the building materials and chemicals, or due to the pollution make the materials hazardous (toxic).
- Some C&DW become hazardous by contact with hazardous/toxic materials; lead-based paint can make the timber, brickwork or plaster hazardous.

Technologies of Processing and Recycling

The wastes originating from demolition of existing structures like old dwelling units, pavements and industrial structures, new constructions and renovation and repairs, are required to be sorted out in following categories;

- Materials which can be recycled in waste recycling plants - concrete, stone, blocks, tiles, etc.,
- Materials which can be disposed as scrap-metals, steel, aluminium, doors and windows, frames, etc., and
- Materials which can be disposed of by other methods-paint, asbestos, glass, electrical, etc.

The schematic of the recycling process, as is common in many countries is depicted in Fig. 3.36. The first step towards recycling is use of excavators along with task-specific attachments to methodically dismantle buildings and to process the material at the site. Jaw crushers, jack hammers, saws, debris buckets, metal shears, etc. are widely adopted. Precast concrete elements and concrete blocks can be reused with little or no processing, if care is taken during demolition to separate them.

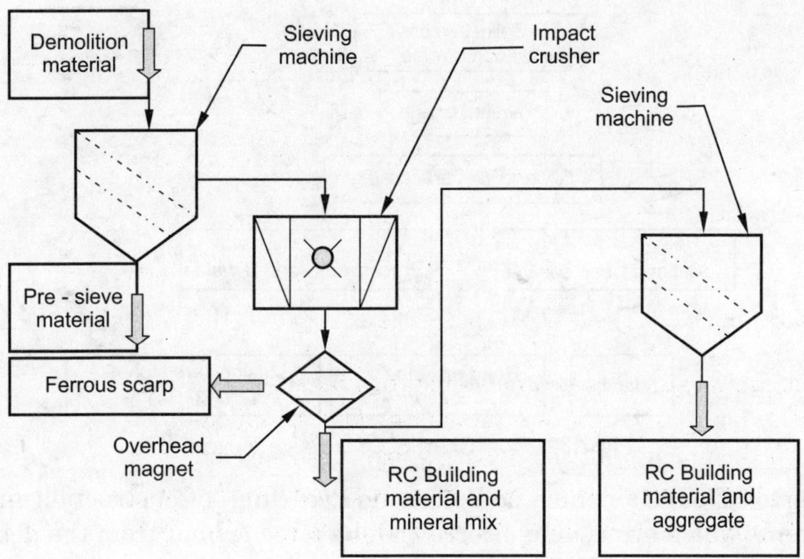

Fig.7.31: General recycling process

on-line and in-plant sorting and processing of C&D wastes are accomplished by use of both mobile and stationary machines. These include:

- Manual sorting lines
- Shredders
- Crushers
- Aggregate sifters
- Separators–water-based density separators, magnetic separators, eddy current separators, air blower, separators, etc.

Table 7.5: How long does it take for these items to perish on the Earth

Banana peel	3-4 weeks
Paper bag	1 month
Cotton cloth	5 month
Woolen cloth	1 year
Wood	5,10,20 years
Leather shoe	40-50 years
Tin can	50-100 years
Aluminium can	200-500 years
Plastic bag	1 million years
Glass	Unknown

Source: Centre for Science and Environment, New Delhi

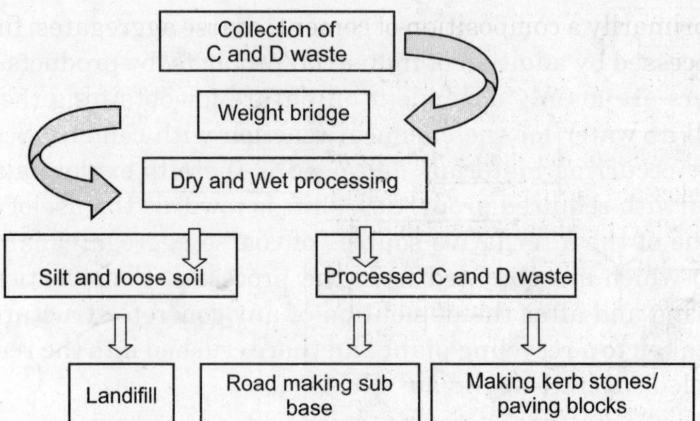

Fig.7.32: Flow Chart of C&D waste recycling

According to CPWD Guidelines on Reuse and Recycling of Construction and Demolition Waste, the items which are usable directly can be screened out from the debris and used without further processing. For example, bricks can be screened out of the demolition debris and used as it is for building wall or can be converted into smaller pieces aggregate or brick bats. Useful products like doors and windows, bricks, reinforcement from RCC components, structural steel can be taken out with little extra efforts and put into reuse.

Fig.7.33: Building Blocks from Recycled Debris, C & D Waste plant, Navi Mumbai

According to CPWD Guidelines on Reuse and Recycling of construction and Demolition waste, the items which are usable directly can be screened out from the debris and used without further processing. For example, bricks can be screened out of the demolition debris and used as it is for building wall or can be converted into smaller pieces aggregate or brick bats. Useful products like doors and windows, bricks, reinforcement, from RCC components, structural steel can be taken out with little extra efforts and put into reuse. Considering 30 % percent of C&D wastes of 12 million tonnes of concrete, and 50 of the concrete as coarse aggregate, the total available recycled concrete aggregate (RCA) in India is of the order of 1.8 million tonnes annually

Concrete is primarily a composition of cement, coarse aggregates, fine aggregates and water, further processed by addition of industrial products/ by products for enhancing the strength. Engineers are mainly dependent on nature for obtaining the Coarse and Fine aggregates as well as water for the chemical reaction with cement. Scarcity is there for all these naturally occurring materials and need is there to explore alternative sources. Even for the water with required properties, shift is towards the use of waste water after due treatment. One of the alternative sources of coarse aggregates is Recycled Concrete Aggregates (RCA) which are obtained from the processed Construction and Demolition (C&D) waste. During and after the demolition of any concrete structure, the demolished concrete waste is taken to a recycling plant and there crushed into the required sizes which is called the Recycled concrete aggregate (RCA).

Sometimes, good sized precast element are also obtained during the demolition, which have potential of being reused or otherwise, these are also crushed and converted into the recycled aggregates. Limitation is there in use of these recycled aggregates. The production of concrete India is governed by BIS and IRC codes i.e. IS: 456, IS: 1343 or IRC: 112. All these codes allow the use of naturally occurring aggregates only conforming to IS: 383. To overcome these limitations, it is required to make a special provision of use of Recycled aggregates in combination with naturally occurring aggregates. Thus, use of recycled aggregates can be there with different value of their share by suitable replacing the component of naturally occurring aggregates. It will help out not only in meeting the situation where there is acute shortage of natural resources, but also a step towards the sustainability.

Bricks: Bricks, after the removal of the stuck up mortar remain reusable for construction work. These can also be laid on as brick pavers or for landscaping or any other artistic creations. Brick paved streets are aesthetically pleasing and rain water also percolates through the pavement. Also, a brick surface is cooler in hot months. These street advantages make bricks a good choice in driveways. Bricks on edge are also sometimes used as economical pavement solutions in smaller compounds.

Construction debris consisting of bricks can be recycled into brick aggregate through screening, crushing, re-screening and blending, which can then be used as pavement base material by proper mix proportions with cement and fly ash. Brick waste which are not suitable for recycling into the pavement base materials can be used in construction/land fill. Concrete prepared from crushed brick aggregate has good engineering and also better thermal properties but has greater shrinkage than ordinary concrete. Sometimes, during the manufacturing of bricks, due inadequate burning, or sometimes due to over burning, whole lot is turned into the production waste. Though this waste is different from construction and demolition waste, but it can also be recycled like C& D waste and can be suitable used for production of precast elements like paver blocks, kerb stones, interlocking tiles by mixing with cement and using as a concrete mix.

Tiles: The tiles extracted from the wall in good shape and size can be reused after the removal stuck up mortar and then glued with suitable adhesives. Creative items like artefacts, table tops, special effects in drive ways, pedestrian subways etc. be made by recycled tiles.

Timber: The waste timber which has not been recovered in its original form or non-usable in same shape and size can be recycled into new particleboard, medium density fibre boards, animal bedding or used to make renewable energy. Timber used for recycling as to be free from any other demolition products like concrete, mortar, aggregates, sand, bricks, plastic, metals, tiles etc. Wood chips are produced from good quality wooden waste such as large size lumbers. Some of the particleboard producing companies and the pulp and paper producing companies are still using the recycled chip for their products.

Metals: Amongst the metals, steel and aluminium are the two major products obtained as waste during the construction as well during the demolition of a building. Structural steel obtained during the demolition of a steel structure or left over steel during the construction can be reused directly without much processing. The members can be resized as op the requirement and can be reused directly. Aluminium scrap can be put into reuse by the solid bonding process. If a care is taken in initial stages i.e. during designing with a valid deconstruction plan, then the reusable scrap can be increased to a much greater extent like house hold appliances, without taking the routing the scrap through a foundry. Reusing a steel beam its existing form is better than re-melting it and rolling a new steel beam, i.e. the energy used to re-melt the beam is saved.

Waste from the manufacture of steel construction products can be easily collected and segregated for recycling. Steel generates almost nil wastage on the construction site. Waste steel which is reusable is equally good in durability criteria and the quality is also well maintained while making products like fire hydrants, steel furniture and also ecologically sustainable.

Aluminium can be recycled into various items. Recycling scrap aluminium requires very less energy in comparison to the energy requirement of new aluminium. Because aluminum is infinitely recyclable, it can be reused in applications vastly different from its previous purpose, and it can also be recast into its original form. These properties make aluminium an ideal material for use in premium applications, even after being recycled many times.

Plastics: Scrap or waste plastic recovered from demolition or construction site is reprocessed and transformed into the entirely different useful products. When compared to other materials like glass and metals, plastic polymers require greater processing to be recycled. The most-often recycled plastic HDPE (high-density polyethylene is reduced to plastic lumber, tables, roadside curbs, benches, truck cargo liners, stationery and other plastic products. Other application of recycled plastic is in the preparation of a road surface that includes recycled plastic aggregate, bitumen (asphalt) with plastic that has been shredded and melted at a temperature below 220° C (428 °F) to avoid pollution. Such road surfaces are very durable and monsoon rain resistant.

Asbestos: Generally, asbestos is disposed as hazardous waste. The demolition of buildings containing large amounts of asbestos based materials have to be deconstructed piece by piece or the asbestos has to be removed carefully before the structure can be demolished. Workers must wear heavy protective equipment. It is important to ensure that asbestos waste has been wetted and sealed in heavy-duty plastic prior to transportation to an approved landfill.

Excavated Materials: Excavated materials are many times contaminated and require special handling and disposal. It may include hazardous as well as non-hazardous material. Excavated contaminated material that can be re-used will be decontaminated prior to re-use, or if not suitable for re-use will be transported to appropriate treatment facilities or approved landfill sites. If, the excavated materials is found suitable for reuse as base material for filling the road, then the cut and fill quantities will be balanced to avoid any off-site disposal. Excavated soils will be retained on site for re-use as backfill while hard rubble will be crushed and re-used on site. Unsuitable material for engineering fill can be used for landscaping.

From Cradle-to-Grave to Cradle

Rather than seeing materials as a waste, as in the cradle to grave system, cradle-to-cradle design is based on the closed loop cycle in which there is no waste. This approach seeks from the start to create buildings, communities and systems that generate wholly positive effects on human and environmental health. It seeks the 'rematerialization' of safe productive materials in systems powered by the sun. This technical metabolism is designed to mirror nature's biological metabolism in which synthetic resources circulate in an endless cycle of production, recovery and reuse.

A large range of building products can be recycled for use if demolition of buildings is carefully planned. This has special significance in dismantling precast concrete elements. Demolished concrete can be recycled as aggregate. Materials like steel and timber have huge potential for reuse.

Recycling Process involves pulverization of materials which are found in demolition, renovation, repairs, etc. Old plaster, bricks bats, concrete blocks, stone waste etc. are prepared to the feeder size to be fed in to the pulveriser and the pulverized material is stabilised by mixing cement to make different products by stabilization.

The processed debris can be used in lean concrete, as mortar for masonry, for making precast products, for rendering plaster and as bedding mortar for pavement tiles. The products which can be developed from the debris include cement debris bricks. Interlocking pavers and concrete M25 can replace sand. Debris banks can help the citizens to deposit their construction and demolition waste which can be recycled to form the building products. Recycling banks at various locations will reduce the transportation of debris and reduce load on the natural resources and provide building products at a lesser cost.

Industrial Waste

There has been an unprecedented growth of industries in the country. Large industries are usually located in the Industrial Estates, while thousands of micro and small scale industries co-exist with residential and commerical areas. These produce huge quantities of wastes, which are hardly segregated. Industrial Waste can broadly be classified as bio-degradable waste, non bio-degradable waste and hazardous waste. The waste is generated in all three forms of Solid, Semi-solid and Liquid Waste. Industrial solid waste, often un-segregated, is dumped in landfills. The waste in these landfills leads to generation of greenhouse gases such as methane and carbon dioxide. This not only contributes to climate change but can also cause fires/explosions and collapse at landfills, as methane is highly inflammable.

Untreated liquid, solid and gaseous waste can seep into air, water aquifers and soil to cause a plethora of ecological and health problems. Particulate matter in the air, polluted water with chemical effluents, hazardous waste, etc. are increasing the urban pollution levels. This problem is worsened in industrial areas, many of which are critically polluted. The environmental impacts of certain elements of the waste streams (such as plastics, metals and glass, etc.) as well as certain waste streams themselves (such as e-waste) stem not only from the waste treatment and disposal itself, but also due to the loss of resources from the supply chain. The use of virgin materials (often non-renewable) contributes to the overall depletion of natural resources. The demand of natural resources makes waste management a major issue.

Already several regulatory requirements exist mandating specific ways for management of industrial waste. These entail additional expenses and overheads. In a resource constrained world, where energy prices and raw material prices continue to be on the rise, it is time to reduce the waste and recycle it to the extent possible.

Various technologies for different kind of industrial waste materials are listed below:

- Bio-methanation: Relevant for Liquid and Semi-solid waste.
- Gasification/Pyrolysis: Relevant for Solid and Semi-solid waste.
- Incineration/Combustion: Relevant for Solid and Semi-solid waste.

An analysis of the WTE process options based on technology systems, environmental aspects, resources recovery and commercial aspects presents the following results:

- Bio-methanation has emerged as a favoured technology for industrial waste
- Gasification and pyrolysis have a distinct promise, as the technology gets more mature.
- Incineration for energy recovery from industrial wastes has been successfully commercialized in the developed countries. However, it represents an increasingly expensive option due to rigorous environmental compliance requirements.

According to the studies conducted by MNRE, C-kinetics and Assocham, there is a potential to reuse the industrial waste and also use them for Waste to Energy (WTE) projects.

Table 7.6: Areas of Reuse of Industrial Waste

Industry	Waste	Areas of Use
Thermal Power Plants	Fly ash	1. Feedstock in the cement industry 2. Making of bricks 3. Structural fill for roads, construction sites, land reclamation 4. Stabilization of soil
Iron and Steel	Blast Furnace Slags	Structural fill for roads, construction sites
Paper and Pulp, Sugar	Lime Sludge	1. As a sweetener for lime in cement manufacture 2. Manufacture of lime pozzolana bricks/ binders 3. For recycling in parent industry 4. Manufacture of building lime 5. Manufacture of masonry cement

Aluminium Smelter	Red Mud	1. As a corrective material 2. As a binder 3. Making construction blocks 4. As a cellular concrete additive 5. Coloured composition for concrete 6. Making heavy clay products and red mud bricks 7. In the formation of aggregate 8. In making floor and wall tiles 9. Red mud polymer door
	Spent Pot Lining	1. Energy savings in the brick and cement industries 2. Beneficial fluxing properties in the brick, cement and steel industries 3. Enhances strength development from cement 4. Recycle into cathodes and anodes for use in the aluminium ramming paste industry 5. Fluorides recovery (cryolite).
Cement	Kiln Dust (Particulate Matter), Waste Heat	1. Feedstock in the cement industry 2. As a soil stabilizer, to create raw materials for road construction 3. Treatment of other waste such ad flue gas and sewage sludge

Source: Assocham, Waste to Wealth, undated, New Delhi

Waste to Energy

The major indicator of potential usefulness of waste is the proportion of organic and inorganic matter, which correlates well with the energy potential in terms of the calorific value. A waste with good proportion of organic matter can be a good substrate for recovery of bio-energy potential (from biodegradable fraction of the organics) or for the recovery of thermal energy potential (equivalent to calorific value observed).

As a result of the conversion process, different kinds, of fuels can be generated, as given below:

Table 7.7: Conversion Processes

Waste to Solid Fuels	Waste to Liquid Fuels	Wate to Gaseous Fuels
Briquettes	Bio-diesel	Biogas
Pellets	Bio-ethanol	Syngas
Charcoal/Bio-char	Bio-Oil	

Several proprietary anaerobic processes have been developed for energy recovery as biogas from various liquid and solid wastes of industrial processes. Anaerobic technology has emerged as a mature technology for industrial applications.

The type of technology processes enabling conversion of waste to energy can be summarized as given below:

- Bio-methanation
- Gasification/Pryolysis, Incineration/Combustion

The Table below indicates key characteristics of the waste materials from various industries and their energy potential:

Table 7.8: Key Characteristics of Indian Industrial Liquid Waste

Sector	Waste Generated (m^3/Tonne)	BOD (mg/L)	COD (mg/L)	Total Dissolved Solids (mg/L)	Indiacative Biochemical Energy Potential (Nm^3 of biogas/m^3)
Distillery	25	45000-50000	90000-100000	7000-90000	25
Paper	15-30	4000-9000	12000-25000	10000-15000	5
Dairy	4-4.5	1000-1200	1800-2500	600-900	0.8
Abattoir	40-50	3500-4000	6000-8000	2500-3000	0.25
Maize	15	12000-12650	10000-20000	4000-6000	6
Tapioca	30	4600-5200	5600-6400	3500-4000	2.5
Tannery	30-40	1200-2500	3000-6000	14000-20000	1
Sugar	0.3-0.5	1250-2000	2000-3000	1000-1200	1

Source: Assocham/CPCB, New Delhi

Waste materials in certain sectors can also be used as inputs for other sectors, e.g. fly ash from thermal power plants are usable as inputs for the construction sector. Under the National Master Plan for Conversion of Waste to Energy, the Ministry of New and Renewable Energy (MNRE) has identified the following industries which offer a high potential for conversion of waste to energy:

- Sugar industry
- Distillery sector
- Dairy sector
- Pulp & paper industry
- Poultry sector
- Tanneries
- Slaughter houses
- Cattle Farms
- Maize starch waste
- Tapioca starch waste

Table 7.9: Key Characteristics of Indian Industrial Solid Waste

Waste Sector	Moisture (%)	Total Solids (%)	Inerts (%)	Organics (Volatile) % TS	Thermal Potential kcal/kg (Dry basis)
Poultry	75-80	20-25	25	75	1000-14000
Cattle Farm	80-90	20-Oct	20	80-85	3700

Bagasse	50	50	5-8	80-90	4000
Pressmud	75-80	20-25	10-20	75-80	4000
Abattoir	75-80	20-25	NA	75-85	NA
Tannery Flashings	75-80	20-25	NA	75-85	NA
Corn cods	10-15	85-90	<5	95	3500
Tapioca peelings	10-15	85-90	5-10	85-90	3000
Rice Husk	5-8	70-78	20-25	75-80	3000
Coal ++	8-10	90	25-30	70-75	4500

Adapted from indiasolar. com; ++ Value for Coal is indicated only for comparison purpose

The Ministry of New and Renewable Energy (MNRE) has identified key sectors generating industrial wastes with high energy potential. The Table below summarizes the industry sectors, the key raw materials, the major waste streams and the potential:

Table 7.10: Industrial Waste Energy Potential

Sectors	Raw Materials	Major Waste	Power generation potential (in MW)		
Distillery	Molasses	Liquid	Solid	2012	2017
		Spend Wash	–	628	785
Dairy	Milk, Cheese	Washings, Whey	–	77	96
Pulp & Paper	Bagasse/Stsaw	Black Liquor	Pith	72	90
Poultry	Chicken	–	Litter	81	102
Tanneries	Raw Hide	Toxic (complex)	Flesh/Hair	8	10
Slaughter house	Animals	Blood	Fresh/Bone	117	146
Cattle farm waste	Cattle	–	Fresh Waste	–	–
Sugar	Sugar cane	Waste water	Pressmud	453	567
Maize Starch	Maize	Steep Liquor	Pith	132	164
Tapioca Starch	Topioca	Waste water	Pith, Peelings	30	37

Source: Ministry of New and Renewable Energy/Assocham/CPCB, New Delhi

The most significant waste-to-energy technologies are based on biological or thermal methods. Biological method involves bio-methanation producing methane enriched bio-gas, which can be used as fuel, whereas thermal method (incineration) involves combustion of organic wastes as fuel with heat energy recovery.

Advanced thermal conversion involves destructive heating of organic materials with a limited supply of oxygen (gasification) or without any oxygen (pyrolysis) to produce a combustible gaseous product consisting of simple hydrocarbons and hydrogen.

The table below presents the types of conversion process and the associated end-products:

Table 7.11: Types of Conversion Processes

Technology	TypeProcess	End Products
Thermo-chemical Conversion	Combustion	Electricity & heat
	Gasification	Methnol, Hydrogen, Gasoline
	Phrolysis	Methanol, Hydrogen, Gasoline, Naptha, Bio-diesel
	Liquefaction	Methanol, Hydrogen, Gasoline
Bio-chemical Conversion	Anaerobic Degestion	Biogas
	Fermentation	Bio Ethanol, Hydrogen
Chemical Conversion	Trans-esterification	Bio-diesel

Thermo-chemical Conversion by Combustion

Combustion produces flue gas from waste at high temperatures. The higher the temperature, higher is the electricity output. However, extreme temperatures in the boilers will cause corrosion.

The following are the key combustion processes:

- Co-combustion of coal and waste
- Dedicated residual derived fuel (RDF) plant
- Incineration

Gasification: Gasification is the process of converting a carbon-based material (industrial, agricultural or municipal waste) into a combustible gaseous product by the supply of a small amount of gasification agent (typically oxygen).

Table 7.12: Advantage and Disadvantages of Various Processes

	Advantages	Disadvantages
1.	Combustion leads to a reduction in the waste that ultimately needs to be disposed off by up to 90%. This significantly lowers the pressure on limited land resources.	The combustion/incineration plants require strict emission control, as they release tars, dioxins, furans and char into the atmosphere. However, the ash released, if captured, can be used in building roads.
2.	Waste dispose in landfills emits Green House Gases, in particular Methane. By offsetting this emission, combustion leads to carbon savings.	Only organic matter can be combusted, while the inorganic matter is released as ash, which need landfill disposal
3.	The initial investment coast of combustion plants are lower than other technologies for Waste to Energy.	

Plasma Gasification Technology resolves the problem of waste management and also provides clean energy. It helps counter the problem of hazardous waste disposal and the need of landfills for municipal solid waste (MSW). Plasma Gasification Technology should not be confused with incineration. Some of the benefits are as follows:

- Gasification instead of combustion of waste ensures less pollution.

- As opposed to oxidation of materials during incineration, organic matter is gasified, inorganic matter is vitrified and metal is melted during plasma gasification. This again ensures that the pollutants created during the process are not toxic.
- Hazardous waste is converted into energy rich syngas.
- Tar, dioxin and furans are not formed.
- Volume of GHG produced is much lesser.
- Toxic ash is not generated, thereby eliminating the problem of its disposal.

Source: *SMS Infrastructure Ltd., Nagpur*

Pyrolysis: Pyrolysis is the process of converting the carbon based solid residue to gas in the absence of any oxygen (or any other agent). The combustible gas produced is called Syngas and contains Hydrogen, Carbon Monoxide and Carbon dioxide. It can be used for running internal combustion engines, substitute furnace oil as also to produce methanol, which is useful both as fuel for heat engines as well as chemical feedstock for industries. These technologies can be applied to most sectors, including Dairy, Poultry, Paper and Pulp, Slaughter house, Sugar and Distillery; as all of these sectors produce waste with a high Biological Oxygen Demand (BOD).

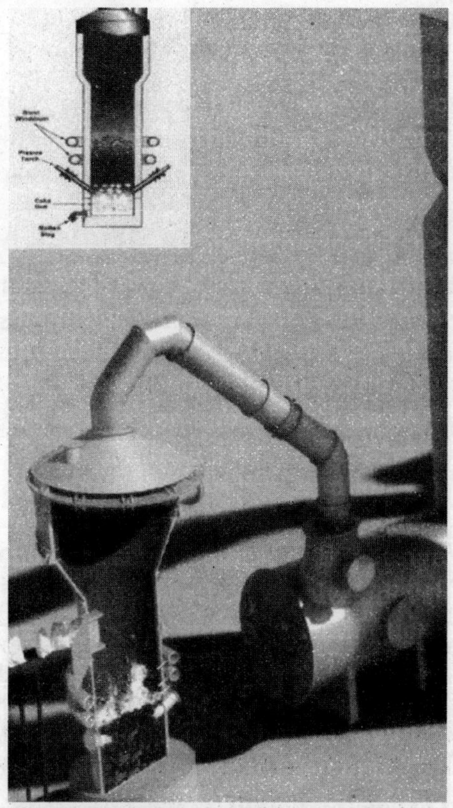

Fig.7.34: A gasification plant

source: SMS Infrastructure Ltd. Nagpur

Liquefaction: Liquefaction is the process of converting biomass and organic materials into hydrocarbon oils and by-products using high pressure (generally up to 200 atm) and temperature (generally up to 350°C). The resulting intermediates are convertible to hydrocarbon fuels and commodity chemicals to generate end-products similar to those produced from petroleum. The difference between Gasification/Pyrolisis and Liquefaction is in the state of the output (*i.e.* liquid fuel instead of gaseous). Advantages and disadvantages of this process are the same as Gasification/Pyrolysis.

Table 7.13: Advantages and Disadvantages of Pyrolysis Process

1.	As gasification occurs at high temperatures, as compared to combustion, the residual effluents such as chloride and potassium are refined and rendered chemically stable. Thus gas is produced with relatively low emissions.	The initial investment required for gasification plants is high and thus larger scale is required to bring economic sustainability.
2..	The feedstock used in gasification is relatively flexible; it can contain wood, plastics, aluminum, agricultural and industrial wastes, sewage sludge, etc.	Waste processing in gasification plants consumes high amounts of power and pure oxygen, which offsets the efficiency of the energy it produces (electricity)
3.	The processes are typically seen as producing a more useful product than standard incineration -gases, oils and solid char can be used as a fuel, or purified and used as a feedstock for petrochemicals and other applications. Syngas may be used to generate energy more efficiently, if a gas engine (and potentially a fuel cell) is used, whilst incineration can only generate energy less efficiently via turbines.	Studies have demonstrated that it may not be possible to maximize recycling prior to treatment by gasification and pyrolysis, due to the requirement for a fairly specific composition of waste, including combustibles, in order for the process to work effectively.

Source: Assocham

Bio-chemical Conversion: The bio-chemical conversion of biomass occurs when different micro-organisms convert biomass into gases or liquids typically under anaerobic conditions and by addition of heat.

Anaerobic Digestion or Bio-methanation: Anaerobic digestion is the naturally occurring process of breakdown of organic material by micro-organisms in the absence of oxygen. This process can be replicated and accelerated under artificial conditions leading to the generation of biogas.

Table 7.14: Pros and Cons of Bio-methanation

Advantages	Disadvantages
As compared to incineration technologies, Anaerobic Digestion does not release any gases into the atmosphere and is a 'net', 100% renewable source of energy.	The initial investment required for anaerobic digestion is high and thus scale is required to achieve commercial viability. Additionally, the maintenance cost is also high, owing to the involvement of bacteria. Only organic matter is treated in this process, and thus the waste waster produced could contain metals and other materials.

This process can easily be introduced into Sugar, Slaughter house and Distilleries for waste to energy projects.

Fermentation: Organic wastes can be converted to ethanol, the alcohol found in beverages, through bacterial fermentation (enzymes may be used to speed up the process), which converts carbohydrates in the feedstock to ethanol. Feedstock for this process includes forestry and agricultural wastes, such as molasses or waste starch. The applicability, advantages and disadvantages are similar to Anaerobic Digestion.

Chemical Conversion—Trans-esterification: Trans-esterification is a process that uses alcohol (like methanol) and reacts it with the triglyceride oils contained in vegetable oils, animal fats, or recycled greases, forming fatty acid alkyl esters (biodiesel) and glycerine. This is a useful technology for converting waste to energy. Depending on the fatty acids' content of feedstock, some pre-treatment of feedstocks may be required before the trans-esterification process. Feedstocks with less than 4% free fatty acids, which include vegetable oils and some food-grade animal fats, do not require pre-treatment. Feedstocks with more than 4% free fatty acids, which include inedible animal fats and recycled greases, must be pre-treated in an acid esterification process.

Table 7.15: Pros and Cons of Trans-esterification

	Advantages	Disadvantages
1.	Bio-diesel, is more combustible than diesel, and hence a more efficient fuel.	The process, if used on fresh edible oil, may contribute to supply changes and re-pricing/increased plantation and resultantly more deforestation.
2.	The process does not have any harmful emissions.	The process may change depending feedstock content; hence it is important to ensure that the feedstock is of the required constitution.

Source: Assocham

Electronic Waste

Anything that has a wire, a plug, a switch, a battery or runs on electricity, which is not going to be used again, comprises Electronic Waste. The term "e-waste" has been designated to describe discarded electronic devices such as computers, CRT, printers, cell phones, pagers, photocopier, T.V., LCD, digital photo and music devices, toys, etc.

The toxic chemicals most likely to be found in electronic waste are some of the most notorious in terms of contributing to serious health issues. These include lead, cadmium, mercury, polybrominated diphenyl ethers (PBDEs) and chromium. Electronic waste, e-waste, e-scrap, or Waste Electrical and Electronic Equipment *(WEEE)* describe loosely discarded, surplus, obsolete, or broken electrical or electronic devices. Environmental groups claim that the informal processing of electronic waste in developing countries causes serious health and pollution problems. Some electronic scrap components, such as CRTs, contain contaminants such as lead, cadmium, beryllium, mercury, and brominate flame-retardants. Recycling, disposal of e-waste may involve significant risk to workers and communities, and great care must be taken to avoid unsafe exposure in recycling operations and leaching of material such as heavy metals from landfills and incinerator ashes.

In addition, uncontrolled fires may arise at landfill sites. When exposed to fire, metals and other chemical substances, such as extremely toxic dioxins and furans (TCDD-tetrachloride dibenzo-dioxin, PCDDs - polychlorinated dibenzo-dioxins. PBDDS-polybrominated dibenzo-dioxin and PCDFs - polychlorinated dibenzo furans) from halogenated flame retardant products and PCB containing condensers can be emitted. The most dangerous form of burning e-waste is the open-air burning of plastics in order to recover copper and other metals. The toxic fall-out from open-air burning affects both the local environment and global air currents, depositing highly toxic by products in the air.

According to the e-Waste (Management and Handling) Rules, 2011, which come into effect from 1st May, 2012, every producer, consumer or bulk consumer involved in the manufacture, sale, purchase and processing of electrical and electronic equipment or components, collection centre, dismantler and recycler of e-waste is obliged to maintain a record of collection, sale, transfer, storage and segregation of wastes and make these records available for inspection by the State Pollution Control Board and follow the guidelines published by the CPCB.

The Rules mandate that every producer of electrical and electronic equipment (listed in schedule I) shall ensure that new electrical and electronic equipment does not contain Lead, Mercury, Cadmium, Hexavalent Chromium, polybrominated biphenyls or polybrominated diphenyl ethers.

Only a maximum concentration value of 0.1% by weight in homogenous materials for lead, mercury, hexavalent chromium, polybrominated biphenyls and polybrominated diphenyl ethers and of 0.01% by weight in homogenous materials for cadmium is permitted.

It is estimated that 75% of electronic items are stored due to uncertainty of how to manage it. These lie unattended in houses, offices, warehouses, etc. and are mixed with household waste, which is finally disposed of at landfills. This necessitates implementable management measures. In industries, management of e-waste should begin at the point of generation. This can be done by waste minimization techniques and by sustainable product design. Waste minimization in industries involves adopting:

- inventory management,
- production-process modification,
- volume reduction,
- recovery and reuse.

Bio-based plastics made with plant-based chemicals or plant-produced polymers, as compared to those from petro-chemicals and bio-based toners, glues and inks are less hazardous and toxic. Electronic product materials are usually non-renewable, as such it should be ensured that the product is built for re-use, repair and/or upgradeability.

Waste-minimization techniques can be broken down into three categories: (*i*) Improved operating and maintenance procedures, (*ii*) Material change and (*iii*) Process equipment modification. Improvements in the operation and maintenance of process equipment can result in significant waste reduction. This can be accomplished by reviewing current operational procedures or lack of procedures and examination of the production process for ways to improve its efficiency. Instituting standard operating procedures can optimise the use of raw materials in the production process and reduce the wastes.

Table: 7.16: Effect of E-Waste on Health

Source of e-wastes	Constituent	Health effects
Solder in printed circuit boards, glass panels and gaskets in computer monitors	Lead(PB)	• Damage to central and peripheral nervous system, blood system and kidney damage. • Affects brain development of children.
Chip resistors and semi-conductors	Cadmium (CD)	• Toxic irreversible effects on human health. • Accumulates in kidney and liver. • Causes neural damage. • Teratogenic.
Relays and switches, printed circuit boards	Mercury (Hg)	• Chronic damage to the brain. • Respiratory and skin disorders due to bioaccumulations in fish.
Corrosion protection of untreated and galvanized steel plates, decorator or hardner for steel housings	Hexavalent chromium	• Asthmatic bronchitis. • DNA damage.
Cabling and computer housing	Plastics including PVC	• Burning produces dioxin. It causes reproductive and developmental problems; • Immune system damage; • Interfere with regulatory hormones
Plastic housing of electronic equipments and circuit boards. Front panel of CRTs	Brominated flame retardants (BFR) Barium (Ba)	• Disrupts endocrine system functions • Short term exposure causes: • Muscle weakness; • Damage to heart, liver and spleen.
Motherboard	Beryllium (Be)	• Carcinogenic (lung cancer) • Inhalation of fumes and dust. Causes chronic beryllium disease or beryllicosis. • Skin diseases, such as warts.

Source: *Ramachandra T.V., Saira Varghese K. Environmentally Sound Options for E-Wastes Management, SPA Journal, New Delhi*

Companies should adopt waste minimization techniques, which will make a significant reduction in the quantity of e-waste generated, thereby lessening the impact on the environment. It is a "reverse production" system that designs infrastructure to recover and reuse every material contained within e-wastes - metals such as lead, copper, aluminum and gold, and various plastics, glass and wire. Such a "closed loop" manufacturing and recovery system offers a win-win situation and environmental sustainability.

Barriers in Solid Waste Treatment/Disposal and Management Reforms

The cities and villages are complex organisms and making them clean and hygienic involves multiplicity of stakeholders, reforms in policy, legal, financing and institutional structures and enforcement. There are number of barriers in ensuring efficient and effective waste management.

Policy and Regulatory Barriers: There is a lack of state policies and guidelines for waste collection and processing. The role of informal sector, that is, waste pickers, informal recyclers are not integrated within the regulatory frameworks. There is lack of clarity on roles, goals and targets for waste collection and processing for various stakeholders. The role of municipalities in regulating the type of waste collection is ambiguous. Policy regarding decentralized/local waste management practices is ad-hoc and disjointed.

There is limited awareness in the urban local bodies about suitability of different waste processing and disposal technologies. No performance guarantee reports are available. Land for developing sanitary landfilling facilities is not available.

Financing Barriers: There is lack of funds with urban local bodies for waste management. There is no standardized information about the financial viability of processing technologies. No clear framework for Public-Private Partnership (PPP) models exist across the waste value chain. There is a limited market for sale of compost and products from waste processing, affecting the financial viability of the waste treatment.

Operational Barriers: Indian cities generate more than 100 million tonnes of solid waste a year. It is estimated that up to 40 percent of municipal waste in India remains uncollected. A large number of cities do not have any processing facilities and the people tend to haphazardly dump wastes all over. Lack of storage and collection facilities, segregation of waste, pollution and contamination due to dumping of garbage on roads, open transportation in trucks, space constraints for landfills, social and environmental issues associated with landfills and lack of reusable and recyclables are the key issues in solid waste management. Cities are running out of space and are unable to provide sufficiently big landfill areas. It is estimated that India has to manage 0.573 MMT of municipal solid waste per day of which about 60 per cent is organic waste, amounting to 0.292 MMT/d. There are only 110 facilities in the country for treating which account for hardly 50 per cent of the organic waste generated. Detailed plans are not prepared by municipalities and Urban Local Bodies for MSW, sanitation, sewage, waste water recycling and drainage. Many municipalities are facing dearth of qualified staff for municipal waste management. There is lack of awareness and practice for segregation of waste at source. There is an absence of sound business models and the business as usual happens because of lack of enforcement.

The setting up waste processing plants entail a high capital expenditure and their financial viability is dependent on financing structure of their capex, along with operating expenditure vis-a-vis the return generated from recoveries.

There are hardly any examples of operational waste processing plants in India which are currently financially viable. There is limited operational data available on the financial performance and viability of waste processing plants in the Indian context and this remains a key challenge for increased adoption of the existing technologies in waste treatment.

Analysis of the financial viability of an indicative waste to energy plant (using bio methanation technology) with 500 TPD waste processing capacity operating for 12 hours daily shows that with existing capital subsidy provided by MNRE (max. of Rs. 10 crore per plant), prevailing average electricity selling price (Rs. 3.5/kWh), zero tipping fee and considering 80:20 capitalization with debt @14% interest rate for 7 year term, the plant will not generate a positive rate of return and thus is not financially sustainable. (ASSOCHAM, 2015).

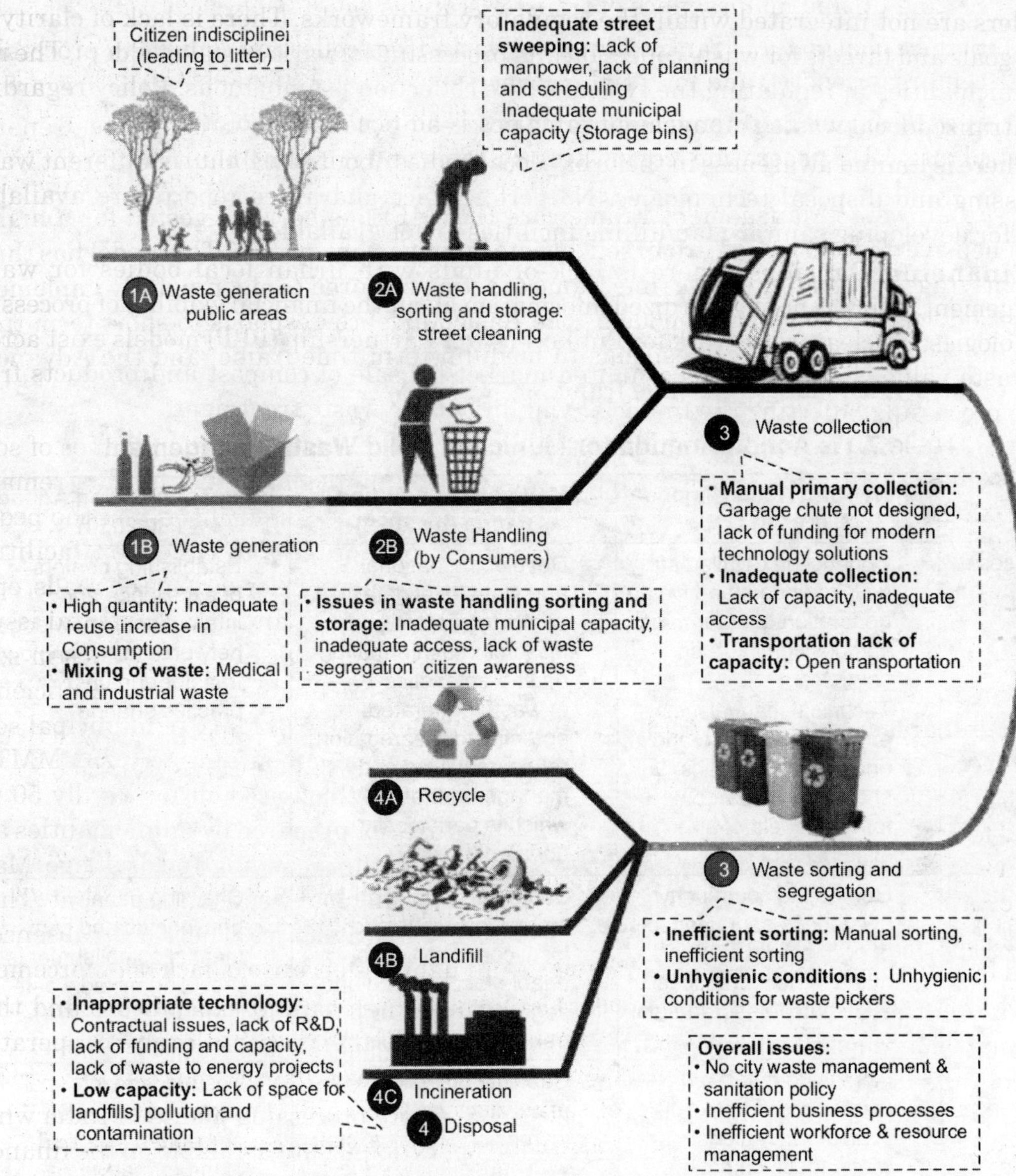

Fig.7.35: Current state of solid waste management

Source: NASSCOM (2015) Integrated ICT and Geospatial Technologies, New Delhi

The Options

In this scenario, it is necessary to explore various options and mobilise the public and private resources in urban waste management. These comprise the following:

(a) Service contract for waste collection at source

(b) Collection of recyclable materials for sale in local market

(c) Setting up and maintenance of transfer stations

(d) Service contract for transportation of waste from collection centers to processing centers for energy and material recovery

(e) Setting up and maintenance of waste processing plants (including waste to energy plants, RDF generation, composting systems and recycling plants).

An urban waste Management Framework (UWMF) has been suggested for a highly comprehensive and integrated urban solid waste management system. The model has three different- stages depending on the time frame of action required to meaningfully implement urban solid waste management models - the Basic Model corresponds to short term time frame, Intermediate Model corresponds to medium term time frame, and the Advanced Model corresponds to long term time frame.

Table 7.17: Action Agenda for Municipal Solid Waste Management

	Short Term Action Agenda (by 2017)	Medium Term Action Agenda (by 2025)	Long Term Action Agenda (By 2050)
For access to data	• Conducting detailed surveys to understand waste generation quantum and composition in urban centres. • Technical manuals for each type of waste and end use applications should be prepared for the benefit of all stakeholders.	• Developing national and state level data banks to capture and disseminate information on characteristics of waste generated, potential for segregation, reuse, recycling and management of MSW which is periodically updated.	• Establishing centers of excellence and international knowledge networks for technology transfer and best practice sharing.
For increasing segregation	• Generating community awareness about impacts of poor solid waste management practices – to improve segregation of waste	• Setting up sorting sheds by municipalities and local communities to enable local sorting and segregation reducing transportation costs	• Charging penalty from households and commercial establishments providing mixed waste
For improved waste collection	• Increasing municipal waste collection to 100% in all urban regions with a focus on source segregation. • Providing color coded waste bins in adequate numbers in commercial areas, markets etc. for collection of segregated waste Transportation of waste through covered garbage trucks and/or compactors	• Capacity building of existing staff and increasing technical staff with sound knowledge in technology and waste management system design. • Promoting informal waste collector's cooperatives for better integration of the informal sector within the value chain.	• Ensuring the informal sector is completely embedded as a part of the formal waste management value chain. • Developing integrated centralized and decentralized waste collection systems based on the type and composition of waste.

	• Creating guidelines for setting up waste collector's cooperatives and holding consultations with NGOs and SHGs to bring the informal waste collection players under the formal purview. • Collecting data on informal waste picking sector to enable designing of better collection models • Consultations and awareness sessions with retailers to stop sale of non-policy compliant plastic bags	• Ensuring safety and well-being of informal sector engaged in waste collection and sorting by providing requisite facilities and safety gear. • Enforcing prohibition of open burning of garbage, leaves and other wastes. • Providing waste trans-portation systems such as hand- carts, tricycles and other simple vehicles which are non-polluting in their localities and mu-nicipality sub- divisions • Setting targets for plastic manufacturers for Extended • Producer Responsibility.	• Deploying automated vacuum based waste collection systems in the new cities being developed in India. • Setting up sensor based collection mechanisms which inform waste collectors when garbage bins are filled.
For appropriate waste processing and disposal	• Creating technical guidelines for enabling municipalities to choose the correct waste processing options based on the composition and quantum of waste • Creating guidebooks to enable better understanding of financial viability of various processing technologies for funders and technology providers • Identifying sanitary landfilling sites in the high priority regions by urban local bodies	• Setting up demonstration units for waste processing technologies (especially plastic) • Designing waste processing plans corresponding to the local context to ensure adequate operational capacity Providing authorization of sites and plans for waste processing and sanitary landfilling • Promoting composting as a decentralized waste processing method especially for households • Engaging fertilizer companies to market the compost to leverage their existing network of distribution channels up to the village level. • Developing co-marketing policies for compost with chemical fertilizers to make them competitive in the agricultural market.	• Incorporating land required for waste processing and disposal in the land use plans at the district and state levels for each individual municipality.

For better policy framework	• Ensuring the development of State Policies • for municipal waste management • Amendment of MSW Rules to incorporate waste reducing, reusing and recycling methods • Building capacities of the local staff for planning, tendering, supervision and monitoring • Prioritizing/categorizing cities/towns based • on population and quantum of waste generation with special emphasis on hilly, coastal and tourist towns and cities for planned implementation of MSW Rules	• Amendment of MSW and Plastic Waste Management Rules for setting targets and timelines for achieving reduction in generation of waste • Increasing inter-governmental coordination between urban, agriculture, and energy departments for clear and targeted policies and cohesive implementation. • Ensuring the development of Urban Local Bodies in all urban centres across India • Developing national guidelines for the siting and • operation of composting facilities in conjunction with compost quality standards • Creating specific guidelines for management of special waste (such as sanitary napkins and diapers)	• Strict adherence to compost quality standards for protecting public health and developing confidence of farmers. • Developing registration and certification mandates for composting facilities to raise industry standards. • Creating a Waste to Energy corporation supported by the Government, focused on R&D, data collection as well as catalyzing finance for WTE projects in India
For better financing environment	• Including waste to energy projects for financing through the National Clean Energy Fund and clearly defining the modalities for getting waste to energy projects funded through the clean energy fund. • Reducing the subsidies provided to chemical fertilizers to enable the uptake of compost generated locally through solid waste. • Providing tax concessions (Excise/VAT) on products such as RDF and C&D products	• Developing output based incentives schemes for compliant processing operations • Developing feed-in tariffs to facilitate the sale of biogas and RDF as alternative energy sources. • Developing buy-back and basket approaches for composting • Using carbon financing to support the financial sustainability of waste to energy projects. • Allocating a dedicated Viability Gap Funding vehicle for implementation of de-centralized waste management projects in the country	• Developing schemes for providing incentives and disincentives to local bodies to promote better implementation of MSWM Rules

Source: *Assocham & C-kinetics, 2015, Vow-value out of waste, New Delhi*

Institutional Arrangements

The Yes Bank Business Council for Sustainable Development in its report E-Waste Management has proposed that the institutional framework may be cantered on the concept of Producer Responsibility Organization (PRO), an entity which is intended to have the physical responsibility for the waste recycling mechanism by providing forward and backward linkages with all other actors involved in the process. Such PROs have been formed in other countries as third party organizations to manage and collect the end-of-life products in lieu of each producer establishing its own separate system. The establishment of PROs and their functioning may be as follows:

- An NGO, private firm or Urban Local Body (ULB) may form a PRO. The selection would be through a bidding process wherein the prospective agencies would be evaluated based on proposals (technical and financial) submitted by them in response to a call for proposals.
- It is envisaged that there could be several such PROs throughout the country (at least one per region) with in-built flexibility for each PRO to design and operate their collection and recycling mechanism according to the region's needs.
- The PROs would enter into contract with collection agents and establish a system (financial and physical) for assuring a reliable flow of waste to the recycling facilities.
- The PRO may deploy mobile technology for collection and aggregation of waste from dispersed sources. Appropriate minimum collection targets may be agreed and written into the contracts of the collection agents. The system may be incentivized to enhance collection beyond the minimum targets. The PRO/collection agents may integrate the informal sector into their operations.
- The PRO would enter into a contract with a recycling facility for recovering different components of waste. The recycler may charge a gate rate to the PRO based on the volume of waste delivered. The gate rate may be fixed to cover the O&M costs, whereas other government measures (such as capital subsidy, land at concessional rates) may be undertaken to assure the viability of the recycling facility. The recycling facility may be incentivised by sale of recovered materials to prospective buyers on a profit sharing basis between the PRO and the recycler.
- A monitoring system would be put in place at the state level to monitor the functioning and utilization of funds by the PROs.
- PROs may also have various functions extending beyond the management of the CSR mechanism, such as for education and training of producers and consumers.

The Ministry of Urban Development in its Atal Mission for Rejuvenation and Urban Transformation (AMRUT, 2015) have also recommended establishment of dedicated Special Purpose Vehicle within the urban local body for integrated, efficient and professional management of urban waste. The model requires integration of activities between the informal and formal sectors, thereby bringing them into the mainstream of waste recycling activity.

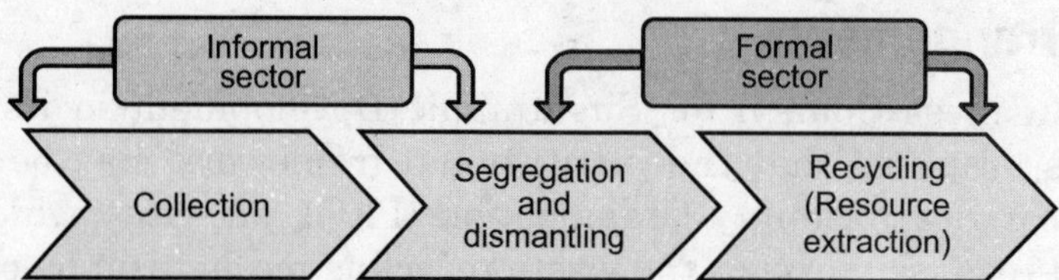

Fig.7.36: Integration of informal sector in managing waste

Source: Adapted from GIZ, 2010

Through dovetailing the informal sector with mainstream recycling units, it is envisaged that the informal sector will serve as a feeder system for the formal recycling industries. While the informal sector carries out dismantling activities in a controlled manner, it is imperative that operational standards are prescribed to ensure that a mix of manual and mechanical operations. A monitoring mechanism should also be put in place so that all operations are environmentally sound and safe for health of workers.

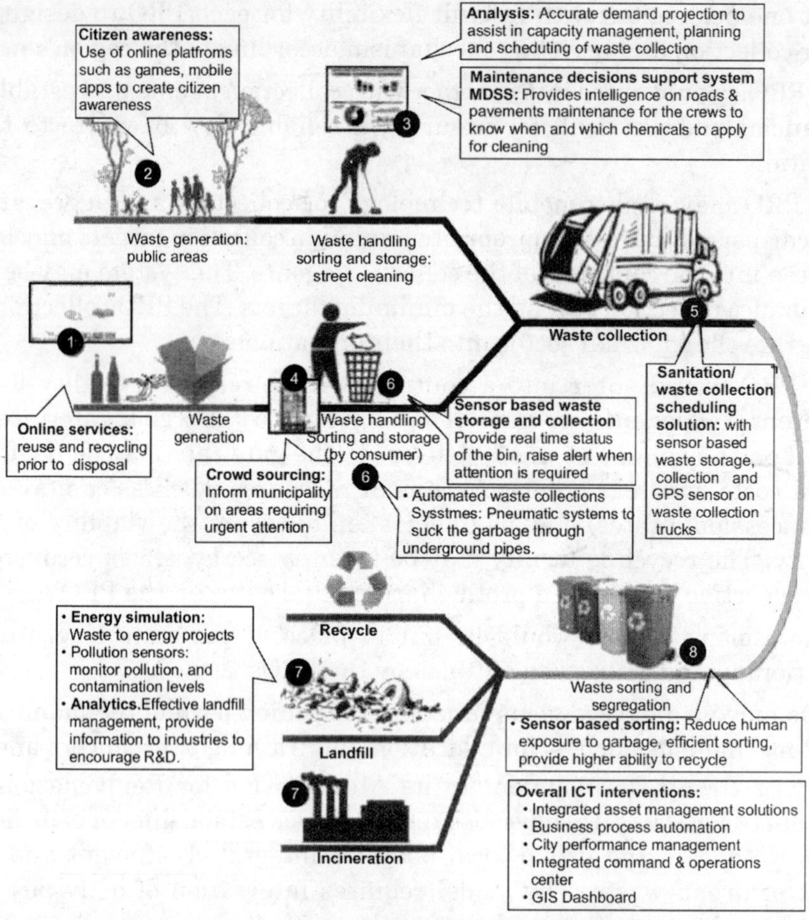

Fig.7.37: Smart solid waste management

Source: *NASSCOM (2015). Integrated ICT and Geospatial Technologies, New Delhi*

Smart Solutions

Technology plays an important role by providing smart, performance based and efficient solutions for digital planning, sanitation, sewerage and garbage collection, resource optimisation, asset management, maintenance, pollution controls, air quality measurements, etc. There is a need to need to redefine the waste management processes based on circular economy concepts, which aim to reduce impact on environment and leverage waste to generate value. Technology can also help involving citizens in waste management and monitoring by using mobile and web channels.

The following ICT enablers can be invoked for Solid Waste Management:

- Online platforms: Online platforms provide options and alternatives to the user to look into reusing old stuff. The existing user is also encouraged to look for options to sell and regain value from the product before discarding the product as waste.
- Analytics: Accurate projections on total waste generated, waste type and identification of high waste generation areas enable effective planning and management of solid waste, management services. Use of analytics during events with large citizen involvement such as festivals and fairs can ensure smooth collection and transport of waste.
- Crowd-sourcing: Citizens can be encouraged to report waste-related activities which need urgent attention from the authorities.
- Sensor-based waste collection: Sensor-based waste bins to identify status of waste bins if it is empty or filled so as to customise the waste collection schedule accordingly and save costs.
- Automated waste collection system: Automated Waste Collection System (ACS) is a long-term solution and can take care the conventional methods like door-to-door, curb-side, block, community bins collection and transportation via chute system from high rise buildings with waste sucked though pipes and minimal human intervention.
- GPS devices and sensors on waste truck: GPS technology to route the waste collection trucks to optimise the collection efficiency and ensure contractors dump waste in designated places. It will also give a clear picture of waste generated per ward.
- Sensor-based sorting: Sorting waste material with the use of sensor technology helps in smart sorting. The sensor technology can recognise materials based on their visible spectrum or colour with infra-red/ultra-violet spectra or based on their specific and unique spectral properties of reflected light, or atomic density or conductivity/ permeability or atomic characteristics.
- Pollution sensors: Leverage the pollution sensors to gauge pollution levels at landfills.
- Energy simulation (waste to energy): Use of energy simulation software and analytics can provide accurate projections of waste generation and energy production from waste.

- Analytics-based landfill management: Accurate waste generation and collection. projections along-with break-up of type of waste can enable smart landfill management.
- Integrated asset management solutions: Integrated asset management of all waste. infrastructure assets including the associated data, processes, information systems and governance for manageable operations and higher sustainability.
- Business process automation: Re-engineer, optimise and automate business processes using business process management solution to have a fully integrated and policy-driven set of automated business processes that increase efficiency and reduce service delivery costs.
- Workforce and resource management: Leverage the workforce and resource management solutions to improve workforce engagement and task management. Optimise the workforce with the help of management solutions like planning, forecasting and scheduling, shift management, mobile applications to execute tasks and efficient performance management tools.
- City performance management: Monitor the performance of city subsystems through the use of digital technologies and big data analytics to manage city governance, efficient performance and proactive crisis management.
- Integrated command and operations centre: Leverage integrated command and operations center to monitor city services on real-time. Improve/synchronise maintenance activities to reduce downtime and improve maintenance effectiveness.
- Geospatial dashboard: Bin locations, landfill locations, waste management assets can be mapped by geospatial system.

Source: NASSCOM (2015), Integrated ICT and Geospatial Technologies, New Delhi

DEBRIEFING QUESTIONS

1. What are the main types and quantities of urban wastes in India?
2. What are the critical issues in solid waste management?
3. Highlight the major stipulations in 'Municipal Solid Waste (Management and Handling) Rules, 2000.
4. Why open dumps of solid waste in landfill sites is an environmental disaster?
5. List various solid waste treatment technologies.
6. Write short notes on:
   i. Recycling
   ii. Solid Waste and Energy, Recycling Facility Technology
   iii. Four Bin Method of Waste Segregation
   iv. Incineration Technology
   v. Pyrolysis/Gasification
   vi. Composting

vii. Vermi Composting

viii. Pneumatic Waste Collection

7. Write notes on:

   i. C & D Wastes

   ii. Industrial Waste

   iii. Electronic Waste

8. Write a note on smart/digital tools for efficient solid waste management.

8 Energy Conservation and Efficiency

Cities account for three quarters of the total global energy demand and produce almost 80% of CO_2 emissions. Half of the Indian population is expected to become city dwellers by around 2050. Concern is growing over the long-term availability and security of fossil fuel supplies. For these reasons, the cities of the future need to powered by renewable energy to a greater extent.

	Fuel	Resource Life	Global Warming	Acid Rain	Photochemical Smog	Particulates	Toxics	Risks	Others	Comments
Fossil fuels										
Coal	●	●	●	●	●	●	●	●	●	Modem large-scale coal burning can be much cleaner than small boilers and fires
Oil	●	●	●	●	●	●	●	●	●	Toxic wastes associated with oil extraction
LPG / CNG	●	●	●	●	●	●	●	●	●	Almost clean, but transport is usually by road.
Electricity (Grid, with current supply mix)	●	●	●	●	●	●	●	●	●	The energy mix in the grid of coal, nuclear, oil, hydro, wind, etc. will hopefully get cleaner over time
Renewables										
Solar, Wind, etc.	●	●	●	●	●	●	●	●	●	Some evacuted tube collectors may contain ozone depleting gases, also PV Toxics such as cadmium
Wood and other biomass	●	●	●	●	●	●	●	●	●	Worst scenario, Modern stoves/boilers may be cleaner

Fig.8.1: A conceptual assessment of various sources of energy

India's per capita energy and electricity consumption is about 479 kg of oil equivalent (kgoe) and 704.2 kWh respectively. Large differences exist in the 'energy consumption' between the urban and rural areas. Demand for energy is rapidly growing due to industrialization, prosperity and urban growth, resulting in higher demands for lighting, heat, mechanical power, transport, air- conditioning, communications and services. Energy usage for lighting is comparatively declining as against a growing consumption for the following end-uses prompted by increasing income levels:

- Space heating
- Water heating
- Refrigeration
- Air conditioning
- Cooking and other appliances such as computers.

To make the transition to low carbon development, it is necessary to reduce the carbon footprint of the built environment-homes, offices, public buildings, factories, transport, and the carbon intensity of energy supplied. Unless cities are environmentally sustainable they will not be economically or socially sustainable. Making the transition from high to low carbon cities and buildings can be done by proper planning and partnership among the governments, city authorities, consumers and businesses.

Primary energy use in India is dominated by coal (40 percent), followed by fuel wood (34 percent) and petroleum fuels (15 percent). The share of fossil fuels increased from around 60 percent during 1995-1996 to 85 percent of total energy use by 2010. The growth in fossil fuel consumption results in increased greenhouse gas (GHG) emissions, local and regional environmental problems (such as air and water pollution, and land degradation). Over 95,000 villages and 60 percent of the households still have no access to electricity. Kerosene lamps are the main source of lighting in these households. Where available, electricity is characterized by erratic supply and fluctuating voltage. The extraction of biomass low efficiency of devices and high human drudgery associated with cooking characterize the energy consumption pattern. The demand of fossil fuels is increasing exponentially and thus becoming more expensive due to the scarcity of fuel. Additionally, the environmental and social impacts of the consumption of fossil fuels generally include the pollution, global warming, waste disposal problems, land degradation, depletion of natural resources and sustainability and affordability of energy.

CLIMATE CHANGE, ENERGY AND BUILDINGS

The IPCC Fourth Assessment Report reported that fossil fuels provide 85% of the total primary energy (2008). Furthermore, the combustion of fossil fuels accounted for 56.6% of all anthropogenic GHG emissions (CO_2 eq) in 2004. To maintain both a sustainable economy that is capable of providing essential goods and services to the citizens and to maintain a supportive global climate system, requires a major shift in how energy is produced and utilized. However, renewable energy technologies, which release much lower amounts of CO_2 than fossil fuels, are growing.

GHG emissions associated with the provision of energy services are a major cause of climate change. The IPCC concluded that "Most of the observed increase in global average temperature since the mid-20th century is very likely due to the observed increase in anthropogenic greenhouse gas concentrations." Concentrations of CO_2 have grown to about 390 ppm CO_2 or 39% above pre-industrial levels by the end of 2010. The global average temperature has increased by 0.76°C (0.57°C to 0.95°C) between 1850 to 1899 and 2001 to 2005, and the warming trend has increased significantly over the last 50 years. While the

IPCC report focuses on the energy sector, forest and land use change, the release of non- CO_2 gases from industry, commerce and agriculture also contribute to global warming. Since about 1850, global use of fossil fuels (coal, oil and gas) has increased to dominate energy supply, both replacing many traditional uses of bioenergy and providing new services. The rapid rise in fossil fuel combustion (including gas flaring) has produced a corresponding rapid growth in CO_2 emissions.

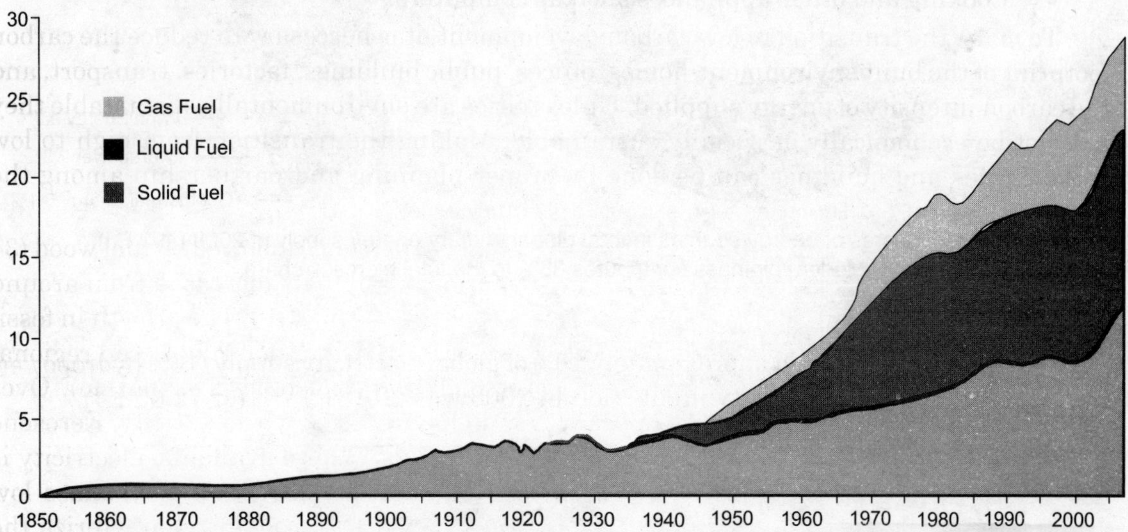

Fig.8.2: Global CO_2 emissions from fossil fuel burning, 1850 to 2007. Gas fuel includes flaring of natural gas. All emission estimates are expressed in Gt CO_2.

Source: Boden and Marland, 2010

Fossil fuel reserves and resources (oil, gas and coal) have the potential to add quantities of CO_2 to the atmosphere— if burned over coming centuries—that would exceed the range of any of the scenarios considered. Reserves refer to what is extractable with today's technologies at current energy prices. Resources represent the total amount estimated to be available without regard to the technical or economic feasibility of extracting it.

In recent years (2000 to 2007), increases in carbon intensity have mainly been driven by the expansion of coal use by both developed and developing countries, although coal and petroleum use have fallen slightly since 2007, partly due to the financial crisis. Since the early 2000s, the energy supply has become more carbon intensive, thereby amplifying the increase resulting from growth in GDP per capita.

Renewable Energy (RE) Option

On a global basis, it is estimated that RE accounted for 12.9% of the total 492 EJ of primary energy supply in 2008. The largest RE contributor was biomass (10.2%), with the majority (roughly 60%) of the biomass fuel used in traditional cooking and heating applications in developing countries. Hydropower represented 2.3%, whereas other RE sources accounted for 0.4% .

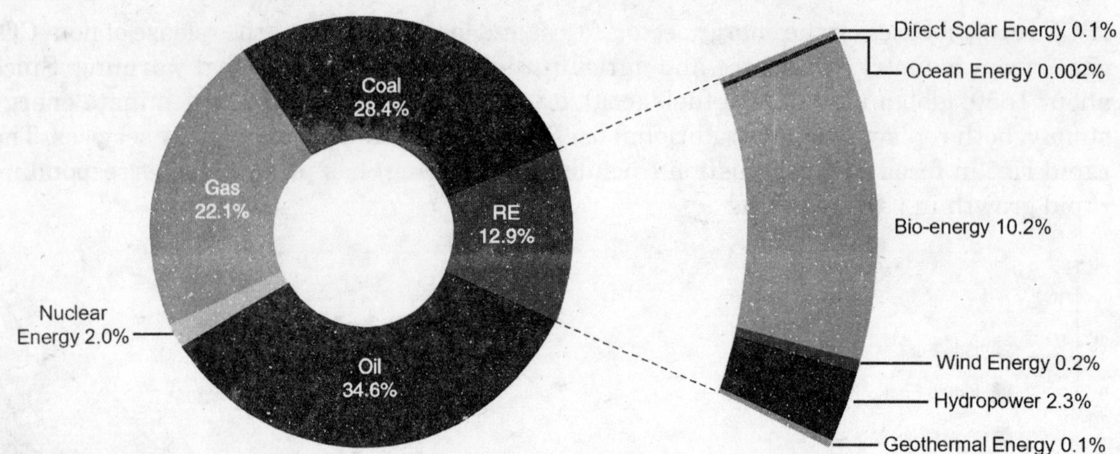

Fig.8.3: Shares of energy sources in total global primary energy supply in 2008 (492 EJ). Modern biomass contributes 38% to the total biomass share.

Source: IEA (2010).

In 2008, RE contributed approximately 19% of global electricity supply (16% hydropower, 3% other RE). Global electricity production in 2008 was 20,181 TWh (or 72.65 EJ).

Fig.8.4: Share of primary energy sources and renewables in world electricity generation

Source: IEA, 2010

The IPCC identified a number of ways to lower heat-trapping emissions from energy sources in providing energy services. They include:

- Improve supply side efficiency of energy conversion, transmission and distribution including combined heat and power.
- Improve demand side efficiency in the respective sectors and applications (*e.g.*, buildings, industrial and agricultural processes, transportation, heating, cooling, lighting).

- Shift from high GHG energy carriers such as coal and oil to lower GHG energy carriers such as natural gas, nuclear fuels and RE sources.
- Utilize carbon capture and storage (CCS) to prevent post-combustion or industrial process CO_2 from entering the atmosphere. CCS has the potential for removing CO_2 from the atmosphere when biomass is burned.
- Change behaviour to better manage energy use or to use fewer carbon- and energy intensive goods and services.

Two additional means of reducing GHGs include enhancing the capacity of forests, soils and grassland sinks to absorb CO_2 from the atmosphere and reducing the release of black carbon aerosols and particulates from diesel engines, biomass fuels and from the burning of agricultural fields. Additional reductions in non- CO_2 heat-trapping GHGs (CH_4, N2O, hydrofluorocarbons, sulphur hexafluoride) can also reduce global warming.

Application of renewable energy

Renewable energy is the form of energy from solar, geophysical or biological sources that is replenished by natural processes at a rate that equals or exceeds its rate of use. RE is obtained from the continuing or repetitive flows of energy occurring in the natural environment and includes resources such as biomass, solar energy, geothermal heat, hydropower, tide and waves and ocean thermal energy, and wind energy. However, it is possible to utilize biomass at a greater rate than it can grow, or to draw heat from a geothermal field at a faster rate than heat flows can replenish it.

Table 8.1: Overview of Energy Sources

Energy source	Select energy technologies	Primary energy sector (electricity, thermal, Mechanical, transport)
Bioenergy	Traditional use of fuel wood/charcoal	Thermal
	Cook stoves (Primitive and advanced)	Thermal
	Domestic heating systems (pellet based)	Thermal
	Small and Large-scale boilers	Thermal
	Anaerobic digestion for Biogas production	Electricity/Thermal/Transport
	Combined heat and power (CHP)	Electricity/Thermal
	Co-firing in fossil fuel power plant	Electricity
	Combustion-based power plant	Electricity
	Gasification-based power plant	Electricity
	Sugar and Starch-based crop ethanol	Transport
	Plant and Seed oil-based bio-diesel	Transport
	Lignocellulose sugar based biofuels	Transport

	Pyrolysis-based biofuels	Transport
	Aquatic plant-derived fuels	Transport
	Gaseous biofuels	Thermal
Direct Solar	Photovoltaic (PV)	Electricity
	Concentrating PV (CPV)	Electricity
	Concentrating solar thermal power (CSP)	Electricity
	Low temperature solar thermal	Thermal
	Solar cooling	Thermal
	Passive solar architecture	Thermal
	Solar cooking	Thermal
	Solar fuels	Transport
Geothermal	Hydrothermal, condensing flash	Electricity
	Hydrothermal, binary cycle	Electricity
	Engineered geothermal systems (EGS)	Electricity
	Submarine geothermal	Electricity
	Direct use applications	Thermal
	Geothermal heat pumps (GHP)	Thermal
Hydropower	Run-of-river	Electricity/Mechanical
	Reservoirs	Electricity
	Pumped storage	Electricity
	Hydrokinetic turbines	Electricity/Mechanical
Ocean Energy	Wave	Electricity
	Tidal range	Electricity
	Tidal current	Electricity
	Ocean currents	Electricity
	Ocean thermal energy conversion	Electricity/Thermal
	Salinity Gradients	Electricity
Wind energy	Onshore, large turbines	Electricity
	Offshore, large turbines	Electricity
	Distributed, small turbines	Electricity
	Turbines for water pumping/other mechanical	Mechanical
	Wind kites	Transport
	Higher-altitude wind generators	Electricity

Source: IPCC, SRRES CCM, 2010

The rate of utilization of direct solar energy has no bearing on the rate at which it reaches the Earth. Fossil fuels (coal, oil, natural gas) do not fall under this definition, as they are not replenished within a time frame that is short relative to their rate of utilization.

Table 8.2: Renewable Energy Theoretical Potential Expressed as Annual Energy Fluxes

Renewable source	Annual Flux (Ej/yr)	Ratio (Annual energy flux/2008 primary energy supply)	Total reserve
Bioenergy	1,548	3.1	—
Solar energy	3,900,000	7,900	—
Geothermal energy	1,400	2.8	—
Hydropower	147	0.30	—
Ocean energy	7,400	15	—
Wind energy	6,000	12	—
Annual primary energy source	**Annual use 2008 (Ej/yr)**	**Lifetime of proven reserve (year)**	**Total (Ej)**
Total fossil fuels	418	112	46,700
Total uranium	10	100-350	1,000-3,500
Total RE	64	—	—
Primary energy supply	492 (2008)	—	—

Source: IEA (2010)

The full utilization of variable renewable sources such as wind and solar power can be enhanced by energy storage. Storing energy as heat is commonly practised today, and multiple means of storing electricity have been developed. Pumped water storage is a well-developed technology that can utilize existing dams to provide electricity when variable sources are not providing. Other technologies include flywheel storage of kinetic energy, compressed air storage and batteries. If electric vehicles become a major fraction of the fleet, it is possible to utilize their batteries in a vehicle-to-grid system for managing the variability of RE supply.

It is necessary to examine the total cost of end-use efficiency measures plus RE technology, and then determine whether there is rebound effect for a specific case. Furthermore, carbon leakage may also reduce the effectiveness of carbon reduction policies. Carbon leakage is defined as the increase in CO_2 emissions outside of the countries taking domestic mitigation action divided by the reduction in the emissions of these countries. If carbon reduction policies are not applied uniformly across sectors and political jurisdictions, then it is possible for carbon emitting activities that are controlled in one place to move to another sector or country where such activities are not restricted.

Reducing energy needed at the services delivery stage is an important means of conservation of energy. Because RE sources usually have a lower power density than fossil or nuclear fuels, energy savings at the end-use stage are often required to utilize a RE technology for a specific energy service. For example, it may not be possible to fuel all vehicles on the planet with biofuels at their current low engine efficiencies, but if vehicle fuel efficiency were greater, a larger fraction of vehicles could be run on biofuels. Similarly, by lowering demand, the size and cost of a distributed solar system may become competitive. The importance of end-use efficiency in buildings and renewable technology are becoming viable options. Furthermore, electricity distribution and management is simplified and system balancing costs are lower if the energy demands are smaller. Energy efficiency at the end-use stage thus facilitates the use of RE.

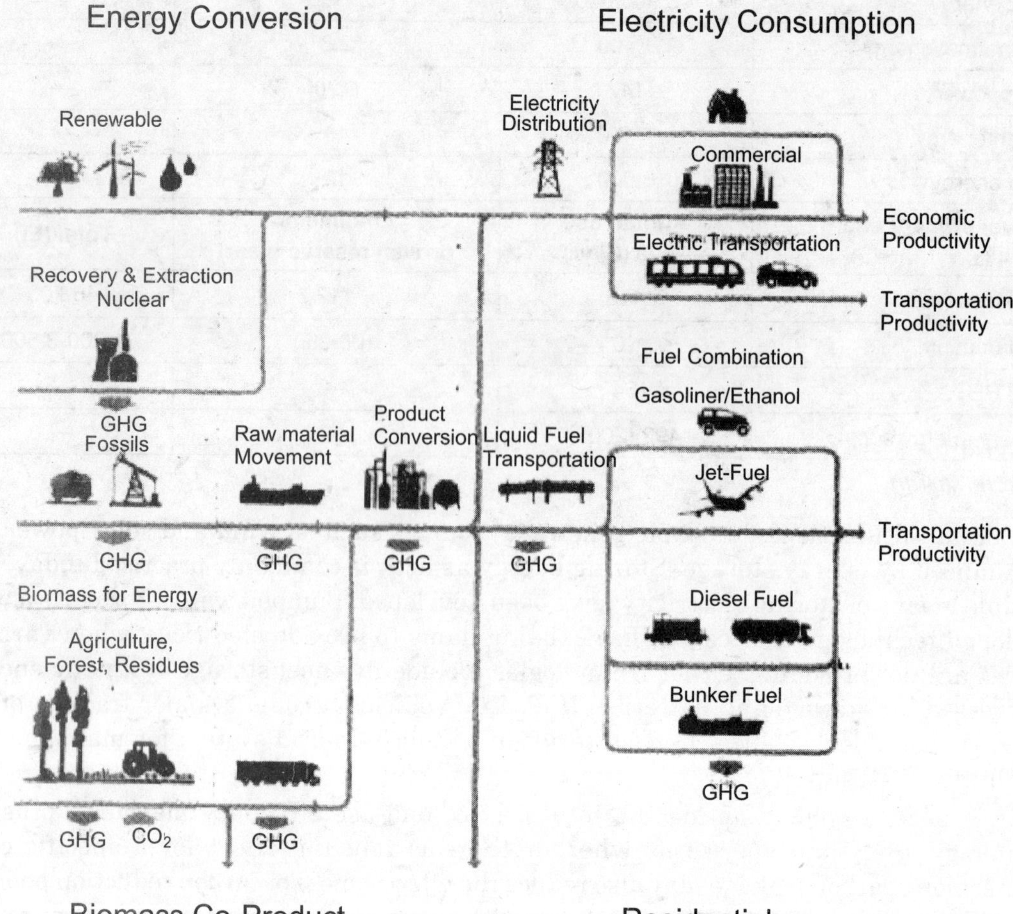

Fig.8.5: Illustrative system for energy production and use illustrating the role of RE along with other production options. A systemic approach is needed to conduct life cycle systems analysis.

Source: IPCC, 2010, Special Report on Renewable Energy Sources and Climate Change Mitigation

Often the lowest cost option is to reduce end-use energy demand through efficiency measures, which include both new technologies and more efficient practices. Examples can be found in efficient appliances for lighting, heating and cooling as well as heating and cooling in the building sector. For example, compact fluorescent or light-emitting diode lamps use much less electricity to produce a lumen of light than does a traditional incandescent lamp. Properly sized variable-speed electric motors and improved efficiency compressors for refrigerators, air conditioners and heat pumps can lower primary energy use in many applications. Efficient houses and small commercial buildings such as the Passivhaus design from Germany are so air tight and well insulated that they require only about one-tenth the energy of a conventional dwelling (Passivhaus, 2010). Energy efficient design of high-rise buildings in tropical countries could reduce emissions from cooling at a substantial cost savings

Globally, per capita incomes as well as broader indicators such as the Human Development Index are positively correlated with per capita energy use, and economic growth can be identified as the most relevant factor behind increasing energy consumption in the last decades. As economic activity expands and diversifies, demands for more sophisticated and flexible energy sources arise. Economic development has therefore been associated with a shift from direct combustion of fuels to higher quality electricity.

Particularly for developing countries, the link between social and economic development and the need for modern energy services is evident. Access to clean and reliable energy constitutes an important prerequisite for fundamental determinants of human development, contributing, inter alia, to economic activity, income generation, poverty alleviation, health, education and gender equality. Because of their decentralized nature RE technologies can play an important role in fostering rural development.

Lifecycle assessments facilitate a quantitative comparison of 'cradle to grave' emissions across different energy technologies. Alongside the commonly known CO_2 production pathways from fossil fuel combustion, natural gas production (and transportation) and coal mines are a source of methane, a potent greenhouse gas, and uncontrolled coal mine fires release significant amounts of CO_2 to the atmosphere.

Traditional biomass use results in health impacts from the high concentrations of particulate matter and carbon monoxide, among other pollutants. Long-term exposure to biomass smoke increases the risk of a child developing an acute respiratory infection and is a major cause of morbidity and mortality in developing countries.

In this context, non-combustion-based RE power generation technologies have the potential to significantly reduce local and regional air pollution and lower associated health impacts compared to fossil-based power generation. Improving traditional biomass use can reduce negative impacts on sustainable development, including local and indoor air pollution, GHG emissions, deforestation and forest degradation.

Impacts on water resources from energy systems strongly depend on technology choice and local conditions. Electricity production with wind and solar PV, for example, requires very little water compared to thermal conversion technologies, and has no impacts on water quality. Limited water availability for cooling thermal power plants decreases their efficiency, which can affect plants operating on coal, biomass, gas, nuclear and concentrating solar power. There have been significant power reductions from nuclear and coal plants in recent years.

Sustainable Energy and Buildings

Energy efficiency and renewable energy are the twin pillars of sustainable energy policy which can reduce the level of energy demand and slow down the rate at which resources are depleted. According to the International Energy Agency, improved energy efficiency in buildings, industrial processes and transportation could reduce the world energy needs in 2050 by one third, and help control the emissions of greenhouse gases. The integration of renewable energies in buildings offers an economically viable way to meet building power/thermal loads by superior environmental performance with high efficiency and low greenhouse gas emissions. In order to improve and increase energy efficiency, use of renewable energy and related technologies (biomass, solar hydrogen, geothermal, wind, and mini-hydraulic energy sources, waste recycling), the government has worked out several policy reforms and programmers. Development and implementation of several legislative and policy reforms as given below over the years are indicative of the changing trends and priorities for the energy sector for buildings and human settlements.

Integrated Energy Policy 2006: This policy has identified key areas in the building sector for energy efficiency measures. This includes building design, construction, heating, ventilation, air conditioning, lighting, and household appliances. It provides for mandatory periodic energy audits for all buildings with loads above 1 MW, and also for all government buildings. It also provides for energy benchmarking of buildings to be done by BEE.

Mission under the national climate action plan: The Enhanced Energy Efficiency Mission has provided for market transformation for energy efficiency. This has recommended mandatory labelling of appliances and equipment, mandatory maximum energy efficiency norms per square feet for new buildings, existing building (through retrofits) to be ECBC compliant, replacement of inefficient appliances and fiscal instruments.

The National Mission on Sustainable Habitat also recommended strategies for mitigation that include harmonising energy building code with the national building code, implementation strategies and incentives, building performance and rating systems, demonstration projects, consumer awareness programmes, enhance appliance standards and labelling, etc.

National Solar Mission is relevant from the perspective of the encouragement of the application of the rooftop solar PV and small power plants to replace diesel gensets, etc. There are direct and indirect benefits of utility based incentive programmes.

National Building Code: Building by-laws are under the state governments. The Bureau of Indian Standards developed the National Building Code (NBC) in the 1980s that guides municipalities and development authorities on building by-laws. The voluntary code covers most aspects of building design and construction, with a small part dedicated to energy efficiency. NBC was revised in 2005. The Code provides guidance on aspects of energy conservation related to day lighting, natural ventilation, energy-efficiency aspects, artificial lighting requirements and HVAC design standards.

Since a degree of climate change is inevitable, adaptation to climate change and conservation of energy are essential components of sustainable development. Adaptation can be either anticipatory or reactive to an altered climate. Some RE technologies may assist in adapting to change, and are usually anticipatory in nature. The linkage between climate mitigation (reducing emissions of GHGs), climate adaptation and buildings has been clearly established as given below:

- Active and passive solar cooling of buildings helps counter the direct impacts of rising mean temperatures,

- Solar PV and wind require no water or fuel for their operation, and hence may become increasingly important as droughts and high river temperatures limit the power output of thermal power plants;
- Water pumps in rural areas, remote from the power grid, can utilize PV or wind for raising agricultural productivity during climate-induced increases in dry seasons and droughts; and tree planting and forest preservation along coasts and riverbanks is a key strategy for lessening the coastal erosion impacts of climate change. With suitable choice of species and silvicultural practices, these plantings can also yield a sustainable source of biomass for energy.

Energy Conservation Act and Energy Conservation Building Code (ECBC)

The Energy Conservation Act, 2001 obliges the Central/State Governments to facilitate and enforce efficient use of energy and its conservation. Bureau of Energy Efficiency (BEE) was established in 2001 for the purpose, and to take suitable steps to prescribe guidelines for EC, energy audit, etc. Accordingly, it has prescribed Energy Conservation Building Code (ECBC). The Code defines norms and standards of energy consumption expressed in terms of per square meter of the area wherein energy is used and includes the location of the building (section 2(j) of the Act).

The Bureau of Energy Efficiency is mandated to take suitable steps to prescribe guidelines for ECBC under clause (p) of Section 14 of the Act (Section 13(2) (d) of the Act). In addition, the Central Government including the State Governments can amend the ECBC to suit regional and local climatic conditions (Section 14(q) and Section 15(a) of the Act) as well as direct every owner or occupier of the building or building complex, being designated consumer to comply with the provisions of the ECBC for efficient use of energy and its conservation (Section 14 (r) and Section 15 (b) of the Act). Further, the Central Government and the State Governments can direct, any designated consumer, if considered necessary for efficient use of energy and its conservation, to get energy audit conducted by an accredited energy auditor in such manner and at such intervals of time as may be specified by regulations (Section 14(s) and 15(c) of the Act).

This Code covers building aesthetics, building envelope, mechanical systems and equipment, including heating, ventilating, and air conditioning (HVAC) system, interior and exterior lighting system, service hot water, electrical power and motors including thermal comfort in non-centrally air conditioned/ heated buildings. The ECBC has been finalized after extensive consultation with all stakeholders by a Committee of Experts constituted by the Bureau of Energy Efficiency.

The implementation of ECBC for design, construction and performance of buildings with proper site layout, design, use of natural light with minimum solar heat gain, comprehensive procedures and energy efficient materials aims to ensure substantial reduction in electricity.

Compliance Requirements

As per the code, all the buildings or building complexes with a connected load of 500 kW or greater or a contract demand of 600 kVA or greater have to comply with the Code. Buildings with 1,000 m^2 or more of conditioned area are likely to fall under the above load conditions.

The following sections which deal with mandatory and prescriptive requirements of new and existing buildings are related to this specified threshold area. It is important to mention here that these mandatory and prescriptive requirements are applicable only where the building has a connected load of 500 kW or more or contract demand of 600kVA or more.

i. *Mandatory Requirements*: Compliance with the requirements of the Code shall be mandatory for all applicable buildings mentioned under the scope.

ii. *New Buildings*: The Code compliance procedure requires the new building to fulfill a set of mandatory provisions related to energy use as well as show compliance with the specified requirements stipulated for the different building components and systems. Its requirements can be met by following one of two methods:

(a) **Prescriptive Method** specifies prescribed minimum energy efficiency parameters for various components and systems of the proposed building. The prescriptive requirements are covered in Chapter 4 through Chapter 8 dealing with the building envelope, HVAC systems, service hot water and pumping, lighting systems, and electric power respectively. To use the building envelope section as an example, designers can choose the prescriptive method that offers flexibility in selecting insulation for roof that meets specified thermal characteristic (*e.g.* R-value), in place of meeting prescriptive requirements of U-factor of the roof assembly.

(b) **Whole Building Performance** (WBP) is an alternative method to comply with the Code. This method offers considerable design flexibility. It allows for Code compliance to be achieved by optimizing the energy usage in various building components and systems (envelope, HVAC, lighting and other building systems) in order to find the most cost-effective solution. WBP method requires an approved computer software program to model a Proposed Design, determine its annual energy use and compare it with the Standard Design of the building.

Table 8.3: ECBC Building Envelope Requirements: An Overview

Building component	Mandatory requirements	Prescriptive requirement
Opaque Construction (Roofs and walls)	Building Envelope sealing Requirements	Maximum U-factor and Minimum R-values of roofs and walls (ECBC 4.3.1. and 4.3.2.) Cool roof specifications(ECBC 4.3.1.1)
Fenestration (door, windows and skylights)	Calculation of U-factors and Solar heat gain coefficient (SHGC) of glazing (ECBC 4.2.1 and 4.2.1.2.) Air leakage Maximum limits (ECBC 4.2.1.3)	Maximum U-factors and SHGC Wall-windows ratio (WWR), and Visible transmission (VLT) of glazing (ECBC 4.3.3) Skylight roof ratio (SSR) Maximum U-factor and SHGC of SHGC of glazing [ECBC 4.3.4)

iii. *Additions to Existing Buildings:* The Code also applies to additions in existing buildings. The requirements are triggered when new construction is proposed in the existing building.

As per the Code:

Where the addition plus the existing building exceeds the conditioned floor area of 1,000 m^2 or more, the additions shall comply with the provisions of Section 4 through Chapter 8, and is demonstrated in either of the following ways:

- The addition alone shall comply with the applicable requirements, or
- The addition, together with the entire existing building, shall comply with the requirements of this Code that would apply to the entire building, as if it were a new building

Exception to above:

When space conditioning is provided by existing systems and equipment, the existing systems and equipment need not comply with this Code. However, any new equipment installed must comply with specific requirements applicable to that equipment.

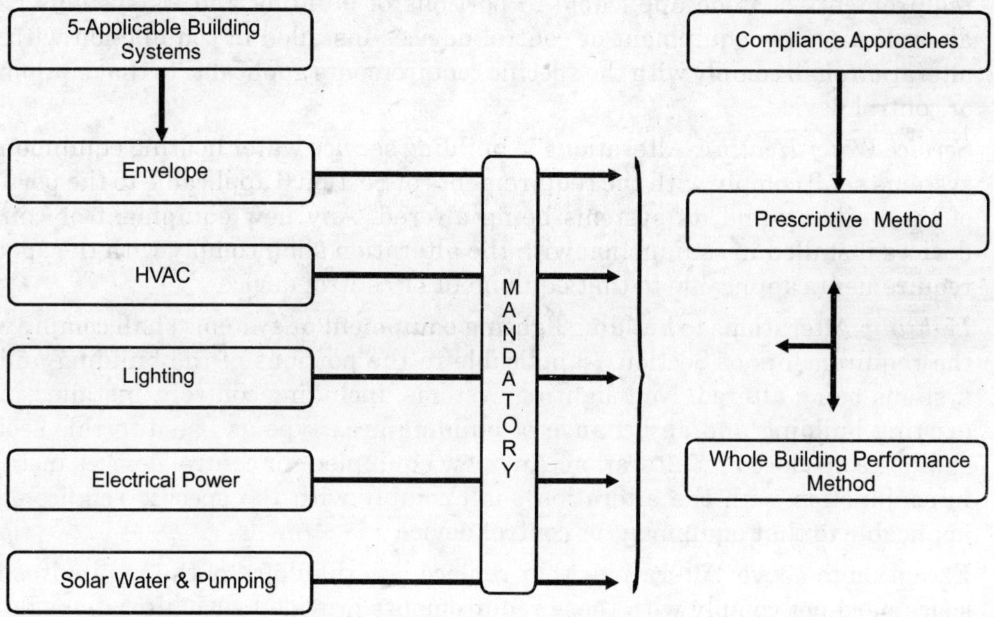

Fig.8.6: ECBC Compliance Process

iv. *Alterations to Existing Buildings*: When making alterations to an existing building, the portions of a building and its systems that are being altered must be made to comply with mandatory and prescriptive requirements.

As per the Code:

Where the existing building exceeds the conditioned floor area threshold (of 1000 m^2 or more), portions of a building and its systems that are being altered shall meet the provisions of Section 4 through Chapter 8 (of the Code). The specific requirements for alterations are given in the following subsections.

v. *Building Envelope*: Alterations to the building envelope shall comply with the requirements of Code or fenestration, insulation, and air leakage applicable to the portions of the buildings and its systems being altered.

Exception to above: The following alterations need not comply with these requirements provided such alterations do not increase the energy usage of the building:

- Replacement of glass in an existing sash and frame, provided the U-factor and SHGC of the replacement glazing are equal to or lower than the existing glazing
- Modifications to roof/ceiling, wall, or floor cavities, which are insulated to full depth with insulation
- Modifications to walls and floors without cavities and where no new cavities are created

vi. *Heating, Ventilation, and Air Conditioning*: Alterations to building heating, ventilating, and air-conditioning equipment or systems shall comply with the requirements of Code applicable to portions of building and its systems being altered. Any new equipment or control devices installed in conjunction with the alteration shall comply with the specific requirements applicable to that equipment or control device.

vii. *Service Water Heating:* Alterations to building service water heating equipment or systems shall comply with the requirements of Section 6 applicable to the portions of the building and its systems being altered. Any new equipment or control devices installed in conjunction with the alteration shall comply with the specific requirements applicable to that equipment or control device.

viii. *Lighting*: Alterations to building lighting equipment or systems shall comply with the requirements of Section 7 applicable to the portions of the building and its systems being altered. New lighting systems, including controls, installed in an existing building and any change of building area type as listed in this section shall be considered an alteration. Any new equipment or control devices installed in conjunction with the alteration shall comply with the specific requirements applicable to that equipment or control device.

Exception to above: Alterations that replace less than 50% of the luminaires in a space need not comply with these requirements provided such alterations do not increase the connected lighting load.

ix. *Electric Power and Motors:* Alterations to building electric power systems and motors shall comply with the requirements of Section 8 applicable to the portions of the building and its systems being altered. Any new equipment or control devices installed in conjunction with the alteration shall comply with the specific requirements applicable to that equipment or control device.

Compliance Approaches

Every building project is different, each building has its own site that presents unique opportunities and challenges, each building owner or user has different requirements, and climate and microclimate conditions can vary significantly among projects. Architects and engineers need flexibility in order to design buildings that address these diverse requirements. The Code provides this flexibility in a number of ways. Building components and systems have multiple options to comply with the Code requirements. To use the building envelope section as an example, designers can choose the Prescriptive Method that requires roof insulation be installed with a minimum R-value. Alternatively, the other options allow the designer to show compliance with the thermal performance (U-factor) of roof construction assembly. In addition, building envelope trade-off option covered in Section-4 permits trade-offs among building envelope components (roof, walls, and fenestration) for Code compliance. If more flexibility is needed, the Whole Building Performance Method is also available.

(a) *Prescriptive Method*: The Code specifies a set of prescriptive requirements for building systems and components. Compliance with the Code can be achieved by meeting or exceeding the specific levels described for each individual element of the building systems, covered in Section 4 through Chapter 8. For building envelope, the Code provides a Trade-Off option that allows trading off the efficiency of one envelope element with another to achieve the overall efficiency level required by the Code.

(b) *Whole Building Performance Method*: Use of energy simulation software is necessary show ECBC compliance via the Whole Building Performance Method. Energy simulation is a computer-based analytical process that helps building owners and designers to evaluate the energy performance of a building and make it more energy-efficient by making necessary modifications in the design before the building is constructed.

The computer-based energy simulation programs model the thermal, visual, ventilation, and other energy-consuming processes taking place within the building to predict its energy performance. The simulation program takes into account the building geometry and orientation, building materials, building facade design and characteristics, climate, indoor environmental conditions, occupant activities and schedules, HVAC and lighting system and other parameters to analyze and predict its energy performance. Computer simulation of energy use can be accomplished with a variety of computer software tools and in many cases may be the best method for guiding a building project to be energy-efficient. However, this approach does require considerable knowledge of building simulation tools and very close communication between members of the design team.

The Whole Building Performance Method involves developing a computer model of the Proposed Design and comparing its energy consumption to the Standard Design for that building. Energy consumption in the Standard Design represents the upper limit of energy use allowed for that particular building under a scenario where all the prescriptive requirements of the Code are adopted. Code compliance will be achieved if the energy use in Proposed Design is no greater than the energy used in the Standard Design. Three basic steps are involved:

i. Design the building with energy efficiency measures; the prescriptive approach requirements provide a good starting point for the development of the design.

ii. Demonstrate that the building complies with the mandatory measures.

iii. Using an approved simulation software, model the energy consumption of the building using the proposed features to create the Proposed Design. The model will also automatically calculate the energy use for the Proposed Design.

If the energy use in Proposed Design is no greater than the energy use in the Standard Design, the building complies with the Code.

The biggest advantage of using this approach is that it enables the design and construction team to make comparisons between different design options to identify the most cost-effective and energy-efficient design solution. For instance, the efficiency of the indoor lighting system might be improved in order to justify fenestration design that does not meet the prescriptive envelope requirements. As long as the total energy use of the Proposed Design does not exceed the allowed energy use in Standard Design, the building will be ECBC compliant.

Compliance Documents

(i) General: As per the Code:

Plans and specifications shall show all pertinent data and features of the building, equipment, and systems in sufficient detail to permit the Authority having jurisdiction to verify that the building complies with the requirements of this code. Details shall include, but are not limited to:

Building Envelope

- Insulation materials and their R-values
- Fenestration U-factors, SHGC, visible light transmittance (if using,the trade-off approach), and air leakage
- Overhang and side fin details
- Envelope sealing details.

HVAC

- Type of systems and equipment, including their sizes, efficiencies, and controls
- Economizer details
- Variable speed drives
- Piping insulation
- Duct sealing
- Insulation type and location
- Report on HVAC balancing.

Service Hot Water and Pumping

- Solar water heating system and pumping efficiency

Lighting

- Schedules that show type, number, and wattage of lamps and ballasts
- Automatic lighting shutoff details
- Occupancy sensors and other lighting control details
- Lamp efficacy for exterior lamps.

Electrical Power

- Schedules that show transformer losses, motor efficiency, and power factor correction devices
- Electric check metering and monitoring system details.

The documents submitted should include sufficient detail to allow thorough review by the Authority having jurisdiction for Code compliance. Additional information may be requested by the authority, if needed, to verify compliance.

(ii) Supplemental Information

The Authority having jurisdiction may require supplemental information necessary to verify compliance with this Code, such as calculations, worksheets, compliance forms, manufacturer's literature, or other data.

Building Envelope

The building envelope refers to the exterior facade, and is comprised of opaque components and fenestration systems. Opaque components include walls, roofs, slabs on grade, basement walls, and opaque doors. Fenestration systems include windows, skylights, ventilators, and doors that are more than one-half glazed. The envelope protects the building's interior and occupants from the weather conditions and shields them from other external factors, *e.g.* noise, air pollution, etc.

Envelope design strongly affects the visual and thermal comfort of the occupants, as well as energy consumption and heat transfer (Conduction, Convection and Radiation) in the building.

Heat transfer takes place through walls, windows, and roofs in buildings from higher temperature to lower temperature in three ways-conduction, convection, and radiation. Conduction is the transfer of heat by direct contact of particles of matter within a material or materials in physical contact. Convection is the transfer of heat by the movement of a fluid (air or gas or liquid). Radiation is the movement of energy/heat through space without relying on conduction through the air or by the movement of air.

The surface of the sun, estimated to be at a temperature of about 5500°C, emits electromagnetic waves. These waves are also known as solar radiation or short-wave radiation with wave length in the range of 0.3 to 2.5 microns or 300 nm to 2500 nm, and has three components: Ultra Violet (UV), Visible (the sun light which is visible to human eye) and Solar (or near) infrared.

When the 'Solar Infrared' component of the waves comes in contact with the earth or any object or a building, it transfers its energy to the object/building in the form of heat.

The phenomenon is known as solar radiation heat transfer. Radiation heat transfer can be between any two bodies having different temperatures with heat transfer taking place from the body at higher temperature to the body at the lower temperature. The Figure 8.7 shows all three modes of heat transfer across a building wall facing the external environment.

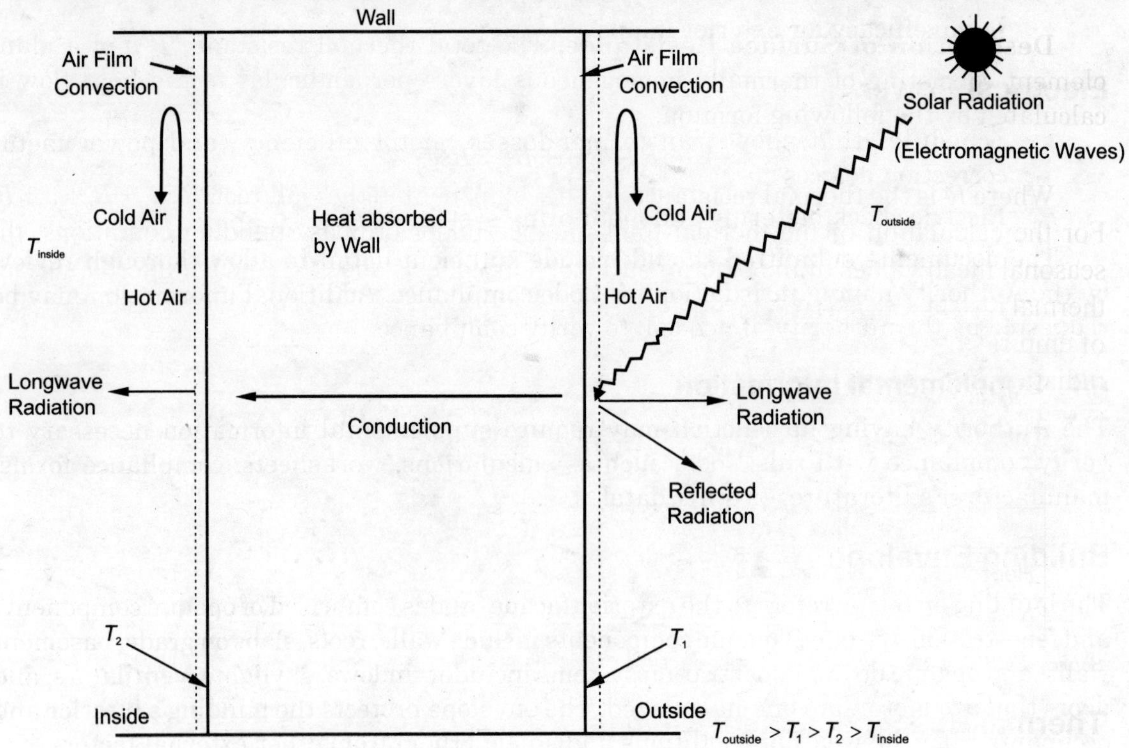

Fig.8.7: Schematic three modes of Heat Transfer

Conductive heat transfer across the envelope also depends upon the conductivity of the building material used. Different materials offer different thermal resistance to the conduction process. Individually, walls and roofs are comprised of a number of layers composed of different building materials. Thus, it is important to establish overall thermal resistance and heat transfer coefficient (U-factor), also termed thermal transmittance.

Conduction: Conduction is heat transfer through a solid medium as a result of a temperature gradient. The heat flow direction is in accordance with the second law of thermodynamics, from a region of higher temperature to that of lower temperature. The Conductivity is the property of material. The rate of heat transfer (q) through a homogeneous medium is given by Fourier's Law of Conduction: $q = kA(dT/dx)[W]$

q : Rate of heat transfer $[W]$

k : Thermal conductivity of the material $IWm^2 - Kr^1]$

A : Area [m²]

T : Temperature $[K]$

x : Distance in the direction of heat flow $[m]$

Resistance: Thermal Resistance is proportional to the thickness of material of construction and inversely proportional to its conductivity. Thus, a lower value of conductivity means less heat flow and so does the greater thickness of material. Together these parameters form the Thermal Resistance to the process of heat conduction.

$$R = d/k[\mathrm{m}^2\,\mathrm{kW}^1]$$

Description of Surface Resistance: The total thermal resistance *R-p* of a plane element consisting of thermally homogeneous layers perpendicular to the heat flow is calculated by the following formula:

$$RT{:}\ Rsi + Rt + Rse$$

Where R_t is the thermal resistances of the component in the wall/roof. $R_t{:}R_1 + R_2 + ... + R_n$ For the calculation of the thermal transmittance under ordinary building conditions, the seasonal mean values of the exterior surface thermal resistance (R^ and the interior surface thermal resistance ($R_s j$) can be obtained from Table given below. These values are the result of empirical studies and represent magnitudes of order. They consider both convection and radiation influences.

Table 8.4: Values of Surface Film Resistance Based on Direction of Heat Flow

R_{si}			R_{se}		
Direction of Heat Flow			Direction of Heat Flow		
Horizontal	Up	Down	Horizontal	Up	Down
0.13	0.10	0.17	0.04	0.04	0.04

Source: ECBC, 2008

Thermal Resistance of an Element Consisting of Homogenous Layers

A building element is usually composed of a number of different materials. When materials are placed in series, their thermal resistances are added so that the same area will conduct less energy for a given temperature difference. Formation of air film at the surface of wall or roof, due to convection movements of air, also provides resistance to the heat flow, similar to the construction material The total resistance of the wall or roof includes all of the resistances of the individual materials that make it up as well as both the internal and external air-film resistance.

$R_1 R_2, ... R_n$ are the thermal resistance of each layer. Thermal Resistance is also depicted as reciprocal of Thermal Conductance (U): $R = 1/U$

Thermal Resistance of Unventilated Air Layers

Table below gives the thermal resistances of unventilated air layers (valid for remittance of the bounding surfaces > 0.8). The values under "horizontal" should be used for heat flow directions ± 30° from the horizontal plane; for other heat flow directions, the values under "up" or "down" should be used.

Table 8.5: Thermal Resistance of Unventilated Air Layers Between Surfaces with High Emittance

Thickness of Air Layer (mm)	Thermal Resistance (m^2 K W'1) Direction of Heat Flow		
	Horizontal	Up	Down
5	0.12	0.10	0.10
7	0.12	0.12	0.12
10	0.14	0.14	0.14
15	0,16	0.16	0.16
25	0.18	0.17	0.18
50	0.18	0.17	0.20
100	0.18	0.17	0.20
300	0.18	0.17	0.21

R -Value Calculations for Cavity Wall Construction

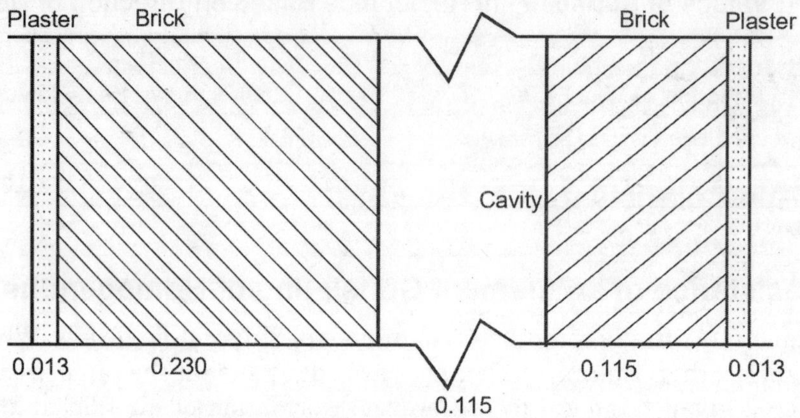

Fig.8.8: Typical Cavity Wall Construction

Table 8.6: Thermal Resistance of Various Building Materials

R1: Resistance for Layer 1 (13 mm Gypsum Plaster)	0.056 Ktn²/W
R2: Resistance of Layer 2 (230 mm brick wall, density = 1920 kg/m³)	d_2/k_2 = 0.230/0.81 = 0.284 ICm²/W
R3: Resistance of Layer 3 (115 mm air gap)	1.8 Km²/W
R4: Resistance of Layer 4 (115 mm brick wall, density = 1920 kg/m³)	d4/^4 = 0.115/0.81 =0.1426 Km²/W
R_t: Minimum R-value for the composite wall	*R*1 + *R*2 + *R*3 + *R*4 = 0.056+ 0.2840 + 0.18 + 0.1426 = 0.6626 I < *tn*²/W
Rx: *Rsi + Rt + Rse*	0.1 + 0.6626 + 0.04 = 0.8026
Maximum U-factor for the composite wall: $U_{m}j_n$	$1/R_t$ = 1/0.8026 = 1.246 W/m²-K

Source: ECBC, Table 11.4, 2008

The design of the building envelope is generally the responsibility of an architect. The building designer is responsible for making sure that the building envelope is energy-efficient and complies with the mandatory and prescriptive requirements of the Code. It also provides 'trade-off options' which allows flexibility to the designer to 'trade-off' prescriptive requirements of building components, while meeting the minimum energy performance requirements of the envelope.

Fenestration

Heat transfer across glazing products or fenestration (windows, door, and skylights) is similar to the heat transfer that takes place across walls and roofs through conduction and convection. So, U-factor of glazing is analogous to the U-factor of wall assembly. In addition, direct solar radiation contributes to the solar heat gain through the fenestration system.

Fenestration and doors must be rated using procedures and methods specified in the ECBC. Three fenestration performance characteristics are significant in the ECBC: U-factor, SHGC, and Visible Light Transmittance (VLT).

The U-factor of fenestration is very important to the energy efficiency of buildings, especially in cold climates. The U-factor must account for the entire fenestration product, including the effects of the frame, the spacers in double glazed assemblies, and the glazing. There are a wide variety of materials, systems, and techniques used to manufacture fenestration products, and accurately accounting for these factors is of utmost importance when calculating the U-factor. According to ECBC, Fenestration U-factors must be determined in accordance with ISO15099.

Solar Heat Gain Coefficient (SHGC) and U-Factor

Conduction heat flow through the fenestration (*e.g.* glass windows) is similar to the process discussed for walls and roofs. However regardless of outside temperature, heat gain through the fenestration is also dependent on direct and indirect solar radiation. The ability to control this heat gain is characterized in terms of SHGC. SHGC is the ratio of the solar heat gain that passes through the fenestration to the total incident solar radiation that falls on the fenestration. The solar heat gain includes directly transmitted solar heat and absorbed solar radiation, which is then re-radiated, converted, or conducted into the interior space. SHGC indicates how well the glazing/glass and fenestration products insulate heat caused by sun falling directly on the glass.

In hot climates, SHGC is more important than the U-factor of the glazing. A lower SHGC means that lesser heat can pass through the glazing. The SHGC is based on the properties of the glazing material, whether the window has single, double, or triple glazing, and the window operation (either operable or fixed). Glazing units with a low SHGC help to reduce the air conditioning energy use during the cooling season.

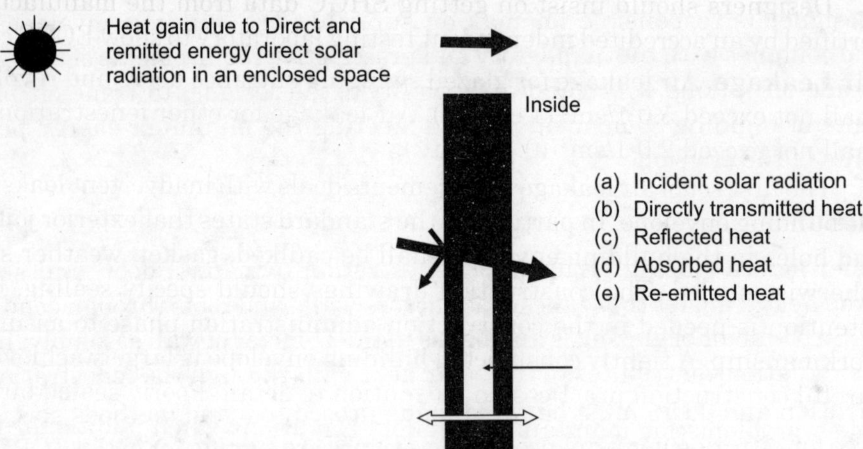

Fig.8.9: Direct and Indirect Solar Radiation

i. **U-factor:** Clear glass, which is the most common type of glass, has no significant thermal resistance (*R*-value) from the pane itself. It has a value of *R*-0.9 to *R*-1.0 due to the thin films of air on the interior and exterior surfaces of the glass. The *U*-factor (thermal conductance), must account for the entire fenestration system including the effects of the frame, the spacers in double glazed assemblies, and the glazing. There are a wide variety of materials, systems, and techniques used to manufacture fenestration products, and accurately accounting for these factors is of utmost importance when meeting the fenestration requirements. The Code also specifies *U*-factor for sloped glazing and skylights, and minimum *U*-factors for unrated products.

   ECBC has used W/m^2 -C as the unit for *U*-factor. Since differences in temperature are always denoted in *K* in physics literature, ECBC User Guide has used W/m^2 K as the unit of *U*-factor, where, *K* is the difference in temperatures. *U*-factors for fenestration systems (including the sash and frame) are required to be determined in accordance with ISO-15099 by an accredited independently laboratory and labelled and certified by the manufacturer or other responsible party.

   In the U.S, the fenestration U-factors are determined in accordance with the National Fenestration Rating Council (NFRC) Standard 100. NFRC is a membership organization of window manufacturers, researchers, and others that develops, supports, and maintains fenestration rating and labeling procedures. Most fenestration manufacturers have their products rated and labelled through the NFRC program. Certified products receive an 8½ by 11 inch NFRC label that lists the U-factor, SHGC, and the visible transmittance.

ii. **Solar Heat Gain Coefficient:** The ECBC requires that SHGC be determined in accordance with ISO-15099 by an accredited independent laboratory, and labelled and certified by the manufacturer or other responsible party. SHGC has replaced Shading Coefficient (SC) as the preferred specification for solar heat gain through fenestration products.

Designers should insist on getting SHGC data from the manufacturers duly certified by an accredited independent testing laboratory to show ECBC compliance.

iii. **Air Leakage.** Air leakage for glazed swinging entrance doors and revolving doors shall not exceed 5.0 $1/sm^2$ (1 cfm/ft^2). Air leakage for other fenestration and doors shall not exceed 2.0 $1/sm^2$ (0.4 cfm/ft^2).

The first set of air leakage requirements deals with inadvertent leaks at joints in the building envelope. In particular, the standard states that exterior joints, cracks, and holes in the building envelope shall be caulked, gasket, weather stripped, or otherwise sealed. The construction drawings should specify sealing, but special attention is needed in the construction administration phase to assure e proper workmanship. A tightly constructed building envelope is largely achieved through careful construction practices and attention to detail. Poorly sealed buildings can cause problems for maintaining comfort conditions when additional infiltration loads exceed die HVAC design assumptions. This can be a significant problem in high-rise buildings due to stack effect and exposure to stronger winds.

iv. **Opaque Construction.** U-factors shall be determined from data or procedures contained in the ASHRAE Fundamentals, 2005.

v. **Building Envelope Sealing.** Air leakage can also occur through opaque construction. Apart from adding cooling or heating load in the building, air leakage can cause condensation within walls and roof can damage insulation material and degrade other building materials.

Building sealing is more important in air-conditioned buildings. In naturally ventilated buildings, the concept of building sealing and tight envelope runs counter to conventional and traditional wisdom.

Building Envelope Sealing and Air Leakage

Air leakage is the passage of air through a building envelope, wall, window, joint, etc. Leakage to the interior is referred to as infiltration and leakage to the exterior is referred to as ex-filtration. Excessive air movement significantly reduces the thermal integrity and performance of the envelope and is, therefore, a major contributor to energy consumption in a building.

A tightly constructed building envelope is largely achieved through careful construction practices and attention to detail Building envelopes should be carefully designed to limit the uncontrolled entry of outdoor air into the building. Air leakage introduces sensible heat into conditioned spaces. In climates with moist outdoor conditions, it is also a major source of latent heat, which is removed by the air-conditioning system at considerable expense. In addition to causing energy loss, excessive air leakage can cause condensation to form within and on walls. This can create many problems including reducing insulation R-value, permanently damaging insulation, and seriously degrading materials.

It can rot wood, corrode metals, stain brick or concrete surfaces and in extreme cases cause concrete to break, bricks to separate, mortar to crumble and sections of a wall to fall, jeopardizing the safety of occupants. It can corrode structural steel re-bar, and metal hangars and bolts with very serious safety and maintenance consequences. Moisture accumulation in building materials can lead to the formation of mould that may require extensive remedying the situation.

Virtually anywhere in the building envelope where there is a joint, junction or opening, there is potential for air leakage. Air leakage will cause the HVAC system to run more often and longer at one time, and still leave the building uncomfortable for its occupants,

All openings in the building envelope, including gaps, joints and other openings that are potential sources of air leakage, should be to be sealed to minimize air leakage. Areas in the building envelope where attention should be paid to infiltration control include:

i. Joints around fenestration and door frames
ii. Openings at penetrations of utility services through roofs, walls, and floors.
iii. Site-built fenestration and doors.
iv. Building assemblies used as ducts or plenums.
v. Joints, seams, and penetrations of vapor retarders.
vi. All other openings in the building envelope.

Prescriptive Requirements

For envelope component-based compliance approach, ECBC sets requirements for:

- Exterior roofs and ceilings
- Cool roofs
- Opaque walls
- Vertical fenestration
- Skylights

Roofs

In roofs, the U-factor for the overall assemblies or minimum R-values for the insulation must be complied with the provisions of the Code. In real practice, the heat gains through the walls, roof, and fenestration depends upon the climate zone in which the building is located. The National Building Code of India, 2005 has divided the country in five climate zones (Hot Dry; Warm-Humid; Composite; Temperate/Moderate; and Cold), and the air temperature and humidity variations need to be considered while designing the building.

Building Comfort Requirements

The ECBC building envelope requirements are based on and vary with the climate zone in which the building is located. Based on the characteristics of climate, the thermal comfort requirements in buildings and their physical manifestation in architectural form are also different for each climate zone (Table 8.7). These physical manifestations, in turn, dictate the ECBC requirements for the envelope, as well as other building components.

Table 8.7: Comfort Requirements and Physical Manifestations in Buildings

Thermal Requirements	Physical Manifestation
1. Hot and Dry Climate Zone **Reduce Heat Gain**	
Decrease exposed surface area	Orientation and shape of building
Increase thermal resistance	Insulation of building envelope
Increase thermal capacity (Time lag)	Massive structure
Increase buffer spaces	Air locks/lobbies/balconies/verandahs
Decrease air exchange rate (ventilation during day-time)	Smaller window openings, ventilation
Increase shading	External surfaces protected by overhangs, fins and trees
Increase surface reflectivity	Pale colour, glazed china mosaic tiles, etc.
Promote heat loss	
Increase air exchange rate (Ventilation during night-time)	Courtyards/wind towers/arrangement of openings
Increase humidity levels	Trees, water ponds, evaporative cooling
2. Warm and Humid Climate Zone **Reduce Heat Gain**	
Decrease exposed surface area	Orientation and shape of building
Increase thermal resistance	Roof insulation and wall insulation Reflective surface of roof
Increase buffer spaces	Balconies and verandas
Increase shading	Walls, glass surfaces protected by overhangs, fins and trees
Increase surface reflectivity	Pale color, glazed china mosaic tiles, etc.
Reduce solar heat gain	Use glazing with lower SHGC and provide shading for windows. Minimize glazing in East and West
Promote heat loss Increase air exchange rate (Ventilation throughout the day)	Ventilated roof construction, courtyards, wind towers and arrangement of openings
Decrease humidity levels	Dehumidifiers/desiccant cooling
3. Moderate Climate Zone **Reduce Heat Gain**	
Decrease exposed surface area	Orientation and shape of building
Increase thermal resistance	Roof insulation east and west wall insulation
Increase shading	East and west walls, glass surfaces protected by overhangs, fins and trees
Increase surface reflectivity	Pale colour, glazed china mosaic tiles, etc
Promote heat loss	
Increase air exchange rate (Ventilation)	Courtyards and arrangement of openings

4. Cold (Cloudy/Sunny) Climate Zone **Reduce heat loss**	
Decrease exposed surface area	Orientation and shape of building. Use of trees as wind barriers
Increase thermal resistance	Roof insulation, wall insulation and double glazing
Increase thermal capacity (Time lag)	Thicker walls
Increase buffer spaces	Airlocks/lobbies
Decrease air exchange rate	Weather stripping and reducing air leakage
Increase surface absorptive	Darker colours
Promote heat gain	
Reduce shading	Walls and glass surfaces
Trapping heat	Sun space/green-house/Trombe wall, etc.
5. Composite Climate Zone **Reduce heat gain in summer and reduce heat loss in winter**	
Decrease exposed surface area	Orientation and shape of building. Use of trees as wind barriers
Increase thermal resistance	Roof insulation and wall insulation
Increase thermal capacity (Time lag)	Thicker walls
Increase buffer spaces	Air locks/Balconies
Decrease air exchange rate	Weather stripping
Increase shading	Walls, glass surfaces protected by overhangs, fins and trees
Increase surface reflectivity	Pale color, glazed china mosaic tiles, etc.
Reduce solar heat gain	Use glazing with lower SHGC and provide shading for windows. Minimize glazing in East and West
Promote heat loss in summer / monsoon	
Increase air exchange rate (Ventilation)	Courtyards/wind towers/arrangement of openings
Increase humidity levels in dry summer	Trees and water ponds for evaporative cooling
Decrease humidity in monsoon	Dehumidifiers/desiccant cooling

Source: ECBC, 2008

Exterior roofs can meet the prescriptive requirements in two ways:

- Use the required R-value of the insulation (this R-value does not apply to building materials or air film. It should be referred exclusively for insulation), or
- Use a roof assembly U-factor that meets the maximum U-factor criterion for thermal performance. The U-factor takes into account all elements or layers in the construction, assembly, including the sheathing, interior finishes, and air gaps, as well as exterior and interior air films.

As per the Code:

- The roof insulation shall not be located on a suspended ceiling with removable ceiling panels.

The Code requirements for the U-factor and R-values for 24 hours use buildings and daytime use buildings for five climate zones are shown in Table below.

Table 8.8: Roof Assembly U-Factor and Insulation R-value Requirements

Climate Zone	24-Hour use buildings Hospitals, Hotels, Call Centres, etc.		Daytime use buildings, Other Building Types	
	Maximum U-factor of the overall assembly (W/m^2 K)	Minimum R-value of insulation alone (m^2 K/W)	Maximum U-factor of the overall assembly (W/m^2 K)	Minimum R-value of insulation alone (m^2 K/W)
Composite	U-0.261	R-3.5	U-0.409	R-2.1
Hot and Dry	U-0.261	R-3.5	U-0.409	R-2.1
Warm & Humid	U-0.261	R-3.5	U-0.409	R-2.1
Moderate	U-0.409	R-2.1	U-0.409	R-2.1
Cold	U-0.261	R-3.5	U-0.409	R-2.1

Source: ECBC, 2008

INSULATION

The mandatory requirements of ECBC address the proper installation and protection of insulation materials. It is recommended that insulation materials be installed according to the manufacturer's recommendations and in a manner that will achieve the rated insulation R-value. Compressing the insulation reduces the effective R-value and the thermal performance of the construction assembly.

It is recommended that insulation be installed in a permanent manner and in substantial contact with the inside surface of the construction assembly. If the insulation does not entirely fill the cavity, the air gap should be on the outside surface. Maintaining substantial contact is particularly important for insulation installed between floor joists. Without proper support, gravity will cause the insulation to fall away from the floor surface, leaving an air gap above the insulation. Air currents will ultimately find their way to the gap, and when they do, the effectiveness of the insulation will be substantially reduced.

It is not a good practice to install insulation directly over suspended ceilings with removable ceiling panels. This is because the insulation's continuity is likely to be disturbed by maintenance workers. Also, suspended ceilings may not meet the ECBC's infiltration requirements unless they are properly sealed. Many building codes consider the space above the ceiling to be an attic and require that it be ventilated to the exterior. If ventilated to the exterior, air in the attic could be quite cold (or hot) and the impact of the leaky suspended ceiling could be a risk.

Cool Roofs

Depending on the material and construction, a roof will have different properties that determine how it conducts heat to the inside of the building. "Cool roofs" are roofs covered with a reflective coating that has a high emissivity property that is very effective in reflecting the sun's energy away from the roof surface. These "cool roofs" are known to stay 10°C to 16°C cooler than a normal roof under a hot summer sun; this quality greatly reduces heat gain inside the building and the cooling load that needs to be met by the HVAC system. Fig. 8.11 indicates how solar heat radiation is reflected, absorbed and emitted from the roof and the need to develop cool roofs.

The heat transfer process in the roof is similar to the heat transfer that takes in a wall. Heat transfer across the roof is more prominent compared to the wall because of higher incidence of solar radiation. Depending on the properties of the roof material and construction, the roof reflects part of the solar radiation back to the environment, and absorbs the other part of the heat in the roof. Finally, portion of the absorbed heat in the roof is emitted as long-wave radiation back to the environment and the remaining part of the absorbed heat is conducted inside of the building. This heat transfer process is governed by the Solar Reflectance and Emissivity (Thermal Emittance) properties of the roof material, apart from the thermal conductivity of the materials used in the roof.

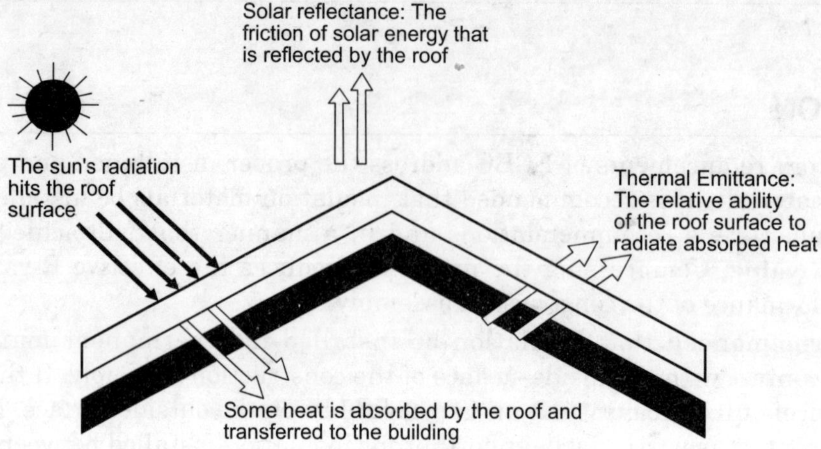

Fig.8.10: Heat Transfer through Roof

Source: ECBC, 2008

The solar reflectance is the fraction of solar radiation reflected by roof. The complement of reflectance is absorbance; whatever radiant energy incident on a surface that is not reflected is absorbed in the roof. The reflectance and absorbance of building materials are usually measured across the solar spectrum, since these are exposed to that range of wavelength.

Reflectance is measured on a scale of 0 to 1, with 0 being a perfect absorber and 1 being a perfect reflector. Absorbance is also rated from 0 to 1, and can be calculated from the relation: Reflectance + Absorbance = 1.

Emissivity (or thermal remittance) of a material (usually written e) is the ratio of energy radiated by a particular material to energy radiated by a black body at the same temperature. It is a measure of a material's ability to radiate the absorbed energy. A true black body would have an $e = 1$ while any real object would have < 1. Emissivity is a dimensionless quantity (does not have units). In general, the duller and blacker a material is, the closer its emissivity is to 1. The more reflective a material is, the lower its emissivity. The emissivity of building material, unlike reflectance, is usually measured in the far infrared part of the spectrum.

An ideal exterior surface coating of a building in hot climate and under indoor cooling would have reflectance near 1, and absorbance near zero, and emissivity near 1 to radiate absorbed heat back to the sky. If designing a cool roof, requirements for minimum solar reflectance and initial emitance levels are specified in the ECBC (4.3.1.1).

Roofs with slopes less than 20° shall have an initial solar reflectance of no less than 0.70 and an initial emittance no less than 0.75. Solar reflectance shall be determined in accordance with ASTM E903-96 and emittance shall be determined in accordance with ASTM E408-71 (RA 1996).

In hot climates, cool roofs (or high emissivity or thermal emittance roof surfaces) are an effective way to reduce solar gains and cut building energy costs. Because cool roofs gain less heat than normal roofs, they reduce the need for air conditioning and make buildings more comfortable to the people inside. The light colour reflects sunlight and heat away from the building and the high emissivity or emittance allows heat to escape to the atmosphere when the surface becomes heated. Although some surfaces, such as galvanized metal, have a high reflectance, they have a low emittance. These surfaces reflect heat but heat that is absorbed cannot escape. Other surfaces, such as dark paint, have a high emittance but a low reflectance. These surfaces allow heat to escape, but do a poor job of reflecting heat that strikes the surface.

Most cool roof materials for low-sloped roofs are white or another light color. For steep sloped roofs that are often visible from the ground, roofing material manufacturers have developed popular roof colors other than white that still reflect solar radiation or emit the sun's energy away from the building. Cool roofs have other benefits in addition to reducing operating absorbed heat costs. For building owners, they can cut maintenance costs and increase the life expectancy of the roof. Cool roofs can also help to reduce the urban heat island effect and slow down global warming that makes our cities hotter and produces unhealthy air.

Urban Heat Island Effect

An Urban Heat Island is an urban area, which is significantly warmer than its surroundings. As urban centres grow in size, they tend to have a corresponding increase in average temperature. Scientists refer to this phenomenon as the "Urban Heat Island Effect". The two main causes of the urban heat island is modification of the land surface by urban development and waste heat generated by energy usage. One consequence of urban heat island is the increased energy required for air conditioning and refrigeration in buildings.

Roofing

Products for low-slope roofs, found on commercial and industrial buildings fall into two categories: single-ply materials and coatings. Single-ply materials are large sheets of pre-made roofing that are mechanically fastened over the existing roof and sealed at the seams. Coatings are applied using rollers, sprays, or brushes, over an existing clean, leak-free roof surface. Products for sloped roofs are currently available in clay or concrete tiles. These products stay cooler by the use of special pigments that reflect the sun's infrared heat. In India, lime coats, white tiles grouted with white cement, special paints, etc. are used as cool roofing materials.

Opaque Walls

Opaque walls can meet the prescriptive requirements by either using a construction that has an assembly U-factor lower than the specified value as shown in Table below, or by using insulation with R-value more than the prescribed value. R-value is for the insulation alone and does not include building materials or air films.

Table 8.9: Opaque Wall Assembly U-Factor and Insulation R-value Requirements

Climate Zone	Hospitals, Hotels, Call Centers (24-Hour)		Other Building types (Day time)	
	Maximum U-factor of the overall assembly (W/m^2 -K)	Minimum R-value of insulation alone (m^2 K/W)	Maximum U-factor of the overall assembly (W/m^2 K)	Minimum R-value of insulation alone (m^2 K/W)
Composite	U-0.440	R-2.10	U-0.440	R-2.10
Hot and Dry	U-0.440	R-2.10	U-0.440	R-2.10
Warm and Humid	U -0.440	R-2.10	U-0.440	R-2.10
Moderate	U -0.440	R-2.10	U-0.440	R-2.10
Cold	U-0.369	R-2.20	U-0.352	R-2.35

Source: ECBC, 2008

Vertical Fenestration

The ECBC limits the area of vertical fenestration, under the prescriptive approach, to a maximum of 60% of the gross wall area.

The ECBC addresses energy losses through fenestration by specifying the following requirements: maximum U-factor (or thermal transmittance) and maximum SHGC, for the following window to wall ratio (WWR):

- WWR up to 40% and
- WWR in the range of more than 40% and up to 60%

Vertical fenestration should meet the requirements for maximum area weighted U-factor and maximum area weighted SHGC. The U-factor and SHGC requirements of the rated (labeled) fenestration for two WWR ranges for Code compliance are given in Table below.

Table 8.10: Vertical Fenestration U-factor (W/m^2*K) and SHGC Requirements

Climate	Maximum U-factor	Maximum U-factor (WWRd 40%)	Maximum SHGC (40%<WWRd 60%)
Composite	3.30	0.25	0.20
Hot and Dry	3.30	0.25	0.20

Warm and Humid	3.30	0.25	0.20
Moderate	6.90	0.40	0.30
Cold	3.30	0.51	0.51

Source: ECBC, 2008

Table 8.11: Defaults for Unrated Vertical Fenestration (Overall Assembly including Sash and Frame)

Frame Type	Glazing Type	Clear Glass			Tinted Glass		
		U-factor (W/m^2 K)	SHGC	VLT	U-Factor (W/m^2 K)	SHGC	VLT
All frame types	Single Glazing	7.1	0.82	0.76	7.1	0.70	0.58
Wood, vinyl, fiber glass frame	Double Glazing	3.3	0.59	0.64	3.4	0.42	0.39
Metal and other prime type	Double Glazing	5.1	0.68	0.66	5.1	0.50	0.10

Source: ECBC, 2008

Energy Efficient Fenestration

Windows are affected by many factors, which in turn affect the comfort and energy performance of buildings. Understanding these factors is critical in designing buildings that meet the needs of building owners and users. Once these factors are identified, a designer can then apply the appropriate technology to address them. A fenestration product is comprised of three areas: the vision area, the glazing, and the opaque area or the frame. In a window, glazing is generally 90-95% of the total area and therefore, it is the most important part to address for achieving energy efficiency. However, the frame becomes important to optimize the overall energy efficiency of the window.

The energy efficiency of a fenestration product is affected by:

- Films which are applied to improve energy efficiency

Table 8.12: SHGC "M" Factor Adjustments for Overhangs and Fins

Project Location	Orientation	Overhang "M" Factors for 4 Projection Factors				Vertical Fin "M" Factors for 4 Projection Factors				Overhang + Fin "M" Factors for 4 Projection Factors			
		0.25	0.5	0.75	1.00	0.25	0.50	0.75	1.00	0.25	0.50	0.75	1.00
		–	–	–	+	–	–	–	+	–	–	–	+
		0.49	0.74	0.99		0.49	0.74	0.99		0.49	0.74	0.99	
North latitude 15° or greater	N	.88	.80	.76	.73	.74	.67	.58	.52	.64	.51	.39	.31
	E/W	.73	.65	.56	.50	.80	.72	.65	.60	.60	.39	.24	.16
	S	.79	.64	.52	.43	.79	.69	.60	.56	.60	.33	.10	.02
Less than 15° North latitude	N	.83	.74	.69	.66	.73	.65	.57	.50	.59	.44	.32	.23
	E/W	.80	.67	.59	.53	.80	.72	.63	.58	.61	.41	.26	.16
	S	.78	.62	.55	.50	.74	.65	.57	.50	.53	.30	.12	.04

Source:ECBC, 2008

Minimum Visible Transmission of Glazing for Vertical Fenestration

Vertical fenestration product shall have the minimum Visual Light Transmittance (VLT), defined as function of Window Wall Ratio (WWR), where Effective Aperture > 0.1, equal to or greater than the minimum VLT requirements.

Table 8.13: Minimum VLT Requirements

Window/Wall Ratio	**Minimum VLT**
0 - 0.3	0.27
0.31-0.4	0.20
0.-11-0.5	0.16
0.51-0.6	0.13

Source: ECBC, 2008

Day lighting, Visual Light Transmittance and Effective Aperture

Visual Light Transmittance (also known as Visual Transmittance VT) is defined as the ratio of light that passes through the glazing to the light passing through perfectly transmitting glazing. In other words it also refers to the fraction of visible light transmitted through the glazing. VLT is concerned with the visible portion of the solar spectrum as opposed to SHGC, which takes into account the entire solar radiation. VLT affects energy consumption in building by providing daylight that creates the opportunity to reduce electric lighting and its associated cooling loads. Glazing with low SHGC generally has a low VLT; however, if the VLT is too low, the outside view from inside the building will be impaired. With lower VLT, the daylighting in the interior may also reduce to a level that may require supplemental electrical lighting for some occupants' functions, or to make the environment productive and enjoyable to the occupants. Thus, buildings with lower window to wall ratios (WWR) may need higher VLT.

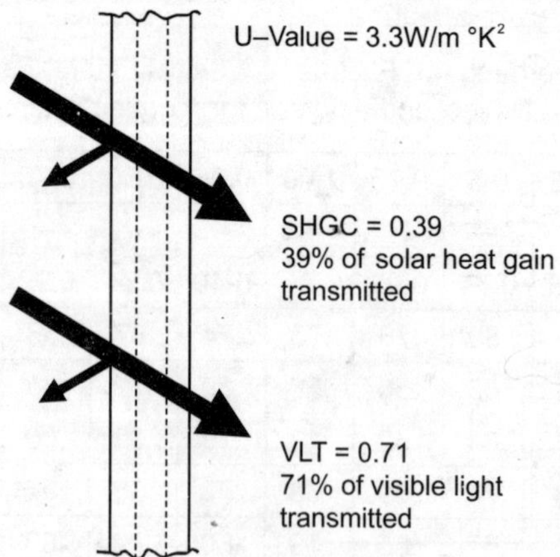

Fig.8.11: Illustration showing U-factor, SHGC, and VLT

Effective Aperture of Glazing

As the area of an aperture/opening in the building envelope increases, the amount of daylight received in the building space also increases. However the glazing material within that aperture can effectively reduce the amount of visible light that enters the space. Therefore, aperture size alone is not an effective determinant to measure illumination levels. If the glazing in an opening is a perfectly transparent material the effective aperture size would be equal to the area of the opening (because the visible transmittance of the glazing would be one). If however, the glazing has a VLT of 0.5, the opening will transmit only half of the light striking it, and the effective aperture will be half of the actual size of the opening. The Effective Aperture (EA) or light admitting potential of a glazing system is determined by multiplying the Visible Light Transmittance of the glazing by the window-to-wall ratio of the building. The window to wall ratio is the ratio of the net window area to the exterior wall area.

Skylights

A skylight is a fenestration surface having a slope of less than 60 degrees from the horizontal plane. Other fenestration, even if mounted on the roof of a building, is considered vertical fenestration.

Skylights can be installed into a roof system either flush-mounted or curb-mounted (including site built). In order to create a positive water flow around them, skylights are often mounted on "curbs" set above the roof plane. However, these curbs, rising 6 to 12 inches (15 to 30 centimeters) above the roof, create additional heat loss surfaces right where the warmest air of the building tends to collect. Portions of roof that serve as curbs that mount the skylight above the level of the roof are part of the opaque building envelope.

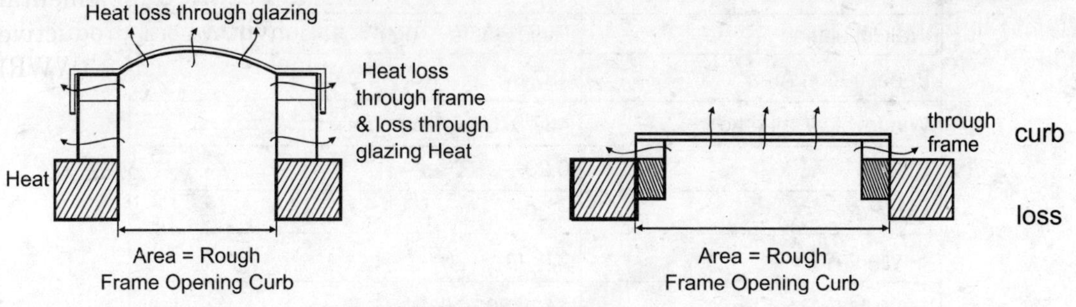

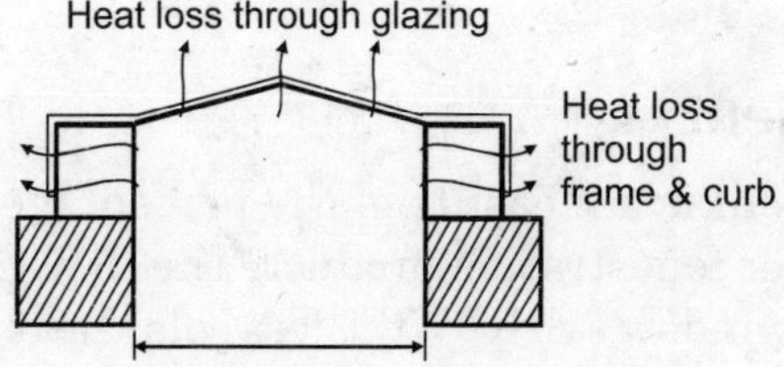

Fig.8.12: Skylight Installations

Source: ECBC, 2008

Skylights shall comply with the maximum D-factor and maximum SHGC requirements. Skylight area is limited to a maximum of 5% of the gross roof area for the prescriptive requirement.

Table 8.14: Skylight U-Factor and SHGC Requirements

Climate	Maximum U-Factor		Maximum U-SHGC	
	With curb	Without curb	0-2% SRR	2–1.5% SRR
Composite	11.24	7.71	0.40	0.25
Hot and Dry	11.24	7.71	0.40	0.25
Warm and Humid	11.24	7.71	0.40	0.25
Moderate	11.24	7.71	0.61	0.4
Cold	11.24	7.71	0.61	0.4

Source: ECBC, 2008

Table 8.15: Prescriptive Requirements for Skylights (an example)

Location	Chennai
Climate Zone	Warm-Humid
Building Type	Daytime Use Building
Roof Area	1,863 m^2
Roof Insulation	Rigid Board 1 inch R- 2.1 m^2-°K/W
Wall Area	3,706 m^2
Wall Insulation	Rigid Board 1 inchR= 1.41 m^2-°K/W
Fenestration Area	487 m^2
Window to Wall ratio	487/3706 = 13%
SHGC	0.20
U-factor	3.30
Skylight Area	112 m^2
Skylight to Roof Area	112/1863= 6%

Source: ECBC, 2008

Solar Heat Gain Coefficient

The Solar Heat Gain Coefficient is a measure of the percentage of heat from the sun that gets through a window or other fenestration product. The SHGC is expressed as a number between 0 and 1. The lower a window's SHGC, the less solar heat it transmits to the interior of the building. SHGC can also refer to shading so the lower the SHGC the more effective the product is at saving the heat gain from entering the interior.

Heating, Ventilation and Air Conditioning

Heating, Ventilation and Air Conditioning (HVAC) refer to the equipment, distribution systems, and terminals that provide, either collectively or individually, the heating, ventilation, or air-conditioning requirement to a building or a portion of building. The HVAC system accounts for a significant (nearly 50%) portion of a commercial building's energy use. HVAC energy use in a commercial building can increase/decrease significantly depending on how efficiently the combination of air side systems and central plant operates. The requirements generally vary from 40% to 90% of full load. Proven technologies and design concepts can be used to build energy efficiencies in the system, generate significant energy and cost savings besides efficient performance.

HVAC systems affect the health, comfort, and productivity of occupants. Issues like user discomfort, improper ventilation, lack of air movement and poor indoor air quality, and poor acoustic design are linked to HVAC system design and operation. In many existing buildings, envelope upgrades are often necessary to improve comfort and energy efficiency, through improvements such as reducing envelope leakage. Generally, upgrading an existing building envelope is expensive. Other strategies such as central plant, airside or control system upgrade may be necessary to improve occupant comfort and energy efficiency.

The best HVAC design considers all the interrelated building systems while addressing indoor air quality, thermal comfort, energy consumption, and environmental benefits. Optimizing both the design and the benefits requires that the architect and HVAC system designer address these issues early in the schematic design phase and continually revise subsequent decisions throughout the remaining design process. It is also essential that a process be implemented to monitor proper installation and operation of the HVAC system throughout construction. An effective commissioning plan for each of the systems at full load and part load including controls calibration and commissioning is essential to the optimal performance of the building.

Air Conditioning System

Basic components of the system include a compressor, condenser (air-cooled or water cooled), evaporator and an expansion device, similar to that of a domestic refrigerator.

Room air is drawn across an indoor coil called the evaporator that cools and dehumidifies the air during the cooling cycle. The condenser condenses the refrigerant and transforms the high pressure vapor into high pressure liquid. Heat is rejected via air drawn across the condenser coils using fans (air-cooled condenser) or using a shell and tube heat exchanger in conjunction with a condenser water reticulation system and cooling towers (water cooled condenser). The expansion device transforms the high pressure high temperature liquid refrigerant to low pressure low temperature mixture of refrigerant liquid and vapor. This mixture fully evaporates in the evaporator absorbing the heat from the water (cooling the water in a chilled water system) or cooling the air drawn across the coil (direct expansion system). The compressor then raises the pressure and temperature of the refrigerant and the cycle continues on.

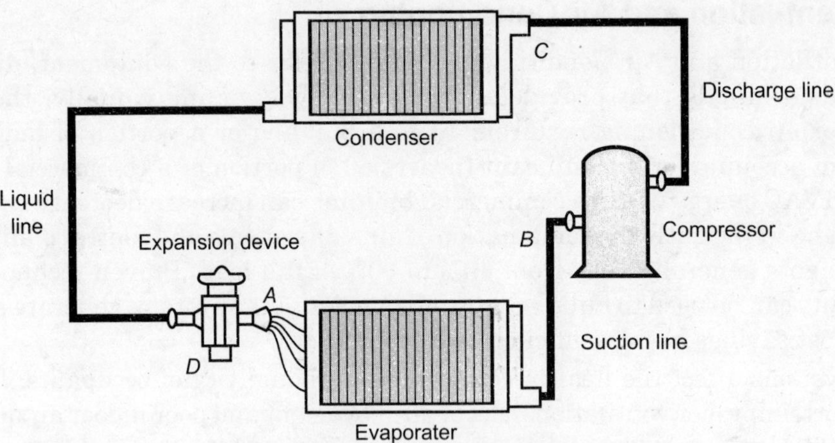

Fig.8.13: An Air-conditioning System

Heating Systems

Heating system types can be classified by the heating equipment type. The heating equipment used in buildings includes boilers (oil and gas), furnaces, (oil, gas, and electric), heat pumps, and space heaters.

Boiler-based heating systems have steam and/or water piping to distribute heat. Boilers can be self-contained units, or they can be packaged units which are factory-built systems, disassembled for shipment, and reassembled at the site. The heated water may serve preheat coils in air handling units; reheat coils, and local radiators. Systems that circulate water or a fluid are called hydraulic systems. Heating water may also be used for heating of service water and other process needs, depending on the building type. Some central systems have steam boilers rather than hot water boilers because of the need for steam for conditioning needs (humidifiers in air-handling units) or process needs (sterilizers in hospitals, direct-injection heating in laundries and dishwashers, etc.). The remaining heating systems include heat pumps and space heaters that heat directly and require little or no distribution.

Mandatory Requirements

The Code contains mandatory requirements for the following elements of the HVAC system:

- Natural Ventilation
- Controls
- System Balancing
- Economizers
- Equipment Efficiency
- Piping and Ductwork
- Condensers
- Hydronic Systems

Natural Ventilation

Natural ventilation of buildings shall comply with the design guidelines provided for natural ventilation in the National Building Code of India 2005,(NBC 2005) Part 8, 5.4.3 and 5.7.1 These guidelines from NBC, 2005 are reproduced below. However, the exact relevance of these general guidelines in the design of commercial buildings needs to be critically examined. The following are the general guidelines for ventilation of buildings:

i. Building need not necessarily be oriented perpendicular to the prevailing outdoor wind; it may be oriented at any convenient angle between 0° and 30° without losing any beneficial aspect of the breeze. If the prevailing wind is from east or west, building may be oriented at 45° to the incident wind so as to diminish the solar heat without much reduction in air motion indoors.

ii. Inlet openings in the buildings should be well distributed and should be located on the windward side at a low level, and outlet openings should be located on the leeward side. Inlet and outlet openings at high levels may only clear the top air at that level without producing air movement at the level of occupancy.

iii. Maximum air movement at a particular plane is achieved by keeping the sill height of the opening at 85% of the critical height (such as head level) for the following recommended levels of occupancy:
   - For sitting on chair 0.75 m
   - For sitting on bed 0.60 m
   - For sitting on floor 0.40 m

iv. Inlet openings should not, as far as possible, be obstructed by adjoining buildings, trees, sign boards or other obstructions or by partitions inside in the path of air flow.

v. In rooms of normal size having identical windows on opposite walls the average indoor air speed, increases rapidly by increasing the width of the window up to two-thirds of the wall width; beyond that the increase is in much smaller proportion than the increase of the window width. The air motion in the working zone is maximum when window height is 1.1 m. Further increase in window height promotes air motion at higher level of window, but does not contribute additional benefits as regards air motion in the occupancy zones in buildings.

vi. Greatest flow per unit area of openings is obtained by using inlet and outlet openings of nearby equal areas at the same level.

vii. For a total area of openings (inlet and outlet) of 20% to 30% of floor area, the average indoor wind velocity is around 30% of outdoor velocity. Further increase in window size increases the available velocity but not in the same proportion. In fact, even under most favorable conditions the maximum average indoor wind speed does not exceed 40% of outdoor velocity.

viii. Where the direction of wind is quite constant and dependable, the size of the inlet should be kept within 30 to 50% of the total area of openings and the building should be oriented perpendicular to the incident wind. Where direction of the wind is quite variable, the openings may be arranged so that as far as possible there is approximately equal area on all sides. Thus no matter what the wind direction is, there would be some openings directly exposed to wind pressure and others to air suction and effective air movement through the building would be assured.

ix. Windows of living rooms should open directly to an open space. In places where building sites are restricted, open space may have to be created in the buildings by providing courtyards.

x. In the case of rooms with only one wall exposed to outside, provision of two windows on that wall is preferred to that of a single window.

xi. Windows located diagonally opposite to each other with the windward window near the upstream comer give better performance than other window arrangements for most of the building orientations.

xii. Horizontal louvers, that is a sunshade, atop a window deflects the incident wind upward and reduces air motion in the zone of occupancy. A horizontal slot between the wall and horizontal louver prevents upward deflection of air in the interior of rooms. Provision of inverted L type louver increases the room air motion provided that the vertical projection does not obstruct the incident wind.

xiii. Provision of horizontal sashes inclined at an angle of 45° in appropriate direction helps to promote the indoor air motion. Sashes projecting outward are more effective than projecting inward.

xiv. Air motion at working plane 0.4 m above the floor can be enhanced by 30% using a pelmet type wind deflector.

xv. Roof overhangs help promoting air motion in the working zone inside buildings.

xvi. Verandah open on three sides is to be preferred since it causes an increase in the room air motion for most of the orientations of the building with respect to the outdoor wind.

xvii. A partition placed parallel to the incident wind has little influence on the pattern of the air flow, but when located perpendicular to the main flow, the same partition creates a wind shadow. Provision of a partition with spacing of 0.3 m underneath helps by augmenting air motion near floor level in the leeward compartment of wide span buildings.

xviii. Air motion in a building unit having windows tangential to the incident wind is accelerated when unit is located at end-on position on downstream side.

xix. Air motion in two wings oriented parallel to the prevailing breeze is promoted by connecting them with a block on downstream side.

xx. Air motion in a building is not affected by constructing another building of equal or smaller height on the leeward side; but it is slightly reduced if the leeward building is taller than the windward block.

xxi. Air motion in a shielded building is less than that in an unobstructed building. To minimize the shielding effect, the distances between two rows should be 8H for semidetached houses and 10H for long rows houses. However, for smaller spacing the shielding effect is also diminished by raising the height of the shielded building.

xxii. Hedges and shrubs deflect air away from the inlet openings and cause a reduction in indoor air motion. These elements should not be planted at a distance of about 8m from the building because the induced air motion is reduced to minimum in that case. However, air motion in the leeward part of the building can be enhanced by planting a low hedge at a distance of 2m from the building.

xxiii. Trees with large foliage mass having trunk bare of branches up to the top level of window, deflect die outdoor wind downwards and promotes air motion in the leeward portion of building.

xxiv. Ventilation conditions indoors can be ameliorated by constructing buildings on earth mound having a slant surface with a slope of 10° on the upstream side.

xxv. In case of industrial buildings the window height should be about 1.6m and the width about two- thirds of wall width. These should be located at a height of 1.1m above the floor. In addition, openings around 0.9 m high should be provided over two-thirds of the length of the glazed area in the roof lights.

xxvi. Height of industrial buildings, although determined by the requirements of industrial processes involved, is generally kept large enough to protect the workers against hot stagnant air below the ceiling as also to dilute the concentration of contaminant inside. However, if high level openings in roof or walls are provided, building height can be reduced to 4m without impairing the ventilation performance.

Stack Effect

Natural ventilation by stack effect occurs when air inside a building is at a different temperature than air outside; Thus, in heated buildings or in buildings wherein hot processes are carried out and in ordinary buildings during summer nights and during pre-monsoon periods, when the inside temperature is higher than that of outside, cool outside air will tend to enter through openings at low level and warm air will tend to leave through openings at high level. It would, therefore, be advantageous to provide ventilators close to ceiling. Ventilators can also be provided in roof as chimney, stack, cowl, vent pipe, covered roof and ridge vent.

Energy Conservation by Ventilation System

Maximum possible use should be made of wind-induced natural ventilation. This may be accomplished by following the design guidelines:

i. Adequate number of circulating fans should be installed to serve all interior working areas during the summer months in the hot dry and warm humid regions to provide necessary air movement when ventilation alone does not afford sufficient relief.

ii. The capacity of a ceiling fan to meet the requirement of a room with the longer dimension D meters should be about 55D m^3/min.

iii. The height of fan blades above the floor should be $(3H + W)/4$, where H is the height of the room, and W is the height of the work plane.

iv. The minimum distance between fan blades and the ceiling should be about 0.3 meters.

v. Electronic regulators should be used instead of resistance type regulators for controlling the speed of fans.

vi. When actual ventilated zone does not cover the entire room area, then optimum size of ceiling fan should be chosen based on the actual usable area of room, rather than the total floor area of the room. Thus smaller size of fan can be employed and energy saving could be achieved.

vii. Power consumption by larger fans is obviously higher, but their power consumption per square meter of floor area is less and service value higher. From the point of view of energy consumption, the number of fans and the optimum sizes for rooms of different dimensions are given in the National Building Code of India 2005.

Optional Methods for Determining Acceptable Thermal Conditions in Naturally Conditioned Spaces in Tropical Countries

The adaptive model of thermal comfort is derived from a global database of 21,000 measurements taken primarily in office buildings in the tropical climate. The allowable operative temperature limits may not be extrapolated to the outdoor temperature above and below the end points of the curves in this figure. If the mean monthly outdoor temperature is less than 10°C or greater than 33.5°C, this option may not be used. Occupant-controlled naturally conditioned spaces are those spaces where the thermal conditions of the space are regulated primarily by the occupants through opening and closing of windows. Field experiments have shown that occupants' thermal responses in such spaces depend in part on the outdoor climate and may differ from thermal responses in buildings with centralized HVAC systems primarily because of the different thermal experiences, changes in clothing, availability of control, and shifts in occupant expectations. This optional method is intended for such spaces.

For this optional method, the space must be equipped with operable windows that open to the outdoor and that can be readily opened and adjusted by the occupants of the space. Mechanical ventilation with unconditioned air may be utilized, but opening and closing of windows must be the primary means of regulating the thermal conditions in the space. The space may be provided with a heating system, but this optional method does not apply when a heating system is in operation. It applies only to spaces where the occupants are engaged in near sedentary physical activities, with metabolic rates ranging from 1.0 met to 1.3 met. This optional method applies only to spaces where the occupants may freely adapt their clothing to the indoor and/or outdoor thermal conditions.

Table 8.16: Limits on Temperature Drifts and Ramps

Time period	0.25 h	0.5 h	1 h	2 h	4 h
Max. operative temperature change allowed	1.1° C (2.0°F)	1.7°C (3.0°F)	2.2° C (4.0°F)	2.8°C (5.0°F)	3.3°C (6.0°F)

Source: ECBC, 2008

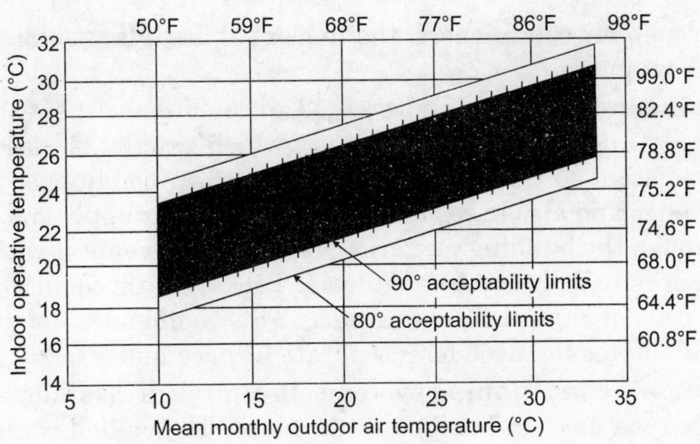

Fig.8.14: Acceptable operative temperature ranges for naturally conditioned spaces.

Source: ECBC, 2008

Allowable indoor operative temperature for spaces that meet these criteria may be determined from the Figure 8.14. This figure includes two sets of operative temperature limits—one for 80% acceptability and one for 90% acceptability. The 90% acceptability limits may be used when a higher standard of thermal comfort is desired.

1 met= 58W/m^2: for typical office activity, one person is likely to produce 100- 125 watts of heat. Minimum equipment efficiencies are required to be met for all HVAC equipment. These include chillers, unitary air conditioner split air conditioner, packaged air conditioner, and boilers.

Type of Air-conditioning Systems

There are primarily two main types of air conditioning systems:

i. A direct expansion or"DX" type system in the form of room air conditioners split system air conditioners and packaged air conditioners. Heat exchange takes place directly from the refrigerant within the copper tubes to the air being drawn across the finned coil by an evaporator fan. The DX type systems offer localized solutions for a building's heating and cooling needs. These systems are typically appropriate for smaller (single-zone) buildings.

ii. Central plant system uses chilled water recirculation. Compared to a DX systems, a central plant HVAC will be able to provide better thermal comfort and flexibility.

Direct Expansion Systems or DX Systems

Unitary air conditioners: These are normally used for cooling individual rooms and provide cooling only when needed. Room air conditioners house all the components of an air conditioning system in one casing. Their efficiency is generally lower than that of central plant systems.

Split-system air conditioning systems: This consists of an outdoor metal cabinet that contains the condenser and compressor and an indoor cabinet that contains the evaporator.

In many split-system air conditioners, the indoor cabinet also contains a furnace or the indoor part of a heat pump.

Packaged air conditioners: In a packaged air conditioner, the evaporator, condenser, and compressor are all located in one cabinet, which usually is placed on a roof or on a concrete slab adjacent to the building. This type of air conditioner is typical in small commercial buildings and also in residential buildings. Air supply and return ducts come from indoors through the building's exterior wall or roof to connect with the packaged air conditioner, which is usually located outdoors. Packaged air conditioners often include electric heating coils or a natural gas furnace. This combination of air conditioner and central heater eliminates the need for a separate furnace indoors.

Central plant air conditioning systems: In central air-conditioning systems, chilled water is generated via a central chilled water plant. The chilled water is distributed to airhandling units or fan-coil units via a chilled water reticulation system consisting of chilled water pipes, valves, fittings, and pumps. The chillers used in central chilled water plants include air-cooled chillers as well as water-cooled chiller systems that work in conjunction with cooling towers.

Table 8.17: An overview of Central HVAC and DX system

	Central Chilled Water Systems	Direct Expansion or "DX" Systems
Building space requirements	Requires separate building space to house the chillers, boilers, pumps, AHU's, distribution networks and control panels. In addition, space is required outdoors for condensing unit for air-cooled machines and cooling tower for water-cooled machines. The building structure should be designed to take the weight of equipment. Suitable vibration control must be considered and adequate load bearing beams and columns must be available for such equipment.	No separate plant room space is required as the refrigeration package is integral to the package unit/condensing unit which is generally located outdoors. Evaporator units are generally located indoors. The local systems are smaller in size and are less bulky.
Aesthetics	Central systems are generally designed as concealed systems and the visible distribution grills, etc. can be easily blended with the aesthetics	The appearance of local units can be unappealing and may not necessarily blend well with the aesthetics.
Zoning	Central HVAC system may serve multiple thermal zones and have their major components located outside the zone(s) being served, usually in some convenient central location. This system can provide better flexibility in terms of zoning	A local HVAC system typically serves a single thermal zone and has its major components located within the zone itself or directly adjacent to the zone. Multiple units are required for multiple zones. This system is less flexible to zoning requirements

Air quality	The quality of air conditioning is comparatively superior, with better control over temperature, relative humidity, air filtration, and air distribution. Best suited for applications demanding close control of temperature, humidity, and cleanliness and can be customized as per the design conditions.	The air quality is not comparable to central systems. These systems typically cannot provide close humidity control or high efficiency filtration. The compact systems, being standard factory items, typically cannot be modified to suit the required design conditions all the times
Controls	Central HVAC systems require a control point for each thermal zone. The controls are field wired and are integrated to a central control panel. The controls are complex and depend on the type of system. Constant air volume (CAV) systems alter the temperature while keeping the constant air delivery. CAV systems serving multiple zones rely on reheat coils to control the cooling. This incurs a lot of energy wastage due to simultaneous cooling and heating. Space temperature control can also be achieved by applying a variable air volume (VAV) system, which primarily alters the air delivery rates. The VAV system may or may not have a reheat coil, which provides additional heat when the space does not need to be cooled or needs less cooling. Proper zoning using "face zoned" AHUs in conjunction with downstream VAV boxes provides energy efficient cooling.	Local units are off-shelf items complete with integrated controls. They usually have a single control point which is typically only a thermostat. The room-by room or "zone" control minimizes over cooling typical of central air-conditioning systems. With the zone-control ability of the compact systems, only occupied spaces are maintained at a comfort level, and conditioning for the rest of the building is turned down or shut off. Some DX systems have limited capacity and have limited capability to reduce airflow during low load situations. Hence, there is a limitation in saving fen energy in DX systems with some types of DX systems only having on/off control for the compressors that can result in considerable hunting and space temperature fluctuation.
Efficiency	Central systems usually cannot be isolated for complete shut down. In a central system, the individual control option is not always available. For individual control the system shall be designed as variable air volume system with localized thermostats. Central systems designed for VAV system is based on block load calculations, which allows the system to borrow air from areas with low load. By variable speed drive on air handling units, it is possible to achieve excellent savings in power. Proper zoning can avoid conflicting demands for heating and cooling.	In a building where a large number of spaces may be unoccupied at any given time, such as a dormitory or a motel, local systems may be totally shut off in the unused spaces, thus providing huge energy saving potential. As a self-contained system, a local HVAC system may provide greater occupant comfort through totally individualized control options — if one room needs heating while an adjacent one needs cooling, two local systems can respond without conflict. As the compact systems are small, they are designed for full peak load and the standard rooftop or package units are not typically available with variable speed option. This type of system therefore has very limited potential to operate efficiency during part load situations. The VRF system addresses to this problem.

Refrigerant Containment	Central plant systems provide an excellent means to contain all refrigerant within the chiller housing and plant room. It is possible to detect any minor leaks within the localized plant room and take remedial action to arrest the leak.	Unlike central systems, DX systems pose a greater risk of refrigerant leaks to the atmosphere. With DX systems installed in several localized areas it may be very difficult or impossible to detect these leaks, especially in split systems with long pipe runs using high pressure refrigerant.
Operations and maintenance (O&M)	Large central systems can have a useful life of 25 years. Central, systems allow major equipment components to be kept isolated in a mechanical room; Grouping and isolating key operating components allows maintenance to occur with limited disruption to building functions.	Local systems can have a useful life of up to 15 years. Local systems maintenance may be relatively simple.
Cost	The initial purchasing and installation cost of a central air- conditioning system is much higher than a local system. These systems can offer higher system efficiencies and pay back the elevated initial, costs through reduced costs operations within a few years. Cost benefits can also be achieved by energy efficiency measures like thermal heat recovery, economizers, energy storage systems, etc.	Packaged and split units have much lower first costs than a central system. The operating costs of unitary systems is usually higher due to lower efficiency ratings and lower part load performance values. The potential for adoption of energy efficiency measures is limited

Source ECBC, BIS. ASHARE and CPWD

Chillers

A chiller is essentially a packaged vapor compression cooling machine. The chiller rejects heat either to condenser water (in the case of a water-cooled chiller) or to ambient air (in the case of an air-cooled chiller). Water-cooled chillers incorporate the use of cooling towers, which improve heat rejection more efficiently at the condenser, than air-cooled chillers. For a watercooled chiller, the cooling tower rejects heat to the environment through direct heat exchange between the condenser water and cooling air. For an air-cooled chiller condenser fans move air through a condenser coil. As heat loads increase, water-cooled chillers are more energy-efficient than air-cooled chillers. A typical chiller is rated between 15 to 1000 tons (53 to 3,500 kW) in cooling power.

Chillers are classified according to compressor type. Electric chillers for commercial comfort cooling have centrifugal, screw, scroll, or reciprocating compressors. Centrifugal and screw chillers have one or two compressors. Scroll and reciprocating chillers are built with multiple, smaller compressors.

- Centrifugal chillers are the quiet, efficient, and reliable workhorses of comfort cooling. Although centrifugal chillers are available as small as 70 tons, most are 300 tons or larger.
- Screw chillers are up to 40% smaller and lighter than centrifugal chillers, so are becoming popular as replacement chillers.

- Scroll compressors are rotary positive-displacement machines, also fairly new to the comfort cool-ing market. These small compressors are efficient, quiet, and reliable. Scroll compressors are made in sizes of 1.5 to 15 tons.
- Use of CFC free refrigerant is becoming essential. To replace chillers using ozone depleting substances, new chillers must be compliant to environmental protocols. Generally MoEF gives subsidy for replacement of chillers using ozone depleting substances/ refrigerant. This aspect may be kept in view to comply with the requirements to get subsidy.

The energy efficiency of cooling and heating systems in terms of Coefficient of Performance (COP), Energy Efficiency Ratio (EER) and Integrated Part-Load Value as specified by the Code are given below.

Coefficient of Performance (COP) - Cooling

The ratio of the rate of heat removal to the rate of energy input, in consistent units, for a complete refrigerating system or some specific portion of that system under designated operating conditions.

Coefficient of Performance (COP) - Heating

The ratio of the rate of heat delivered to the rate of energy input, in consistent units, for a complete heat pump system, including the compressor and, if applicable, auxiliary heat, under designated operating conditions.

Energy Efficiency Ratio (EER)

The ratio of net cooling capacity in BTU/hr to total rate of electric input in watts under designated operating conditions.

Integrated Part-Load Value (IPLV)

A single number figure of merit based on part-load EER, COP, or KW/ton expressing partload efficiency for air-conditioning and heat pump equipment on the basis of weighted operation at various load capacities for the equipment.

As per the Code:

Cooling equipment shall meet or exceed the minimum efficiency requirements presented in Table below. Heating and cooling equipment not listed here shall comply with ASHRAE 90.12004.

Table 8.18: Chillers Specifications

Equipment Class	Minimum COP	Minimum IPLV	Test Standard
Air Cooled Chiller<530 kW (<150 tons)	2.90	3.16	ARI 550/590-1998
Air Cooled Chiller>530 kW (>150 tons)	3.05	3.32	ARI 550/590-1998
Centrifugal Water Cooled Chiller < 530 kW (< 150 tons)	5.80	6.09	ARI 550/590-1998

Centrifugal Water Cooled Chiller>530 and<1050 kW (>150 and <300 tons)	5.80	6.17	ARI 550/590-1998
Centrifugal Water Cooled Chiller > 1050 kW (>300 tons)	6.30	6.61	ARI 550/590-1998
Reciprocating Compressor, Water Cooled Chiller all sizes	4.20	5.05	ARI 550/590-1998
Rotary Screw and Scroll Compressor, Water Cooled Chiller <530 kW (<150 tons)	4.70	5.49	ARI 550/590-1998
Rotary Screw and Scroll Compressor, Water Cooled Chiller >530 and <1050 kW (>150 and <300 tons)	5.40	6.17	ARI 550/590-1998
Rotary Screw and Scroll Compressor, Water Cooled Chiller > 1050 kW (> 300 tons)	5.75	6.43	ARI 550/590-1998

Source: ECBC, 2008

As per the Code: Unitary Air Conditioner shall meet IS 1391- Part 1, split air conditioner shall meet IS 1391 - Part 2, packaged air conditioner shall meet IS 8148 and boilers shall meet IS 13980 (the standard specifics the procedure with above 75% thermal efficiency).

Table 8.19: Power Consumption Ratings for Unitary Air Conditioners Under Test Conditions

Rated Cooling Capacity		Maximum power Consumption (kW)
Kcal/h	kW	
1,500	1.7	1.1
2,250	2.6	1.4
3,000	3.5	1.6
3,750 -	4.4	2.0
4,500	5.2	2.4
6,000	7.0	3.2
7,500	8.7	4.25
9,000	10.5	5.2

Source: BSI Code No.: IS 1391 (Part-1): 1992 (amendment No. 2, Dec.2006)

Table 8.20: Power Consumption Ratings for Split Air Conditioners Under Test Conditions

Rated Cooling Capacity		Maximum power Consumption (kW)
Kcal/h	kW	
3,000	3.5	1.7
4,500	5.2	2.6
6,000	7.0	3.4
7,500	8.7	4.5
9,000	10.5	5.4

Source BSI: Code No.: IS 1319 (Part-2): 1992 (amendment No. 2, Dec.2006)

Table 8.21: Power Consumption Rating for Packaged air Conditioners under Test Conditions

Cooling Capacity		Maximum Power Consumption in Watts Cooled Air Cooled	
Watts	Tons of Refrigeration	Water Cooled	Air Cooled
10,000	3	3,750	4,750
17,500	5	6,000	7,000
26,250	7.5	9,000	10,000
35,000	10	11,500	13,500
52,000	15	17,000	20,000

Source: BSI Code No.: IS 8148: 2003

Selection of individual equipment efficiency should be considered in the context of the whole HVAC system. In a chilled-water system, for example, although the chiller is at the core of the system and typically is the single largest energy user, simply selecting a high-efficiency chiller does not guarantee high performance. Auxiliary equipment (such as fans and blowers) and design decisions (such as "approach temperatures") can have substantial effects on overall efficiency. Thus, attention to overall system design and auxiliary components is critical to achieving optimal performance and comfort. Even in packaged air-conditioning systems, leaky ductwork, improper sizing, refrigerant charge, and air flow rates can considerably affect energy performance.

Controls

Controls are one of the most critical elements for improving efficiency of any HVAC system. Controls determine how HVAC systems should operate to meet the design goals of comfort, efficiency, and cost-effective operation. In this context, the Code specifies the use of time clocks, temperature controls/thermostats, and two-speed or variable speed drives for fans.

1. Time Clock Control

As per the Code: All mechanical cooling and heating systems shall be controlled by a timeclock that:

— Can start and stop the system under different schedules for three different day-types per week

— Is capable of retaining programming and time setting during loss of power for a period of at least 10 hours, and

— Includes an accessible manual override that allows temporary operation of the system for up to 2 hours *Exceptions to the above are:*

— Cooling systems < 28 kW (8 tons)

— Heating systems < 7 kW (2 tons)

2. Temperature Control

As per the Code: All heating and cooling equipment shall be temperature controlled. Where a unit provides both hearing and cooling, controls shall be capable of providing a temperature dead band of 3°C (5°F) within which the supply of heating and cooling energy to the zone is shut off or reduced to a minimum. Where separate heating and cooling equipment serve the same temperature zone, thermostats shall be interlocked to prevent simultaneous heating and cooling.

It is important to clearly establish design conditions and ensure adequate dead band between cooling and heating set points to avoid conflicting thermostat control conditions. Increasing the dead band can make the system more stable and efficient.

3. Controls for Cooling Towers and Closed Circuit Fluid Coolers

As per the Code: All cooling towers and closed circuit fluid coolers shall have either two speed motors, pony motors, or variable speed drives controlling the fans.

A variable speed drive (VSD) is an electronic device that controls the rotational speed of a piece of motor-driven equipment (*e.g.* a blower, compressor, fan, or pump). Speed control is obtained by adjusting the frequency of the voltage applied to the motor. This approach usually saves energy for varying load applications.

Through the application of VSD on the cooling tower fan, the fan speed can be reduced during lower ambient conditions for reducing energy consumption. However condenser water reset strategy may require condenser fan speeds to be maintained to improve chiller efficiency by lowering condenser water temperature. At lower loads the overall system efficiency should be the driver as pumping and tower fan energy form significant proportion of overall chiller plant energy.

4. Energy and Water Efficiency in Cooling Towers.

Water-based HVAC systems offer significant energy savings due to the ability of water to transport large quantities of heat over relatively long distances, more efficient than air-based systems. Additionally, they offer advantages such as smaller equipment size and cost, along with reduced maintenance and extended life of mechanical equipment. However, for water scarce urban areas, the viable installation and operation of cooling towers will require balancing energy efficiency and water conservation. Air-cooled chillers are a good option to conserve water, however, these chillers consume more energy to provide the same amount of cooling when compared to water cooled machines with cooling towers.

Energy Efficiency Measures

- Proper site selection and sizing of the tower can reduce fan speed, capacity, and sound and help to conserve energy
- Centrifugal fans in favor of lower energy axial fans can reduce horsepower by 50% ormore for the same capacity
- Fan control through two-speed motors, pony motors, or variable speed motors can save energy as well.

Water Efficiency Measures

- An optimized bleed rate for the tower should be maintained to regulate water consumption. The evaporation rate is dependent on the load, which can vary widely and a constant bleed rate usually discharges more water than required. A properly operating conductivity meter can automatically control bleed to the proper amount required to maintain the desired tower chemistry in the system at all times.

- Contaminant induction should be minimized and a proper blow down rate should be maintained. Water treatment regimens are effective for water conservation, keeping the cooling loop cleaner, saving energy, reducing maintenance, and improving reliability of the entire cooling system.
- New technological solutions like hybrid wet-dry cooling tower designs, which combine wet and dry cooling can be adopted to reduce water use, as much as 70% compared to conventional towers. Typically, a dry finned coil section is combined in series with an evaporative section in these units. The dry finned section handles as much of the load as possible, with the unit able to operate completely dry at reduced ambient both open and closed circuit versions are available.

Piping and Ductwork

Pipe Insulation: To minimize heat losses, the Code requires that piping of heating and cooling systems, (including the storage tanks) must be insulated. The Code specifies required R-values of insulation for heating and cooling systems based on the operating temperature of the system. These are as shown in Tables below:

Table 8.22: Insulation of Heating Systems

Designed operating temperature of piping	Insulation with minimum R-value (m^2 K/W)
60°C and above	0.74
Above 40°C and below 60°C	0.35

Table 8.23: Insulation of Cooling Systems

Designed operating temperature of piping	Insulation with minimum R-value (m2 K/W)
Below 15°C	0.35
Refrigerant suction piping	
Split system	0.35

Source: ECBC, 2008

Duct Sealing. Duct sealing is critical to avoid air leaks that prevent the HVAC system from functioning as designed and operated. The ASHRAE 90.1 energy code can be referred to for appropriate seal levels for all ductwork in order to minimize energy losses from the HVAC system. ASHRAE 90.1 specifies sealing requirements based on the duct location, static pressure classification, and type of the duct (exhaust or return) as given below:

Table 8.24: Minimum Duct Sealing Level

Duct Location	Duct Type Supply		Exhaust	Return
	≤ 498.2 Pa	≥ 498.2 Pa		
Outdoors	A	A	C	A
Unconditioned spaces	B	A	C	B
Conditioned spaces	C	B	B	C

Source: ASHRAE 90.1 (2004)

Table 8.25: Duct Seal Levels Requirements

Seal Level	Sealing Requirements
A	All transverse joints and longitudinal seams, and duct wall penetrations. Pressure-sensitive tape shall not be used as the primary sealant.
B	All transverse joints and longitudinal seams. Pressure-Sensitive tape shall not be used as the primary sealant
C	Transverse joints only.

Source: ASHRAE (2004)

System Balancing

System balancing is a process for maintaining the performance of an HVAC system, and for providing the occupants with a comfortable conditioned space. Balancing an air or water-based HVAC system of buildings will make it more energy-efficient, provide improved thermal comfort, extend the life of the building equipment and reduce the cost of operating it. Balancing is achieved by optimizing the air/water distribution rates for the HVAC system.

As per the Code: Construction documents shall require that all HVAC systems be balanced in accordance with generally accepted engineering standards.

Construction documents shall require that a written balance report be provided to the owner or the designated representative of the building owner for HVAC systems serving zones with a total conditioned area exceeding 500 m^2.

Construction documents provide vital information to the building owners on how to properly operate and maintain a system that has been properly balanced. Verify during final inspection that an operations manual has been passed on to the building owner and that it contains the following information at a minimum.

- HVAC equipment capacity
- Equipment operation and maintenance manuals
- HVAC system control maintenance and calibration information, including wiring diagrams, schedules, and control sequence descriptions.
- A complete written narrative of how each system is intended to operate.

Air System Balancing

Air System Balancing refers to the adjustment of airflow rates through air distribution system devices, such as fans and diffusers. This is done by adjusting the position of dampers, splitter vanes, extractors, etc., manually or by using automatic control devices, such as constant air volume or variable air volume boxes.

Balancing is necessary to verify that each space served by a system receives the air volume designed for that space. Proper means for air balancing should be installed at each supply air outlet and zone terminal device. These include balancing dampers or other means of supply-air adjustment provided in the branch ducts or at each individual duct register, grille, or diffuser. Installation in the duct system of all devices used for balancing, shown on the approved mechanical plans, typically, on the ductwork layout, should be verified.

As per the Code: Air systems shall be balanced in a manner to minimize throttling losses. Then, for fans greater than 0.75 KW (1.0 HP), fans must then be adjusted to meet design flow conditions.

Air Handling Unit Concepts

An air handler is responsible for moving air throughout the duct work in an air conditioning system. All air handlers contain a blower motor and squirrel cage blower housing which facilitates the movement of air. Most air handlers also include system controls, which are connected to the thermostat. Depending on the type of system, an air handler can also be integrated with a gas, oil, electric furnace, heat pump, and cooling coils (or evaporator coil for the air conditioning).

High-efficiency air distribution systems can substantially reduce fan power required by an HVAC system, resulting in dramatic energy savings. The largest gains in efficiency for air distribution systems are realized in the system design phase for new constructions or major retrofit projects. Passive or natural air transport systems have the highest efficiency, and successful, modern examples of this approach are steadily accumulating. For buildings that require mechanical ventilation, innovative design approaches and a methodical examination of the entire air system can greatly improve efficiency and effectiveness.

All these factors can be manipulated to reduce the energy consumption of the system. Air flow has a dominant effect on energy consumption because it shows up twice in the energy equation: as the first term and as a squared function in the second term (pressure). The pressure a fan must work against depends on twc primary factors: the flow and duct design features such as diameter, length, surface treatment, and impediments such as elbows, filters, and coils. Typical pressure losses are on the order of 2 to 6 inches water gauge, (wg); an efficient system operates at less than 1.5" wg. A fan's duty factor is the number of hours per year that it operates, sometimes presented as a percentage: Many large fans spin at full speed continuously (8,760 hours per year). Using simple or complex controls, duty factors can often be reduced to about 3,000 hours per year or less by limiting fan operation to occupied periods. The mechanical efficiency of the fan and its drive system can typically be raised from the 40 to 60% range to the mid-80% range. Wire to air efficiency will need to be considered and is function of motor efficiency × mechanical efficiency. Design options for improving air distribution efficiency include:

- Variable-air-volume (VAV) systems
- VAV diffusers

- Low-pressure-drop duct design
- Low-face-velocity air handlers
- Fan sizing and variable-frequency-drive motors
- Displacement ventilation systems.

Hydronic System Balancing

Hydrolic System Balancing refers to the adjustment of water flow rates through distribution system devices, such as pumps and coils, by manually adjusting the position of valves, or by using automatic control devices, such as flow control valves.

When something is balanced, it is even on both sides. Therefore, a balanced hydronic system is one that delivers even flow to all of the devices on that piping system. Each component has an effective equal length of pipe on the supply and return. And when a system is balanced, all of the pressure drops are correct for the devices. When that happens, the highest efficiencies are possible in the system. One need not have to change system supply temperatures to accommodate one zone only. The system has the least amount of pressure drop possible, which translates into reduced pumping costs. A balanced hydronic system is one that is efficient. If a system that is not delivering the water to the right devices in the right amounts, then the system is out of balance.

Hydronic systems shall be proportionately balanced in a manner to first minimize throttling losses; then the pump impeller shall be trimmed or pump speed shall be adjusted to meet design flow conditions.

- Impellers need not be trimmed nor pump speed adjusted for pumps with pump motors of 7.5 kW (10 hp) or less,
- Impellers need not be trimmed when throttling results in no greater than 5% of the nameplate horsepower draw, or 2.2 kW (3 hp), whichever is greater.

Condensers

A condenser is a heat exchanger that liquefies refrigerant vapor through heat removal The typical condensing unit houses a compressor, a condenser fan motor, and coils, along with controls which make all the components work sequentially.

Care shall be exercised in locating the condensers in such a manner that the heat sink is free of interference from heat discharge by devices located in adjoining spaces and also does not interfere with such other systems installed nearby.

Condensers should be located in such a manner that there is no restriction to the air flow around the condenser coils, there is no short-circuiting of discharge air to the intake side, and the heat discharge of other adjacent equipment is not anywhere the air intake of the condenser.

Condenser water treatment is important to eliminate mineral buildup in condensers and chilled water systems. Mineral deposits create poor heat transfer situations there by reducing the efficiency of the unit.

As per the Code: All high-rise buildings using centralized cooling water system shall use soft water for the condenser and chilled water system.

Prescriptive Requirements

The prescriptive requirements apply only if the HVAC system in the building meets the following criteria:

(a) Serves a single zone

(b) Cooling (if any) is provided by a unitary packaged or split-system air conditioner or heat pump

(c) Heating (if any) is provided by a unitary packaged or split-system heat pump, fuelfired furnace, electric resistance heater, or baseboards connected to a boiler

(d) Outside air quantity is less than 1,400 l/s (3,000 cfm) and less than 70% of supply air at design conditions.

Other HVAC systems shall comply with ASHRAE 90.1-2004.

As per the Code: HVAC Zone is a space or group of spaces within a building with heating and cooling requirements that arc sufficiently similar so that desired conditions (*e.g.*, temperature) can be maintained throughout using a single sensor (*e.g.*, thermostat or temperature sensor).

Heat Pump: A heat pump consists of one or more factory-made, assemblies that normally include indoor conditioning coil, compressor, and outdoor coil, including means to provide a heating function. Heat pumps provide the function of air heating with controlled temperature, and may include the functions of air cooling.

Economizers: Economizers allow the use of outdoor air to cool the building when the outside temperature is cooler than that inside. An economizer consists of dampers, sensors, actuators, and logic devices that together decide how much outside air to bring into a building. Under the right conditions, sensors and controls shut down the compressor and bring in the outside air through the economizer louvers. A properly operating economizer can cut energy costs by as much as 10% of a building's total energy consumption, depending mostly on local climate and internal cooling loads.

As per the Code: Each individual cooling fan system that has a design supply capacity over 1,200 l/s (2,500 cfm) and a total mechanical cooling capacity over 22 kW (6.3 tons) shall include either:

(a) An air economizer capable of modulating outside-air and return-air dampers to supply 100% of the design supply air quantity as outside-air; or

(b) A water economizer capable of providing 100% of the expected system cooling load at outside air temperature of 10°C (50°F) dry-bulb/7.2°C (45°F) wet-bulb and below.

Exceptions to above are:

(a) Projects in the hot-dry and warm-humid climate zones arc exempt

(b) Individual ceiling mounted fan systems < 3,200 l/s (6,500 cfm) are exempt.

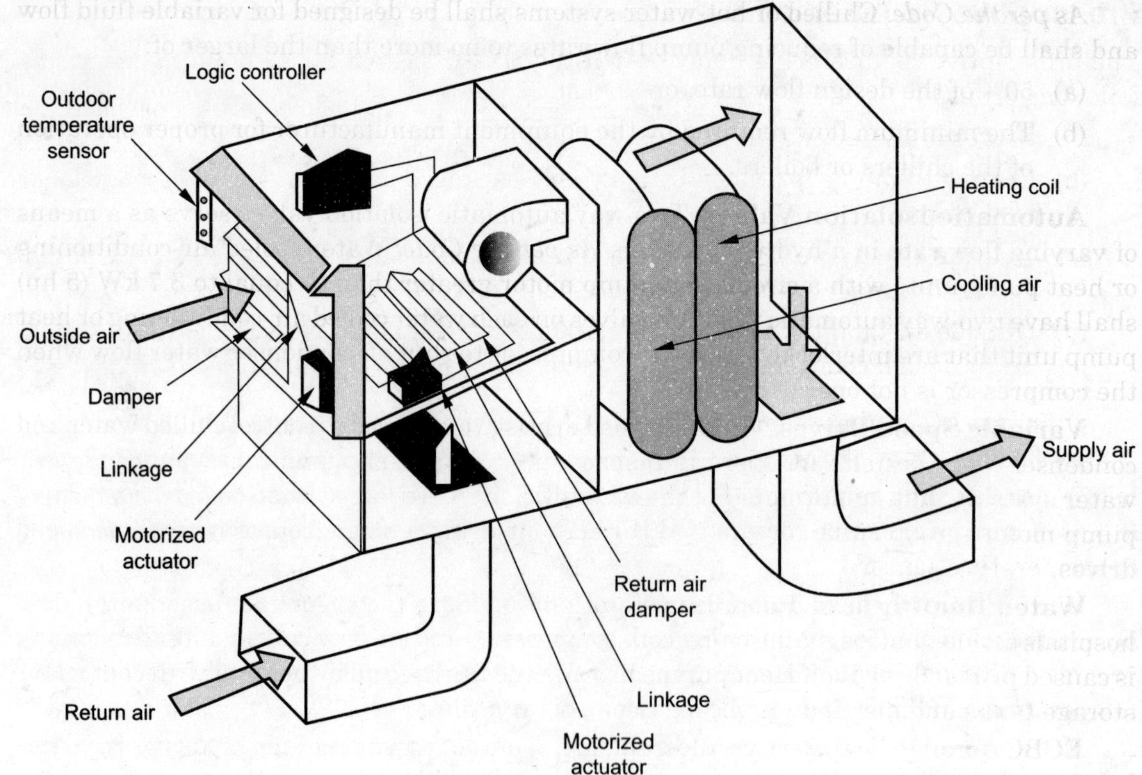

Fig.8.15: Air-Side Economizers

Partial Cooling: One can use the building's intrinsic thermal mass to reduce peak cooling loads by circulating cool night-time air to pre-cool the building prior to daily occupancy in the cooling season. The building control system can operate ventilation fans in the economizer mode on a scheduled basis. Care should be taken to prevent excessive fan operation that would offset cooling energy savings. It is also ensured that night humidity does not preclude the use of this strategy.

As per the Code: Economizers shall be capable of providing partial cooling even when additional mechanical cooling is required to meet the cooling load.

Testing of Air-Side Economizer: *As per the Code:* Air-side economizers shall be tested in the field following the requirements in Appendix F (of the Code) to ensure proper operation.

Exception to above: Air economizers installed by the HVAC system equipment manufacturer and certified to the building department as being factory calibrated and tested as per the procedures in Appendix F of the Code.

Variable Flow Hydronic Systems: Fluid from the heating or cooling source is supplied to heat transfer devices, such as coils and heat exchangers, and back through the hydronic system. The Code specifies the type of equipment and capabilities in such a way as to reduce pump energy. Variable fluid flow, automatic isolation valves, and variable speed drives enable the system to operate below design flow when needed.

As per the Code: Chilled or hot-water systems shall be designed for variable fluid flow and shall be capable of reducing pump flow rates to no more than the larger of:

(a) 50% of the design flow rate, or

(b) The minimum flow required by the equipment manufacturer for proper operation of the chillers or boilers.

Automatic Isolation Valves: Two-way automatic isolation valves serve as a means of varying flow rate in a hydronic system. *As per the Code:* Water cooled air-conditioning or heat pump units with a circulation pump motor greater than or equal to 3.7 kW (5 hp) shall have two-way automatic isolation valves on each water cooled air-conditioning or heat pump unit that are interlocked with the compressor to shut off condenser water flow when the compressor is not operating.

Variable Speed Drives: Variable speed drives are required to control chilled water and condenser water systems for energy efficiency. As per the Code Chilled water or condenser water systems that must comply with either §5.3.2.1 or §5.3,2.2 (of ECBC) and that have pump motors greater than or equal to 3.7 kW (5 hp) shall be controlled by variable speed drives.

Water Heating and Pumping: For some building types such as large hotels and hospitals service water heating can be major energy consumer. Inefficiency in water heating is caused primarily by inefficiency of the heating equipment and by heat loss from hot water storage tanks and distribution piping network.

ECBC through mandatory requirements seeks to minimize energy usage in water heating systems by:

- Utilizing solar water heating
- Specifying heating equipment efficiency
- Maximizing heat recovery and minimizing electric heating
- Insulating hot water storage tanks and pipelines
- Reducing standby losses
- Reducing heat and evaporation losses in heated swimming pools

Mandatory Requirements for Solar Water Heating

Residential facilities, hotels and hospitals with a centralized system shall have solar water heating for at least 1/5th of the design capacity.

Exception to above: Systems that use heat recovery for at least 1/5th of the design capacity.

There are two types of solar water heaters. Passive heaters collect and store solar thermal energy for water heating applications and do not require electrical energy input for recirculating water through solar collector. Active heaters collect and store solar thermal energy for water heating applications and require electrical energy input for operation of recirculation pumps or other components.

An exception is provided by the Code for systems that use heat recovery systems for at least one-fifth (20%) of the design capacity. For example, heat is rejected from the air conditioner's condenser to the atmosphere. By recovering this waste heat and utilizing it to heat water, wherever feasible, it is possible to substantially reduce water-heating costs.

Storage Gas: A gas water heater designed to heat and store water at less than 80°C. Water temperature is controlled with a thermostat.

Storage Electric Water Heater: An electric water heater designed to heat and store water at less than 80°C. Water temperature is controlled with a thermostat. Storage electric water heaters have a manufacturer's specified capacity of at least two gallons.

Storage Heat Pump: An electric water heater that uses a compressor to transfer thermal energy from one temperature level to a higher temperature level for the purpose of heating water. It includes all necessary auxiliary equipment such as fans, storage tanks, pumps or controls.

Instantaneous Gas: A gas water heater controlled manually or automatically by water flow activated control or a combination of water flow and thermostatic controls.

Water Heating Equipment

Instantaneous Electric: An electric water heater controlled automatically by a thermostat, instantaneous water heaters are not generally designed for use with solar water heating systems or as heat sources for indirect fired water heaters. They are also typically inappropriate for use with recirculation systems.

Indirect Gas: A water heater consisting of a storage tank with no heating elements or combustion devices, connected via piping and recirculating pump to a heat source consisting of a gas or oil fired boiler, or instantaneous gas water heater.

Passive Solar Systems which collect and store solar thermal energy for water heating applications and do not require electricity to recirculate water through a solar collector.

Active Solar Systems which collect and store solar thermal energy for water heating applications requiring electricity to operate pumps or other components.

Service water heating equipment shall meet or exceed the performance and minimum efficiency requirements as per the Indian Standards:

(a) Solar water heater shall meet the performance/minimum efficiency level mentioned in IS 13129 (Parts 1 & 2).

(b) Gas Instantaneous Water heaters shall meet the performance/minimum efficiency level mentioned in IS 15558 with above 80% thermal efficiency.

(c) Electric water heater shall meet the performance/minimum efficiency level mentioned in IS 2082.

The mandatory requirements of the Code include minimum efficiencies prescribed in relevant Indian Standards for the various water heating equipments, such as electric and gas heaters, instantaneous heaters, boilers, and pool heaters.

For Solar water heating systems, IS 13129 (Part 1) provides information on the 'Performance Rating Procedure Using Indoor Test Methods', and IS 13129 (Part 2) provides the information on the 'Procedure for System Performance Characterization and Yearly Performance Prediction. These standards however, do not provide any performance/minimum efficiency levels.

For Gas Instantaneous Water Heaters, IS 15558 describes the procedure for the measurement of thermal efficiency of the heaters. As per this Standard, thermal efficiency of the water heaters (under test conditions) shall not be less than:

- 84 percent for water heaters with a nominal heat input exceeding 10 kW
- 82 percent for water heaters with a nominal heat input not exceeding 10 kW. However, the Code specifies thermal efficiency of 80% or more.

For Electric Water Heaters, IS 2082 (Part 1) covers the safety and performance requirements of heaters with rated capacities in the range of 6 liters to 200 liters. In these heaters, certain amount of energy is consumed to keep the water hot while it is not being used. This consumption of electricity is called standing loss (or standby loss).

Standby Losses. These losses account for energy lost while storing heated water. This includes any heat losses through the water heater tank wall, fittings, flue, and any pilot light energy. Standby loss depends on the design and insulation of the water heater, as well as the difference between the temperature of the water and that of the air around the tank. Water heating energy can be reduced by decreasing standby losses.

IS 2082 (Part 1) specifies the standing loss in the heaters. For hot water temperature difference of 45°C, in no case the standing loss should exceed the values specified in the following Table.

Table 8.26: Standing Loss in Storage Type Electric Water Heaters

Rated Capacity in Litres	Loss in kWh/day for 45° Difference
6	0.792
10	0.990
15	1.138
25	1.386
35	1.584
50	1.832
70	2.079
100	2.376
140	2.673
200	2.970

Source: IS 2082 (Part 1): 1993 (Reaffirmed 2004 Edition) Stationary Storage Type Electric Water Heaters-Specification (Fourth Revision)

Supplemental Water Heating System: *As per the Code:* Supplemental Water Heating System shall be designed to maximize efficiency and shall incorporate and prioritize the following design features:

(a) Maximum heat recovery from hot discharge system like condensers of air conditioning units

(b) Use of gas-fired heaters wherever gas is available

(c) Electric heater as last resort

Piping Insulation

As per the Code: Piping insulation shall comply with §5.2.4.1. The entire hot water system including the storage tanks, pipelines shall be insulated conforming to the relevant IS standards on materials and applications, as reproduced below.

Table 8.27: Insulation of Hot Water Piping

Heating System	
Designed Operating	**Insulation with Minimum**
Temperature of Piping	R-value (m^2K/W)
60°C and above	0.74
Above 40°C and below 60°C	0.35

Source: ECBC, 2008

Guidelines for Temperature Controls

Water heating systems are required to have controls that are adjustable down to a 49°C setpoint or lower. An exception is made where a higher setting is recommended by the manufacturer to prevent condensation and possible corrosion. To comply with this requirement, the water heater must have thermostatic control with an accessible setpoint. This setpoint must be adjustable down to whichever is lower: 49°C or the minimum manufacturer's recommended setting to prevent condensation. Both standby and distribution losses will be minimized by designing a system to provide hot water at the minimum temperature required. In addition to the potential energy savings, maintaining water temperature as low as possible reduces corrosion and scaling of water heaters and components. Another important benefit is improved safety with respect to scalding. Accidental scalding from temperatures as low as 60°C is responsible for numerous deaths each year. The bacteria that cause Legionnaire's disease has been found in service water heating systems and can colonize in hot water systems maintained below 46°C. Careful maintenance practices can reduce the risk of contamination. In health-care facilities or service-water systems maintained below 60°C, periodic flushing of the fixtures with high temperature water or other biological controls may be appropriate.

Lighting

Lighting accounts for an average 15% of total energy consumption in India. Lighting is an area that offers many energy efficiency opportunities in almost any building facility, existing or new.

People need light for different reasons. Different tasks and building facilities require different amounts and types of light. For example, a surgeon in an operation theater needs lots of light with low glare and excellent color rendering, restaurant owners and diners often want low light levels, corporate boardrooms call for lighting that reinforces a feeling of importance and success while adapting to audio-visual presentations; retail outlets want to make their merchandise sparkle so that it draws the customers. An office executive needs modest ambient lighting level, good task lighting on work surface, and minimal glare to effectively read and work on computers.

While energy efficiency is an attractive goal for many reasons, lighting designers also need to consider a host of other factors, including the effect of quality of light on the visual comfort and productivity of the occupants. Small improvement in lighting quality can improve productivity of the user substantially. In practice, the right quality and quantity of light can be provided efficiently (with less energy) by using the right technology and its effective integration with daylight.

Using efficient lighting equipment and controls is the best way to ensure lighting energy efficiency while maintaining or even improving lighting conditions. For instance, modern fluorescent lighting, such as electronically ballasted T-8 systems, can provide the same quantity of light as older fluorescent lighting while consuming as little as two-thirds of the energy. Compact fluorescent sources are three to four times more efficient than the traditional incandescent lamps.

An energy-efficient lighting design involves sensitive integration of many requirements and considerations that include building orientation, interior building layout, task illumination, daylight strategies, glazing specification, choice of lighting system and controls, etc. The designer is also responsible for making sure that lighting complies with the Code, meeting both mandatory and prescriptive requirements.

The lighting requirements in the Code apply to:

(a) Interior spaces of buildings

(b) Exterior building features, including facades, illuminated roofs, architectural features, entrances, exits, loading docks, and illuminated canopies

(c) Exterior building grounds lighting that is provided through the building's electrical service.

Exceptions to above:

(d) Emergency lighting that is automatically off during normal building operation and is powered by battery, generator, or other alternate power source (*b*) Lighting in dwelling units.

Lighting Controls

Lighting controls allow lighting to be turned down or completely off when it is not needed—the simplest way to save energy. Maximizing the use of controls involves developing a set of strategies complying with the Code requirements for various devices, including on-off controls, dimming controls, and systems that combine the use of both types of equipment. These controls can be quite sophisticated, but in general, they perform two basic functions:

1. They turn lights off when not needed, and
2. They modulate light output so that no more light than necessary is produced.

The equipment required to achieve these functions varies in complexity from simple timers to intricate electronic dimming circuits, each applicable to different situations. Controls include time clocks, occupant and motion sensors, automatic or manual daylighting controls, and astronomical time switches (the automatic switches that adjust for the length of the day as it varies over the year).

Manual lighting controls range from a single switch to a bank of switches and dimmers that are actuated by toggles, rotary knobs, push buttons, remote control, and other means. Manual controls can be cost-effective options for small-scale situations. However, as the lighting system becomes larger, automated systems become more cost-effective and are better at controlling light. Manual controls may not give the desired results in real situations because the decision to shut off the lights when they are not needed is based entirely on human initiative. It is worthwhile to determine the amount of local vs. central control that is needed from the lighting control system.

Although there is no simpler way to reduce the amount of energy consumed by lighting systems than to manually turn lights off whenever not needed, this is not done as often as it could be. In response to that problem, the Code requires several automatic switches that either work on time schedule or sense the presence of occupants.

Several different approaches can be used to control electric lighting. The control hardware and design practices are discussed below:

(a) *Scheduling Control*: Use a time scheduling device to control lighting systems according to predetermined schedules.

(b) *Occupancy Sensing*: Control lights in response to the presence or absence of people in the space.

(c) *Daylighting*: Switch or dim electric lights in response to the presence or absence of daylight illumination in the space.

(d) *Lumen Maintenance*: Gradually adjust electric light levels over time to correspond with the depreciation of light output from aging lamps.

Interior lighting systems in buildings larger than 500 m^2 shall be equipped with an automatic control device. Within these buildings, all office areas less than 30 m^2 enclosed by walls or ceiling-height partitions, all meeting and conference rooms, all school classrooms, and all storage spaces shall be equipped with occupancy sensors. For other spaces, the automatic control device shall function on either:

(a) A scheduled basis at specific programmed times. An independent program schedule shall be provided for areas of no more than 2,500 m and not more than one floor; or,

(b) Occupancy sensors that shall turn the lighting off within 30 minutes of an occupant leaving the space. Light fixtures controlled by occupancy sensors shall have a wall mounted, manual switch capable of turning off lights when the space is occupied.

Exception to above: Lighting systems designed for 24-hour use.

Programmable timing, also known as automatic time scheduling, is the oldest form of automatic lighting control. Time scheduling manages the on and off times of a building's lighting systems. Scheduling systems function by turning off all or some of the lights when a building space is unoccupied. In the most basic time-scheduling scheme, a time switch switches lighting circuits on or off based on programmable schedules. For example, exterior lighting is usually switched on to correspond to sundown and is switched off again at daybreak. By contrast, time scheduling of interior lighting systems is based, for the most part, on occupancy schedules. In some cases, time switches are used to energize additional lighting control systems, such as daylighting controls, which are held off during unoccupied periods. Time scheduling systems employ the following components:

- A central processor is usually capable of controlling several output channels, each of which may be assigned to one or more lighting circuits.
- Relays are series-wired to lighting control zones and are controlled by the central processor.
- Overrides are required to accommodate individuals who use the space during scheduled off hours. Individuals can activate manual switches or use telephone overrides to regain temporary control of the lights in a given space.

Occupant-sensing devices are an alternative to scheduling controls and an acceptable means of meeting the requirement for automatic shutoff. The designer is free to arrange occupant sensing controls in any manner that makes sense for the building design. In office spaces, each room or space might have an occupant sensor. Of course, the smaller the area controlled, the greater the energy savings will be. In open office areas, several occupant sensors may be connected so that the lights remain on if any one of the sensors detects occupants. However, it is necessary that all the general lighting be controlled by one or more occupant sensors.

Occupancy sensors are automatic scheduling devices that detect motion and turn lights on and off accordingly. Most devices can be calibrated for sensitivity and for the length-of-time delay between the last detected occupancy and extinguishing of the lights. The: most energyefficient, occupancy sensors, known as "manual- on,automatic-off" require that the users manually switch on the lights when entering a controlled zone (the "lights off" function is still automatic).

Occupancy sensor systems typically consist of a motion detector, a control unit, and a relay, Usually, two or more of the components are integrated into one package. Most systems also require a power supply in the form of a transformer, which steps down the building voltage to 24V. The detector collects information, then sends it to the controller, where it is processed. Output from the controller activates the relay, which in turn switches the light circuit.

There are two major types of occupancy controls:

- Wall box units are designed to fit into a standard wall switch box and operate on the building voltage (*i.e.*, a separate power supply is not required). They are excellent, inexpensive replacements for standard wall switches. Their main limitation is their relatively short range. Consequently, they tend to be used in small offices and meeting rooms.
- Wall and Ceiling units typically contain an integrated sensor/controller unit wired (Class 2) to a switch pack containing the relay and power supply. They are far more popular than wall box units and have very few application limitations.

Occupancy-sensing lighting controls represent a refinement of the technology to detect intruders for security applications. With lighting control, two different means of detecting occupancy are used:

- Passive Infrared (PIR) sensors perceive and respond to the heat patterns of motion. This same technology is used in most residential and commercial security systems. The chief advantage of PIR sensors is that they are relatively inexpensive and reliable. They sometimes may raise "false trigger" (that is, respond to non-occupant motion in a space). The major limitation of PIR sensors is that they are strictly line-of sight devices, unable to see around corners or partitions.
- Ultrasound (US) detectors radiate ultrasonic waves into a space, then read the frequency of the reflected waves. Motion causes a slight shift in frequency, which the detector interprets as occupancy. They are more sensitive than PIR sensors, which is both an advantage and a disadvantage. They are often used very effectively in partitioned spaces but are also more prone to false triggering due to their sensitivity to air movement. Proper design and installation minimizes this problem.

Space Control: Along with controls for individual lights or sets of fixtures, master controls are required for each space which can shut off all the lights within the space. For example, the last person leaving the office is much more likely to use a master switch than to go through the office turning off every switch. A master switch can easily turn lights off at the end of a working day.

Each space enclosed by ceiling-height partitions is required to have at least one control device to independently control the lighting within the space. The device can be a switch that is activated either manually or automatically by sensing an occupant sensor. Each control device, regardless of type, shall have the following functions:

(a) Control a maximum of 250 m^2 for a space less than or equal to 1,000 m^2, and a maximum of 1,000 m^2 for a space greater than 1,000 m^2.

(b) Be capable of overriding the required shutoff control [ECBC 7.2.1.1] for no more than 2 hours

(c) Be readily accessible and located so that the occupant can see the control.

Table 8.28: Space Area and Lighting Controls

S.No.	Space Area	Coverage Area for each control device
1	Up to 1000 m^2	250 m^2 max.
2	More than 1000 m^2	1000 m^2 max

Exception to above: The required control device may be remotely installed if required for reasons of safety or security. A remotely located device shall have a pilot light indicator as part of or next to the control device and shall be clearly labelled to identify the controlled lighting.

Control in Daylight Areas. Luminaire in daylight areas greater than 25 m^2 shall be equipped with either a manual or automatic control device that:

- is capable of reducing the light output of the Luminaires in the daylighted areas by at least 50%
- controls only the luminaries located entirely within the day lighted area.

As per ECBC: A complete lighting unit consisting of a lamp or lamps together with the housing designed to distribute the light, position and protect, the lamps, and connect the lamps to the power supply.

Exterior Lighting Control. Lighting for all exterior applications not exempted in §7.3.5 (of the Code) shall be controlled by a photo sensor or astronomical time switch that is capable of automatically turning off the exterior lighting when daylight is available or the lighting is not required. The following lighting applications shall be equipped with a control device to control such lighting independently of general lighting:

(a) *Display/Accent Lighting:* Display or accent lighting greater than 300 m^2 area shall have a separate control device

(b) *Case Lighting*: lighting in cases used for display purposes greater than 300 m^2 area shall be equipped with a separate control device

(c) *Hotel and Motel Guest Room Lighting:* Hotel and motel guest rooms and guest suites shall have a master control device at the main room entry that controls all permanently installed Luminaries and switched receptacles

(d) *Task Lighting*: Supplemental task lighting including permanency installed undershelf or under- cabinet lighting shall have a control device integral to die Luminaries or be controlled by a wall-mounted control device provided the control device complies with §7.2.1.2 (of ECBC)

(e) *Non-visual Lighting:* Lighting for non-visual applications, such as plant growth and food-warming, shall be equipped with a separate control device

(f) *Demonstration Lighting*: lighting equipment that is for sale or for demonstrations in lighting education shall he equipped with a separate control device accessible only to authorized personnel.

Many special lighting applications must be controlled separately, including display lighting in retail stores, case lighting, hotel/motel guest rooms, task lighting, non-visual lighting applications, and demonstration lighting.

Lighting used to highlight artwork or merchandise in retail stores, art galleries, lobbies, and other spaces must have a separate lighting control. This additional control can save considerable energy since the hours required for display lighting are generally fewer than the hours that the space is occupied. In a retail store, for instance, employees typically arrive one to two hours before the store opens in order to prepare the store, and often employees need to stay for an hour or two after the store closes. Without a separate control for display lighting, the display lighting would have to be operated for two to four hours each day when it is not needed. Controls for display lighting can be situated in remote locations, but it is advisable that they have indicator lights and be clearly marked to indicate which lighting is controlled.

Case Lighting: Lighting is frequently installed in closed casework for the display of jewelry and other fine merchandise. Such lighting is required to have a separate control from that used for general illumination of the space. The reason for this requirement is the same as for display lighting: the case lighting is only needed during store hours, not during the entire occupancy period of the space. Usually, the control for case lighting is integral to the case.

Hotel/Motel/ Guest Room Lighting: A master lighting control is required at the entry door of hotel and motel guest rooms to control all permanently installed luminaries and switched receptacles. The control is usually a three-way device wired in combination with local controls. In multiple-room suites, a single master control must be located at the main entrance. This master lighting control allows guests or the housekeeping staff to turn off all permanently installed luminaries when they are exiting the room.

Task Lighting: All supplemental task lighting in a space shall have a separate control. Desk lamps will inherently meet this requirement, but the requirement also applies to permanently installed under-shelf or under-cabinet lighting. Such lighting can have a switch integral to the luminaries or be controlled by a wall-mounted control device, provided the control device is accessible and the controlled lighting can be observed when the switch is toggled.

Non-Visual Lighting: Lighting needed for non-visual purposes, such as plant growth or food warming, must have a separate control. This is because such lighting is likely tube needed at different times than the general lighting.

Demonstration Lighting: Lighting on display in retail lighting stores and lighting that is being demonstrated in classrooms and lighting education facilities must have a separate control. The justification is that such lighting is operated on a separate schedule from the general lighting.

Exit Signs

Internally illuminated exit signs shall not exceed 5W per face. Electrically powered exit signs normally use incandescent bulbs. Most LED and some CFL exit signs can meet ECBC requirement. Due to their low power consumption, LED exit signs can be purchased with builtin back-up power supplies (*i.e.*, batteries). With an estimated service life of 10 years or more, LEDs require significantly fewer lamp replacements than exit signs equipped with either incandescent lamps or CFLs.

Energy Efficient Lighting Controls

High-efficiency lighting components, such as T-8 fluorescent lamps and electronic high frequency ballasts, make a significant impact on lighting energy and its associated costs by reducing the kW required to illuminate buildings. Lighting controls, on the other hand, affect lighting energy by directly reducing lighting's time of use. Some lighting control techniques, such as using photocell controls in building spaces that incorporate day lighting, not only reduce lighting time of use but also decrease lighting power and may even reduce the average cost of electricity by eliminating some lighting kW during peak demand periods.

Source: ASHRAE Manual 2004.

Interior Lighting Power

The prescriptive lighting requirements limit the installed electric wattage for interior building lighting. These lighting power requirements are minimum requirements. Designers working on specific projects may be able to design more efficient lighting systems.

For interior lighting power requirements, the installed lighting power used by luminaires, including lamps, ballasts, current regulators, and central devices (except as specifically exempted in 7.1) is first calculated using the procedure discussed under 7.3.4.1 of ECBC. Calculated installed power is then compared with the maximum permissible Interior Light Power Densities, specified for various building types (Building Area Method) or building space functions (Space Function Method) given in the ECBC.

Interior lighting includes all permanently installed general and task lighting shown on the plans. Interior lighting, for a building or a separately metered or permitted portion of a building, shall not exceed allowed power limits.

The building area method is the simplest method to follow since fewer calculations are required. However, if the project applies to only a portion of the entire building, is not listed as a building type, or has more than one occupancy type, the space function method should be used to determine compliance. Trade-off of lighting power allowances are not permitted between portions of a building where different methods are used.

Building Area Method

This method provides the procedure of calculating total watts per square meter for the entire building based on its type. The sum of all the interior lighting power for various areas of the building cannot exceed the total watts to be in compliance. The first step is to identify the allowed power lighting density for appropriate building area types listed in Table 8.29 of ECBC. If more than one listed type applies to the area, the more general building area type should be used.

The second step is to calculate the gross lighted floor area for each of the building area types (this can be done using the building plans). Finally, the last step is to multiply the allowed watts per square meter listed for each selected building type by the corresponding lighted floor areas to determine the allowed light power allowance.

Table 8.29: Interior Lighting Power- Building Area Method

Building Area Type	LPD (W/m^2)	Building Area Type	LPD (W/m^2)
Automotive facility	9.7	Multi-family Residential	7.5
Convention Center	12.9	Museum	11.8
Dining: Bar, Lounge/Leisure	14.0	Office	10.8
Dining: Cafeteria/Fast food	15.1	Parking Garage	3.2
Dining: Family	17.2	Performing Arts Theater	17.2
Dormitory/Hostel	10.8	Police/Fire Station	10.8
Gymnasium	11.8	Post Office/Town Hall	11.8

Healthcare-Clinic	10.8	Religious Building	14.0
Hospital/healthcare	12.9	Retail/Mall	16.1
Hotel	10.8	School/University	12.9
Library	14.0	Sports Arena	11.8
Manufacturing Facility	14.0	Transportation	10.8
Motel	10.8	Warehouse	8.6
Motion Picture Theater	12.9	Workshop	15.1

Source: ECBC: 2008/ASHRAE Manual, 2004

In cases where both-general building area type and a specific building area type are listed, the specific building area type shall apply.

Space Function Method

Similar to the building area method, the first step of the space function method is to identify the appropriate building type and their allowed lighting power densities, which varies according to the function of the space. These are listed in ECBC Interior Lighting Power - Space Function Method.

Table 8.30: Interior Lighting Power- Space Function Method

Space Function	LPD (W/m^2)	Space Function	LPD (W/m^2)
Office-enclosed	11.8	Reading Area	12.9
Office-open plan	11.8	Hospital	
Conference/Meeting/Multipurpose Hall	14.0	Emergency	29.1
Classroom/Lecture/Training	15.1	Recovery	8.6
Lobby	14.0	Nurse Station	10.8
Hotel	11.8	Exam. Treatment	16.1
Performing Arts Theater	35.5	Pharmacy	12.9
Motion Picture Theater	11.8	Patient Room	7.5
Audience/Seating Area	9.7	Operating Room	23.7
Gymnasium	4.3	Nursery	6.5
Convention Center	7.5	Medical Supply	15.1
Religious Building	18.3	Physical Therapy	9.7
Sports Arena	4.3	Radiology	4.3
Atrium-first three floors	6.5	Laundry - Washing	6.5
Atrium-each, additional floor	2.2	Automotive - Service Repair	7.5
Lounge/Recreation	12.9	Manufacturing Facility	
Dining Area	9.7	Low Bay (< 8 m ceiling)	12.9
Hotel	14.0	High Bay (> 8 m ceiling)	18.3

Motel	12.9	Detailed manufacturing	22.6
Bar lounge/leisure dining	15.1	Equipment room	12.9
Family dining	22.6	Control room	5.4
Preparation	12.9	Convention Center-Exhibit Space	14.0
Laboratory	15.1	Library	
Transportation	5.4	Card file & Cataloging	11.8
Airport-concourse	6.5	Stacks	18.3
Air/train/bus-baggage area	10.8	Hotel/Motel Guest rooms	11.8
Ticket-counter terminal	16.1	Dormitory/Living rooms	11.8
Restrooms	9.7	Museum	
Dressing/Locker/Fitting room	6.5	General exhibition	10.8
Corridor/Transition	5.4	Restoration	18.3
Hospital	10.8	Bankoffice-banking activity area	16.1
Manufacturing facility	5.4	Retail	
Stairs-active	6.5	Sales Area	18.3
Active storage	8.6	Mall concourse	18.3
Hospital	9.7	Sports Arena	
Inactive storage	3.2	Ring Sports Area	29.1
Museum	8.6	Court Sports Area	24.8
Electrical/Mechanical Facility	16.1	Indoor Field Area	15.1
Workshop Library	20.5	Parking garage-garage area Warehouse	2.2
Fine material storage	15.1		
Medium/Bulky material storage	9.7		

Source: ECBC: 2008

For each space that is enclosed by partitions, which are 80% or greater than ceiling height, the gross interior floor area must be determined. This applies to all space area types except for retail. The gross interior floor area should be calculated by measuring to the center of the partition walls and must also include spaces allotted to balconies or other projections.

Finally, the individual lighting power allowances for each space is determined by multiplying its gross lighted floor area by the allowed lighting power density for that space. The lighting power allowances are summed to equal the Interior Lighting Power Allowance for the building.

Installed Interior Lighting Power

The installed interior lighting power calculated for compliance with §7.3 of the Code shall include all power used by the Luminaires, including lamps, ballasts, current regulators, and control devices, except as specifically exempted in §7.1 of the Code.

Exception to above: If two or more independently operating lighting systems in a space are controlled to prevent simultaneous user operation, the installed interior lighting power shall be based solely on the lighting system with the highest power.

Luminaire Wattage

The Code requires that luminaire wattage be incorporated into the installed interior lighting power calculation as follows:

(a) The wattage of incandescent luminaires with medium base sockets and not containing permanently installed ballasts shall be die maximum labeled wattage of the luminaires.

(b) The wattage of luminaires containing permanently installed ballasts shall be the operating input wattage of the specified lamp/ballast combination based on values from manufacturers' catalogs or values from independent testing laboratory reports.

(c) The wattage of all other miscellaneous luminaire types not described in (*a*) or (*b*) shall be the specified wattage of the luminaires.

(d) The wattage of lighting track, plug in bus way, and flexible-lighting systems that allow the addition and/ or relocation of luminaires without altering the wiring of the system shall be the larger of the specified wattage of the luminaires included in the system or 135 W/m (45 W/ft). Systems with integral overload protection, such as fuses or circuit breakers, shall be rated at 100% of the maximum rated load of the limiting device.

Exterior Lighting Power

Lighting power limits are specified for building exterior lighting applications in the Code. The connected lighting power for these applications must not exceed these allowed limits. In addition, trade-offs between applications are not permitted.

Exemptions are allowed for the following lighting applications, only if they are equipped by an independent control device:

(a) Specialised signal, directional, and marker lighting associated with transportation

(b) Lighting used to highlight features of public monuments and registered historic landmark structures or buildings

(c) Lighting that is integral to advertising signage

(d) Lighting that is specifically designated as required by a health or life safety statute, ordinance, or regulation,

Any exterior lighting applications are required to comply with the mandatory requirements in the Code. This requires luminaires operating at greater than 100W to contain lamps with minimum efficacy of 60 lm/W unless the luminaire is controlled by a motion sensor.

Table 8.31: Exterior Lighting Building Power

Exterior Lighting Application	Power Limits
Building entrance (with canopy)	13 W/m^2 (1.3 W/ft^2) of canopied area
Building entrance (without canopy)	90 W/lin m (30 W/lin f) of door width
Building exit	60 W/lin m (20 W/lin f) of door width
Building facade	2 W/m^2 (0.2 W/ft^2) of vertical facade area

Source: ECBC: 2008

Electrical Power

ECBC has mandatory requirements for electric power systems installed in buildings. These provisions are related to distribution transformers, electric motors, power factor, and distribution losses. The mandatory requirements of the Code cover the following electrical equipment and systems of building:

- Transformers
- Energy- Efficient Motors
- Power Factor Correction
- Electrical Metering and Monitoring
- Power Distribution Systems

Transformer Efficiency and Losses

Transformers are of two types — Dry type and Oil filled. Fire safety and environmental concerns associated with transformers are important. Oil filled transformers are not installed at fire hazardous places. Dry type transformer is used when it has to be located near the load center which may be in a hazardous place. As there is no oil used in the transformer and special type of fire resistant insulation are used for the windings, the fire risk is considerably reduced in these transformers.

The efficiency of transformers normally varies anywhere between 96 to 99 percent. The efficiency not only depends on the design, but also on the effective operating load. Transformer losses consist of two parts: No-load Loss and Load Loss.

No-load Loss (also called core loss) is the power consumed to sustain the magnetic field in the transformer's steel core. Core loss occurs whenever the transformer is energized; and it does not vary with load. Core losses are caused by two factors: hysteresis and eddy current losses. Hysteresis loss is that energy loss caused by reversing of the magnetic field in the core as the magnetizing alternating current rises and falls and reverses direction. Eddy current loss is a result of induced currents circulating in the core. Load Loss (also called copper loss) is associated with full-load current flow in the transformer windings. Copper loss is power lost in the primary and secondary windings of a transformer due to the ohmic resistance of the windings.

Energy-Efficient Motors

Electric motors convert electrical energy into mechanical energy. Induction motors are the most commonly used prime mover for various equipment in buildings. In induction motors, the induced magnetic field of the stator winding induces a current in the rotor. This induced rotor current produces a second magnetic field, which tries to oppose the stator magnetic field, and this causes the rotor to rotate.

The 3-phase squirrel cage induction motor is the workhorse of most applications; it is rugged and reliable, and is by far the most common motor type used. These motors drive pumps, blowers and fans, compressors, conveyers and production lines. The 3-phase induction motor has three windings each connected to a separate phase of the power supply.

Induction Motor: The induction motor is the dominant motor in use today with over 90 percent of the installed horsepower—largely because it is rugged, has no brushes or slip rings, and is simple, reliable, and cheap.

The induction motor has stator and rotor magnetic fields that interact to spin a shaft which provides torque for performing useful work. The stator and rotor magnetic fields in an induction motor are each created differently. The stator field results from applying AC voltage directly to copper windings distributed around the circumference of the stator, causing the magnetic field to rotate with varying input power. The speed at which the stator field rotates—the motor's synchronous speed—depends on the construction of the motor. The induction motor is unique in how it develops its rotor field. Typically, the rotor has bars that circulate currents that generate the rotor field. Instead of connecting the bars to a power supply as for the stator field, or otherwise creating a field using a permanent magnet, the rotor field is created by current induced in the rotor bars by the rotating stator field. The rotor field thus rotates at the same speed as the stator field; however, it is offset by a certain angle that increases as more motor torque is required.

Although induction motors are normally considered fixed speed machines, one potential disadvantage is that their operating speed will change with both load and voltage. In fact, it is difficult to find two induction motors that will operate at exactly the same speed, even when operated at identical voltage and frequency, and driving identical loads.

There are two major classes of induction motors, varying in winding pattern and construction: squirrel-cage motors—in which conductive bars mounted in the laminated rotor and held by heavy end rings that short- circuit them at both ends, form the rotor windings - and wound rotor motors, in which coils of wire form the rotor windings. In wound-rotor motors, the windings are generally connected via slip rings and brushes to a variable resistance which can vary the motor's speed and torque.

Induction motors can be constructed so that different pole pairs can be connected and disconnected as desired at the terminal box, producing a variety of speeds. Such squirrel-cage induction motors are multispeed (usually two speed), but are not variable-speed if operating from a supply at fixed frequency. In another variation, part-winding starts induction motor to energize part of the winding when starting and the rest later, in one or more steps, so as to reduce starting torque or current. Induction motors can be constructed for single or polyphase operation with most motors over 20 horsepower designed for three-phase power.

Source: E- Source Technology Atlas Series Volume IV, Drive Power

Drive Power: Motor Efficiency is the ratio of the useful mechanical power output to the total electric power input to the motor. Like all electro-mechanical equipment, motors consume some "extra" energy in order to make the conversion. Efficiency reflects how much total energy a motor uses in relation to the rated power delivered to the shaft.

A motor's nameplate rating is based on output horsepower, which is fixed for continuous operation at full load. The amount of input power needed to produce rated horsepower will vary from motor to motor, with more- efficient motors requiring less input wattage than less-efficient models to produce the same output. Electrical energy input is measured in watts, while output is given in horsepower. Output power for motors may also be stated in watts or kilowatts. One horsepower is equivalent to 746 watts.

Motors shall comply with the following:

(a) All permanently wired polyphase motors of 0.375 kW or more serving the building and expected to operate more than 1,500 hours per year and all permanently wired polyphase motors of 50 kW or more serving the building and expected to operate more than 500 hours per year shall have a minimum acceptable nominal full load motor efficiency not less than IS 12615 for energy-efficient motors.

(b) Motors of horsepower differing from those listed in the two pole energy efficient induction motor shall have efficiency greater than that of the four pole energy efficient induction motors.

(c) Motor horsepower ratings shall not exceed 20% of the calculated maximum load being served.

(d) Motor nameplates shall list the nominal full-load motor efficiencies and the full-load power factor.

(e) Motor users should insist on proper rewinding practices for any rewound motors. If the proper rewinding practices cannot be assured, the damaged motor should be replaced with a new, efficient one rather than suffer from the energy efficiency by rewind practices.

Certificates shall be obtained and kept on record indicating the motor efficiency. Whenever a motor is rewound, appropriate measures shall be taken so that the core characteristics of the motor is not lost due to thermal and mechanical stress during removal of damaged parts. After rewinding, a new efficiency test shall be performed and a similar record shall be maintained.

When motors are oversized and operate for extended periods at the base motor efficiency is assumed to be 90%. significantly less than full load, there are three significant operational penalties—reduced efficiency, reduced slip (important, if the load is a cube-law type), and reduced power factor.

Depending on the motor, efficiency will typically peak at somewhere between 75% load and full load. The larger the motor and the higher its peak efficiency, the more likely it will have a relatively flat efficiency curve between 50% load 5 and full load; with a hump at 75% loads some 0.3 to 1% points higher than at full load. Efficiency drops precipitously below 50% load, with the average 100-hp energy-efficient, motor losing over two points between 50 and 25% load and the average 100-hp standard efficiency induction motor dropping some 5.5 points over the same range. Smaller motors lose even more, particularly at lower efficiencies.

A general rule-of-thumb is that a one percentage point increase in efficiency is equivalent to about a one-third-point increase in slip - a decrease in slip can therefore quickly negate even a significant energy efficiency improvement.

Building Management System (BMS) / Intelligent Buildings

The use of Building Management System (BMS) for operation and control of building, development design, construction and operation, optimizes its resources efficiency and performance, and helps to reduce consumption of energy.The concept of intelligent building further extends the concept of energy saving to other building services, creating a total package for energy management. This improves maintenance and reduces dependence on operators, who may not perform efficiently.

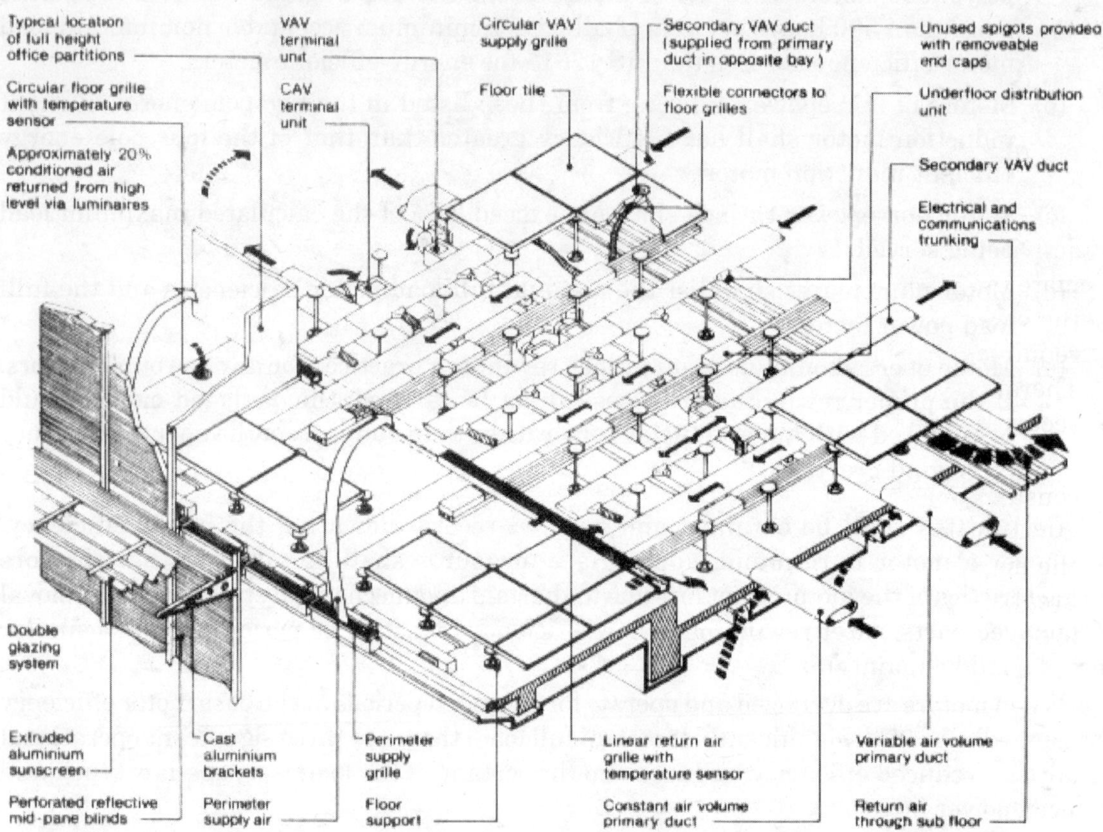

Fig.8.16: Hong Kong and Shanghai Bank: Isometric view of modular sub-floor services layout.

Source: Ove Arup and Partners, Engineers; Foster Associates, Architects.

The concept of intelligent building is the one in which all the building services, safety and security aspects are controlled and monitored by microprocessor based controls, monitoring devices and accessories. Basically such systems are "hands free", requiring very little human interface for operation, and control. The various functions which can be performed are listed as follows:

(a) Control and monitoring of the HVAC System.
(b) Control and monitoring of internal and external lighting.
(c) Monitoring of the incoming electric supply, DG sets and other electrical equipment
(d) Controlling and/or monitoring of the water supply system
(e) Monitoring of the sprinkler system
(f) Monitoring the operation and functioning of fire detection system, which should preferably be intelligent type.
(g) Monitoring the functioning of firefighting equipment *i.e.* fire pumps and accessories.
(h) Controlling entry to a building through an access controlled system, to prevent unauthorized movement.
(i) Monitoring fire exits and other exit points of the building
(j) Enhancing security of the building by installing closed circuit TV (CCTV) monitor.
(k) Providing passive infrared and other sensing devices to detect unauthorized entry around or into the building.
(l) Automatic management of car park systems.

All the above functions are performed by microprocessor-based controllers and their activities are monitored by personal computers (PC) or dedicated control panels. One or more elaborate software programmes are customized and fed into-the system to perform the various tasks. These programmes try to anticipate all possible options, which may be required to be executed. In case of any trouble, an audio visual alarm alerts the supervisor. Thereupon, he can go to the monitor, which clearly displays the faults or any breach of security or danger to the building or its occupants.

The programme helps to optimize the operation of various equipments and reduce consumption of all types of energy. It is estimated that the saving can be between 12% and 20%. The system also provides for maintenance checks and necessary advice. This helps the maintenance staff to carry out preventive maintenance, thus reducing the chances of breakdowns and consequent cost of repairs. The functions of various systems and devices are given below, which relate to energy efficiency in an intelligent building.

- Water Chilling Units
- Pumps
- Cooling Towers
- Heating Equipment
- Air Handling Units

These systems can be made intelligent and computer controlled by Building Management System (BMS). This facilitates the systems to correlate all available data at a very high speed, and take various decisions, like:

i. The time when plants should be started,

ii. Duration and the number of plants which should run to meet the requirement,

iii. Measuring run time to rotate the operation of all the plants and the standby plant in a sequence. This will allow all the equipment to operate for nearly the same duration and equalize the wear and tear. The BMS and intelligent building system, helps to coordinate the operation of numerous services which have become too complex to be handled efficiently and safely manually.

Whole Building Design Approach and High Performance Building

The whole-building design approach considers site, energy, materials, indoor air quality, acoustics, natural resources and their interrelation. This approach brings together building design, energy efficiency, and solar technologies to boost energy savings and make the most of building's elements. It reduces the amount of energy required to operate a building compared to conventional buildings. It improves the comfort the building occupants by using aesthetically pleasing architectural designs to brighten up work areas using sunlight rather than electricity, without causing excess glare. These aim at high performance building design.

High performance buildings are energy-efficient, durable, environmentally responsible and comfortable. They aim to improve the comfort, productivity and efficiency of the people and organisations that occupy them. Ultimately, a high performance building becomes a strategic asset that helps an organisation achieve its primary mission and pays for itself many times during its occupied life.

Following are some of the basic attributes to be considered while designing a high performance building.

- Minimize resource consumption
- Maximum resource reuse
- Use renewable or recyclable resources
- Protect the natural environment
- Create a healthy, non-toxic human environment
- Apply life cycle cost analysis
- Pursue quality in creating the built environment

Technology enabled and knowledge based approaches that are holistic in nature form the basis of high performance buildings (HPB). Intelligent services continuously monitor critical building systems and use sophisticated, analytic and diagnostic tools to identify potential problems and enable building owners and operators to make real time decisions. Advanced technologies with focus on energy efficiency, sustainability and cost factor enable HPBs to perform better.

Over time, automation or smart concepts made rapid strides into the domain of housing. Today, automation has extended to wider systems of a home such as light controls, access control (egress and entry to a house), fire detection, electricity monitoring and control, security information, including video camera (CCTV), burglar alarm system, control of home appliances such as refrigerators, microwave, washing machine, geysers (electric water heaters), garage door operation, entrance gate operation, etc. Pre-programmed retractable sunshades over window openings have also made their entry wherein depending on the direction of the sun rays, they either retract or extend so as to offer shelter against direct sun rays and glare, yet optimising natural-light harvesting. Therefore, the modern home automation can broadly be classified under the following categories based on usage:

- Home security and safety
- Automation and comfort products
- Energy efficiency
- Low maintenance need

Versatile technologies have emerged and new technologies, such POE (Power Over Ethernet), wireless cameras, routers, customised software, specific-purpose protocols, gas sensors, etc. have made these possible.

Water Supply Services : For water supply services, the system can automatically operate and monitor, with built in intelligent controls. The BMS can also be used only to monitor the performance, leaving the operation to their own control systems. The water services include:

- Raw Water Pumps
- Treated Water Pumps
- Hydro-pneumatic System Pumps
- Water Treatment Plants
- Sewage Treatment Plants (STP)
- Cooling towers

The functioning of HVAC also involves water services to an extent.

Electrical Services: The internal lights in a building are controlled on a time basis, so that these are switched on automatically at the right time in the morning and again switched off in the evening. There is a provision for extending working time of lights on a zonal basis.

The external lights can be controlled by a light (Lux) meter which senses light level to switch lights. This helps to conserve energy by adjusting working hours as per natural lights. The BMS is also used to keep track of the functioning of the electric supply and DG sets, so that continuity of power is assured.

Fire and Energy Safety and Security: A building should react automatically to external climate influences. It should also protect people, and business process against intrusion, fire and provide maximum comfort. Whether in industrial systems or in buildings, every technical system depends on the reliable supply of electric power. Even a short failure can have serious consequences. There is a demand for building automation with integrated energy service which saves energy, lowers operating costs and maximizes energy efficiency. Instead of solutions to individual problems all-in-one packages generate a real return on investment throughout the building's entire life cycle.

Voice controlled evacuation systems, which guarantee gradual priority-based evacuation of danger zones and also serve as internal communication systems for customers and employees and as a means of transmitting information during everyday operations. Modern extinguishing systems, are geared to the specific needs of the location and react quickly to fires while doing as little damage as possible to the surroundings.

The biggest threat to building safety comes from fire. A majority of buildings use conventional fire detection system, in which danger signals from various types of detectors are received at a zonal panel and from these on to a central panel. A common signal is received from a group of devices. Hence such system only gives general location of a fire. In the last few year advanced intelligent fire detection systems have become very popular. In these systems every device has an independent address, which is known to the microprocessor based main panel. The microprocessor also constantly checks the health of each device, so one can be sure, that each device is always fully functional. Hence, these types of system enhance safety and also indicate the exact location of fire.

The fire-fighting equipments consist of wet riser sprinklers and fire pumps. The BMS or intelligent fire panels monitor the functioning these devices, so that they do not fail in times of need. Another aspect of safety and security is controlled access to a building. The access control system reduces the load on reception and security staff. All entitled members of staff are provided with a magnetic ID card, which is inserted or flashed near a card reader. If the person is acceptable, the door lock is opened, allowing entry. The system is also used to restrict entry to certain areas or on holidays to certain specified category of staff members. If required exit from a building can also be controlled by use of Access Control Cards. In buildings, where access control is not used for exit, than these gates are connected to the BMS, by which every door opening is monitored, so that a check on exits is kept by the BMS.

Passive Infrared Sensors and/or Door-break Sensors are commonly used to detect any authorized movement. The CCTV are used to record the person's identity. The devices are also used to instantly switch on the lights which stun the intruder and also activate alarms. These are the important functions which all add up to provide a building with artificial intelligence to operate its services and provide safety and security.

Parking Management: An automatic controlled access and exit to and from car parks is a new concept in India, which is now being implemented in some buildings. In this system an access control system is used to identify an authorized user and then allow a motorized barrier to open automatically. The exit from car park is similarly controlled by card readers. Hence, no car can enter or leave the car park without authorization. Several sophistications such as bay or car location identifications, can be added.

Passive Building Design: "Passive design", "Energy-conscious design", "bio-climatic approach" and now "green building" are the buzz words of the architects and engineers. All these strive towards more comfortable, energy-efficient and non-toxic built environment. Passive building design is a systematic process comprising various measures, such as orientation, envelope, form, day lighting, openings, shading, renewable energy and cooling, Table 8.32 gives a broad matrix of these factors.

Table 8.32 Passive Measures Affecting Energy Efficient Design

Passive Design Measure	Sub factor for passive design option	Benefit of Evaluation
Orientation	• Latitude • No. of storeys • Siting • Building form	Helps to evaluate the best direction for orientation
Envelope material	• Thermal transmittance value • Thickness of envelope • Time- lag of material	Calculate the conductance of envelope & assess the heat gain or loss
Building form & effect of surrounding building	• Climate type • Vol. of the mass to be cooled • Surface area, no. of storeys	Increase in surface area increases incident radiation & thus heat gain. Analyze effect of mutual shading with surrounding buildings.
Natural lighting/Day lighting design	• Area of openings • Room height • Aspect ratio of openings • Cill height as compared toht. of working plane • Location of windows • Reflection • Co-efficient of walls • Sky conditions	Determine impact of natural lighting on cooling load due to reduction in the need of artificial lighting
Glazing area of Openings	• Daylight requirement • Sky conditions • Climate type • Glazing area decided by heat gain/ heat loss	Extent of natural lighting supplemented & also heating in colder regions
Efficiency of Shading device	• Height of opening • Depth of overhang • Solar altitude • Solar wall - azimuth	Effective blocking of transmission of radiation, thus protecting from morning & afternoon sun
Solar energy utilization	• Sky conditions • Climate type • Panel size	Minimizing external loads on system & providing backup to electricity requirements
Cooling approach	• Building type & occupancy type • Cost of project • Climate type • Building type & occupancy type • Orientation of openings • Climate type • Location & size of openings	Passive approach can be a substitute to active systems of cooling & thus reduce the load on the system. Evaluation of thermal conditions indoors

Source: ECBC, 2008

The evaluation of passive measures reveals substantial impact on energy performance of building, as given below:

i. Orientation has the maximum impact on the performance
ii. Orientation can increase heat gain up to 41%
iii. Materials and building form influence heat gain or loss can lead to energy performance by 18%.

Accordingly, a weightage can be assigned to each component of design in terms of energy efficient, green building:

Table 8.33: Average Weightage of Passive Measures

Passive measure	Average weightage (Generic)
Orientation	41%
Envelope material	18.5%
Building form	18%
Shading due to neighbouring building	9.5%
Day lighting	4%
Efficiency of shading devices	4%
Cooling approach	3%
Solar energy utilization	2%

Source: ECBC, 2008

Appliance standards and labeling

Energy efficient appliances (such as air-conditioners, refrigerators, ceiling fans and CFLs) have a large potential to save energy, for which the Bureau of Energy Efficiency has introduced energy labelling with minimum energy performance standards (MEPS). There is nearly 17% energy saving by moving up one star (*i.e.* a 4 star vis a vis a 3 star refrigerator has 17% annual energy saving potential). With the current annual growth rate (of 8%) in the production of refrigerators that is likely to continue till 2020, the projected energy savings that can be achieved, according to the BEE, would be 17602 million kWh/Year.

BEE labelled air conditioners have a large energy saving potential. A shift from EER 7 (unlabelled) to 10 (5 star) can save 30% in annual energy consumption. A projected savings that can be achieved by using energy efficient air conditioners according to the BEE is 8682 million kWh/Year by 2020. Use of energy efficient ceiling fans can contribute to an energy savings of 48408 million kWh/Year by 2020 and use of CFLs can contribute to an energy savings of 13081 million kWh/Year by 2020. The energy savings that can be achieved by using energy efficient appliances by 2020 form an important part of energy saving in the building sector.

In a study conducted by TERI, implementation of energy efficiency measures would contribute to a 30% energy savings in new residential buildings and 40% savings in new commercial buildings. In case of existing buildings, the energy saving potential for residential buildings is estimated to be about 20% and that for commercial buildings to be about 30%.

Environment Impact Assessment Process for Building and Energy (MOEF)

As per EIA process, energy aspects for townships and building projects cover the following:

1. **Power:** Requirement: sources of supply, backup sources, consumption and reduction per sq. foot of built up area.
2. **Application of Glass in Building**
   (a) Window openings, specifications of thermal characteristics related to short wave and long wave radiation due to solar energy
   (b) Construction specification of building insulation U values/K values of glass emissivity and thermal characteristics
3. **Renewable Energy**
   Passive solar architecture features layout of streets and buildings to maximize the potential for solar energy, street lighting, emergency lighting, solar hot water system and non renewable energy technologies.
4. **Lighting, Ventilation, Space Conditioning, etc.**
   (a) Including transformers and motors efficiencies, lighting intensity and air-conditioning CFC and HCF facilities,
   (b) Shading of roof and walls to reduce heating/cooling loads and green roofs.
   (c) Thermal characteristics of building envelope (roof, external walls, and fenestration materials used) U values and K values, and rate of heat infiltration into the building.
5. **Impact on micro-climate, and surrounding ecology**
6. **Safety against *the* hazards and emergency.**

Energy Efficient Building Design.

Based on the above parameters, certain generic principles of energy efficient building design can be worked out which comprise the following:

- Integration of energy concept from project outset, demand management
- Compact shape, minimize power requirement
- Daylight design
- Energy efficient lighting, ventilation and space conditioning, site planning and landscape design
- Shading, green roof, building envelope
- Fenestration and window/openings design, glazing design
- Efficient use of passive solar energy, renewable energy
- Energy-efficient, easy-to-use technical systems

- Low water use sanitary ware
- Low-energy electrical appliances
- Use of low embodied energy, recyclable construction materials
- Increased insulation, elimination of thermal bridges wherever possible
- Safety against fire and other hazards.

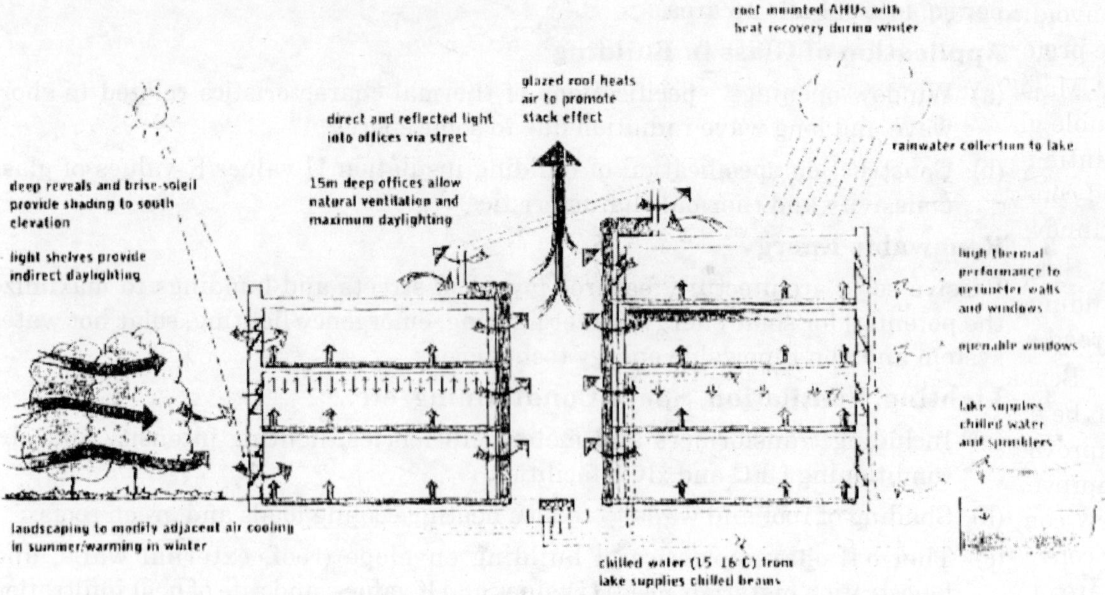

Fig.8.17: A conceptual translation of principles of Low Energy Buildings Design.

Building design involves a synergy of various levels of planning, design, construction and maintenance, leading to a sustainable and energy efficient regime:

Site Planning: Ensure that the proposed building is appropriately oriented and sensitive to the natural features and microclimate of the site. Assess its micro-climatic character taking into account exposure, shelter, natural shading of buildings, interaction of buildings, solar access through the seasons, atmospheric pollution, water and drainage, noise gradients across the site. This gives a basic picture of the qualities of the site and its potential. Minimise earth movements and excavations where possible. Respect ground water levels and design to manage surface water through natural processes. Avoid the formation of heat island and inversion effect by layout planning. Avoid the formation of heat island and inversion effect by layout planning.

Landscape development of the site should improve the shaded areas, micro-climate and visual amenity by the greenery, green roofs, climbing plants on walls, window boxes and balcony gardens, the type of trees to be planted along the roads and near buildings need careful attention. Deciduous trees which shed their leaves during winter permit penetration of sunshine onto building facades and help in heating of buildings. Proper landscaping reduces direct sun from striking and heating up building surfaces. Landscaping creates different airflow patterns and can *be* used to direct or divert the wind advantageously by causing a pressure difference. Additionally, the shade *created* by trees and the effect of grass and shrubs reduce air temperatures adjoining the building and provide evaporative cooling. Trees are the primary elements of an energy-conserving landscape.

Form and Orientation: The objective is to minimize solar heat during the summer and maximise solar gain during the winter so as to reduce the need for cooling, heating and lighting, thus reducing the demand for energy. Design to reduce surface area for heat transfer by avoiding elongated thin forms, spread out, and low density development. Compact forms are preferred with grouping of the buildings for multiple use zoning.

Major components of passive architecture are orientation of building, sun shades, double glazed windows, smart glazing window overhangs, thermal storage wall/roof, roof painting, ventilation, evaporative cooling, day lighting, wind towers, earth air tunneling and construction materials. The active and passive systems can be used simultaneously in innovative ways to make an integrated system.

Sunshine is one of the desired aspects in the buildings, especially the residential buildings. As a matter of urban planning and design as well as building bye-laws, the ratio of road width to building height becomes a very important factor.

Besides use of energy efficient building materials, the energy demand in buildings can be saved substantially by proper designing of walls and roofs, windows and lighting. Improved insulation of walls and roof can reduce the heating and cooling load by 25 per cent. Improved multi-pane windows can reduce the air conditioning heating load significantly. New compact fluorescent lamps require less than a quarter of the electricity consumed by conventional lighting bulbs. These are replacing old fixtures and capturing significant share of market.

Wall to Window Ratio, U Factor, SHGC and VLT Values: ECBC recommends a maximum U-factor of 3.3 w/m2 ok, SHGC of 0.25 for a wall to window ratio (WWR) of upto 40 %, and SHGC of 0.2 for WWR of 40-60 %. A glazing area in excess of 60% of gross external wall area is not recommended. The VLT value linked to WWR should be as per ECBC. SHGC value is particularly critical for South, East and West facade. For glazing U factor and Solar Heat Gain Coefficient (SHGC) should be minimized, whereas VLT (Visible Light Transmittance) should be maximum,

Volume and Building Envelope: Generally, avoid oversized, higher interior spaces in designing for specific functions. Thermal character of building envelope, roof, walls and fenestration should be compatible to U and R values.

Internal Layout: To reduce need for artificial light and for optimum heat efficiency cluster the activities uses which need similar environmental conditions. Avoid open plan to allow for better control of services by the users.

Windows/Doors: Consider the percentage of fenestration on different facades and plan to minimise different temperature zones. Use southerly orientation for passive solar gain. Consider type of glazing and summer/winter ventilation. Use blinds, curtains, shutters draught lobbies and air curtains. Use energy efficient glass with spectrally selective coating to take the maximum advantage of daylight.

Cooling and Air-conditioning: Use natural ventilation, where possible. Consider the use of atria to achieve some of these requirements and to provide amenity space for building users. Minimise the use of air conditioning. Consider the interaction between energy and ventilation strategies to balance potentially conflicting demands. Avoid the use of wet cooling where air conditioning is installed. Specify low-energy, high efficiency plant and fittings, sensor controls and appliances, as per BEE certification. Explore new methods of air-conditioning, such as Passive Energy Draft Cooling (PEDC), High Efficiency Chilling System, earth embedded cooling, thermal storage system, etc. As a thumb rule following standards can be adopted for energy efficiency:

- Lighting power density is less than 15-20 W/m^3
- Lighting performance index lies in the range 21-28 kWh^2/hr
- Air conditioning performance index for different climate zones are: Warm & humid: 195 kWh/m^2 per year (24 hours operation)

  Moderate: 105 kWh/m^2 per year (10 hours operation)

  Composite: 144 kWh/m^2 per year (10 hours operation)

  Square metres per tonne of refrigeration (sqmt/TR) lies in the range 32-42.

Lifts: Consider building forms and heights to economise the reliance on lifts subject to meeting the needs of people with mobility problems.

Resource Recovery: Specify the reuse of materials arising from demolition on site and recycled materials brought in from other sites locally. Shuttering etc. should be reused where possible rather than destroyed on completion. Allocate space for the future recycling of waste glass and a composting facility where appropriate.

Avoid over-designed structures, footings, etc. that may result in the waste of materials. Consider alternative foundations and structures, where appropriate. Specify timbers from sustainable forests. Minimise the use of toxic chemicals and hazardous materials.

Indoor Air Quality: The building design should help in enhancement of indoor air quality for human safety, comfort and health-use of non-toxic paints, glues and building materials to prevent additional environmental impact and high waste treatment/disposal costs during demolition, maximum use of locally available renewable, non-radioactive recyclable and durable material/resources, for example, fly ash bricks, hollow bricks, stabilized mud blocks, etc. for building construction.

Plant location: To reduce distribution loses to a minimum; locate plant close to areas of high energy consumption. Lag pipe runs to high specification, use low temperature storage, time controls and intelligent systems.

Waste heat recovery: Consider the use of a heat exchanger if the building is mechanically ventilated.

Building fabric: Specify insulation standards above the current regulation, where possible. Avoid the use of CFC blown insulation.

Controls: Employ controls that can respond to internal and external conditions. Time and temperature should be sensor controlled according to need of the occupants.

Decorative Finishes: Light coloured finishes improve lighting conditions and reduce the intensity of light required.

Details/Standards of Work: Joint ceiling, insulation, ventilation and thermal installations should be checked to ensure that the work has been carried out to a high standard.

Operation: A user-friendly manual for occupants should be provided to explain the efficient operation of building and equipment.

Commissioning: Before occupancy, a building should be flushed to remove solvents, gases, vapours, smells, etc. that could affect future users. Check the performance of the components, machinery and the equipment against standards and to put right any defects found that could have major long term effects on the energy consumption of the building.

Incentives: Incentives (like bonus, floor area ratio), awards, training and users participation for energy efficiency should be encouraged.

Building Envelope and Surface to Volume Ratio

For any given building volume, the more compact the shape, the less wasteful it is in gaining/losing heat. Also, the building form determines the airflow pattern around the building, directly affecting its ventilation. The depth of a building also determines the requirements for artificial lighting-greater the depth, higher the need for artificial lighting. A proper envelop design should be followed by an efficient lighting and space conditioning design to save energy.

The building envelope and its components are key determinants of the amount of heat gain and loss and wind that enters inside. The primary elements affecting the performance of a building envelope are:

- Materials and construction techniques
- Roof
- Walls
- Fenestration and shading
- Finishes

The following measures can be considered to reduce heat gain by envelope design.

- Select high performance glazing with low U-value, low Shading Coefficient, and high VLT (Visual Light Transmittance).
- Insulate the wall. The options for insulation materials can be extruded polystyrene, expanded polystyrene (thermocol), polyurethane, glass wool, and the likes.
- Brick wall with air cavity can also significantly reduce the heat ingress.
- Hollow blocks, fly ash bricks, and AAC (Autoclaved Aerated Concrete) blocks are also good insulators.
- The heat ingress through the roof can be as high as 12%-15%. Heat reflective paints and garden roofs result in substantial heat reduction.
- Consider shading devices for window openings and skylights.

The choice of materials also helps to maximize indoor comfort.

Thermal Insulation: Insulation is of great value when a building requires mechanical heating or cooling. Insulation helps reduce the space-conditioning loads.

Roof: The roof receives significant solar radiation and plays an important role in heat gain/ losses, day lighting, and ventilation. The roof can also be used advantageously for effective ventilation and day lighting by incorporating vents and skylights.

Walls: Walls are a major part of the building envelope and receive large amounts of solar radiation appropriate thermal insulation and air cavities in walls reduce heat transmission into the building, which is the primary aim in a hot region.

Fenestration and Shading: Proper location, sizing, and detailing of windows and shading form an important part of bioclimatic design as they help to keep the sun and wind out of a building or allow them when needed.

Finishes: The external finish of a surface determines the amount of heat absorbed or rejected by it. Lighter colour surfaces have higher emissivity and should be ideally used for warm climate.

Building Skin Design

The building skin is the first layer of protection for the occupants against sun, wind, and rain and at the same time it has to provide comfort along with:

- Natural ventilation
- Better acoustic insulation - improve noise protection
- Reducing heating energy requirements
- Reducing cooling energy requirements.
- Exploiting solar power

To achieve energy efficiency the following measures can be adopted:

- Minimize exposure on the south and west
- Use simulation tools and techniques that can help in designing the building orientation to minimize heat energy.
- Designing lighting levels for the expected activities keeping into consideration the type of activity, its duration and spaces where they are needed
- Designing with effective and efficient luminaires and using them in conjunction with fenestration/windows with controls for sunlight and skylight.
- Taking care in choosing the colour and finish on ceilings, walls, floors and furnishings. Having flexibility in controlling the lighting - separate and convenient switching devices/dimmers, etc. for areas that need varying lighting levels.
- Designing fenestration/windows in a manner that reduces heat producing radiation entering spaces, and choosing lighting devices/fixtures that have low heat/radiation.

The following types of building skins can be adopted:

Light Weight Flexible Skins: For an architect looking for adaptable and flexible skin systems ETFE (Ethylene tetraifluoro-ethylene) is becoming a popular choice. ETFE is a thin polymer membrane with better insulation and corrosion resistance properties. With its high melting point, it can withstand wide temperature variations and in case of fire, it exhibits an excellent self extinguishing property by shrinking away from the fire creating natural ventilation. The Burj Al Arab Hotel, Dubai, United Arab Emirates, the world's tallest structure, incorporates PTFE (Poly-tetra-fluoro-ethylene based) 15000 sq. m., 200 m high membrane facade with excellent resistance against the extremes of UV radiation, large temperature variation, sandstorms and fire.

Energy Generating Skin/Façade: Integrating Photovoltaics (PV) into building envelopes is not only limited to generating electricity, but can also play a major part in energy-saving designs. With reference to daylight control, in particular semi-transparent photovoltaics provide a wide range of facade applications. Transparency characteristic of crystalline based solar modules is the main determining factor when actual implementation comes into picture.

The LEDs and LCD can display building facade into a flickering screen. In addition to its function as an advertising and communication medium, they also became a new type of ornament for the building, making them stand out in a crowded urban landscape. These media facades transform the building envelope into a huge animated, illuminated screen, especially during the night, attracting the attention of the onlookers and at the same time spurring the vivid spectrum of colours.

Ventilation: Natural ventilation conditions inside a building have direct influence on health, comfort and well-being of building users. Natural ventilation may contribute to the comfort levels in internal spaces especially during hot weather by increasing the rate of evaporation.

The airflow also adds to the general micro-climate by deflecting of channelizing the wind flow pattern. Generally two factors affect the general airflow in built spaces, (*a*) distribution of pressure gradient around a building, and (*b*) thermal forces caused by temperature difference between outdoor and indoor spaces. The first deflects the building up of pressure zone as a result of wind deflection around, within and above the building. Where the wind pressure is lowest, the suction effect takes place. The difference in outdoor and indoor pressure zones would determine the airflow velocity through the internal space, and hence these factors need to be considered while designing the fenestration/ windows in buildings.

Dr. Givoni has classified the ventilation into three clear functions: *"health ventilation,* which maintains air quality in the building by replacing indoor air with fresh outdoor air; *thermal comfort ventilation,* whose purpose is to facilitate the heat exchange from the body to the surrounding air; and *structural ventilation,* which is a means by which the structure is cooled when the indoor temperature is higher than the outdoor temperature."

The ventilation requirements tend to vary with general climate and more specifically with micro-climate conditions during different seasons. Therefore, the analysis of different ventilation requirements needs to be undertaken prior to the deciding the location of windows/fenestration, opening mechanisms and overall building orientation. The surrounding vegetation and landscape also have profound impact on wind flow and its resultant impact on ventilation in the buildings. Various forms of building can be considered for natural ventilation according to micro-climatic conditions.

Fig.8.18: Numerous Jharokha of Hawa Mahal, Jaipur make a unique architectural statement. As the name suggests, the envelope of this building was designed for wind and ventilation, besides privacy and view.

Wind Tower: In a wind tower, the hot air enters the tower through the openings in the tower, gets cooled, and thus becomes heavier and sinks down. The inlet and outlet of rooms induce cool air movement. The tower walls absorb heat during daytime and release it at night, warming the cool night air in the tower. Warm air moves up, creating an upward draft, and draws cool night air through the doors and windows into the building. Wind tower works well for individual units, and not for multi-storeyed apartments.

Courtyard: Due to incident solar radiation in a courtyard, air gets warmer and rises. Cool air from the ground level flows through the louvered openings of rooms surrounding a courtyard, thus producing air flow. If the roof surfaces are sloped towards the internal courtyard, the cooled air sinks into the court and enters the living space through low-level openings, gets warmed up, and leaves the room through higher-level openings.

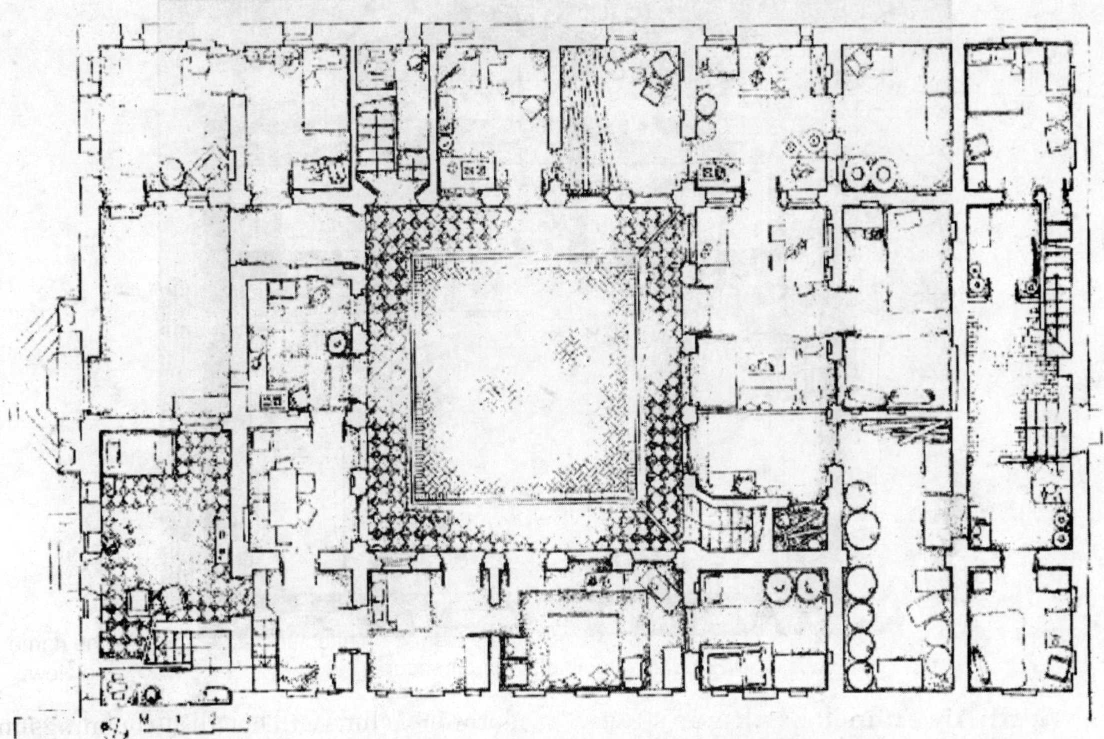

Fig.8.19: Courtyard in a Haveli is the centre piece of climatically comfortable and interactive living. It acts like a sink of cool air and is shaded most of the day. Similar principle had been adopted for streets in Jaipur, Jodhpur, Shahjahanbad and other cities.

Source: Sunand Prasad, Haveli: Urban in Paul Oliver, 1997, Vernacular Architecture of the World, Cambridge.

Air Scoop: To cool down building in India, Pakistan, Iraq and other hot climates traditionally people resorted to heavyweight solutions. Structures are massive, with dense walls to absorb heat. Buildings may be dug into the ground to use the earth mass for cooling and to increase ventilation. In Hyderabad, Pakistan, buildings are equipped with wind catchers which redirect the airflow down into the living quarters Traditional homes in Baghdad may contain a *badgir*,an air scoop that takes in the north-westerly air stream. A *badgir* is a flue in the wall of a house which rises to the highest point of the roof parapet. It is typically 50 by 20 cm in cross section. *Badgirs* are most effective when built into walls transverse to the prevailing wind with wide openings. Once the air is taken in, it sinks as it takes on moisture and cools in the channel. This is an example of an air conditioning system in its most sophisticated and energy-efficient form.

The air taken in rises through a central space or shaft, which has openings at the top. Courtyards also help to lower the temperature. They act as a vertical vent through the height of a building and create a comfortable microclimate. The temperature differential between ground floor and roof can be as much as 20 degrees Celsius. Verandahs have been an important part of Indian homes. They provide semi-enclosed spaces that are used for socialising, sleeping in the hot summer months, cooking, and storage.

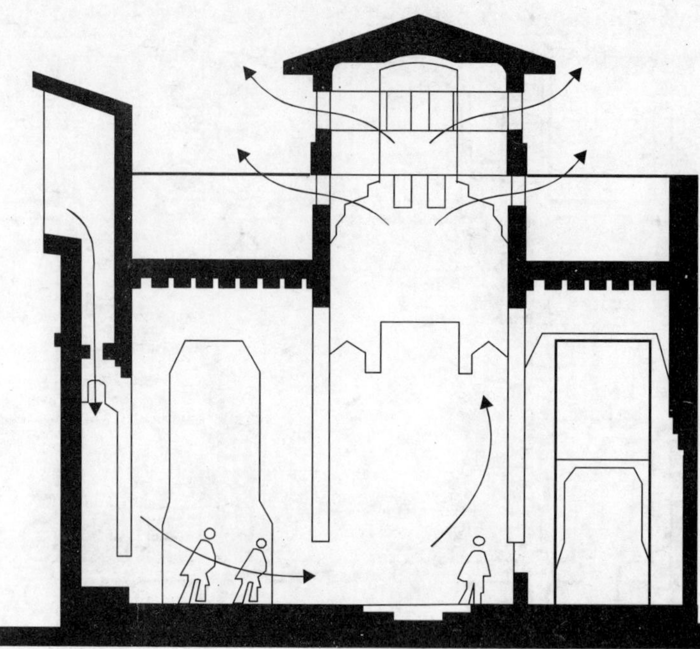

Fig.8.20: Cooling by an air scoop.

Earth Air Tunnel: Earth is an effective moderator of climate. Though surface thermal conditions show patterns of yearly cycles warming the surface during summer and cooling it during winter, at depths below 4-5 meters yet this affect is negligible and the temperature remains more or less constant-equal to yearly mean solar temperature. The moderating characteristic of earth, *i.e.*, variation of temperature with depth, given by the laboratories (1979), is shown in the Table 8.34.

Estimated average earth temperatures for some Indian cities given in the following table show that even in the hottest locations, earth temperatures remain below 30°C. The capacity of the earth to condition the air by effectively damping the ambient air temperature and filtering out the effects of solar radiation can be used in combination with buried pipes or underground tunnels, as were used in Persian architecture.

Table 8.34: Summer Earth Temperatures 4 metres below the Surface in Selected Cities

Sl. No	Name of City	Climatic Zone	Earth Temperature
1.	Bangalore	Composite	24°C
2.	Bombay	Warm and Humid	28°C
3.	Delhi	Composite	25°C
4.	Jodhpur	Hot & Dry	29°C

Source: Kachru, U (1994)

The main barrier to design of buildings is the lack of insulation of the roof slab. Major heat gains in summer are through the roof. In Delhi, for example, 45-50 percent of the total summer solar radiation is transferred through the roof. The envelope system becomes ineffective unless techniques are developed to produce *inexpensive* horizontal cavities to enclose the roof area. A system of promoting wind flow though the building and reducing heat intake is earth berms with wind shafts opening within the rooms. The temperature of subterranean air is almost constant at any time of the year.

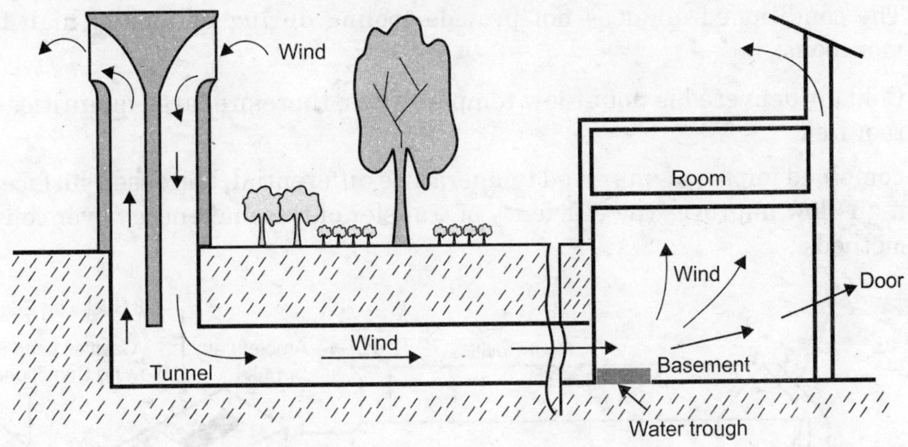

Fig.8.21: Earth-air tunnel with a scoop.

Source: Kachru, Upendra, (1994)

Envelope air-conditioning: At a depth of about 4 m below ground, the temperature remains nearly constant round the year a tunnel or a pipe embedded at a depth of about 4 m below the ground will acquire the same temperature as the surrounding earth. Thus the ambient air ventilated though this tunnel can be used for cooling in summer and heating in winter.

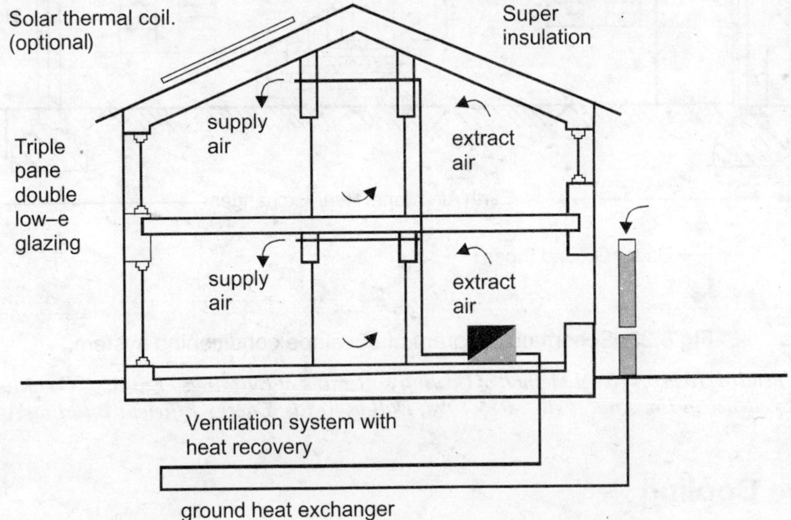

Fig.8.22: Earth-air tunnel with turbulent air-flow.

Source: Kachru, Upendera (1994)

Air drawn through the earth into buildings is cooler and reduces air-conditioning load. However, to provide large cooling loads, the construction costs of the tunnels becomes prohibitive. Other limitations of the system can be the following:

- The ducts provide access to foul smells, fungi, micro-organisms, insects and rodents and thereby create a health hazard;
- The conditioned air does not provide cooling during periods of high humidity, monsoons;,
- Cold air delivered is not at low temperatures, therefore large quantities of air are required.

The combined impact of improved temperature differential, increased surface area and turbulent air-flow improves the efficiency of transfer of thermal energy over conventional passive methods.

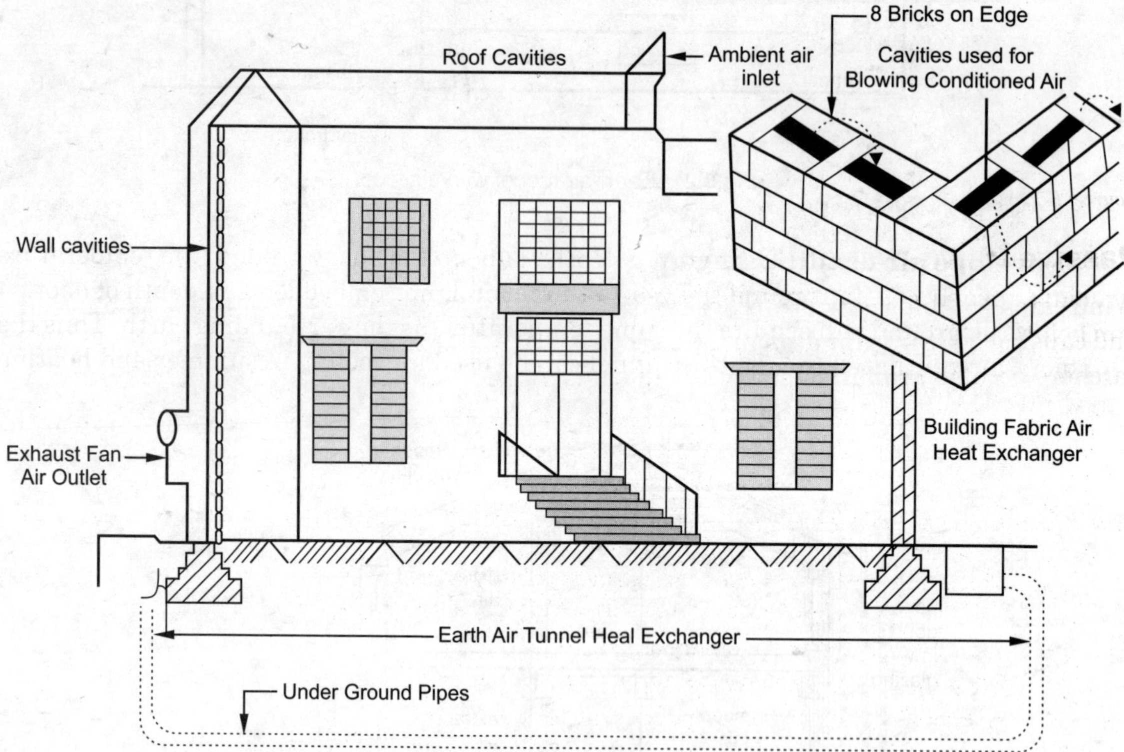

Fig.8.23: Schematic diagram of envelope conditioning system.

Source: Kachru, Upendra (1994), A Novel Method of Cooling & Heating of Buildings - Envelope Conditioning, Proceedings on Conference "Excellence in Housing", Feb. 24-25,1994, Dallas, Texas; Energy Efficient Building Association, Wasua, U.S.A.

Evaporative Cooling

Evaporative cooling lowers indoor air temperature by evaporating water. It is effective in hot and dry climate where the atmospheric humidity is low. In evaporative cooling, the sensible heat of air is used to evaporate water, thereby cooling the air, which in turn, cools the interior space of the building.

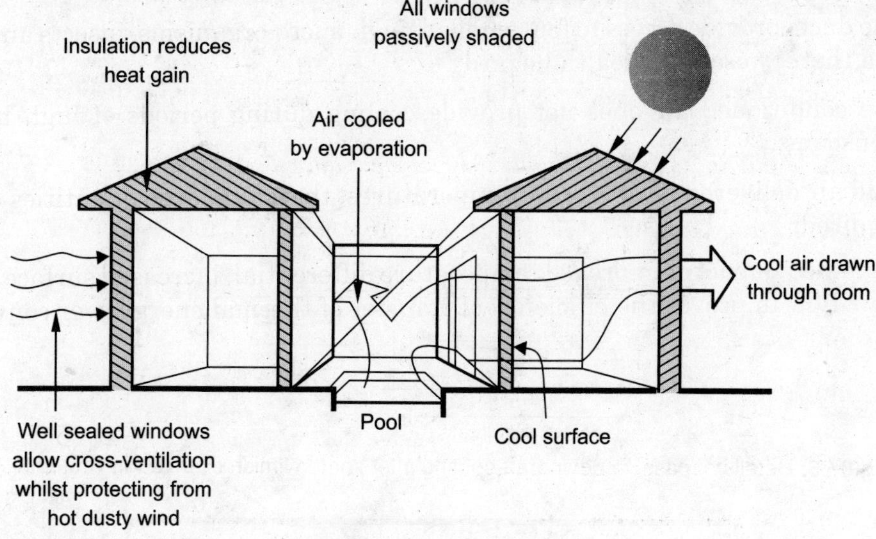

Source: Kachru, Upendra, 1994

Fig.8.24: Evaporative Cooling

Passive Down-draught Cooling

Wind catchers guide outside air over water-filled pots or cooler pads, inducing evaporation and causing *a* significant drop in temperature before the air enters the interior. Such wind catchers become primary elements of the architectural form also.

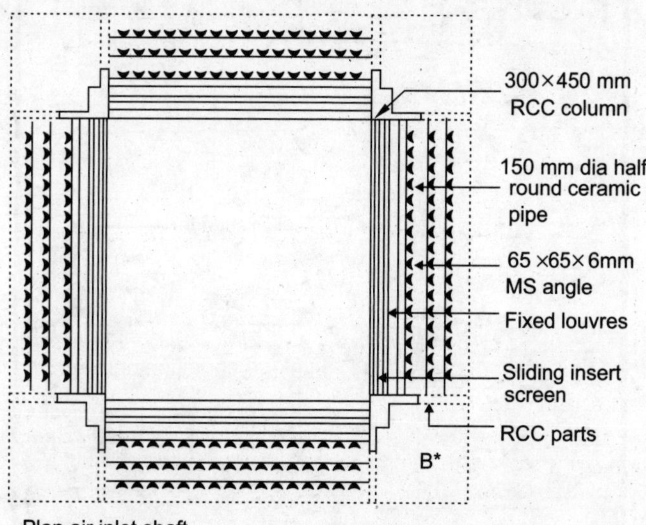

Fig.8.25: Passive down-drought cooling

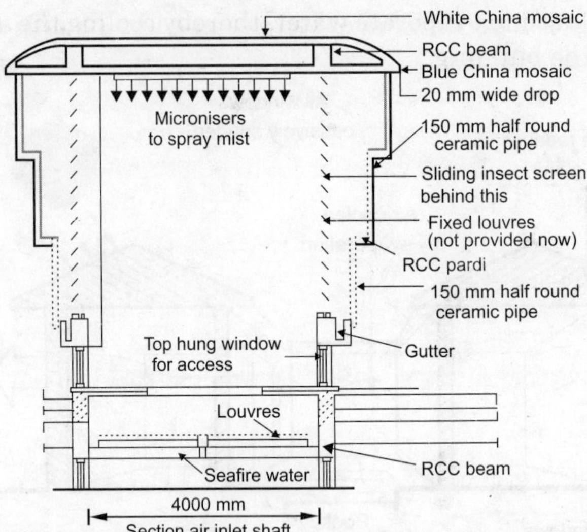

Fig.8.26: Details of passive downdraft cooling inlet shaft, Nimish Patel and Parul Zhaveri.

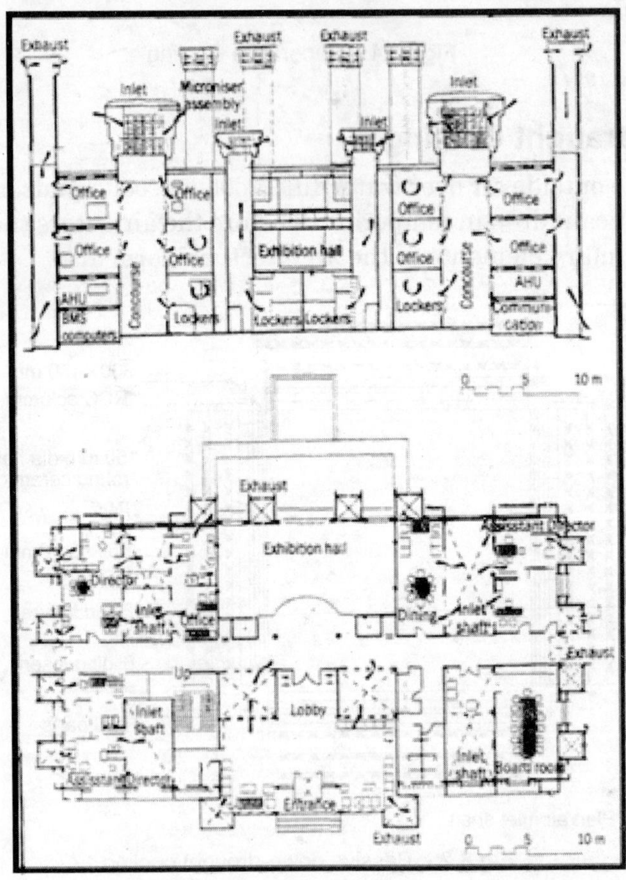

Fig.8.27: Torrent Research Centre, Ahmedabad. Nimish Patel and Parul Zaveri, Architects.

Torrent Research Centre at Ahmedabad designed by Nimish Patel and Parul Zaveri is an exemplary design of Passive Down-draught Evaporative Cooling (PDEC), reducing the use of conventional air conditioning by about 200 tonnes. The shafts as a consequence of their locations, size, height, and their simulated configuration generate the required movement of air in different spaces without using any mechanical or electrical energy. Thus the buildings have a climatically-sealed environment with only designated inlets and outlets, with air cooled at the point of entry using a fine spray of water. In the summer when the outside temperature is 4344°C, the inside temperature remains at 29-30°C, a 13-14°C drop. Six to nine air changes per hour are provided on each floor. Temperature fluctuations inside the building do not exceed beyond 4°C over 24-hours period, while the fluctuations outside are as much as 14-17°C. As such cooler inside temperatures provide comfort to its occupants.

- During the hot season strategy (March-June) passive evaporative cooling micronisers provide a down draft of cool air.
- Insulated building mass and roof and high air change rates (8 – 15) reduce the heat load.
- Air moves across the laboratory through short circulating and the micronisers control automatically the ambient humidity.
- During the monsoon season (July to September), the ventilation rate is increased with micronisers switched off. Ceiling and wall fans induce air movement.
- During the cool season (October - February), the inlets and Exhausts are closed. Insulated walls and roof reduce heat loss.
- During the hot days the building is evaporatively cooled, while the inlets and exhausts are closed at night.

Torrent Research Centre at Ahmedabad designed by architects Parul Zaveri and Nimish Patel, is a good example of energy saving architecture. The pharmaceutical laboratories minimize the use of air conditioning by Passive Down-draft Evaporative Cooling (PDEC) system. This is done by simulated design of inlet and outlet shafts for circulation of air through a fine spray of water that cools it. The buildings work on natural light and have dust control. The PDEC system is adaptable according to seasonal climatic variations.

Lighting Daylight Design

Lighting is the largest single item of running energy cost, particularly in the offices. Also occupants tend to prefer natural light, especially since certain forms of artificial lighting have been implicated as the source of health problems. Energy efficient buildings should make as much use of naturally available light as possible. Lighting is important because of the influence it has over occupant experience.

Principal factors influencing levels of daylight are:

- The view out of the building;
- The effect of daylight on the appearance of interiors;
- Daylight and skylight;

- Combining daylight and artificial light;
- The treatment of sunlight penetrating building;
- The orientation of windows;
- The angle of tilt of windows;
- The obstructions to light admission (*e.g.* nearby buildings);
- The reflectivity of surrounding surfaces.

The following factors should be kept in mind in building design for daylight:

- Windows provide a view out and time orientation for occupants.
- Occupants more readily accept variable illumination when daylight is the light source.
- Natural light gives true perception of colours.
- The amount of glazing obviously influences the amount of daylight available, but greater window area is not always better - it may simply increase contrast.
- Large windows admit light but also lead to heat gain and heat loss routes, and thus potential thermal discomfort.
- The decision regarding which rooms to allow facades should be governed by the activity to take place within them.
- The amount of sky which can be seen from the interior is a critical factor in determining satisfactory day lighting.
- Daylight normally penetrates about 4-6 m from window into the room.
- Roof lights give a wider and more even distributed light, but also permit heat gain, which may cause overheating but provide about three times the benefit of a vertical window of the same size.

Daylight design should reduce dependency on artificial lighting, as well as provide comfortable, glare free light to the interiors. Apart from conventional windows, several options are available for this.

Light Shelves have been in use for some time. Their function resembles the mirror system. They are normally located in the window and provide shade rather like one enlarged louver element. The light shelves can be located outside the window, inside or on both sides. Sun light reflected from the upper surface of the light shelf into the room interior and particularly onto the ceiling where it provides additional, diffused light. In overcast conditions, light shelves cannot increase the lighting level; they operate most effectively when direct sunlight is available. Ceilings are usually designed to be higher than normal to optimize their use.

Prismatic Glazing operates by refracting incoming light. The system consists of a panel of linear prisms (triangular wedges) which refract and spread the incoming light to produce a more diffuse distribution. The view out is substantially restricted, but the system can be used as an alternative to the reflective louver system. Glare can also be somewhat reduced. Maintenance is virtually eliminated if the system is installed between the panes of double-glazed units.

Light Pipes gather incoming sunlight using a solar tracking system. The light is concentrated using lenses or mirrors and is then transmitted to building interiors by 'pipes'. These can be hollow shafts or ducts with reflective internal finishes (similar to working principle of optic fibre cable technology). A special luminaire is required to provide distribution of the light within the building.

Mirror-Based Systems can be used to reflect incoming sunlight and daylight *deeper* into building interiors. Louvres with reflective upper surfaces can be positioned in the building facade at the window opening, and used to shade occupants from unpleasant direct solar radiation while reflecting light onto the ceiling, providing much greater penetration of natural light.

Atria have become much more common as integral components of commercial building designs over the last years. They appear to provide the most attractive combinations, an aesthetically pleasing architectural feature which also has the potential to be environment enhancing and energy efficient. Their use as a buffer space and as a means of providing natural light and ventilation large, deep and complex buildings has been exploited by the designers.

The illumination by day light without compromising on the performance for illumination system needs to address the issue of glare and light controls. Glare can be reduced to a minimum by means of light shelves, overhangs and deeply recessed windows. The interior layouts should place the desks at right angles to windows. Control systems are capable of responding to individual requirements with good switching design (often infra-red) and controls which are user friendly and take into account both-daylight levels and workstation layout Design luminance should be set to achieve about 400 lux with lower levels in circulation areas. There will be good levels of daylight in perimeter work areas without causing excessive glare in the interior. Blinds are easy to operate with a good range of adjustment, which need to be fully closed only in exceptional circumstances. Interior fittings and furnishings should be light in colour and tonal value with tall fittings kept to a minimum. Circulation lighting should be low energy and well planned and controlled with full contribution of daylight. An important added advantage is the external view.

Where lights are operated electronically according to natural light levels, the systems can be either closed or open loop. A closed loop system controls the lighting to top up the daylight to achieve a given minimum acceptable luminance level. Open systems measure external incident daylight to dim lights but with no feedback of the actual levels of realised luminance.

Fenestration/Windows: Windows are very significant building components that affect the energy consumption in buildings, and hence help to conserve energy. The interplay of window size and its impact on illumination levels, solar radiation/heat and ventilation need to be effectively dealt with. Primarily there are three types of lights that enter from windows: (*a*) direct sunlight (*b*) reflected light from the earth's surface, and (*c*) diffused or skylight. Each of these types has their own characteristics and creates different moods. By proper placing of the windows along with appropriate design, glazing, shading devices and sunscreen devices, significant reduction in energy consumption in buildings can be achieved.

The interactions of external climatic conditions and the desired comfort level and type of activities in the internal spaces, define the design objectives of fenestration and windows. The points where these interactions occur become important as this is where the flows of energy do take place such as walls, roofs and floors, window and hence these building components need due attention and consideration.

In the modern buildings with sealed building envelopes, almost all building processes are mechanically run and controlled, leaving little scope for manipulating the external and internal micro-climatic environments. This results in significant loss of energy. Humidity control is another element that need due consideration through the design of heating and cooling system, and the design of fenestration. As a thumb rule, non-domestic buildings, the window area should *be* about 20 per cent of the floor area to provide sufficient light to a depth of about 1.5 times the height of the room. Various switching sensors like timer control, daylight control, occupancy linked control, etc. can be adopted.

Applications of Glass: The deterrent for use of day light is mainly glazing, heat transfer, prone to adverse climatic conditions and noise penetration. With the rapid development in glass technology, one can not only effectively control above factors but can also improve working conditions. Various types of glass are available as described below:

Photo-chromic devices change transmission in response to prevailing radiation levels. Small examples have been in everyday use in the form of sunglasses and spectacles. These react automatically to light levels. There are considerable technical problems to scaling up photo chromic glass to normal window size.

Thermo-chromic glass has changing optical properties in response to temperature variations. It has a laminated structure incorporating a chemical which turns opaque at around 30°C, reducing insolation by about 70 per cent. For this reason an ideal application is as external solar shading.

Electro-chromic glass, is an innovative option, the properties of which can be changed by the application of a small electrical current. Their construction consists of complex multi-layered transparent coatings. The electrical signal reduces the transmission capacity of the electro-chromic layer between two sheets of glass affecting not only daylight but also solar heat.

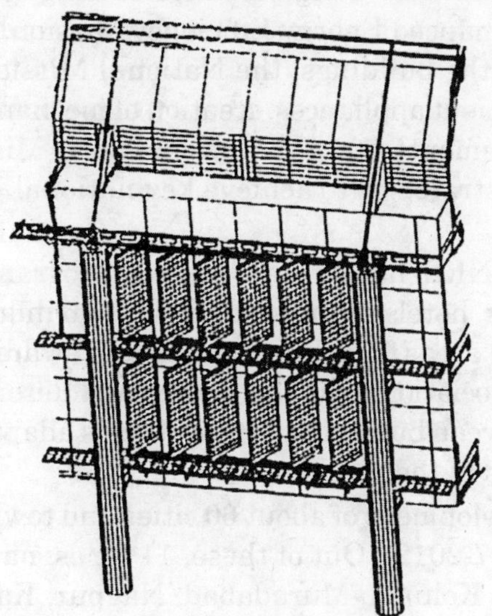

Fig.8.28: The "smart" façade filters sunlight according to internal temperatures, thus regulating solar gain and light effects. The automatically controlled system may be manually overridden.

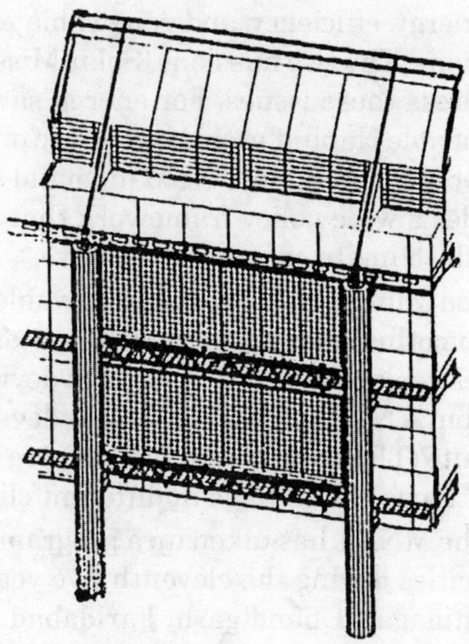

Fig.8.29: Intelligent Building Envelope, the window grids are activated by photo-electric cells to control light at Centre of Islamic Culture in Paris Numerous light sensitive mechanical irises open and close itself to control the lighting inside the building. An ever changing façade added an element of dynamism to the building with automatically controlled louvers and sun blockers to regulate the solar shading. Jean Nouvel, Architect, 1987

Promoting Renewable Energy in India

The harnessing of renewable energy aims not only increasing energy generation, but also helps in a pollution-free environment. It is estimated that India has a potential of generating more than 1,00,000 MW from renewable sources of energy. The potential of renewable energy resources is given in following table.

Table 8.35: Potential of Renewable Energy Sources

Energy Source	Estimated Potential	Cumulative Achievement
1. Biomass power (agro residues)	16,881	683.0 MW
2. Wind power	45,195	9755.85 MW
3. Small hydro power (up to 25 MW)	15,000	2344.67 MW
4. Cogeneration-bagasse	5,000	1033.73 MW
5. Waste to energy	2,700	58.91 MW
6. Solar Power (grid)	2,533	2.12 MW

Source: MNRE report, Jan, 2009.

Energy efficiency and renewable energy sources offer a big scope for cutting carbon emissions. The two missions, Solar Mission and Enhanced Energy Efficiency Mission, seek to address these issues. For energy saving within the buildings, the National Mission on Sustainable Habitat promotes the use of energy-efficient appliances, creation of mechanisms that would help finance DSM [demand side management], etc. The National Solar Mission provides a wide policy framework that integrates strategies to achieve key national goals from the climate change perspective.

The Ministry of New and Renewable Energy (MNRE) has initiated various programmes for promoting solar water heating systems in homes, hotels, hostels, hospitals and industry besides deployment of SPV systems/devices in urban areas for demonstration and awareness creation. A National Rating System developed in association with The Energy and Resources Institute (TERI) for promotion of energy efficient/green buildings. The system is adaptable for all types of buildings in different climatic zones of the country.

The MNRE has taken up a program for the development of about 60 cities and towns as solar cities during the eleventh five year plan (2007-2012). Out of these, 11 cities, namely Gandhinagar, Chandigarh, Faridabad, Dehradun, Kohima, Muradabad, Nagpur, Kalyan and Gurgaon, etc. have been identified for the purpose in the first phase.A solar city is an urban area with an active programme to achieve not only lower reliance on fossil fuel but also sustainable greenhouse gas emission levels too. Thus, a typical solar city programme strives to integrate:

- Energy conservation and energy efficiency measures so as to reduce the energy demand
- Utilisation of locally available resources such as solar and other renewable energy resources to meet the increasing energy demand.

India's first initiative towards the development of solar cities was undertaken by the Gujarat Energy Development Agency (GEDA), Government of Gujarat. It decided to make its capital city Gandhinagar as a Solar City. Accordingly, a Master Plan for the same was prepared by TERI in 2007, which is being implemented.

Solar Radiation: Solar radiation controls our climate. Some radiation does not penetrate to the ground, some is reflected off the atmosphere back into space. The oxygen, ozone, and water vapour in the atmosphere absorb some of the radiation, which warms the atmosphere. Some of this energy reaches the Earth and supports life. The sun is a non-polluting source of renewable energy and is essential in the formation of wind, clouds, thunderstorms, rain, and other weather conditions, some of which can be converted into usable energy. The earth's atmosphere regulates energy flows by filtering through water vapour. Reflection and radiation are essential to the well-being of the planet, which would otherwise overheat.

The radiation of a large area is determined by the amount of solar radiation which falls upon that area during each season. Regions that are fully exposed to the sun for a larger part of the year are predominantly hot; those which receive sunshine only at low angles and for small portions of the year are predominantly cold. Considering the sun-earth relationship, the equatorial regions is the hottest.

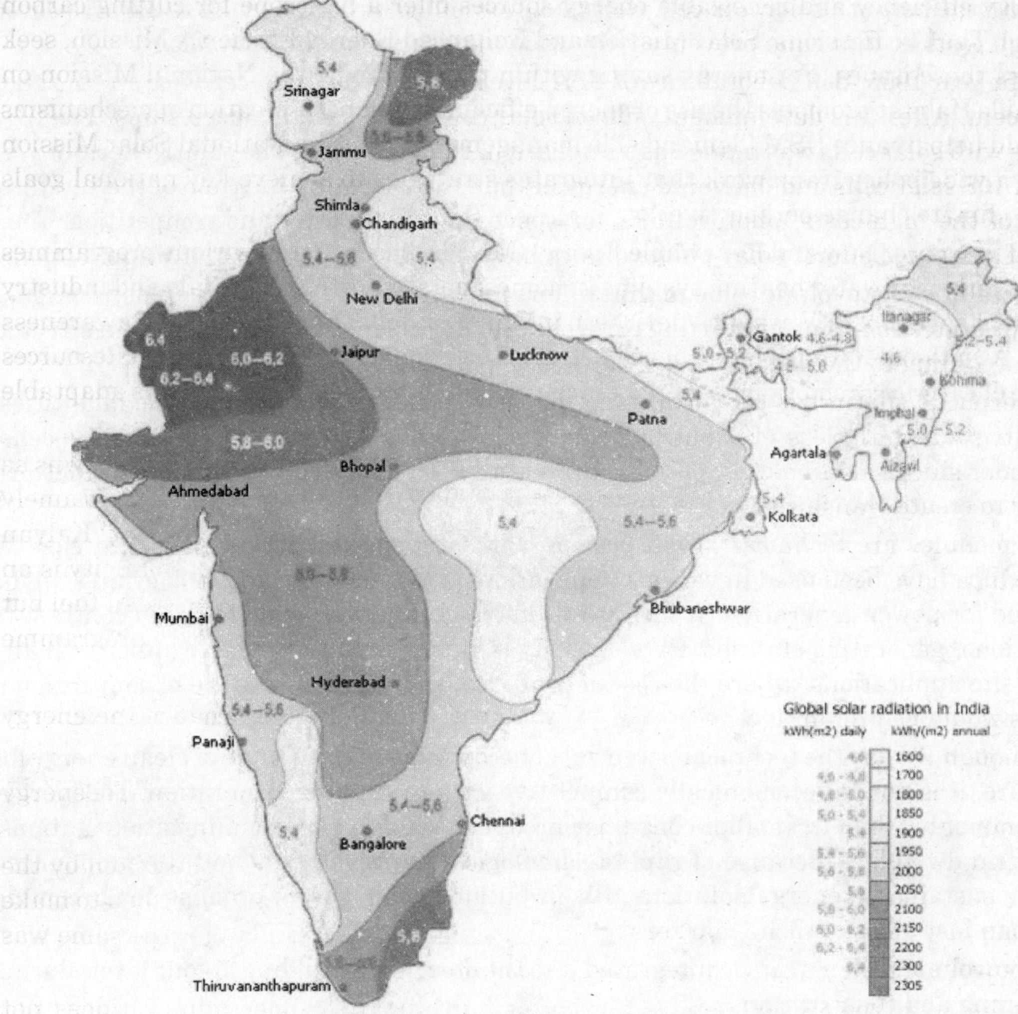

Fig.8.30: Annual solar radiation (kWh/m²) in India.

Source: National Building Code, 2005

Fig. 8.30 gives the global radiation which is the total amount of solar radiation (measured in kilowatt-hours per square meter) received on a horizontal surface in various regions of India. It shows that the regions that receive more solar radiations are not necessarily the hottest. There are other factors which govern the temperature. For a building designer, it is important to know the proportion of direct and diffuse radiation on a horizontal surface and on vertical walls oriented in different directions.

Solar Energy for Buildings: Photovoltaic is the process of converting solar radiation to electrical energy with the help of solar cell. The photovoltaic effect was discovered as early as 1839 by Alexander Becqueral, who was experimenting electrochemically with two metal plates dipped into a dilute acid. He found that the equipment produced more energy when exposed to sunlight. But it was fifty years later that Chutes Fritts proved the photovoltaic effect with the help of a selenium cell. Relatively low efficiency of about 1% to 2% associated with high costs at that time prevented any wide application of the technology.

In the year 1955, Bell Laboratories of USA developed a silicon solar cell with an efficiency of 6 percent. After this development, silicon solar cells were used in space ships. The first satellite with solar cells was named as Vanguard in the year 1958. The solar generator was made of 108 solar cells and delivered energy beyond the operating life time of the satellite. Inspite of the high costs, photovoltaics for space applications had no competition. As a result of sustained efforts solar photovoltaics have found their way in many applications.

Crystalline photovoltaic come in thin wafers and are mostly used under plastic or glass covers for protection, thus permitting external use. Each panel of cells can vary in size and output according to the needs.

Amorphous silicon has the advantage that it is easier to produce in large quantities/areas. It is currently less efficient but considerably cheaper than poly-crystalline cells. Amorphous silicon also makes it possible to create a film, which can be applied in a pattern in order to create translucency.

PV modules are technically well proven, with an expected service life of atleast 30 years; which have been used in various applications. PV is modular technology and can be employed for power generation. It facilitates distributed power generation, in contrast to conventional generating stations. PV electricity is viable and cost-effective option in many remote-site applications, where the cost of grid extension or maintenance of conventional systems would be prohibitive.

Although PV has the technical potential of becoming a major source of clean energy for the future, it is not yet economically competitive with grid power generation. Today, it is most commonly found in smaller consumer products, remote telecommunication stations, and offgrid dwellings. Because of rapid technological improvements and the pronounced need for sustainable energy solutions, PV in building now shows promise for becoming more than just another niche market.

Photovoltaic systems can be integrated into the design of a building at four levels during its planning and construction:

- Design of a building (shape, size, orientation, colour)
- Mechanical integration (multi-functionality of a PV element)
- Electrical integration (grid connection and/or direct use of the power)
- Maintenance and operation control of the PV system integrated with the usual building maintenance and control.

Solar roof is a recent concept which has been adopted throughout the world to produce electricity in buildings for their own use. The basic idea is to use photovoltaic modules for roof construction, thus savings on roof cost to partially cover the cost of the solar electrical generation. The balance of the costs can be recovered from the annual electricity which the system generates.

Renewable Energy Devices and Systems

Rising energy costs along with increased environmental awareness and stricter emission and trading regulations are seeing a shift from fossil fuels to renewable sources to optimise energy supply and consumption. Renewable energy devices, system and conservation measures that are helping bring about reduction in consumption of conventional energy are as follows:

- Solar roof/Photovoltaic Panels
- Solar water heating systems
- Solar cookers (box and dish type)
- Scheffler cookers
- Solar steam generating systems.
- Solar drying/air heating systems
- Solar refrigeration and air conditioning plants
- Solar concentrators for process heat applications
- Solar lanterns
- Solar home lighting systems
- Solar generators
- Street light solar control systems
- Solar hoardings
- Solar street light/garden lights
- Solar traffic lights
- Solar blinkers, Road studs
- Solar power packs
- Building-integrated photovoltaic
- SPV power plants for decentralization applications
- Power projects based on municipal urban waste and industrial waste through combustion/bio-methanation technologies.
- Biomass gasification and co-generation projects in industries
- Biomass gasifier based crematoriums
- Projects on methane utilization for thermal and electrical applications in industries
- Wind turbines for power generation

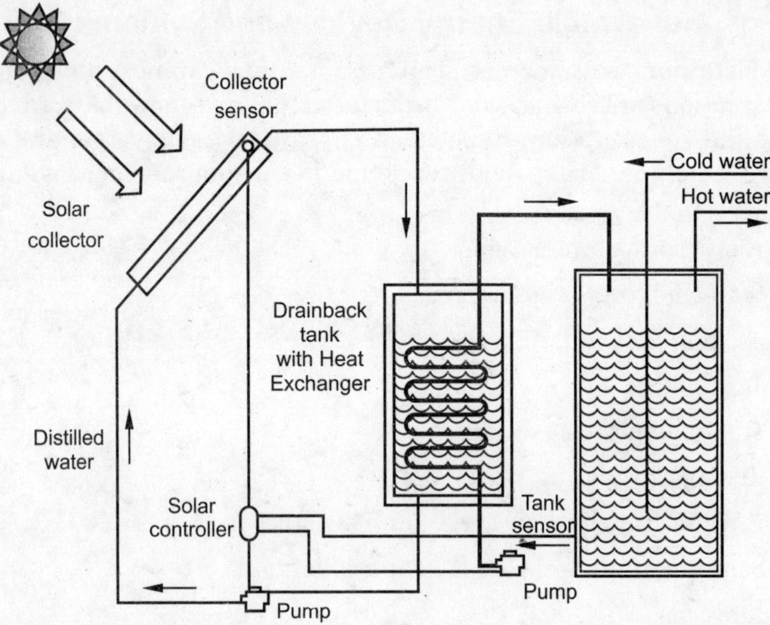

Fig.8.31: Solar water heating system heats water through solar energy by absorption of radiation. The Solar Water Heating System (SWHS) consists of a black coated absorber surface encased in a glass covered and insulated box for absorbing solar radiation. The absorbed solar energy is transferred from absorber to the water flowing in the absorber panel which gets heated and is sent to the hot water and temperatures up to 80°C are achieved,

Solar hot water systems can reduce the operating expenses on fuel for water heating to extent of 70 to 100% depending on season. Typically, a solar hot water system is sized to provide 80-100% of required hot water during sunny winter days. The use of solar energy for water heating helps in reducing greenhouse gas emission. It envisages the adherence to the principles of electro-biology to create an environment that does not change natural ionization, natural magnetic fields and beneficial cosmic and terrestrial radiation, and minimizes technical electromagnetic interference.

Table 8.37: Fuel Saving (for a typical 100 litres capacity solar water heating system at 50 degree centigrade)

Fuel	Calorific val. (kcal/ kg)	Thermal efficiency (%)	Fuel saved (kg/ annum)
Electricity	860 kcal/ kWh	95	1500 kWh
LPG	10800	60	195
Kerosene	10300	50	245
Fuel Oil	9600	75	175
Coal	5000	20	1260
Fire wood	4700	17.3	1550

Assumptions: Solar Insolation 5.5 kWh/ Sq.m./ day Ambient Temperature 25 Degree Centigrade, 300 Sunny days.

Source: "Solar Water Heating" by Dr. Bibek Bandyopadhyay - IREDA News Jan - March, 1998

Organic Solar Cells: Organic solar cells are having potential advantage over traditional inorganic sources, for material diversity, mechanical flexibility, low-temperature processing, and large-area capability. To increase the photovoltaic response of organic solar cells, a number of fundamental issues, such as optical absorption and exciton formation; exciton migration to the donor-acceptor interface; exciton dissociation; charge carrier mobility; and charge collection at the electrodes have to be addressed.

Donor - Acceptor (DA), *i.e.* low band gap polymers are being used in bulk-hetero- junction (BHJ) solar cells due to their wider photon absorption window. These polymers offer the promise of fabricating organic solar cells through solution-processing methods, which can further ease the processing demands and lower production costs. To improve overall device efficiency, itis necessary to understand and develop the structural, electronic, and optical properties of low band gap polymers that serve as the electron-donor and photon collector in bulk hetero -junction solar cells.

Solar streetlights and garden lights featuring battery storage, photovoltaic and optimized solar luminous (LED) are useful, both for remote areas and urban areas which save electricity bill and provide a pleasing ambience. Solar Lantern is both CFL and LED based, with a built in mobile charger unit and also serves as an emergency light.

Parabolic Trough

Parabolic trough power plants use a curved, mirrored trough which reflects the direct solar radiation onto a glass tube containing a fluid (also called a receiver, absorber or collector) running the length of the trough, positioned at the focal point of the reflectors. A change of position of the sun parallel to the receiver does not require adjustment of the mirror.

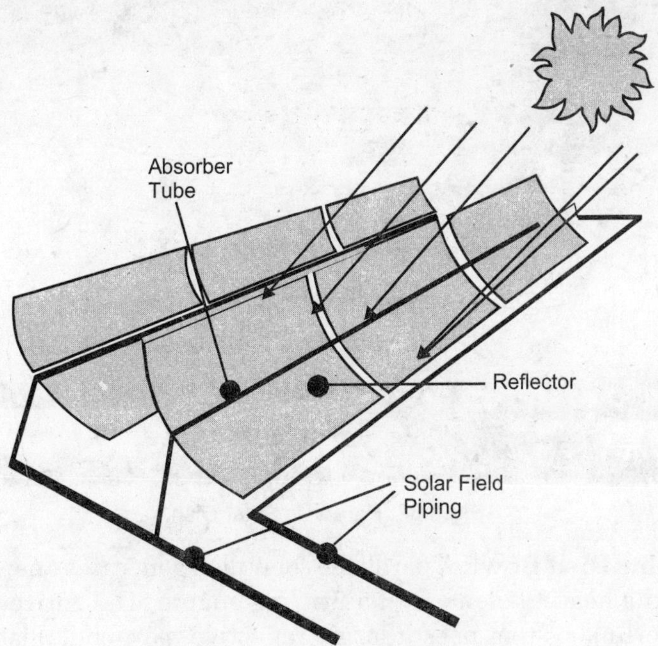

Fig.8.32: Solar Parabolic Trough

Solar Power Tower: Solar Tower mirrors focus the light on the top of the tower. The white surfaces below the receiver are used for calibrating the mirror positions. Power towers (also known as 'central tower' power plants or 'heliostat' power plants) capture and focus of the sun's thermal energy with thousands of tracking mirrors (called heliostats) in roughly a two square mile field. A tower resides in the center of the heliostat field. The heliostats focus concentrated sunlight on a receiver which sits on top of the tower. Within the receiver the concentrated sunlight heats molten salt to over 1,000°F (538°C). The heated molten salt then flows into a thermal storage tank where it is stored, maintaining 98% thermal efficiency, and eventually pumped to a steam generator. The steam drives a standard turbine to generate electricity. This process, also known as the "Rankine cycle" is similar to a standard coal-fired power plant, except it is fueled by clean and free solar energy.

The advantage of this design above the parabolic trough design is the higher temperature. Thermal energy at higher temperatures can be converted to electricity more efficiently and can be more cheaply stored for later use.

Fig.8.33: Parabolic Solar Dish.

Parabolic Solar Dish Bowl: A parabolic solar dish generates energy by concentrating the sun's rays on the heating element of a Stirling engine. The entire unit acts as a solar tracker.A Dish Sterling system uses a large, reflective, parabolic dish. It focuses all the sunlight that strikes the dish up onto a single point above the dish, where a receiver captures the heat and transforms it into a useful energy form. Typically the dish is coupled with a Sterling engine, but sometimes a steam engine is also used. These create rotational kinetic energy that can be converted to electricity using an electric generator.

The advantage of a dish system is that it can achieve much higher temperatures due to the higher concentration of light. Higher temperatures lead to better conversion to electricity and the dish system is very efficient on this point.

Fresnel Reflectors: A linear Fresnel reflector power plant uses a series of long, narrow, shallow-curvature or even flat mirrors to focus light onto one or more linear receivers positioned above the mirrors. On top of the receiver a small parabolic mirror can be attached for further focusing the light. These systems aim to offer lower overall costs by sharing a receiver between several mirrors as compared with trough and dish concepts, while still using the simple linefocus geometry with one axis for tracking.

Rival single axis tracking technologies include the relatively new Linear Fresnel Reflector (LFR) and Compact-LFR (CLFR) technologies. The LFR differs from that of the parabolic trough in that the absorber is fixed in space above the mirror field. The CLFR offers an alternate solution to the LFR problem. CLFR power plants offer reduced costs in all elements of the solar array. These reduced costs encourage the advancement of this technology.

Parabolic Solar Oven: Solar cookers and ovens use no fuel, which means no smoke, no carbon emissions, no ashes or soot, no fire hazard, lesser fuel cost and reduced deforestation. Widespread use of solar cookers gives huge economic and environmental benefits. Solar water heaters/geysers and solar box cookers attain temperatures of up to about 165°C (325° F). They can be used to heat water or cook the food. However, the box cookers are slow and good enough enough for a small family. It does not work during night or cloudy days. For mass cooking parabolic solar ovens are becoming popular.

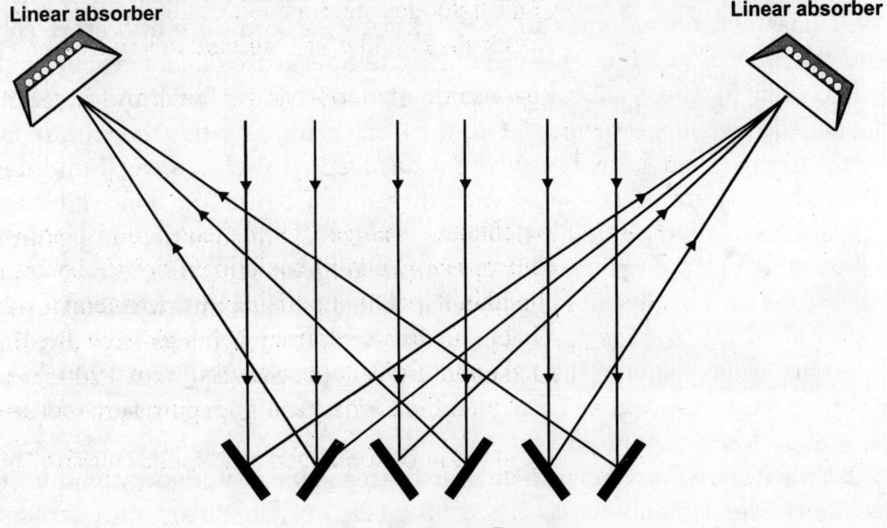

Fig.8.34: Fresnel Reflector

Fig.8.35: The Solar Bowl in Auroville

A hybrid solar oven is a solar cooker equipped with a conventional electrical heating element for cloudy days or night time cooking. A hybrid solar grill consists of an adjustable paraboloid reflector suspended in a tripod with a movable grill surface. These outperform box cookers in temperature range and cooking time. When solar energy is not available the design uses any conventional fuel such as gas, electricity, or wood. Integrated solar cooking combines solar cooker, a fuel-efficient cook-stove and an insulated container / basket to store heated food.

Among various types of solar cookers, solar kettle stands out which uses vacuum tube technology. It uses evacuated glass tubes to capture, accumulate and store solar energy needed to power the kettle. Since the stagnating temperature of solar vacuum glass tubes is a high 220°C (425°F), these solar kettles can also deliver dry heat and function as ovens and autoclaves. As solar vacuum glass tubes work on accumulated rather than concentrated solar thermal energy, solar kettles only need diffused sunlight to work and need no sun tracking at all. Solar kettles use solar vacuum tubes with insulation that keeps heated water hot throughout the night. These are most suitable for hostels, langar and religious establishments which feed large number of visitors/ pilgrims and residents every day. Solar water heaters are available at subsidized price. The system with hybrid (electric) support can be connected to water taps for bathing, washing and kitchen use. Brahm Kumari Ashram at Mount Abu is using high tech solar cookers producing about 1200 meals per day which saves huge consumption of LPG cylinders. Tirumal Tirupati Devasthanam (TTD) has solar steam cooking system for 30,000 meals per day, which saves 200 liters of diesel every day. Sai Baba Sansthan Temple at Shirdi uses solar cooking system for 7000 meals per day with enhanced capacity to cook 20,000 meals.

The Solar Bowl is a unique concentrating technology used by the solar kitchen in Auroville, Pondicherry. Unlike nearly all concentrating technologies that use tracking reflector systems, the solar bowl uses a stationary spherical reflector. This reflector focuses light along a line perpendicular to the sphere's surface and a computer control system moves the receiver to intersect this line. Steam is produced in the solar bowl's receiver at temperatures reaching 150°C and then used for process heat in the kitchen where 1,000 meals per day are served.

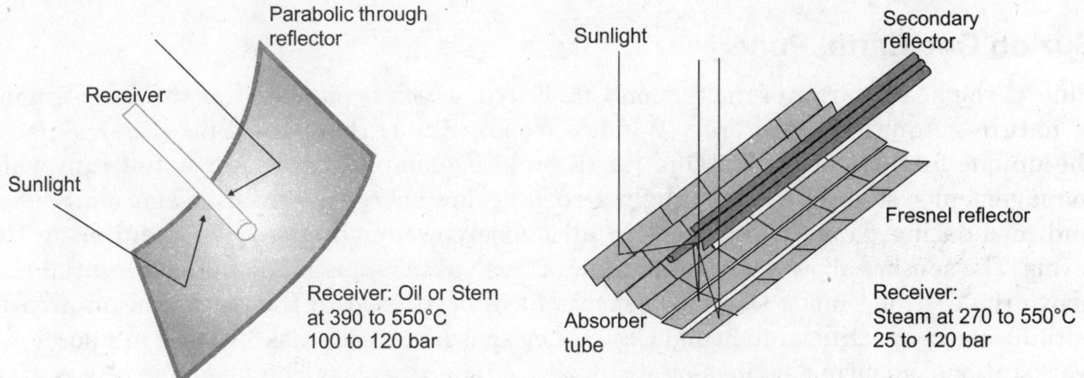

Fig.8.36: Parabolic Trough Solar Energy System

Fig.8.37: Parabolic Solar concentrator at the temple produces steam which powers the cooking procedure.

The Akshardham Temple in Delhi has installed a solar cooker for cooking 4,000 meals every day. The solar concentrator, named ARUN®100, produces steam which powers the cooking process. It works on the principle of a parabola. It uses an ingenious, two-dimensional, fresnel mirror arrangement to get the parabolic effect. The system automatically tracks the sun from its rise to set on both the east-west and north-south axes, intercepting maximum sunlight. The receiver is designed to operate at temperatures up to 400 degrees Celsius and works as a solar boiler, making it an effective replacement of conventional polluting fuels such as furnace oil, diesel, PNG and coal. The dish is mounted on a single pole and occupies a ground area of 3m by 3m and can therefore be erected in locations with space constraints.

Suzlon One Earth, Pune

Pune is the headquarter of the Suzlon One Earth, which is named after the five elements of nature - Aqua (Water), Tree (Wind + Wood), Sky (Ether} and Sun (Fire), Some of the unique features of Suzlon One Earth building complex are a storm and rain water management system, efficient sunlight screening, low energy materials usage, etc. Sensors and modulating dampers are used in all densely occupied spaces for monitoring CO_2 levels. The sensors allow additional inflow of fresh air as soon as the CO_2 differential rises above the certified limit. Daylight sensors have been used in the work station area for optimized use of artificial lighting. Occupancy sensors control task lighting in unoccupied workstations, ensuring savings of about 20% of energy costs. The headquarters is 100% powered by a combination of onsite and off-site renewable energy sources. Out of this, 7% of the total energy consumption comes from 18 onsite hybrid wind turbines, solar panels and photovoltaic cells and 93% of the remaining is from offsite wind turbines. Instead of a regular tall, glass box structure it is spread out in the form of a campus. Three glass chimneys, facilitated by "sky courts" suck air out from the basement.

Fig.8.38: Suzlon One Earth Headquarter, Pune, Architect Christopher Charles Beninger.

Renewable Energy Assisted Pump (REAP), a new technology developed by power distribution company BYPL, and IIT Delhi, has been installed by a number of Delhi government schools in East Delhi. It is easy to install with submersible pump connected to a water tank with a specially designed motor powered by a solar panel.

A Solar House

Frauen-hofer Institute for Solar Energy Systems at Freiburg, designed a solar house which provides 100 percent energy from renewable sources. Its highly transparent insulation system converts sunlight into thermal energy. Controlled ventilation systems combined with heating and heat recycling, high-efficiency solar collectors for supplying warm water, powerful photovoltaic modules placed on the roof for conversion of sun-light into electricity as well as electricity-saving household appliances help in economising energy. The hydrogen-oxygen system also contributes to the seasonal reduction in energy consumption.

The electricity converted from solar cells is used for producing hydrogen and oxygen in a compression electrolyser. Gas tanks store energy required for the season. The gases can be reconverted into the requisite energy. Hydrogen and oxygen are released by electrolysis. This system offers new ways for a closed, pollution-free energy.

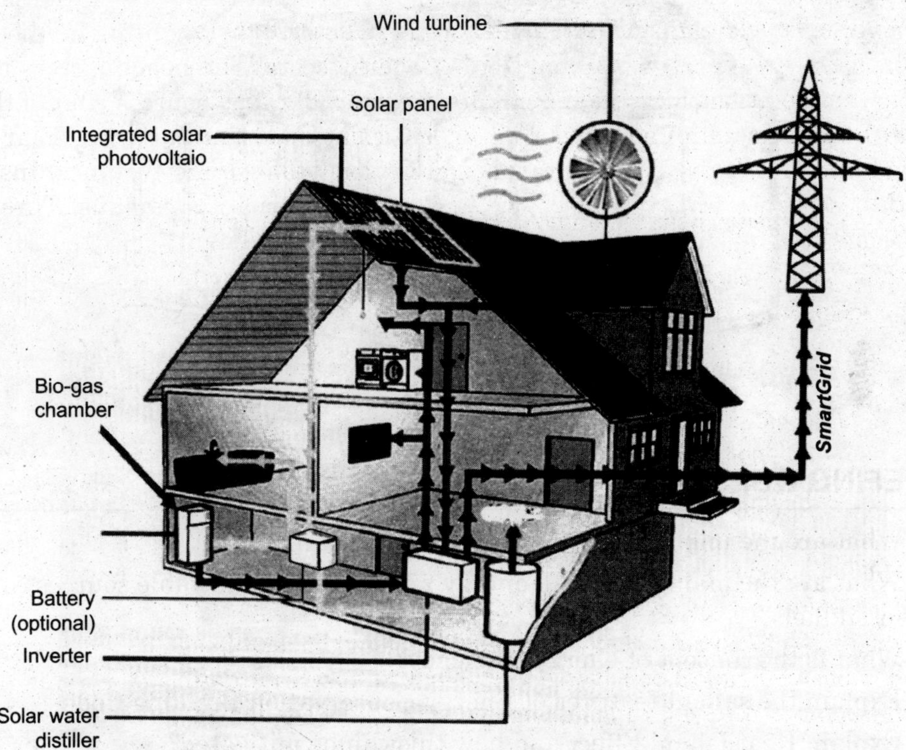

Fig.8.39: A schematic renewable energy house.

Solar Powered Night Shelters

It is estimated that about one percent of urban population is homeless, comprising poorest of the poor, beggers, rickshaw pullers, casual labourers, domestic help, loaders, cleaners, etc. They have a right on the city and need at least a night shelter to sleep. As such, night shelters are provided at transport nodes, construction project sites, markets, etc. These are usually temporary or semi-permanent structures with beds and basic services-water, toilets, electricity. Prefabricated modular structures with solar power can be quickly erected and are economical and comfortable.

Various companies and NGOs have come up with innovative ideas and designs. For example, Indo-Global Social Service Society (IGSSS) and Micro Home Solutions, have designed a modular portable bamboo shelter which can accommodate around 100 people. The porta cabin structures are covered by bamboo from outside. This helps to insulate the walls from the cold. The interiors have bunk beds for better utilization of space. These night shelters are easy to built and are portable. The cost of 18 Beds Night Shelter, covering 270 sqft is around Rs. 1.00 lakh, while the cost of a 36 Beds Night Shelter, covering 540 sqft is about Rs. 1.80 lakh.

Fig.8.40: Solar Powered Prefabricated Night Shelter

DEBRIEFING QUESTIONS

1. What are the implications of energy on environment?
2. What are the main sources of energy and whether renewable sources can replace fossil fuels?
3. What is the concept of energy efficiency?
4. Explain the salient features of Energy Conservation Building Code.
5. Explain Heat Island Effect and how this can be mitigated?
6. Write short notes on:
   i. Co-efficient of Performance
   ii. Energy Efficiency Ratio
   iii. System Balance
   iv. Piping Insulation
   v. Lighting Control and Sensors

7. Explain the concepts of 'Whole Building Approach' and 'High Performance Building'.
8. Explain the concepts of passive measures for achieving energy efficiency.
9. Write short notes on:
   (a) Envelop Air-conditioning System
   (b) Downdraft Cooling
   (c) PV Module
   (d) Parabolic Trough/Solar Dish

9 Water Pollution and Health

Water pollution probably began with the foundation of the earliest cities around 7,000 years ago along the major river systems of the Tigris-Euphrates and Indus. Cities have long used rivers, lakes, and coastal waters as receptacles for diluting and dispersing wastes. The natural processes of water flow help to break down wastes and render them harmless. Ever-increasing urban populations and their growing amounts of wastes, however, have over-taxed the natural recycling capabilities of local rivers and lakes. In cities close to coasts, untreated sewage and industrial effluents flow into the sea and damage beaches and inshore waters.

Table 9.1: Urban Water and Sanitation Coverage

Service	Africa	Asia and Pacific	Middle East	Latin America
Water				
Percent of population covered	68.9	80.9	71.8	91.4
Percent served by:				
House connection	65	48.4	89.7	92
Public standpost	26	24	9.3	3.3
Others	9	27.6	0	4.7
Sanitation				
Percent of population covered	53.2	69.8	60.5	79.8
Percent served by:				
House connection to sewer/septic system	53.0	42.7	100.0	91.2
Pour/flush latrine	3.0	43.1	0	2.1
Ventilated improved pit latrine	13.6	2.7	0	0.9
Simple pit latrine	22.4	8.5	0	5.4
Other	2.6	3.0	0	0.4

Source: G. Waters, Health and Environment, World Health Organization, Geneva, 1995

Although there has been significant progress in controlling water pollution over the past three decades, it has continued to rise in most cities posing a threat to human health and aquatic ecosystems.

Urban-generated pollution comes from both localized and dispersed, or point and non-point, sources. Major point sources include municipal sewage, industrial outfalls, and air emissions from power plants and heavy industries. Non-point sources include silt from earth-moving activities; storm runoff from roads, home gardens, and industrial sites; infiltration from aquifers contaminated with sewage or industrial chemicals, and automobile emissions.

Table 9.2: Water Related Urban Environmental Issues and Options

Problem Area	Effects	Causes	Management Options
Surface Water Pollution • Community • City wide • Regional	• Health problems • Economic costs (additional treatment, new sources of supply, health costs) • Amenity losses	• Pricing policies • Unclear property rights • Poor regulations and/ or enforcement • Municipal & industrial waste disposal practices • Urban runoff • Irrigation practices	• Marginal cost pricing • Regulations, standards, licensing, charges • Improve monitoring and enforcement • Demand management and waste water re-use • Appropriate technology • Land use controls • Waste management
Groundwater Pollution and Depletion • Community • City wide • Regional	• Reduced water quality from saline intrusion, biochemical seepage • Health impacts • Economic costs (damage from land subsidence, health costs, increasing marginal costs of (supply)	• Pricing policies • Unclear property rights • Poor regulations and/ or enforcement • Unsustainable extraction • Sanitation, municipal & industrial waste disposal practices • Poor demand management	• Marginal cost pricing (sustainable extraction, aquifer recharge costs) • Regulations, standards, licensing, charges • Waste management • Appropriate technology • Demand management • Controls on land use and sources of contamination
Coastal/Lake Pollution • Community • City wide • Regional • Transnational	• Health effects due to contaminated seafood and direct contact • Loss or recreational resources & tourism revenues • Damage to fisheries • Amenity losses • Eutrophication	• Unclear property rights • Poor regulations and/ or enforcement • Municipal and industrial waste disposal practices • Disposal of shipboard wastes	• Regulations, standards, licensing, charges • Appropriate technology • Coastal zone management and preservation • Shipping facilities • Waste management • Land use controls

Source: World Health Organisation

Of the many problems associated with urban effluents, nutrient loading, oreutrophication, of local waters is one of the most serious. Nutrients are essential plant foods, but excessive amounts can cause radical plant growth—often massive algal blooms that blocks the sunlight that other organisms need. As plants die and decompose, the dissolved oxygen in bottom waters is depleted—a condition that is deadly for fish and other aquatic life. Those fish and other mobile species that can survive may nonetheless lose critical habitat, their food supplies may be disrupted, or they may be forced into shallow areas where they are subject to increased predation.

Nutrients come from several sources, including run-off from upstream agriculture and urban areas, particularly silt, and air emissions. Nutrient enrichment problems are especially severe in urban estuaries, where water flushing is minimal and inputs, often from numerous cities, are large.

Water quality is closely linked to type of water use and level of economic development. This was clearly seen in developed countries in the mid-18th Century, at the start of industrialisation, when cities expanded, became densely populated with crowded inner city slums and excreta was discharged untreated into urban watercourses. The resultant faecal contamination of surface waters created serious health problems, including choleraand typhoid epidemics. Since drinking water was drawn from the very same rivers or groundwater sources into which excreta was discharged, water borne disease outbreaks became regular occurrence. Aspects of water pollution which have a direct bearing on human health are summarized below.

Sewage

Untreated or inadequately treated municipal sewage is a major source of surface and groundwater pollution throughout the developing world. This is cause for considerable concern since the biological degradation of organic material that is discharged with municipal sewages into watercourses uses substantial amounts of BOD called biochemical oxygen demand. It effects the ecological balance of rivers and lakes. Sewage also carries microbial pathogens; faecal coliform levels may be very high in sewage-contaminated rivers. In developed countries, drinking-water usually receives extensive multi-stage treatment before it is supplied to the consumer. But elsewhere treatment is often unreliable and services intermittent. In many regions therefore peri-urban residents and the rural poor often consume water that is untreated or that is not controlled for drinking-water quality.

Nutrients

Domestic wastewater, agricultural drainage waters and much industrial effluence contain phosphorus and nitrogen. They thus serve as a source of additional nutrients for aquatic organisms and can cause severe eutrophication of lakes and rivers and ultimately of estuaries and coastal waters. A recent survey showed that in the Asia/Pacific region, 54% of lakes are eutrophied; the figures for Europe, Africa, North America and South America are 53%, 28%, 48% and 41% respectively (ILEC and Lake Biwa Research Institute, 1988-1993). Intensive fertilizer use in agriculture has also contaminated groundwater sources with nitrates, with the result that nitrate levels in drinking-water often far exceed the safe levels recommended by WHO.

Synthetic Organics

Although data on water quality is sparse, evidence is growing that many of the 100,000 synthetic compounds in use today find their way into the aquatic environment and accumulate in the food-chain. Persistent organic pollutants (POPs) represent the most harmful group for ecosystems and human health. An international list of the twelve most dangerous POPs has been compiled, including industrial, chemical and agricultural pesticides. These chemicals can bioaccumulate in fish and shellfish to levels that are hazardous to human health. If pesticides are used on a massive scale, groundwater may become contaminated, leading to chemical contamination of drinking water.

Indian Scenario

The total available water in India is estimated at 1,900 billion cubic metres (cu m) per year. About 86 per cent of this is from the surface run-off—rivers, streams, lakes and ponds. Excluding groundwater resources, that still need to be tapped, surface flow represents 97 per cent of the available water.

The Himalayan rivers form a mighty network of snow-fed and perennial waters, including the Ganga and the Brahmaputra river systems and the Indus river tributaries. The Ganga along with its major tributaries the Yamuna, the Ghagra, the Gandak, the Gomti,the Kosi and the Sone rivers has a length of 8,047 km long.

The major rivers of peninsular India are the Mahanadi, the Godavari, the Krishna, the Cauvery, the Narmada and the Tapti. The Godavari basin is the second largest in the country, serving over 10 per cent of its area. The coastal rivers rise in the Western or Eastern Ghats and after a short traverse, flow into the nearby seas. In addition, there are a few other inland rivers like the Luni, the Khan, the Kshipra, etc. The peninsular and other rivers are rainfed, mostly non-perennial and thus their volume fluctuates during the year.

A staggering 70 per cent of the available water in India is polluted, according to scientists at the Nagpur-based National Environmental Engineering and Research Institute (NEERI). "From the Dal lake in the north to the Periyar and Chaliyar rivers in the south, from the Damodar and Hooghly in the east to the Thana creek in the west, the picture of water polluting is uniformly gloomy. Even our large perennial rivers like the Ganga are today heavily polluted, admits the Planning Commission.

Community wastes from human settlements account for four times as much waste water as industrial effluents. Most of these wastes are discharged untreated into the water courses. Out of India's 7936 towns and cities, only 217 have partial or full sewerage and sewage treatment facilities. These cover less than a third of the urban population.

The implications of such massive pollution, for the health of the nation, are extremely serious. According to one estimate, two-thirds of all illnesses in India are related to water borne diseases, such as typhoid, infective hepatitis (jaundice), cholera, diarrhoea and dysentery. Many of them assume epidemic proportions. It is estimated that 73 million work days are lost every year due to water-related diseases.

Many water-borne diseases such as dysentery, botulism, hepatitis and poliomyelitis are the potential health hazards, which owe their origin to polluted water. The types of water pollution are determined both by the medium (surface water, ground water, etc.), the habitat in which they occur (river, lakes, estuaries, coastal waters and open sea), as also the source of type of contamination (nutrient, bacterial and viral, industrial, pesticide, radioactive wastes, etc.).

Table 9.3: Water Related Infections with Estimates of Morbidity, Mortality and Population at Risk

disease	Common Disease Name
Waterborne, Washed and Foodborne	
Cholera	Cholera
Diarrheal diseases	Salmonellosis Shigellosis, Campylobacter, *E. coli* rotavirus Amoebiasis and Giardiasis
Enteric Fever	Paratyphoid, typhoid
Infective Jaundice	Hepatitis A
Pinworm	Enterobiasi
Polio	Poliomyelitis
Leptospirosis Whipworm	Acaviasis Trichuriasis
Water Washed	
(a) Skin and eye infections	
Scabies Schools Sores	Scabies Impetigo
Trachoma Leishmaniasis	Trachoma Leishmaniasis
(b) Others	
Relapsing fever Typhus	Relapsing fever Rickettsia diseases
Water Based	
a. Penetrating Skin b. Ingested Guinea worm	Bilharzia/Schistosomiasis Dracunculiasis
Water Related Insect Vector	
a. Biting fear water Sleeping Sickness b. Breeding at water Filaria Malaria c. River Blindness Yellow Fever Break Bone Fever	African Trypanosomiasis eve Filariasis Malaria Onchocerciasis Onchocerciasis Dengue Fever

Source: Adapted from WHO (1992) Our Planet, Our Health, Report on the World Commission on Health and Environment, Geneva.

About 21 per cent of all communicable diseases in India are water-related. It is estimated that beside premature deaths, about 30.5 million. Disability Adjusted Life Years (DALYs) are lost each year in India due to poor water quality, sanitation and hygiene. Significant urban/rural differences exist in the country relative to water-related environmental health hazards. In rural areas, unsafe drinking water, inadequate excreta disposal, agricultural run-off that contains chemical pesticides are among the main environmental health concerns. In urban areas, environmental health problems are the result of rapid population migrations from rural areas to urban centres and of uncontrolled

Industrialisation. Municipal water supply and waste management services are unable to keep pace with this rapid urban growth and industrialisation.

In the last few decades India has shown remarkable progress in terms of water management, both in terms of provision of drinking water as well as in food security. Though about 90% of the population has been provided with drinking water facilities, the actual situation is not so satisfactory due to lack of equitable distribution of water, water quality issues, high unaccountable flow of water, lack of proper operation and maintenance and a weak monitoring and surveillance mechanism. Apart from this:

- Fresh water aquifers are mined at an unprecedented rate to improve food production, which has resulted in falling water tables.
- Lack of attention to maintenance of vegetation in catchment areas contributing to flooding.
- Poor water quality in rivers and lakes.
- Deforestation and land degradation causing flooding and siltation, salinization and water logging and water contamination by chemicals and fertilisers and human waste are all serious issues.

Table 9.4: Some Sources and Causes of Water Quality Deterioration

Effluent	Factors affecting water quality deterioration
Domestic sewage	BOD, suspended solids, ammonia, nitrate, phosphate
Vegetable processing	BOD, suspended solids, colour
Chemical industry	BOD, ammonia, phelons, non-biodegradable organics, heat
Iron and steel manufacture	Cyanide, phenols, thiocyanate, pH, ammonia, sulphides
Coal mining	Suspended solids, iron, pH, ammonia, sulphides
Metal finishing	Cyanide, copper, cadmium, nickel, pH
Brewing	Suspended solids, BOD, pH
Dairy products	BOD, pH
Oil refineries	Heat, ammonia, phenols, oil, sulphide
Qarrying	Suspended solids, oil
Power generation	Heat

Table 9.5: Point Sources and the Expected Contamination of Groundwater

Source	Inorganic contaminants	Organic contaminants
Urban areas	Heavy metals and salts	Oil products, biodegradable organics
Industrial sites	Heavy metals	Chlorinated hydrocarbons, hydrocarbons, oil products
Land fills	Salts and heavy metals	Biodegradable organics and xenobiotics
Mining disposal sites	Heavy metals, salts and arsenics	Xenobiotics
Dredged sediment and sludge	Heavy metals	Xenobiotics
Hazardous waste sites	Heavy metals (concentrated)	Concentrated xenobiotics
Leaking storage tanks		Oil products (petrol)
Line sources (motorways, railways, sewerage, etc)	Heavy metals, salts	Oil products, pesticides

Chemical contaminants namely fluoride, arsenic and selenium pose a very serious hazard in the country. Apart from this, increase in the concentration of chloride, TDS, nitrate in ground water is of great concern for a sustainable drinking water programme. All these need to be tackled holistically.

With regard to sanitation coverage it is one of the lowest in the Region, with hardly 31% provided with sanitation facilities. Very few cities have been provided with conventional sewerage and sewage treatment facilities with most people, both in urban and rural areas, depending on on-site sanitation. Solid waste management is another major area of concern.

Chemicals in Drinking-Water

Both naturally-occurring and human-made chemical substances in drinking-water can have serious effects on health. A variety of acute and chronic health effects have been reported. Actual risks are determined by the chemical concentration in drinking-water. Thus many toxic substances, if present below a certain threshold level, do not pose a health hazard. When assessing the potential health risk of a chemical constituent, identification and quantification of all the major routes of exposure (*i.e.* the total human exposure), is important. This is because the tolerable health impairments can then be determined.

Fluoride

Fluoride levels of 0.5 to 1 mg/l provide substantial protection against dental caries. However, the margin between beneficial and toxic levels of fluoride is rather narrow and higher levels of fluoride in drinking-water have led to adverse health effects, ranging from unsightly dental fluorosis to crippling skeletal fluorosis. Waters with significant natural fluoride content are usually found at the base of high mountains and in areas with geological deposits of marine origin. Typical examples are the geographical belt running from Syria through Jordan, and Egypt, and from Algeria to Morocco; the Rift Valley running through Sudan and Kenya, and the geographical belt stretching from Turkey through Iraq, the Islamic Republic of Iran, and Afghanistan, to India, northern Thailand and China. Similar areas are found in the USA and Central and South America.

Arsenic

High concentrations of arsenic in drinking-water are found in specific parts of Argentina, Canada, Chile, China, Japan, Mexico, the Philippines and the USA. Adverse health effects, including skin cancer, other skin effects, peripheral neuropathy and peripheral vascular disease ("Blackfoot disease") have been observed in populations ingesting arsenic contaminated drinking-water.

Recently, arsenic contamination of groundwater, the main source of drinking-water, was detected in six districts of West Bengal, India and in several villages in Bangladesh. Levels up to 70 times higher than the national drinking-water standard of 0.05 mg/l have been measured in both countries. The contamination is due to the natural soil composition of the region. While the true extent of the problem is not yet known, it has been estimated that some 30 million people may be at high risk from arsenic exposure. Meanwhile, evidence of chronic arsenic toxicity in the population is accumulating and includes incidence of melanosis (abnormal black-brown pigmentation of skin), hyperkeratosis (thickening) of palm and sole, gangrene of the lower extremities and skin cancer. In West Bengal alone, 2,00,000 people have been reported to be suffering from arsenical skin lesions.

Seven of 16 districts of West Bengal have been reported to have ground water arsenic concentrations above 0.05 mg/ L; the total population in these seven districts is over 34 million (Mandal *et al.*, 1996) and it has been estimated that the population actually using arsenic-rich water is more than 1 million (above 0.05 mg/ L) and is 1.3 million (above 0.01 mg/ L).

Arsenic is a heavy toxic metal and ingestion of amounts beyond 0.05 milligrams per litre of water can pose a serious health hazard, where aquifers are the main source of drinking water. Currently, experiments are being undertaken with a number of arsenic mitigation technologies. The basic methods involve oxidation coagulation-flocculation sedimentation and filtration.

Alternative low-arsenic sources such as rain water and treated surface water may be available and appropriate in some circumstances. Where low arsenic water is not available, it is necessary to remove arsenic from drinking-water:

- The technology for arsenic removal for piped water supply is moderately costly and requires technical expertise. It is inapplicable in some urban areas of developing countries and in most rural areas world-wide.
- New types of treatment technologies, including co-precipitation, ion exchange and activated alumina filtration are being field-tested.
- There are no proven technologies for the removal of arsenic at water collection points such as wells, hand-pumps and springs.
- Simple technologies for household removal of arsenic from water are few and have to be adapted to, and proven sustainable in each different setting.
- Some studies have reported preliminary successes in using packets of chemicals for household treatment. Some mixtures combine arsenic removal with disinfection.

Lead

The pipes, solder, fitting and service connections of some household plumbing systems contain lead and cause contamination of drinking-water supplies. Lead is hazardous to health. It accumulates in the body and its effects on the central nervous system can be particularly serious. The foetus, infants, children and pregnant women are most at risk. Numerous epidemiological studies have demonstrated associations between childhood exposure to lead and a decreased intelligence quotient (IQ), and other health effects.

Nitrates

Drinking-water contaminated with nitrates can lead to serious, even fatal, consequences, especially for infants fed with formula prepared with such water. Most often the nitrate comes from fertilizer added to agricultural soil. In the human body, nitrate is converted to nitrite which then combines with haemoglobin to form methaemoglobin. Since methaemoglobin cannot bind with oxygen, transport of oxygen in the blood is impeded. In very young children severe vyanosis ("blue baby syndrome") follows and can cause death. Technology to remove nitrate from drinking-water is expensive, but good agricultural practice can do much to prevent nitrate contamination. However breast-feeding is by far the best means of avoiding exposure of infants to water contaminated with nitrates.

Pesticides

Groundwater is susceptible to contamination with pesticides that are mobile in atrazine, bentazone, carbofuran, is operaturon, ethylene dibromide and simazine. Pesticide concentrations in many groundwater sources exceed WHO guideline values. Some of these pesticides are believed to be carcinogenic to humans, but are of concern with respect to future generations because of their persistence in soil and water.

Untreated sewage, industrial effluents have and agricultural waste are frequently discharged into inland waterways, lakes and coastal zones, endangering the use of these water bodies for recreational purposes such as swimming, canoeing and windsurfing. Recreational exposure to polluted waters can cause diarrhoea, respiratory infections, skin irritation and other diseases depending on specific pollutants involved. Control of land based pollution sources is evidently the most effective means of preventing discharge of sewage into rivers, lakes, coastal beach areas and other marine waters used for recreation. Additionally, intensive bacteriological monitoring is recommended for public beaches and recreational areas. The economic benefits measures is well proven. In many European tourist resorts, investments have been made in sewage treatment and construction of long ocean outfalls to carry waste water beyond coastal waters used by bathers. The regional and local government of the Italian Adriatic coast, for example, have invested heavily in such infrastructure (WHO, 1994). Coastal bathing water quality in the area has now improved and tourist numbers have accordingly increased.

Water-related Vector-borne Diseases

The aquatic environment provides an essential habitat for the mosquito vectors and intermediate snail hosts of parasites that cause human diseases. Among these diseases, malaria out-ranks all others in severity and distribution. Vector-borne diseases have, of course, always affected humans, even in pristine environments. But the accelerated development of water resources that started in the 1960s has led to habitat modifications and ecological conditions that favour certain disease vectors. Health issue linked to irrigation development have become a particular focus of attention, but increased transmission is also linked to the construction of dams and reservoirs, to changes in land use patterns (with indirect effects on local hydrology and relative humidity) and to poor water management in urban areas.

Lack of financial resources has often been implicated in mismanaged water resources development. In the case of irrigation, rising costs have often obliged planners to omit certain components (most notoriously the drainage component) to make proposed schemes economically viable. Poor intersect oral collaboration has been another contributory factor. Many countries have no established procedures for involving the health sector at the planning and design stages of water resources development. Moreover, prolonged post World-War II reliance on residual insecticides for vector control has resulted in neglect of public health and hygiene issues among civil engineers. This has reduced the apparent need for engineers and public health experts to collaborate on water resources projects. The most important water-related Vector-borne diseases affected by development projects in recent decades include:

Malaria: The mosquitoes thrive in water storage tanks, blocked drains and on construction sites. Can engineers prove the maxim that - in the case of malaria – prevention is better than cure? Malaria is a major public health problem throughout the world; according to the World Health Organization, it kills around 2 million people every year and there are over 300 million new cases each year. Malaria is caused by parasites in the human bloodstream and is transmitted from one human to another by the bite of female anopheline mosquitoes.

In the 1950s and 1960s, there was dramatic success in reducing the incidence of malaria although the original, widely accepted 1958 goal of "eradication" was not achieved. Now, the number of casualties is spiralling as a result of reduced government and agency funding for malaria control mosquitoes' increasing resistance to insecticides, and growing drug resistance among malarial parasites. Not all types of mosquito transmit malaria but all mosquitoes are a nuisance and many can create other health hazards. As limitations to chemically based control emerge, more thought is being given to appropriate technology in mosquito control, and the role of environmental management in the Amazon region due to development in urban areas; in Africa and South Asia due to poor water management; in South-East Asia due to deforestation, afforestation, and in many parts of the lowland tropics due to expansion or intensification of irrigated agriculture.

Schistosomiasis: Mainly in sub-Saharan Africa in association with irrigation schemes and reservoirs.

Dengue fever: In urban but also increasingly in some rural tropical areas; here general solid waste problems have combined with inadequate household storage of drinking-water.

Filariasis: In tropical urban areas dur to the vectors breeding in organically-polluted water, *i.e.* in open sewage canals, blocked drains and in wastewater collected for reuse in peri-urban agriculture.

Japanese encephalitis: Japanese encephalitis is a neurologic infection with a broad range of manifestations. It can range from subtle changes in behaviour to serious problems, including blindness, ataxia, weakness, and movement disorders. Japanese encephalitis is caused by the Japanese encephalitis virus (JEV).

The annual million deaths due to malaria, the 200 million people infected globally with schistosomiasis, and the tens of thousands of dengue fever cases illustrate the cost of water-related vector-borne diseases, both in terms of human suffering and healthcare costs. Yet attributing a specific number of disease cases to poor water management, especially if the disease was endemic before hydrological changes occurred and the relevant vector ecology complex, would be difficult. Solid assessment of hydraulic systems and breeding places which contribute most to disease transmission risks is called for.

Table 9.6: Protection and Development of Water Resources

Are any of the rivers and lakes in the Municipality being polluted upstream?	Regional water resources can only be properly managed at the river basin level. It is therefore necessary to have planning and control for the whole river basin. The Municipality should keep a close watch of developments at that level.
Is there evidence of groundwater pollution in the region?	Groundwater, which is the responsibility of the Department of Mineral Resources, and surface water, for which various government departments share responsibility, should be managed as one system. The regional strategy must work out a coherent management framework for water resources as a whole.
Has the aquifer recharge area of the region been built over, causing groundwater depletion?	Because of the impact which urbanisation and its demand for groundwater can have, aquifer management is a vital aspect of regional water resource management.
Is the groundwater of the region being overexploited?	Water Pollution Control, Domestic Sanitation and Drainage actions to be taken at the municipal level.

Water and Health

Most of the vital functions of the body involve the use of water. Blood which contains 83% water transport nutrients, hormones, enzymes, medicines, oxygen and other life sustaining material into our cells and carry wastes to organs for removal. Water is present the mucous linings of organs and fluids between internal organs provide lubrication and reduce friction within the body, joints, bones etc. Digestive juices help to breakdown food and make transportation through digestive system easy. Body temperature is maintained through perspiration or storage of water. Reactions in the body such as digestion, enzyme reactions and waste removal in kidneys, intestines, respiration etc. also involve water.

One in six people still have no regular access to safe drinking water. More than twice that number lack access to adequate sanitation facilities. Those without access to adequate sanitation are the poorest and most vulnerable. The problem is particularly severe in remote rural and rapidly growing urban areas.

Assessment and Treatment of Dehydration

Assess the degree of dehydration of someone with diarrhoea using the following guide (WHO, 1991):

Table 9.7: Assessing Degree of Dehydration

1.	Look at:			
	Condition	Weli, alert	'Restless, irritable	Lethargic or unconscious
	Eyes	Normal	Sunken	Very sunken and dry
	Ears	Present	Absent	Absent
	Mouth and Tongue	Moist	Dry	Very dry
	Thirst	Drinks normally, not thirsty	'Thirsty, drinks eagerly	Drinks poorly or not able to drink
2.	Lift the skin between two fingers and release	Skin falls back quickly	Skin falls back slowly	Skin falls back slowly
3.	Decide	No sign of dehydration	Two or more signs, there is some dehydration	Two or more signs, there is severe dehydration

Source: World Health Organisation, 1991

The skin pinch may be less useful in people with marasmus (severe wasting) or kwashiorkor or (malnutrition with swelling of the feet, hands and face), or obese people. Tears are a relevant sign only for infants and young children.

Prevention and treatment of dehydration using rehydration drink

To prevent and treat dehydration, give plenty of liquids or rehydration drink. Use small sachets of Oral Rehydration Salts (ORS), if available, for mixing with water. If not available, prepare a rehydration drink as follows:

To 1 litre of water add:

- 2 level tablespoons of sugar (or honey)
- ¼ teaspoon of salt
- ¼ teaspoon of sodium bicarbonate (baking soda).

If there is no baking soda use another ¼ teaspoon of salt. Add a small amount of fruit juice if available. The drink should be no more salty than tears. Sip regularly until urinating is normal. An adult needs at least 3 litres per day and a child at least 1 litre or a glass of liquid for each watery stool passed.

Table 9.8: Control of Environmental Diseases

Type of disease	Personnel measures				
	Drink safe water	Dispose of excreta safely	Wash and clean regularly	Store and cook properly	Dispose of wastewater safely
	Engineering measures				
	Provide safe water (quality)	Control defecation provide latrines	Provide enough water (quantity)	Provide cooking utensils. fuel etc.	Ensure adequate drainage
Faecal-oral:					
°Water-borne*	•••	•••	—	••	—
°Water-scarce* (e.g. diarrhoeas) Skin/eye infections	—	•••	•••	—	—
°Water-scarce*	—	—	•••	—	•
Excreta-related:					
°Water-based* (e.g. schistosomiasis)	••	•••	—	—	—
Water-related* (e.g. malaria)	—	—	—	—	•
Soil-based	—	••	•	••	—
Food-based	—	••	•••	•••	—

**Water-borne: pathogens in water which is swallowed; Water-scarce: personal hygiene affected by lack of water; Water based: pathogens infect humans through contact with water; Water-related: spread by vectors (e.g. mosquitoes) which live near or breed in water; Importance of measures: • • • High; • Low. ° The provision of adequate shelter, including blankets.*

— *The maintenance of a sanitary environment.*

Note that improvements in water and sanitation often reduce the severity of a disease more than its incidence.
Source: WHO, 1991

According to Central Public Health Environmental Engineering Organisation (CPHEEO) estimates, 90% of urban population has access to potable water supply. The average water availability in large cities is 5 to 6 hrs/ day with low water pressure. Another concern is depletion of ground water that supplies to 80% of rural water and 50% of urban and industrial usage. Death toll shave increased recently to 1.5 million children every year by drinking of contaminated water in rural area. 1.51 lakh people suffered from water quality problem such as excess arsenic, fluoride, salinity and iron.

Estimated Morbidity and Mortality

As much as 90 per cent of waste water in India is discharged without treatment into rivers and streams. Unsanitary water, which provides a breeding ground for parasites, amoebas and bacteria, damages the health of 1.2 billion people a year. People suffering from water-borne diseases occupy half the world's hospital beds.

Table 9.9: Estimated Morbidity and Mortality of Water-related Diseases

Diseases	Morbidity (episodes/year or as stated)	Mortality (deaths/ year)	Relationship of diseases to water supply and sanitation
Diarrhoeal diseases	1,000,000,000	3,300,000	Strongly related to unsanitary excreta disposal, poor personal and domestic hygiene, and unsafe drinking water
Infection with intestinal	1,500,000,000[1]	100,000	Strongly related to unsanitary excreta disposal and poor personal and domestic hygiene
Schistosomiasis	200,000,000[1]	200,000	Strongly related to unsanitary excreta disposal and absence of nearby sources of safewater
Dracunculiasis	100,000[1]	—	Strongly related to unsafe drinking water
Trachoma	150,000,000[3]	—	Strongly related to insufficient face washing often in the absence of nearby sources of safewater
Malaria	400,000,000	1,500,000	Related to unsatisfactory water management, water storage, operation of water points and drainage
Dengue fever	1,750,000	20,000	Related to unsatisfactory solid waste management, water storage, operation of water points and drainage
Poliomyelitis	114,000	—	Related to unsanitary excreta disposal, poor personal and domestic hygiene and unsafe drinking water
Trypanosomiais	275,000	130,000	Related to the absence of nearby sources of safe water
Bancroftian filariasis	72,800,000[1]	-	Related to unsatisfactory water management, water storage, operation of water points and drainage
Onchocerciasis	177,000,000[1]	40,000[5]	Related to unsatisfacotry water management in large-scale projects

1. *People currently infected*
2. *Excluding Sudan*
3. *Active trachoma. There are approximately 5,90,000 cases of blindness or severe complications of the disease annually*
4. *Includes an estimated 270,000 blind people*
5. *Mortality resulting from blindness.*

Source: W.H.O.

Pure water does not exist in nature. Natural minerals eroded from soil and rock imparts flavour and an appeasing taste etc. and is good for human consumption. Man made chemicals, discharges from industry, urban dwelling, farm residues, leaches, etc. contaminate water. In its pure state, water is one of the most aggressive solvents known. Called the "universal solvent", water, to a certain degree, will dissolve virtually everything to which it is exposed. Pure water has a very high energy state and, like everything else in nature, seems to achieve energy equilibrium with its surroundings. It will dissolve the quantity of material available until the solution reaches saturation, the point at which no higher level of solids can be dissolved. Contaminants found in water include atmospheric gases, minerals, organic materials (some naturally occurring, others man-made) plus any materials used to transport or store water. The hydrologic cycle illustrates the process of contamination and natural purification.

The impact of the various impurities generated during the hydrologic cycle and/or bacterial colonization depends upon the water user's particular requirements. In order to assess the need for treatment and the appropriate technology, the specific contaminants must be identified and measured.

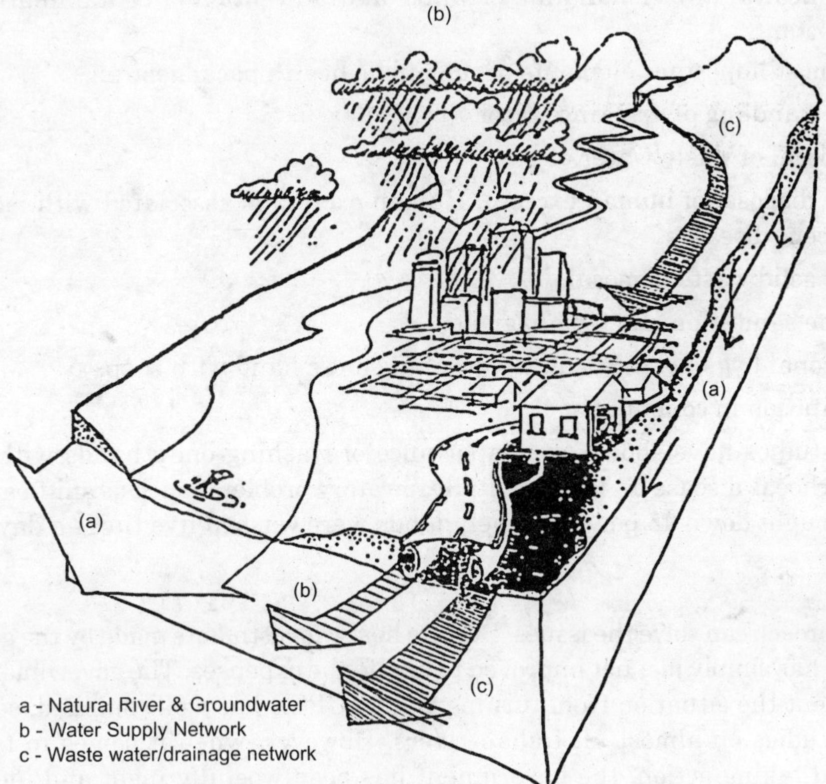

Fig.9.1: Warner's 'three rivers' Metaphor of Urban Hydrology.

Water Supply, Sanitation and Health

Safe drinking water and improved sanitation play a major role in the overall well-being of the people, with a significant bearing on the infant mortality rate, death rate, longevity and productivity. The poor, both in rural and urban areas, bear a disproportionate burden of non- availability of water as well as of poor quality. They often supplement public sources of water with supplies obtained at prices from other sources. Women bear the physical burden of fetching water. Women and children are particularly vulnerable to the effects of water contamination.

70-80 per cent of illnesses are related to water contamination and poor sanitation. The national objectives of reducing morbidity and mortality largely depend on the reduction of diarrhoea and jaundice. In fact, no water supply and sanitation programme can be successful if water-relented illnesses are not reduced. It is a matter of concern that despite the progress made with water supply, level of water-related sickness continues to be high.

Causes of contamination of water are indiscriminate use of chemical fertilisers and chemicals, poor hygienic environment of the water sources, improper disposal of sewage and solid waste, pollution from untreated industrial effluents, over-exploitation leading to quality degradation. Thus, the supply of additional quantity of water by itself does not ensure good health; proper handling of water and prevention of contamination are also equally important.

Among the most important elements of the public health package are:

- Safe handling of drinking water.
- Disposal of waste water.
- Safe disposal of human excreta. Human excreta is associated with more than 50 per cent diseases.
- Safe solid waste disposal.
- Home sanitation and food hygiene.
- Personai hygiene, particularly, washing one's hand with soap.
- Sanitation in community.

Recent studies have shown the importance of washing one's hands with soap as it reduces diarrhoeas disease by 43 per cent. Respiratory problems such as sniffles and coughs were also brought down 45 per cent when hands were washed five times a day.

Strategies

No single approach can solve the issues. Despite heavy investments made by the government, the situation has simply just not improved to justify the expenses. The government has done little to prevent the situation from turning worse. CPCB has well compiled, scientifically competent studies on almost all Indian rivers. However, when it comes to tackling the situation and taking action, the government has been woefully inept and inefficient. In fact, tackling river pollution was a non-issue till the start of the Ganga Action Plan (GAP) in 1985.

Sewage and municipal effluents account for 75 per cent of the pollution load in rivers, while the remaining 25 per cent is credited to industrial effluents and non-point pollution sources. Thus, river cleaning programmes such as GAP, Yamuna Action Plan (YAP), and more recently, the National River Conservation Plan (NRCP), were envisaged. By and large, attempts have been made to tap drains, divert sewage to sewage treatment plants (STPs) and then to let out this supposedly treated sewage into the river.

The biggest drawback of river cleaning programmes such as GAP and YAP is that the failure to envisage who would pay for running the treatment facilities in the long run. With power supplies to most Indian towns being erratic, the facilities which are heavily dependent on power, lie underutilised, undoing all attempts to clean the rivers. NRCP, set up with GAP as its role-model, failed to rectify these shortcomings. Moreover, river cleaning programmes do not address the problem of pollution resulting from agricultural runoffs.

There has been a change in water quality of the rivers since river cleaning schemes have been launched, but it goes unnoticed by a layperson. However, researchers confirm that there has been some improvement in the water quality. Studies carried out by Metro economic, a UK-based environment consultant, corroborate government studies that some parameters indicating water quality have shown marginal improvements. The studies indicate that parameters like dissolved oxygen (DO) and BOD have improved than would have been the situation had there been no attempt to reduce pollution.

But this 'improvement' is hard to perceive for the common man. For most part of the river, the little water that flows in it is dark in colour and has an unfriendly odour. The water can hardly be used without any treatment. However, any river cleaning programme in the world has taken 20-25 years to show results and at a much bigger investment.

Promoting Water Conservation

For most cities, extending water supply coverage to current residents is a major challenge. Yet cities are also facing pressures to expand their municipal water supplies, demand for municipal water could grow by a factor of five or more over the next four decades. Many cities already face critical water shortages and high costs of supply.

The usual response is to increase supplies through expensive investments in new public infrastructure. Yet evidence suggests that cities can also manage the demand for water by reducing wasteful use (through pricing and conservation efforts) and by preventing pollution. A comprehensive strategy would include improving the operation and maintenance of water supply systems, removing subsidies and price distortions that encourage waste, and public education. Demand management is a particularly attractive option for cities in India. Recycling, especially of industrial wastewater, is another strategy, providing companies with a cost-effective and reliable source of water and at the same time protecting the environment.

In India, several cities have been implementing demand management programmes. In Mexico City, for instance, the water utility implemented a new rate structure that charges more per cubic meter as consumption levels increase. It is hoped that this measure will provide metered industries with incentive to conserve water. In the cities, the imposition of effluent charges induced reductions in water demand. About 15 percent of domestic consumption could be saved through measures such as improved efficiency in public facilities, a leakage reduction programme, recycling of cooling water used in air conditioning, and installation of water-efficient flush toilets.

Reducing Water Pollution

By reducing water pollution, cities can reap the double benefit of effectively increasing the water supply while lessening the deterioration of the aquatic environment. As the "pollution shadow" spreads, cities must go further afield to find clean water, which significantly increases the costs of water supply.

Of all the pollutants, urban sewage may be the worst offender in near-urban waters, although industrial pollutants can be a major source. In addition, up to half of the contaminants reaching urban waters come from non-point sources, such as urban runoff.

Controlling urban runoff, although difficult, is essential if cities are to mitigate their impacts on nearby water bodies.

- Regional water resources can only be properly managed at the river basin level. It is therefore necessary to have planning and control for the whole river basin. The municipality should keep a close watch of developments at that level.
- Groundwater, which is the responsibility of the Department of Mineral Resources, and surface water, for which various government departments share responsibility, should be managed as one system. The regional management framework for water resources as a whole.
- Because of the impact which urbanisation and its demand for groundwater can have, aquifer management is a vital aspect of regional water resource management.

National River Conservation Plan

The National River Action Plan (NRAP) was formulated along the lines of the Gang Action Plan (GAP). Under NRAP, work will be accorded priority according to their benefits and implemented in a phased manner according to the availability of resources, Apart from Ganga, Yamuna, Gomti and Damodar, other Indian rivers, whose stretches are badly polluted, were included in the NRAP. Proposed in 1992-93, this plan was renamed the National River Conservation Plan(NRCP).In July 1995, the Central Ganga Authority (CGA),which oversaw GAP, was reconstituted as the National River Conservation Authority (NRCA),and the Ganga Project Directorate (GPD) was redesignated the National River Conservation Directorate (NRCD).

NRCP essentially addresses pollution load from sewer systems of towns and cities, and works towards intercepting and diverting sewage, setting up STPs, electric crematoria, low-cost sanitation, riverfront development, afforestation and solid waste management. Setting up STPs account for 41 per cent of the estimated budget, while intercepting and diverting sewage takes about 30 per cent of the funds allotted. The programme emphasises treating urban sewage because MEF believe that 75 per cent of the pollution load in Indian rivers is due to untreated urban sewage dumped into the rivers.

As for the pollution load from industries flowing into rivers, the responsibility for ensuring that industries, whether large-scale or small-scale, meet pollution standards is that of the respective state pollution control boards (SPCBs).

Water Purification

Conserving the scarce water and recylcing used water is a major challenge in the urban areas of India. For this, a new technology known as root zone cleaning system is feasible. It consists of a receiving tank into which the toilet, kitchen and bath water flows by gravity; a mud area where suitable reeds, rushes and other plants are grown; and an outlet tank in which the water cleaned by the natural process is accumulated. The processed water is clean and without any smell. Fish and amphibians living and growing within it show that the water is clean and the system is successful. It avoids chlorinating, which removes germs and bacteria, but leaves chemical residues that are unfavourable for human consumption.

Urbanisation, removal of forest cover and the addition of impervious surfaces have increased frequency of flooding and changed stream characteristics. By adopting, the natural drainage system, water-table recharges and increases lag time for run-off. Protection of flood plains and drainage sales is imperative for flood control, for regulating stream control and for maintaining water quality.

Advanced Integrated Wastewater Pond System

Treating the incoming waste water through a series of carefully engineered ponds is called the Advanced Integrated Wastewater Pond Systems (AIWPS), which uses bacteria and oxygen-releasing micro-algae to remove parasites, heavy metals, bacterial and viral pathogens from waste water. This has been experimented in cleaning of River Ganges at Varanasi, so as to meet World Health Organisation standards for purity of food crop irrigation water. While pond systems are generally more efficient and less costly than mechanical treatment facilities, they require more land and more time. For this reason, larger municipalities generally favour mechanical systems. The ponds use the ability of bacteria and algae to decompose and oxidize sewage. It greatly diminishes the dependency on electricity to treat the water. Sewage is stored from 20 to 45 days as compared to mechanical facility (where sewage is stored only for one day). Methane produced from the waste is captured at the site and used to meet the meagre energy requirements of the AIWP system.

Table 9.10: Relative Importance and Final Determinant Weightings for the Water Quality Indices

Deteminand	General (G) Rel. imp	Wt	Bathing (B) Rel. imp	Wt	Supply (W) Rel. imp	Wt	Fish (F) Rel. imp	Wt
Dissolved oxygen	1.33	0.30	2.15	0.15	2.13	0.18	1.00	0.34
pH	3.13	0.13	3.10	0.10	2.79	0.15	2.81	0.12
Suspended material	2.57	0.15	1.73	0.19	2.82	0.15	2.41	0.14
Temperature	3.15	0.12	3.78	0.09	3.59	0.12	1.35	0.26
BODs	2.20	0.18	2.23	0.15			2.48	0.14
Faecal coliforms	3.18	0.12	1.00	0.32	1.78	0.24		
Ammonia					2.50	0.16		

Source: Smith, 1987

Table 9.11: Standards for Drinking Water (after Brown et al., 1972)

	European standards	International standards
Biology		
Coliform bacteria	Nil	Nil
Escherchia coli	Nil	Nil
Streptococcus faecalis	Nil	Nil
Clostridium perfringens	Nil	Nil

Virus	Less than I plaque-forming unit per litre per examination in 10 L of water	
Microscopic organisms	Nil	
Radioactivity		
Overall α radioactivity	<3pCil^{-1}	<3pCiL^{-1}
Overall β radioactivity	<3pCil^{-1}	
Chemical elements	(mgL^{-1})	(mgL^{-1})
Pb	<0.1	<0.1
As	<.05	<0.005
Se	<0.01	<0.01
Hexavalent Cr	<0.05	<0.05
Cd	<0.01	0.01
Cyanides (in CN)	<0.05	<0.05
Ba	<1.00	<1.00
Cyclic aromatic hydrocarbon	<0.20	
Total Hg	<0.01	<0.01
Phenol compounds (in phenol)	<0.001	<0.001-0.002
NO_3 recommended	<50	
NO_3 acceptable	50-100	
NO_3 not recommended	>100	
Cu	<0.05	0.05-1.5
Total Fe	<0.1	0.10-1.0
Mn	<0.05	0.10-0.5
Zn	<5	5.00-15.0
Mg if SO_4 > 250 mgL^{-1}	<30	<30
if SO_4 < 250 mgL^{-1}	<125	<125
SO_4	<250	250-400
H_2S	0.05	
Cl recommended	<200	
Cl acceptable	<600	
NH_4	<0.05	
Total hardness	2-10 meqL^{-1}	2-100 meqL^{-1}
Ca	75-200 mgL^{-1}	75-200 mgL^{-1}
F	In the case of fluorine the limits depends upon air temperature:	

Mean annual maximum day time temperature (°C)	Lower limit (mgL^{-1})	Optimum (mgL^{-1})	Upper limit (mgL^{-1})	Unsuitable (mgL^{-1})
10-12	0.9	1.2	1.7	2.4
12.1-14.6	0.8	1.1	1.5	2.2
14.7-17.6	0.8	1.3	1.3	2.0
17.7-21.4	0.7	0.9	1.2	1.8
21.5-26.2	0.7	0.8	1.0	1.6
26.3-32.6	0.6	0.7	0.8	1.4

**No. 100 ml sample to contain E. color more than 10 coliform.*

Table 9.12: Comparison between Surface Aerator and Submersible Aerator

	Surface Aerator	Dry Motor Submersible Aerator
(a)	Large tank surface area required	Deeper tanks save space.
(b)	Water is thrown into air in form of drops	Air is drawn mixed with liquid and released in the form of line bubble.
(c)	Water drops are coming in with contact air before falling back into tank	Air bubbles released at bottom of tank get in contact with water before being released into atmosphere.
(d)	Oxygen is absorbed into water drops before they fail into tank	Oxygen gets absorbed into water during air bubble travel from bottom to top.
(e)	Duration of water drops being in air is short hence oxygen absorption is less effective	Duration of air bubble travelling from bottom to top of tank is more compared to water drops falling into tank. Hence oxygen absorption is more effective.
(f)	Due to propeller motion agitation takes	Due to velocity of liquid mixed with air places at top of tank effectively being forced through nozzle not only agitation but through mixing takes place at bottom of tank.
(g)	Settlement of solid particles present in liquid cannot be avoided	Settlement of solid can be avoided since through mixing takes place at bottom of tank.

Submersible Aerators developed from dry motor submersible pumps is a new concept of aeration, which has several advantages over the conventional surface aerators. These aerators do not require investment in civil engineering works such as bridges and foundations. These aerators could be relocated at will. The oxygenation capacity of this submerged aerator is required to be evaluated for application in sugar industry, paper and pulp, and distilleries. Further studies are required to be done.

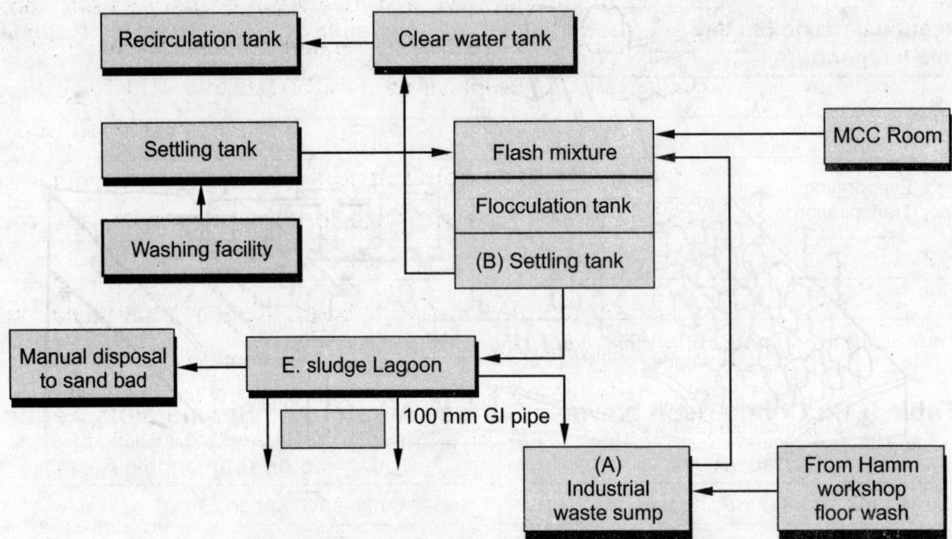

Fig.9.2: Flow Chart of Effluent Treatment Plant

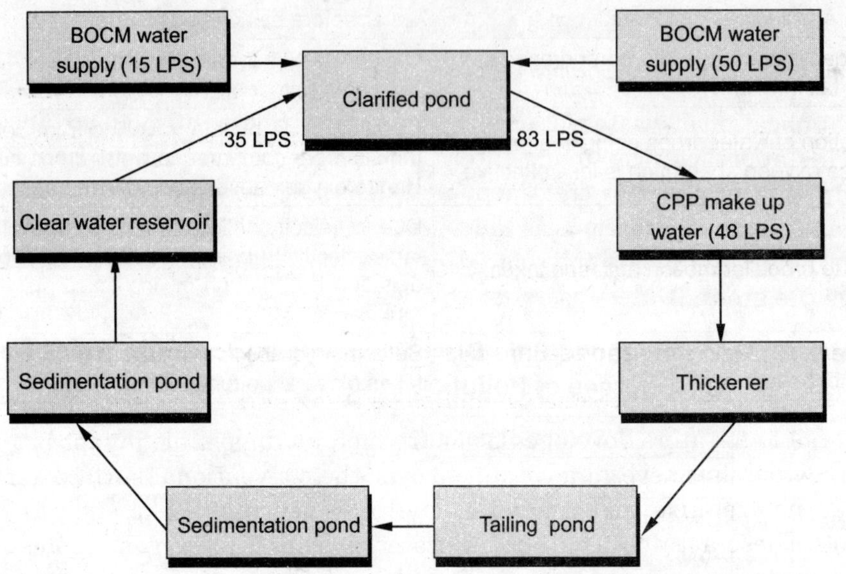

Fig.9.3: Closed Water Circuit for CPP

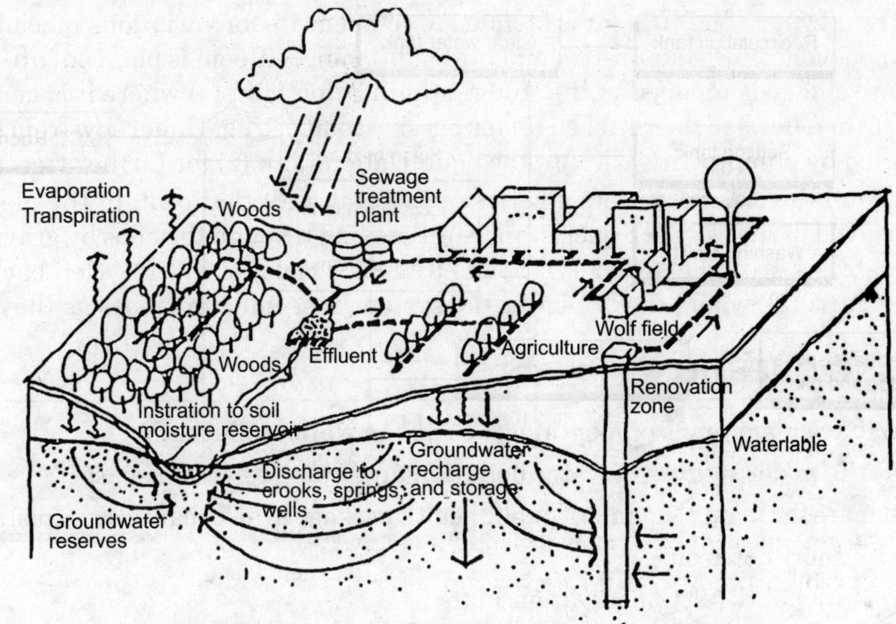

Fig.9.4: The Wastewater Renovation and Conservation Cycle.

Source: William E. Soper, Surface Application of Sewage Effluent and Sludge, Madison, Wisc. ASA-CSSA-SSSA, 1979

Land application of waste water, after secondary treatment, is an advanced treatment method. This approach (the 'living filter') considers effluent and nutrients as resources rather than as a product for treatment and disposal. Treatment is provided by natural biological and chemical processes as the waters, moves through the living filter provided by soil, plants, micro-organisms and related ecosystems. The renovated water then percolates to recharge the groundwater reservoir.

Table 9.13: Recommended Safe Distance between Domestic Water Wells and Sources of Pollution (after Romero, 1970).

Source of pollution	Distance (m)
Septic tank	15
Cess pit	45
Sewage farm	30
Infiltratlon ditches	30
Percolation zones	30
Pipes with watertight joints	3
Other pipes	15
Dry wells	15

As a typical SAS septage treatment plant, effluent is discharged into a headworks (1) consisting of a receiving station and degritter. From there it flows by gravity into in-ground tanks (2) where the effluent is blended to compensate for variations in loads and is preconditioned. Following equalization and preconditioning, effluent is pumped to a clarifier. (3) From here sludge is pumped to the sludge stabilization tank (4) where it is aerobically stabilized. It then flows to the reed bed (5) for passive composting. Underflow from the reed bed is recycled by pumping it back into the equalization tank (2) for further treatment.

Liquid from the clarifier enters a series of solar silos (6) that contain the plants and animals involved in the clean-up operation. After moving through the silos by gravity flow, the liquid enters a second clarifier (7). Solids from this clarifier are recycled back to the equalization tank (2), where they re-enter the system, reseeding microbes as they travel.

DEBRIEFING QUESTIONS

1. What are major environmental issues due to water pollution?
2. List the major sources of water pollution and water related diseases.
3. What are the main elements of public health package in an urban area infrastructure?
4. Write short notes on:
   (a) National River Conservation Plan
   (b) Advance Integrated Waste Water Pond System
   (c) Closed Water Circuit
   (d) Effluent Treatment
   (e) Water Quality Indices
   (f) Wastewater Renovation and Conservation Cycle

10 Wastewater Treatment

Liquid waste is usually discharged from homes, commercial establishments and industries into the sewerage and drainage system. Liquid waste may contain sewage, urine, oils, fats,chemicals, nutrients, etc. Discharge from industries and wastes from the hospitals, labs, etc.is often mixed with municipal waste and sewerage system and discharged into the rivers or streams without treatment.

For example, in the NCT of Delhi, about 22 drains, each having several sub-sewers empty their contents (a total of 3800 MLD) into the river Yamuna and this accounts 80% of the river pollution load. Six of these drains (Najafgarh, Shahdara, Savita Vihar, Maharani Bagh, Barapulla and Sen Nursing Home drains) contribute 90% of the flow and 80% of BOD levels. The total Wastewater generated is about 3800 MLD, but the treatment capacity is only 2460 MLD and the capacity for carrying Wastewater to STPs is only 1556 MLD. Besides domestic sewage, industrial effluents (180 MLD) from planned and unplanned industrial areas also contribute to the pollution load of the river. The solid waste, including construction and domestic waste and fly ash is also dumped into the river.

Table 10.1: Water Quality Status of Drains Joining Delhi's Stretch of Yamuna

Sl.No.	Location	Suspended solids (10 mg/L)	COD (250 mg/L)	BOD (30 mg/L)
1.	Najafgarh drain + Supplementary drain	103	75	25
2.	Magazine Road drain	502	866	262
3.	Sweeper's Colony drain	292	384	144
4.	Khyber Pass drain	33	17	02
5.	Metacafe House drain	50	71	33
6.	ISBT drain+Mori Gate drain	137	224	82
7.	Tonga Stand drain	214	350	148
8.	Civil Mill drain	298	351	148
9.	Delhi Gate drain	541	853	277
10.	Sen Nursing Home drain	126	88	28
11.	Drain No 14	29	27	4
12.	Barapulla drain	79	92	38
13.	Maharani Bagh drain	353	256	99
14.	Tughlaqabad drain	299	209	89

15.	Shahadara drain	95	126	45
16.	Abu Fazal drain	51	45	11
17.	Jaitpur drain	93	160	65

COD—*chemical oxygen demand. This is the capacity of water to consume oxygen during decomposition of organic matter. It determines the amount of organic pollutants in surface water.*
BOD—*bio-chemical oxygen demand.*
Source: *Central Pollution Control Board.*

MAJOR WATER POLLUTANTS

Huge quantities of human, industrial, liquid, tanneries and solid wastes enter into the water bodies, drains and rivers. This needs to be urgently stopped by providing sanitation, sewerage, local treatment facilities and bio-digester based toilets in the areas adjacent to river where people live and congregate. Up-gradation of water quality to prescribed standards requires strict enforcement of Water Pollution Act. The installation of treatment facilities (STPs and ETPs) by polluting units and surrounding urban areas, and their efficient and effective functioning are essential. The pollutants include:

(a) Sediments: Particles of solid and minerals introduced by erosion, unscientific farming, and unregulated construction cause damage throughout the length of the watersheds. Such sediments discolour the water, fill the channels, lakes, reservoirs, and harbours, smother the feeding and breeding grounds of fish, and reduce the aeration of the water by the penetration of sunlight.

(b) Plant nutrients: When large concentrations of nitrogen and phosphates are discharged into the stream by sewage, industrial waste, or the leaching of fertilizers from farm fields, the waters bloom with algae.

(c) Oxygen-reducing wastes: Fish, aquatic plant life, and water quality all require the presence of free oxygen in the water. This vital element is depleted or eliminated by the introduction of the organic wastes from sewage, animal manures, food process plants and many industries. The lower reaches of most rivers are oxygen deficient.

(d) Infectious organisms: Disease-causing agents such as coliform bacteria and viruses are introduced into the stream by sewage and by food processing and packing plants.

(e) Organic chemicals: Many of the new synthetic pesticides, detergents, and industrial chemicals are toxic to the fish, plants, birds and animals that inhabit the waterways. Some are poisonous, even in low concentrations. Others produce stench and discoloration. Most resist conventional water treatment.

(f) Minerals and inorganic chemicals: These are by-products of mining, the petroleum industry, farming, and manufacturing plants. They are composed mainly of traces of metals, salts, acids, oils and other chemical compounds which are detrimental to all users of the contaminated water.

(g) Heat: Thermal pollution reduces the capacity of water to absorb oxygen, break down wastes, or support aquatic life. Most of such heated water is discharged by industrial and power generation plants which require enormous volumes of water for washing and cooling.

(h) Radioactive materials. The most significant sources have been fallout from nuclear weapons and devices, and the use of radioactive substances in industry and research. Radioactive pollutants are relatively new and their long-term effects are not yet understood. It is recognised, however, that they compromise a serious potential threat to the world environment.

The quality of drinking water is determined by following tests:

- Temperature Test: The temperature of the water should be tested. It is desirable to
- have a temperature of 20-25°C.
- pH Test: The water is to be tested by pH value for its acidity or alkalinity. pH above 7 shows alkaline nature below 7 shows acidic nature of the water. Water should have an equal to 7. pH is a logarithmic scale. It is advisable to have pH of water between 6.5 – 8.5.
- Bio-chemical Oxygen Demand (BOD): Organic wastes, when they enter water, begin to break it down by the action of bacteria. This process utilises the dissolved oxygen in the water. The principal of quantitative measure, the organic wastes, are therefore expressed in terms of Bio-chemical Oxygen Demand (BOD). BOD measures the amount of dissolved oxygen that would be depleted by a specified quality of organic wastes in a given time (5 days) at a standard temperature (usually 20 °C). The degradable pollution content of various kind of organic wastes can be commonly expressed in terms of number of pounds of BOD they contain.
- Coliform Bacteria Test: Water is to be tested by Coliform bacteria test. Coliform bacteria, which is the main reason for water borne diseases, has to be at minimum. The standard is 1 mg per litre of water is allowed.
- Turbidity test: Water should be turbid free. Turbid water contains soil particles and organic wastes which can be removed through filtration.
- Phosphorous Test: Water should be tested for phosphorous. Phosphorous is required to human body in a very small quantity. Phosphate more than what is required is dangerous to health. 1 mg of phosphate per litre is desired.
- Nitrate Test: Water is to be subjected to Nitrate test. Nitrates are also required in small amount 10 mg per liter. Anything, more than this is harmful.
- Suspended Particles Test: Water contains lot of suspended particles in the form of inorganic element like Arsenic, cadmium, chromium, cyanide, fluoride, lead, mercury, etc. All types of elements should be at desired levels.

Control of pollution from industrial effluents was also accorded priority in the Ganga Action Plan. To begin with, 68 grossly polluting units on the banks of the river were identified. Earlier, among these industries, only 14 units had necessary effluent treatment plants (ETPs). With the initiative taken through the Pollution Control Boards in union with the programme on control of domestic effluents, 55 industries have been closed down. For control of industrial pollution, a fresh exercise was made and 119 additional units have been enlisted for controlling the discharge of their effluents into the river.

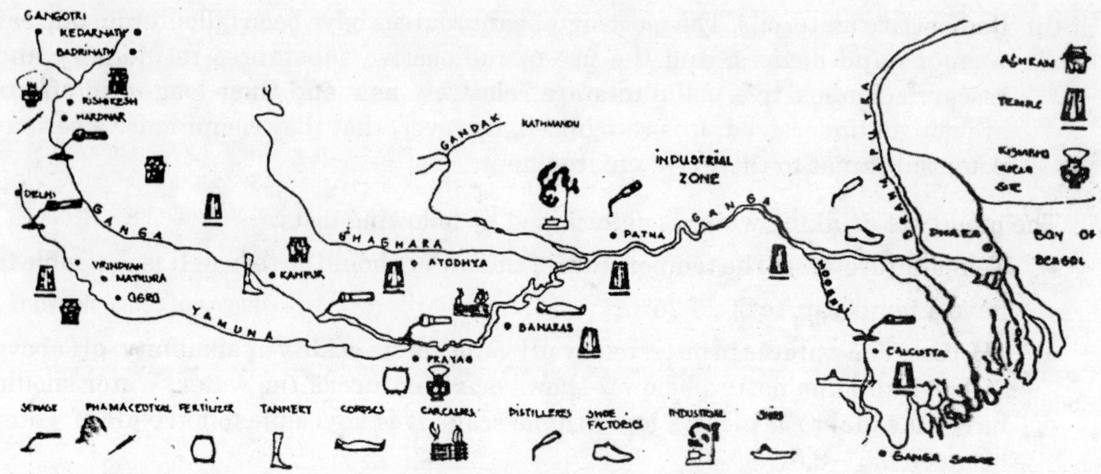

Fig.10.1: The Ganga-river of purity and poison

Source: Worldwatch, 1995

As supporting measures to prevent pollution, besides interception, diversion and treatment of sewage, schemes were taken up for providing low cost sanitation, setting up of electric crematoria, river front facilities and biological regeneration of the river. The other important achievements of the Ganga Action Plan include the following:

Urban-generated pollution comes from both localized and dispersed, or point and non-point, sources. Major point sources include municipal sewage, industrial outfalls, and air emissions from power plants and heavy industries. Non-point sources include silt from earth-moving activities; storm runoff from roads, home gardens, and industrial sites; infiltration from aquifers contaminated with sewage or industrial chemicals; and automobile emissions.

Water pollution includes surface water, groundwater and coastal/lake pollution. As given below, it causes various environment and health impact:

Table 10.2: Water Related Urban Environmental Issues and Options

Problem Area	Effects	Causes	Management options
Water Related Problems			
Surface Water Pollution • Community • City wide • Regional	• Health problems • Economic costs (additional treatment new sources of supply. Health costs) • Amenity losses	• Pricing policies • Unclear property rights • Poor regulations and/or enforcement • Municipal & industrial waste disposal practices • Urban runoff • Irrigation practices	• Marginal cost pricing • Regulations, standards, licensing, charges • Improve monitoring and enforcement • Demand management and wastewater re-use • Appropriate technology • Land use controls • Waste management

Groundwater Pollution and Depletion • Community • City wide • Regional	• Reduced water quality from saline intrusion, biochemical seepage • Health impacts • Economic costs (damage from land subsidence, health costs, increasing marginal costs of (supply)	• Pricing policies • Unclear property rights • Poor regulations and/or enforcement • Unsustainable extraction • Sanitation, municipal & industrial waste disposal practices • Poor demand management	• Marginal cost pricing (sustainable extraction, aquifer recharge costs) • Regulations, standards, licensing, charges • Waste management • Appropriate technology • Demand management • Controls on land use and sources of contamination
Coastal/Lake Pollution • Community • City wide • Regional • Transnational	• Health effects due to contaminated seafood and direct contact • Loss or recreational resources & tourism revenues • Damage to fisheries • Amenity losses • Eutrophication	• Unclear property rights • Poor regulations and/or enforcement • Municipal and industrial waste disposal practices • Disposal of shipboard waste	• Regulations, standards, licensing, charges • Appropriate technology • Coastal zone management and preservation • Shipping facilities • Waste management • Land use controls

Source: Carl Bartone, 1991, Annotated Outline of Strategy Paper on Urban Management and the Environment, UMP Environment Profile

Table 10.3: Hydrological Cycle, Pollution Sources and its Effects

hydrological cycle process	Store & transfer agents of Fresh water	Sources of pollution	Type of pollution	Effects of pollution
1	2	3	4	5
Runoff	River	Sewage	(1) Organic matter (2) Detergents (3) Farm (4) Nitrates (5) Phosphates	Foul Odours (1) oxygen deficiency (1) Excess five-collforms (1) High] growth of floral species (1,4,5) Cessation of aeration and photosynthesis causing harm to Biota (2,3) waterborne diseases like Cholera, Diarrhoea, Hepatitis, Panathoid fever. Typhoid fever by carriers living in or on water (1) Listlessness & blue colour skines in infants (4,5).
		Industries	(1) Mercury (2) Lead (3) Thermal steel, Power plants) (4) Amonia Sulphate (5) Fluorine (6) Cadmium (7) Poly Chlorinated biphenyls (PCBs)	Interference with self purification process (1,2,4,5,6) Kills native flora and fauna (1,2,3,4,5,6) Nervous disorder (1,2) Depression (2) premature hatching of fish eggs and destruction of biota (3) Congenial to spread bacterial diseases (3) Congenial to spread bacterial diseases (3) Corrosive effect on metal (4) Oxygen deficiency (3) Absorbed by grasses, when eaten

			(8) Pathalets-kepon (9) Phathalets-Mirex	by herbivorous causes fluorosis (5) disease minimata & itai by taking contaminated rice and fish (1,6) Brittle bones, kidney failures and high blood pressure (6) Brown biotches on skin, weakness and eye discharge (7) See row VI (8,9)
		Mining and deforestation	(1) Sediments (2) coal deposits leave iron sulphide (3) cadmuim	Sitting (1) Turbidity (1) Turbidity harms flora and fauna by checking photosynthesis and areration (1) Iron sulphide reacts with oxygen in water and forms sulphuric acid and iron oxide which kills the biota. Sulphuric acid and iron oxide which kills the biota. Same as in IInd row (3)
		Transportion	(1) Oil (due to slick, spill, accidents and cleaning of oil tanks (2) lead-gasolin from motor boats.	Harm to fish plant life in water by checking aeration and passage of sun light (1) Fire hazard (1) Lead causes depression and metal disorders (2)
		Tourism	(1) Solid wastes.	Chocking of water surface and harming biota.
	Groundwater	Agriculture	(1) Fertilizer (2) Pesticides (3) Herbicide (4) DDT (5) BHG (6) Phthalates keep on (7) Pathlates-Mirex.	More than 45 mg per lit. harms babies (1) Water Slinity (1) Causes liver, nervous and blood disorders (2,3,4,5) Inter- terence in reproduction of water preacher (6,7) Dizziness, trembling, tremors visual problems, liver cancer (6) Cancer (7)
		Human waste through septic tanks, land fills & drains	(1) Humic acid (2) Soluble fraction of organic degradation (3) Organic particulate (4) Dissolved carbon dioxide in land fill areas.	Water turned brakish and bitter (2) Odours (2) The decomposition in land fill areas may produce CO_2 which gets dissolved in percolating water and increase its aggressiveness to calcium and magnesium based rocks which may echance calcification of soil (4) The particulate matter can clog the pores and check the flow of underground water (3) Increase in tubidity (2,3).
		Industries	(1) Sulphate (2) Amonia (3) Gynide (4) Fluorides (5) Zince (6) Gobait (7) Copper oxide	Increase in water salinity reduces plants ability to absorb water, causing stress and deaths in plants (1,4,5,6). Increase ions in water displaces alkalise ions and increase alkalinity (2)

		Atmospheric pollution due to industries & aeroplanes	(1) Sulpher (2) Nitrogen (3) crabon-monixide (4) Carbon	The smoke when come in contact with the water vapour it produces sulphuric & nitric acid and its percolation into ground water may give adverse effect on skins etc. (1,2)/kills plants by cloging stomata (4).
		Under ground explosion & waste disposal	(1) Radio-activity	Causes mutation in still born flora and fauna reducing its life span (1) cause skin ailments
Storage	Lake	Introduced alien species	(1) Turbidity increased (mud fish) (2) New species may check to extinction the local plant due to abrupt growth.	Hinder sunlight, hampering photosynthesis resolution in death of plant, than animals (1) Extinction of local species (2).
		Tourism savage industries & Transportation		Same as in the Case of River
	Glaciers & Ice-Caps	Atmospheric (pollution)	(1) Dust (storm volcanoes & forest fires (2) Sulphur-dioxide (3) Carbon-monoxide (4)Hydrocarbon	Though the level of pollution of fresh water type (solid) not significant but with the increase use of atomic fuel may increase radio-activity and kill the lickens and algae
Seepage	Soil-Moisture	Industries garbage	Same as associated with such type of pollution.	

Note: To see what type of pollutant gives what affect then see the number of the chemical in column 4 and check the affect in column 5.

Groundwater is especially arsenic to contamination with pesticides that and are mobile in atrazine, bentazone, carbofuran, is operaturon, ethylene dibromide and simazine. Pesticide concentrations in many groundwater sources now exceed WHO guideline values. Someof these pesticides are believed to be carcinogenic to humans, but are of concern with respectto future generations because of their persistence in soil and water.

Other Issues

Untreated sewage, industrial effluents have and agricultural waste are frequently discharged into inland waterways, lakes and coastal zones, endangering the use of these water bodies for recreational purposes such as swimming, canoeing and windsurfing. Recreational exposure to polluted waters can cause diarrhoea, respiratory infections, skin irritation and other diseases depending on specific pollutants involved. Control of land based pollution sources is evidently the most effective means of preventing discharge of sewage into rivers, lakes, coastal beach areas and other marine waters used for recreation.

Additionally, intensive bacteriological monitoring is recommended for public beaches and recreational areas. The economic benefits measures is well proven. In many European tourist resorts, investments have been made in sewage treatment and construction of long ocean outfalls to carry Wastewater beyond coastal waters used by bathers. The regional and local government of the Italian Adriatic coast, for example, have invested heavily in such infrastructure (WHO, 1994). Coastal bathing water quality in the area has now improved and tourist numbers have accordingly increased.

The Central Pollution Control Board (CPCB) report on 'Status of sewage treatment (2009)', states that out of 38,000 MLD of sewage generated, treatment capacity exists for only about 12,000 MLD. Nearly 39 per cent plants do not conform to the general standards prescribed under the Environmental (Protection) Rules for discharge into streams. The waste-water management offers a huge scope for improvement and introducing newer technology in its treatment.

The existing sewage treatment plants (STPs) are designed to treat sewage up to secondary level with Biochemical Oxygen Demand (BOD) and suspended solid (SS) as the effluent quality parameters. The Central Public Health and Environmental Engineering Organisation (CPHEEO) Manual (2012) prescribes stricter standards for disposal than the existing standards provided by the Environment Protection Rules, 1986. It suggests that faecal coliform, a cause of water borne diseases, should be controlled in effluent discharging into surface water.

The Diagnosis

In order to decide the suitable kind of treatment, it is necessary to diagnose the sources and causes of water quality deterioration. Accordingly, the treatment of Wastewater can be taken up in four steps, viz., preliminary treatment for grit removal, primary treatment for removal of raw sludge, secondary treatment (oxidation and activated sludge removal) and Tertiary treatment (disinfection). A major source of pollution of water is because of the nitrogen contents in land due to increasing use of the fertilisers. Various industrial activities with potential to contaminate the ground and aquifers need to be identified and checked accordingly.

PROTECTING THE QUALITY OF WATER

Since rivers and their tributary streams are the potential repositories and carriers of pollution, they must be guarded from contamination at all points and from all sources. The major pollutants present in almost every river, include:

- Sediments: Particles of solid and minerals introduced by erosion, unscientific farming, and unregulated construction cause damage throughtout the length of the watersheds. Such sediments discolor the water, fill the channels, lakes, reservoirs, and harbors, smother the feeding and breeding grounds of fish, and reduce the aeration of the water by the penetration of sunlight.

- Plant nutrients: When large concentrations of nitrogen and phosphates are discharged into the stream by sewage, industrial waste, or the leaching of fertilizers from farmfields, the waters bloom with algae.
- Oxygen-reducing wastes: Fish, aquatic plant life, and water quality all require the presence of free oxygen in the water. This vital element is depleted or eliminated by the introduction of the organic wastes from sewage, animal manures, food process plants and many industries. The lower reaches of most rivers are oxygen deficient.
- Infectious organisms: Disease-causing agents such as coliform bacteria and viruses are introduced into the stream by sewage and by food processing and packing plants. Humans become infected by contact.
- Organic chemicals: Many of the new synthetic pesticides, detergents, and industrial chemicals are toxic to the fish, plants, birds and animals that inhabit the waterways.
- Some are poisonous even in low concentrations. Others produce stench and discoloration. Most resist conventional water treatment.
- Minerals and inorganic chemicals: These are by-products of mining, the petroleum industry, farming, and manufacturing plants. They are composed mainly of trace metals, salts, acids, oils and other chemical compounds which are detrimental to all users of the contaminated water.
- Heat: Thermal pollution reduces the capacity of water to absorb oxygen, break down wastes, or support aquatic life. Most such heated water is discharged by industrial and power generation plants which require enormous volumes of water for washing and cooling.
- Radioactive materials: The most significant sources have been fallout from nuclear weapons and devices, and the use of radioactive substances in industry and research. Radioactive pollutants are relatively new and their long-term effects are not yet understood. It is recognised, however, that they compromise a serious potential threat to the world environment.

The quality of drinking water is determined by following tests:

- Temperature Test: The temperature of the water should be tested. It is desirable to have a temperature of 20-25°C.
- pH Test: The water is to be tested for pH. is a good determinant of acidity or alkalinity of liquid. pH above 7 shows alkaline nature below 7 shows acidic nature of the water. Water should have a equal to 7. pH is a logarithmic scale. It is advisable to have pH of water between 6.5 – 8.5.
- Bio-chemical Oxygen Demand (BOD): Organic wastes, when they enter water, begin to break it down by the action of bacteria. This process utilises the dissolved oxygen in the water. The principle of quantitative measure, the organic wastes, are therefore expressed in terms of Bio-chemical Oxygen Demand (BOD). BOD measures the amount of dissolved oxygen that would be depleted by a specified quality of organic wastes in a given time (5 days) at a standard temperature (usually 20°C). The degradable pollution content of various kind of organic wastes can be

commonly expressed in terms of number of compounds of BOD they contain.

- Coliform Bacteria Test: Water is to be tested by Coliform bacteria test. Coliform bacteria is the main reason for water borne diseases, which has to be minimum. The standard is 1 mg per liter of water.
- Turbidity test: Water should be turbid free. Turbid water contains soil particles and organic wastes which can be removed through filtration.
- Phosphorous Test: Water should be tested for phosphorous. Phosphorous is required to human body in a very small quantity. Phosphate more than what is required is dangerous to health. 1 mg of ph per litre is desirable.
- Nitrate Test: Water is subject to Nitrate test. Nitrates are also required in small amount 10 mg per liter. Anything more than this is harmful.
- Suspended Particles Test: Water contains lot of suspended particles in the form of inorganic element like arsenic, cadmium, chromium, cyanide, fluoride, lead, mercury, etc. All types of elements should be at desired levels.

For protecting the quality of water in the rivers and water bodies, besides curative action, preventive enforcement action should be intiated. Where the hazard of pollution exists, the minimum charge for operating permits should cover the expenses of adequate policing and controls. Mandatory performance bonds and liability insurance should be adequate to pay for all damages plus any corrective measures which might possibly be needed. In cases of violations or incidents of contamination, additional fines of sufficient magnitude to minimize future risks should be levied. The idea is to set the charge high enough to force the polluter to stop.

As a rule, no new development, manufacturing, processing, or operation of any polluting activity should be permitted which may result in the significant degradation of any water resource. Along the cities and coastlines an increasing threat is being posed by the infiltration of Wastewater, sewage and seawater into the aquifers. As the hydrostatic pressure of the landward water table is reduced by drawdown, the normal flow to the oceans is reversed. Preventive techniques include the construction of extensive system of efficient treatment plants, filters, dams and barricades, coupled with the regulation of freshwater consumption.

Table 10.4: Water Pollution by Industrial Sector

Sector	Pollutants
Iron and steel	BOD, COD, oil metals, acids phenols and cyanide
Textiles and leather	BOD, solids, sulphates and chromium
Pulp and paper	BOD, COD, solids, chorinated organic compounds
Petrochemicals and refineries	BOD, COD, mineral oils, phenols and chromium
Chemicals	COD, organic chemicals, heavy metals, SS, and cyanide
Non-ferrous metals	Fluorine, SS
Micro-electronics	COD, organic chemicals
Mining	SS, metals, acids, salts

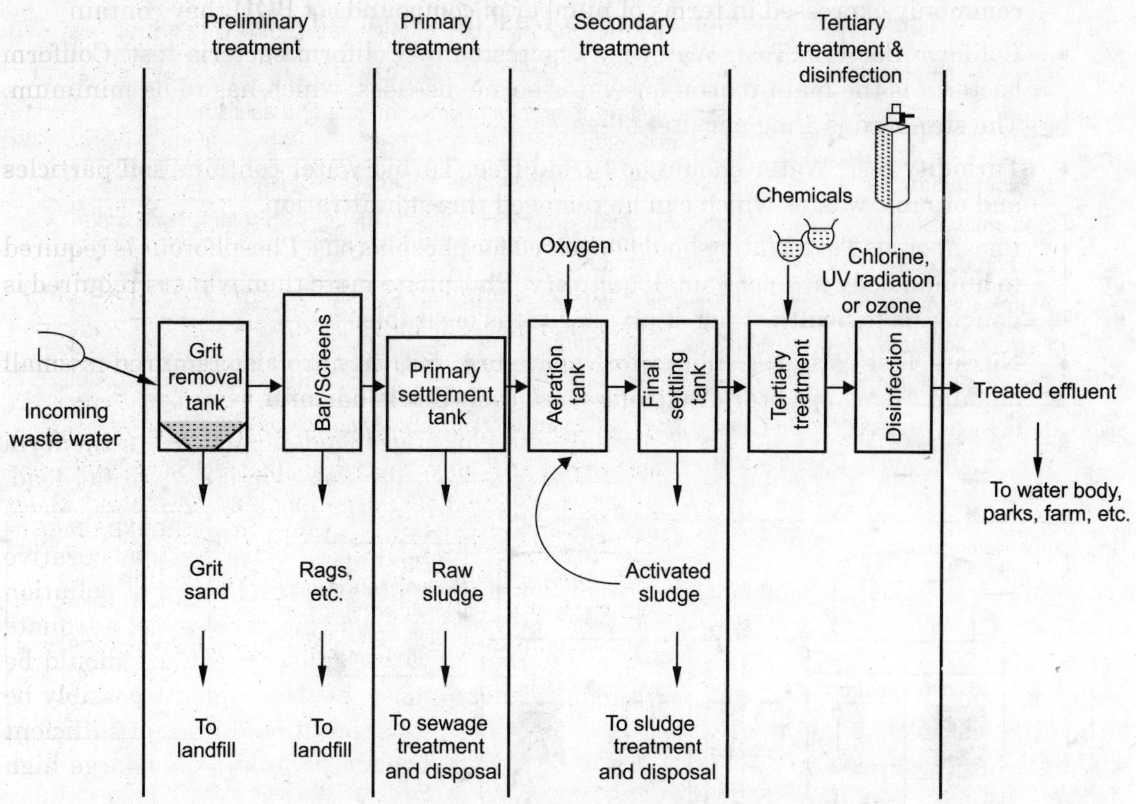

Fig.10.2: Wastewater treatment

Table 10.5: Physical Treatment of Wastewater

Process	Aim	Examples
Screening	Remove coarse solids	Vegetable canneries, paper mills
Centrifuging	Concentrate solids	Sludge dewatering in chemical industry
Filtration	Concentrate fine solids	Final polishing and sludge dewatering in chemical and metal processing.
Sedimentation	Remove settled solids	Separation of inorganic solids in ore extraction, coal and clay production
Flotation	Remove low-specific gravity solids and liquids	Separation of oil, grease and solids in chemical and food industry
Freezing	Concentrate liquid and sludge	Recovery of pickle liquor and non-ferrous metals
Solvent extraction	Recover valuable materials	Coal carbonizing and plastic manufacture
Ion exchange	Separate and concentrate	Metal processing
Reverse osmosis	Separate dissolved solids	Desalination of process and wash water
Adsorption	Concentrate and remove trace impurities	Pesticide manufacture, dyestuffs removal

Source: Harrison (ED) 1990, Pollution: Causes, Effects and Controls, Royal Society and Chemistry.

Table 10.6: Chemical used to Treat Industrial Wastewater

Chemical	Purpose
Calcium hydroxide	pH adjustment, precipitation on metals and assisting sedimentation
Sodium hydroxide	Used mainly for pH adjustment in place of lime
Sodium carbonate	pH adjustment and precipitation of metals with soluble hydroxide
Carbon dioxide	pH adjustment
Aluminium sulphate	Solid separation
Ferrous sulphate	Solids separation
Chlorine	Oxidation

Source: Harrison (ed) 1990, Pollution-Causes, Effect and Controls, Royal Society and Chemistry

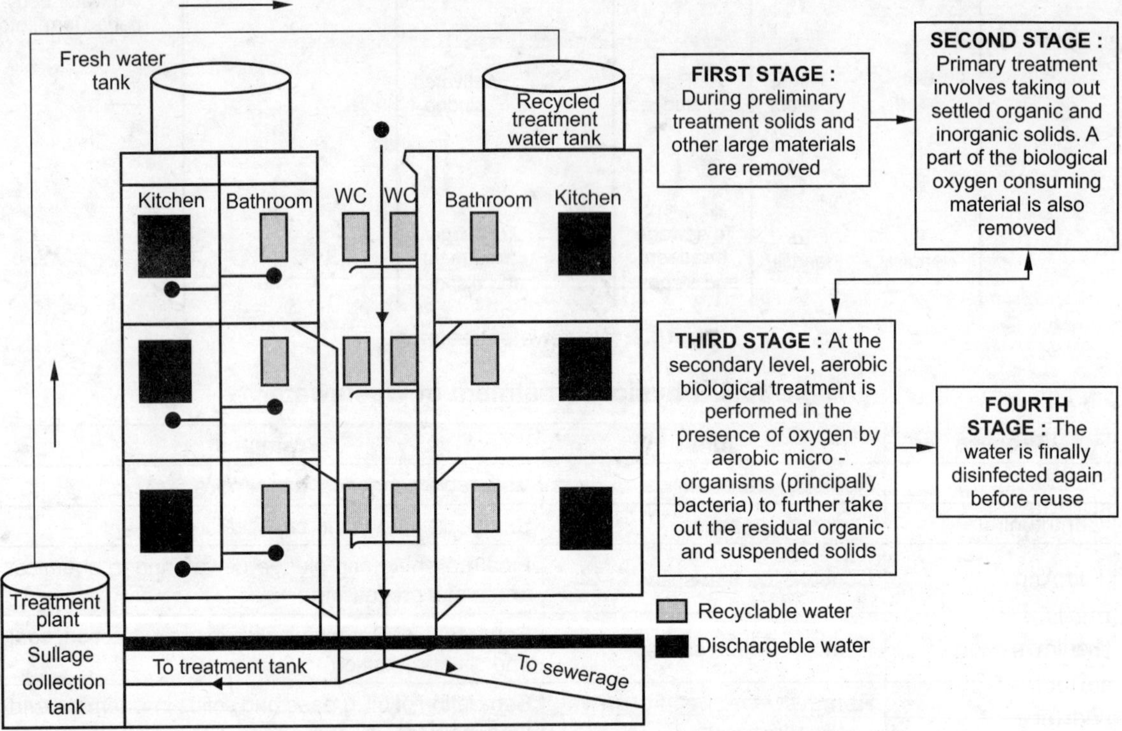

Fig.10.3: Two tier wastewater recycling system

Source: Anon 2001, Total Water Management for Communities, Ion Exchange (India) Ltd. Mumbai

About 50 to 60 per cent of the wastewater generated can be recycled for various purposes,thus reducing much of domestic wastewater loading on the public sewage treatment system. In a house, the water requirement for flushing the toilets is of the order of one-fourth of the total grey water generated. The excess water can also be used for gardening purposes.

In the dual piping system, while one pipe collects kitchen and sink water, toilet waste flows through the other. The sewage is discharged and the remaining water is treated before being reused for non-potable purposes.

Effluent Treatment System

The effluent quality standards prescribed under Environment (Protection) Act, 1986 for disposal of wastewater into inland water bodies and for irrigational uses are presented below:

Table 10.7 : Effluent Quality Standards

Parameter	On land for irrigation	Into inland water bodies
pH	6.0–8.5	5.5–9.0
Total suspended solids, mg/L	200	100
BOD_5, mg/L	100	30
COD, mg/L	—	250
NH_3.N (as N), mg/L	—	50
Phosphates (as P), mg/L	—	5
Boron, mg/L	2	2
Oil and grease, mg/L	10	10
Total coliforme, No/100 mL	NS	NS
Fecal coliforms, No/100 mL	NS	NS

NS : No standards prescribed

Source: *Environment (Protection) Act, 1986, Ministry of Environment and Forests. Govt. of India, Central Pollution Control Board*

These standards have not considered the public health risks that may arise due to the micro-biological characteristic of wastewater used for irrigation, which was clearly established by many UNDP/World Bank and other studies. These studies recommended a standard mean fecal coliform concentration of 1,000 per 100 mL, and an intestinal nematode eggs concentration of $\leq$ 1/100 mL for both restricted and unrestricted irrigation.

As per standard literature, the BOD of raw sewage in India is in the range of 250- 300 mg/L. Field sample analysis of the Central Pollution Control Board, however, has shown the values are lower than stated in the literature. Further, the 'intermediate' wastewater collection technology proposed in this paper could lead to much lower BOD (up to 30-40% reduction) because of the pre-treatment occurring in the household septic tanks. Similarly, the coliform levels also will be lower than in raw sewage. With a conservative assumption of at least 2 log removal in septic tanks, a coliforms level of 104 to 105 can be expected. Hence, only a minimum level of treatment will be required to achieve prescribed BOD standards and the design of wastewater treatment system needs to consider mainly the improvement of micro-biological qualities.

For more than two decades, waste stabilization ponds have been the widely recommended treatment system for developing countries especially due to their excellent coliform removal efficiencies. A manual of guidelines on the design, construction and operation of waste stabilization ponds was published by the Central Public Health Engineering Research Institute (now NEERI) over 30 years ago. Process design procedures for WSP systems have also been described in the 'Manual of sewerage and sewage treatment' prepared by CPHEEO in 1980. Nevertheless, the system has not found widespread application in India. It may be due to either the poor awareness among local authorities on the need for wastewater treatment, lenient view taken by the regulatory authorities in respect of enforcement of wastewater disposal standards or the large area requirement for adoption of this system.

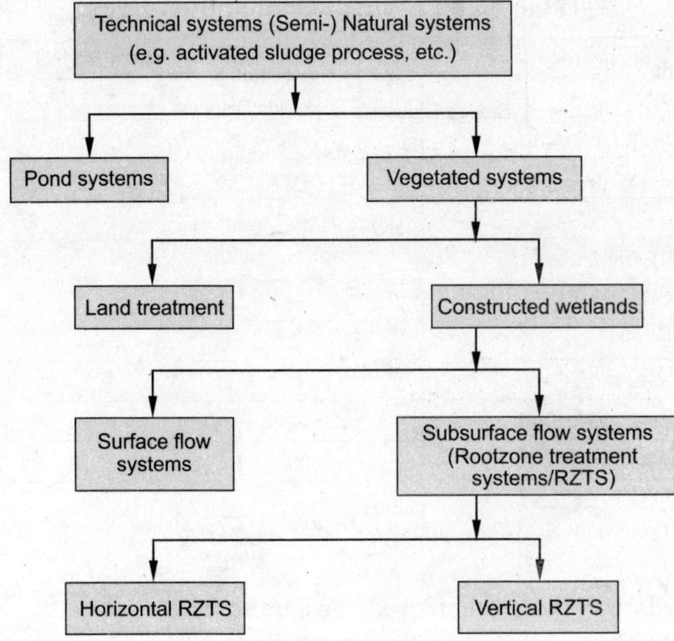

Fig.10.4: Wastewater treatment system

Source: Central Pollution Control Board

Much of the Wastewater can be retrieved,purified and reused after treatment. Based on the type of liquid waste, the following treatment options are available:

- Primary treatment
- Secondary treatment
- Tertiary treatment
- Sludge disposal

Primary treatment may include removal of solids, which uses screens, grit chambers and sedimentation tanks. Sometimes, removal of heavy and bulky materials, grit, fat and grease is included in the pre-treatment process. Secondary treatment includes BOD treatment through aeration tanks, sedimentation tanks/clarifiers and disinfection of the Wastewater. Tertiary treatment may include chemical/nutrient and toxins (pesticides, VOC, metal) removal. After the treatment, Wastewater is normally discharged into the rivers/ streams, which should be stored and reused.

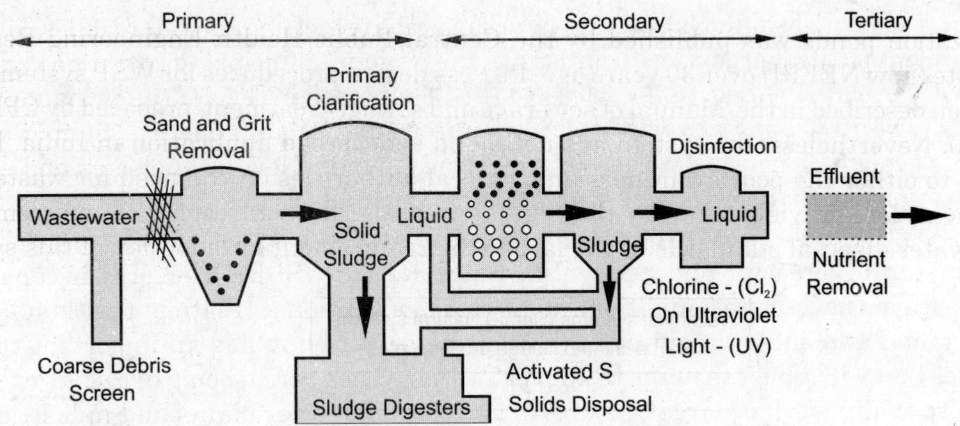

Fig.10.5: Treatment of liquid waste

Source: http://www.sheffy6marketing.com/index.php?page=test-child-page

Domestic Wastewater can be recycled after secondary treatment. Normally, grey water is recycled and used for flushing or for horticulture purposes.

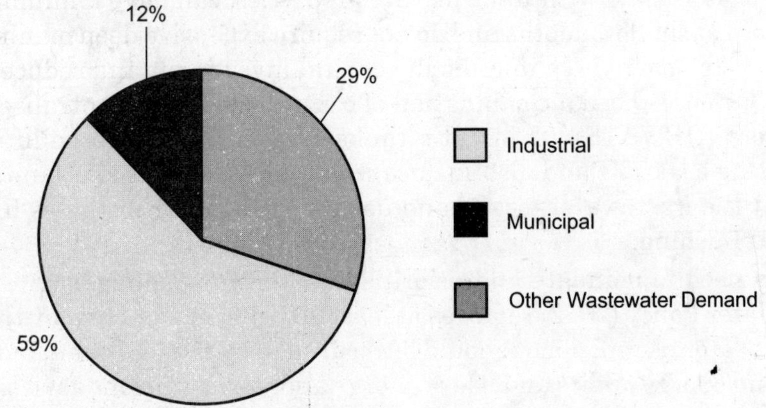

Fig.10.6: Demand for treated Wastewater in India – Share by markets (2009)

Recycled Wastewater, used for horticulture and landscaping does not require secondary or tertiary treatment. Storm water can be used only with pre-treatment process and primary treatment. Such Wastewater can be stored in water bodies or with simple treatment and overflow can be discharged into ground aquifers. Water bodies can be constructed in parks and gardens. Recycled treated water can also be used for flushing or for horticultural purpose.

Recycled Wastewater can also be used as "make up water" for air-conditioning plant, provided it has been treated at higher scale and odour has been removed. Hence tertiary treatment is normally carried out in such Wastewater. Water bodies as storage system require lesser treatment as these have advantage of natural treatment. Recycled water shall have a distinct distribution system, and it is to be ensured that there are no cross connections with potable water system.

Land is extremely difficult to obtain in urban areas for setting up water bodies, landfills and treatment plants. Large water utilities like the Nagpur Municipal Corporation and Delhi Jal Board are exploring Least Lifecycle Cost (LLC) approach in their turnkey management contracts. The LLC approach will reduce requirements of land, energy, manpower and chemicals over the longer term thereby providing savings. New technologies and energy saving equipment that conserve resources are in demand. Some of the new technologies being adopted include Membrane Filtration,

Membrane Bioreactor and Sequential Batch Reactor (SBR). The SBR is capable of producing five times better quality effluent than conventional treatment with 40-50 per cent less land and energy requirement. Wastewater recycling has an important role for industries that will reduce their operating cost with payback period being five years or lesser. This is especially vital for large consumers like thermal power plants and industries like food and beverage, textiles, steel, automobiles and pharmaceutical. Stricter environmental standards, like Zero Liquid Discharge, mandate recycling and reuse of Wastewater.

Indian cities often have centralised wastewater system. The system does not work where per capita water supply is less than 135 lpcd and population density is low. Brazil which faces similar sanitation conditions uses the in-block sewers. These are simplified tertiary sewers that are laid within the private properties, can have minimum diameter of te100 mm, are laid at shallow depths and do not require expensive deep manholes. Such low cost sanitation systems will help in reducing capital investment and reduce maintenance requirements. Our cities need a combination of centralised and decentralised wastewater treatment system (DEWATS). The septic tanks system has led to pollution of ground water from leaking soak pits and also surface water pollution. On-site sanitation systems collect and treat the wastewater locally and use the effluent for purposes like gardening, landscaping and flushing.

The policies need to mandate and prioritise wastewater management and recycling. The National Water Policy (2012) requires at least 20 per cent recycling by the state public water agencies. Water use efficiency should be inculcated by the ULBs and industries which include water and energy audits and use of energy and water efficient devices. Sustainable use of water resources with rain water harvesting and controlled aquifer recharge to replenish groundwater should be accorded priority.

Some of the technical measures in this regard are given below:

Advanced Integrated Wastewater Pond System: Treating the incoming Wastewater through a series of carefully engineered process is called the Advanced Integrated Wastewater pond Systems, (AIWPS) which uses bacterial and oxygen-releasing micro algae to remove parasites, heavy metals, bacterial and viral pathogens from Wastewater. This has been experimented in cleaning of Ganga at Varanasi. Ganges water will meet World Health Organisation standards for the flood crop irrigation water.

While pond systems are generally more efficient and less costly than mechanical treatment facilities, they require more land and more time. For this reason, larger municipalities generally favour mechanical systems. The ponds use the ability of bacteria and algae to decompose and oxidize sewage. It greatly diminished the dependency on electricity to treat the water. Sewage is stored from 20 to 45 days as compared to mechanical facility where sewage is stored only for one day. Methane produced from waste is captured at the site and used to meet the meagre energy requirements of the AIWPS systems.

Removing Arsenic from Drinking water: Arsenic is a heavy toxic metal and ingestion of amounts beyond 0.05 milligram per litre of water can pose a serious health hazard, where aquifers are the main source of drinking water. Currently experiments are being undertaken with a number of arsenic mitigation technologies. The basic methods involve oxidation-coagulation-flocculation sedimentation-filtration.

Bucket Treatment Method: The bucket treatment method is simple and cost effective. Based on the chemical behavior of arsenic in water, the bucket-type chemical treatment was designed to develop a method to remove arsenic from water. The method uses a ratio of chlorine or potassium permanganate and alum to treat ground water for arsenic removal. The water is collected in a bucket and the chemicals are mixed and rigorously stirred. The arsenic-free water is then drained out from the top. Another method is water filter which uses activated alumna in a nylon bag to filter water and enables it free from arsenic.

Assess the offender to pay full corrective costs: Where the hazard of pollution exists, the minimum charge for operating permits should cover the expenses of adequate policing and controls. Mandatory performance bonds and liability insurance should be adequate to pay for all damages plus any corrective measures which might possibly be needed. In cases of violations or incidents of contamination, additional fines of sufficient magnitude to minimize future risks should be levied. The idea is to set the charge high enough to force the polluter to stop.

Prohibit the Dumping of Garbage, Sewage and Other Wastes: This essential mandate is long overdue on both fresh and saltwater bodies. It can be enforced effectively by annual licensing renewal and inspections, as per the requirements of pollution control standards.

Outlaw the Manufacture of Compounds which Damage Water Quality: Prohibition on use alone, being much more difficult to control, is not sufficient. A further prohibition against manufacture for domestic consumption or for export is evidently needed, such as DDT

Preclude Pollution: As a governing rule, no new development, manufacturing, processing, or operation of any polluting activity should be permitted, which may result in the significant degradation of any water resource. Along the cities and coastlines an increasing threat is being posed by the infiltration of Wastewater, sewage and seawater into the aquifers. As the hydrostatic pressure of the landward water table is reduced by drawdown, the normal flow to the oceans is reversed. Preventive techniques include the construction of extensive system of efficient treatment plants, filters, dams and barricades, coupled with the regulation of freshwater consumption within the limits of adequate supply. Common technologies available for cleaning of Wastewater are given below:

Reed Bed Channel

One possible alternative low-cost treatment, hence, could be the 'reed-bed channels', which is an engineered version of constructed wetlands. Constructed wetlands, have been widely recognized as low-cost treatment systems requiring minimum levels of operation and maintenance, for both developed and developing countries. They are particularly suitable for developing countries as an alternative to waste stabilization ponds, since their pathogen removal efficiencies also have been found to be equally effective with flexible design criteria.

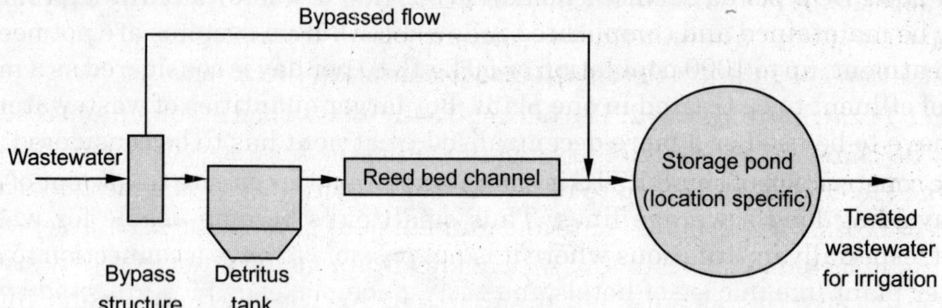

Fig.10.7: Schematic of Proposed Reed Bed System for Wastewater Reuse.

Source: Sundarvadival,M., 2002,Intermidiate **Technologies for Wastewater Management in India, Shelter, Vol.5, No.i. January**

In almost all cities and towns of India, wastewater flowing through long open drains that is either redirected for use in agricultural fields or collected in ponds (meant to store storm water run-off) for later use, is common. By converting these terminal (trunk) drains into gravel media beds supporting commonly found Indian wetland plants such as *Phragmites carca* and *Typha latifolia,* surface flow reed-bed channels for treatment can be developed. During lean flow periods of the day, the beds will function as sub-surface flow reed beds. A detritus tank pre-treatment preceded by storm water overflow structure to bypass the flows in excess of twice the peak dry weather flow are the other requirements. The proposed reed bed channel arrangement for wastewater treatment offers the following advantages:

- treatment will require little or no additional land;
- utilisation of existing drains will reduce system costs;
- due to channel-like feature, there will be no spreading and stagnating wastewater over a large area as in waste stabilization ponds or conventional constructed wetlands;
- running wastewater in the channel will not support mosquito breeding, which has been reported as a public health hazard with waste stabilization ponds or conventional constructed wetlands.

Root Zone Treatment System (RZTS)

Wastewater containing bio-degradable matter is treatable with RZTS after sedimentation of particulate matter. Especially domestic sewage and effluents from agro-based industries are suitable for treatment with RZTS. Wastewater containing not easily bio-degradable substances can be treated in RZTS by appropriate operation of the plant (e.g. prolonging the residence time of the wastewater or changing the conditions in the filter bed between aerobic and anaerobic environment). Storm water can also be treated as well with RZTS after removal of settleable substances. RZTS have to be planned and designed in such a way that they comply with the environmental stipulations, standards and criteria, laid down by the concerned regulatory agencies.

RZT Systems are very suitable for treatment of small quantities of wastewater (say 5 persons or 250 g BOD per day), since a homogeneous flow of wastewater through small RZTS can easily be maintained and complicated inflow distribution systems are not needed. For sewage treatment, up to 1000 population or 50 kg BOD per day is considered as a maximum quantity of effluent to be treated in one plant. For larger quantities of wastewater a series of RZTS have to be used or a more decentralised treatment has to be considered.

As the construction of small RZTS is easy, it is suitable for on-site treatment of domestic sewage, avoiding long sewerage lines. Thus small plots become usable for wastewater treatment, especially in situations where it is not possible to have a connection to a central wastewater plant like in case of hotels and tourist complexes, which are sited away from cities. In case of dispersed settlements, RZTS offers a feasible solution for decentralized sewage treatment.

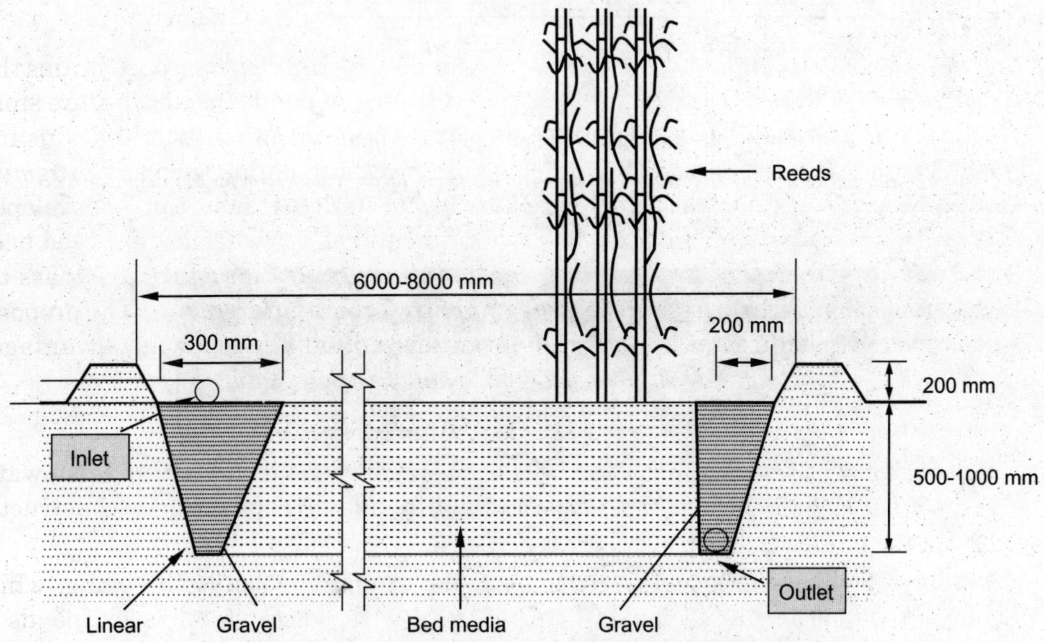

Fig.10.8: Rootzone Treatment Horizontal Flow System

Source CPCB 2001

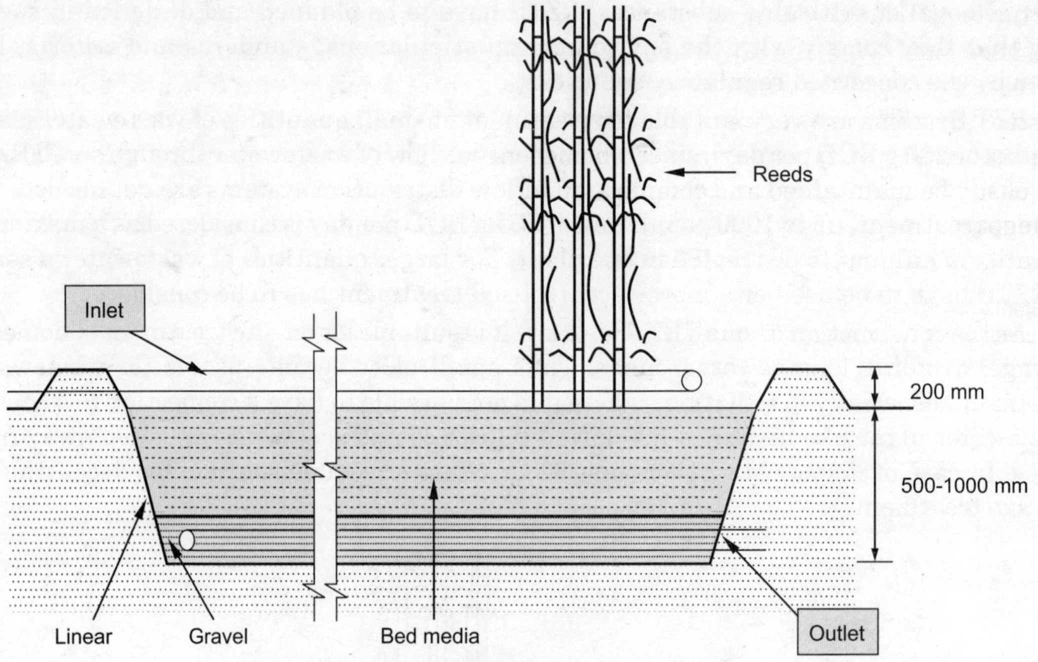

Fig.10.9: Vertical Flow Rootzone Treatment System

Source: *Central Pollution Control Board (2001) Guidelines on Rootzone Treatment System for Municipal and Industrial Wastewater*

Fig.10.10: Floating Wetlands are part of the INTACH's On-channel treatment of Palam Drain; Dwarka, New Delhi. The Project involved treating polluted flows in the Palam drain over 4 km with the techniques of bacterial bioremediation, Daphnia Treatment and Floating Wetlands, which reduced the pollution levels and foul odours.

Source: INTACH

Integrated Water Management includes surface water sources, groundwater aquifers and inter-basin transfer of water. This water mapping, desilting of existing lakes / depressions for augmentation of storage of rain / flood water, reuse and recycling of Wastewater and measures for conservation of water. Major water bodies can be developed as places of tourist interest / picnic spots to ensure their protection. The concept of 'zero run-off drainage', with retention ponds, rain-gardens sediments traps and bio-swale can be adopted, with a segregated wastewater disposal system.

Fig.10.11: A rain garden is a shallow planted depression, which captures rainwater and allows it to seep and move into the ground. This water otherwise flows as surface run-off that causes erosion, flooding and reduced groundwater recharge. A Bio-swale is a channel for storm water planted with water tolerant species.

Table 10.8: Pros and Cons of Decentralized Wastewater Treatment Systems

Type	Kind of treatment	Use for type of wastewater	Advantages	Disadvantages
Septic tank	Sedimentation sludge stabilization	Wastewater of settleable solids, especially domestic	Simple, durable, little space because of being underground	Low treatment efficiency, effluent not odourless
Imhoff tank	Sedimentation sludge stabilization	Wastewater of settleable solids, especially domestic	Durable, little space because of being underground, odourless effluent	Less simple than septic tank, needs very regular desludging
Anaerobic filter	Anaerobic degradation of suspended and dissolved solids	Pre-settled domestic wastewater of narrow COD/ BOD ratio	Simple and fairly durable, high treatment efficiency, little permanent space required because of being underground	Costly in construction because of special filter material, blockage of filter possible, effluent smells slightly despite high treatment efficiency

Baffled Septic tank	Anaerobic degradation of suspended and dissolved solids	Pre-settled domestic wastewater of narrow COD/ BOD ratio,	Simple and durable, high treatment efficiency, less space required because of being underground hardly any blockage, relatively cheap compared to anaerobic filter	Less efficient with weak Wastewater, longer start-up phase than anaerobic filter
Root zone Treatment System	Aerobic facultative-anaerobic degradation of dissolved and fine suspended solids, pathogen removal	Suitable for domestic wasterwater where settleable solids and most suspended solids already removed by pre-treatment	High treatment efficiency, pleasant landscaping possible, no wastewater above ground, no odour	High space requirement, skill and care required during construction, intensive maintenance and supervision during first 1-2 years
Anaerobic pond	Sedimentation anaerobic degradation and sludge stabilization	Domestic and strong and medium waste-water	Simple in construction, flexible in degree of treatment, little maintenance	Wastewater pond occupies open land, there is always some odour, can even be stinky, mosquitoes are difficult to control
Aerobic pond	Aerobic degradation, pathogen removal	Pre-treated domestic wastewater	Simple in construction, reliable in performance high pathogen removal rate, fish farming possible when large in size and low loaded	Large space requirement, mosquitoes and odour can become a nuisance if undersized, algae can raise effluent BOD
Duck-weed Pond	Anaerobic except aerobic at top, Degradation of Suspended and dissolved Solids, Nutrient Removal	Sullage or Pre-treated sewage	Simple in construction, Revenue generation through pisciculture, suitable for rural and semi-rural	High space requirement, possibility of odour can not be ruled out, proper harvesting of duckweed is must.

Source: Central Pollution Control Board, 2001, Decentralised Sewage Treatment System, New Delhi.

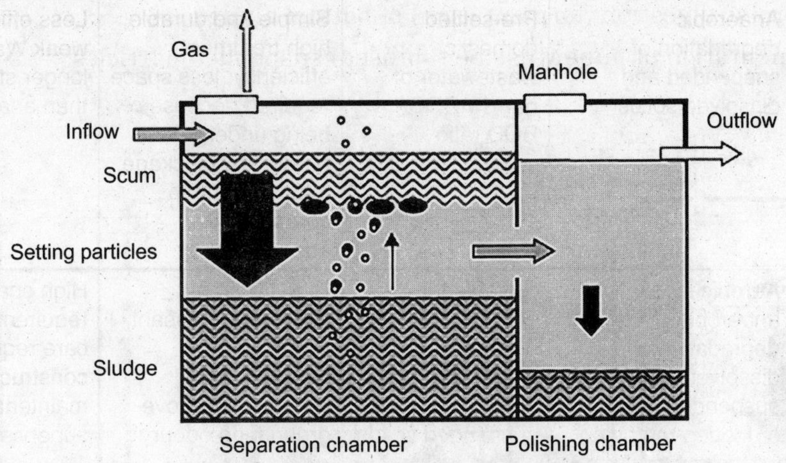

Fig.10.12: Principle of Septic Tank

Source: *Baetens, Tency (2004). Constructed Wetlands for Wastewater Treatment in Tropical and Sub-tropical Regions. Good Governance India, Jan-Feb.*

The septic tank is the most common, small-scale decentralised treatment plant. It is compact and robust in comparison to the cost of its construction and is extremely efficient. It is basically a sedimentation tank in which settled sludge is stabilised by anaerobic digestion. Dissolved and suspended matter leaves the tank more or less untreated.

Two treatment principles, a mechanical treatment by sedimentation and a biological treatment by contact between fresh wastewater and active sludge compete with each other in the septic tank. Optimal sedimentation takes place when the flow is smooth and undisturbed. Biological treatment is optimised by quick and intensive contact between new inflow and old sludge, particularly when the flow is turbulent. The way the new influent flows through the tank decide which treatment effect predominates.

A septic tank consists minimum of 2 or 3 compartments. The first compartment occupies about half the total volume because most of the sludge and scum accumulates here. The following chamber(s) are provided to calm down the turbulent liquid. They are of equal size and occupy the other half of the total volume. Scum accumulates and must be removed regularly every two years. Scum does not harm the treatment process as such, but it does occupy tank volume.

A baffled reactor or baffled septic tank is a design incorporating an initial settling tank followed by at least four and up to ten narrow chambers or compartments, all of them fitted with a baffle wall to force the wastewater in every chamber to flow downwards. The active sludge that is washed out from one chamber is trapped in the next. The compartments placed in series help to digest difficult and slow degradable substances.

The up-flow velocity should never be more than 2 m/h. Equal distribution of inflow and widespread contact between the fresh incoming influent with the active sludge present in the reactor in order to ensure quick inoculation for digestion are important process features.

1. Sedimentation/floatation of solids
2. Anaerobic digestion of suspended and dissolved solids through contact with sludge

3. Anaerobic digestion (fermentation) of bottom sludge
4. Sedimentation of mineralised (stabilised) suspended particles

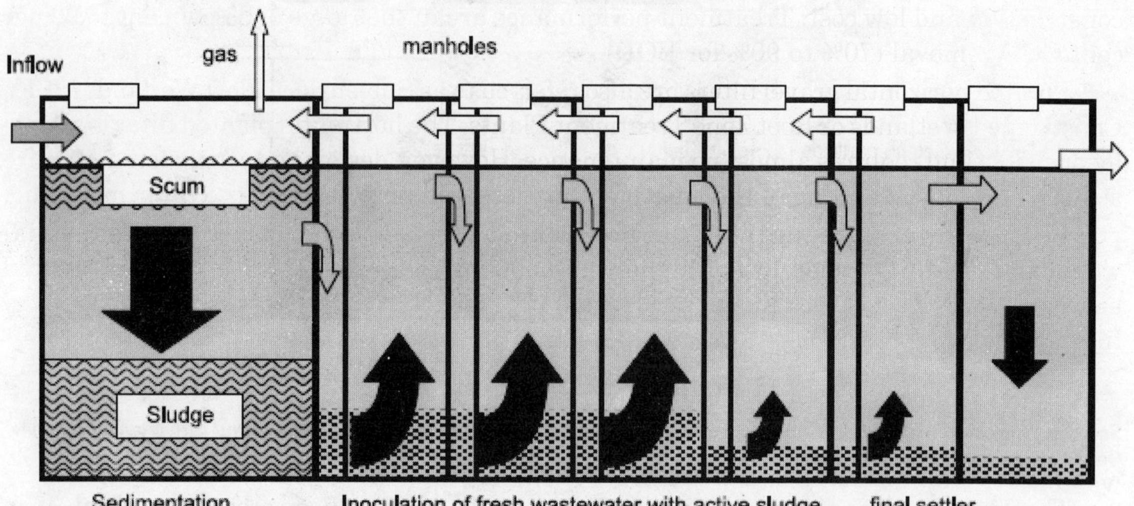

Fig.10.13: Principle of Anaerobic Baffled Reactor

Source: *Baetens Tency, (2000)*

1. Continuous oxygen supply to the upper layers only.
2. Anaerobic conditions in the lower parts of the filter.
3. Roots of plants provide favourable environment for bacteria diversity.

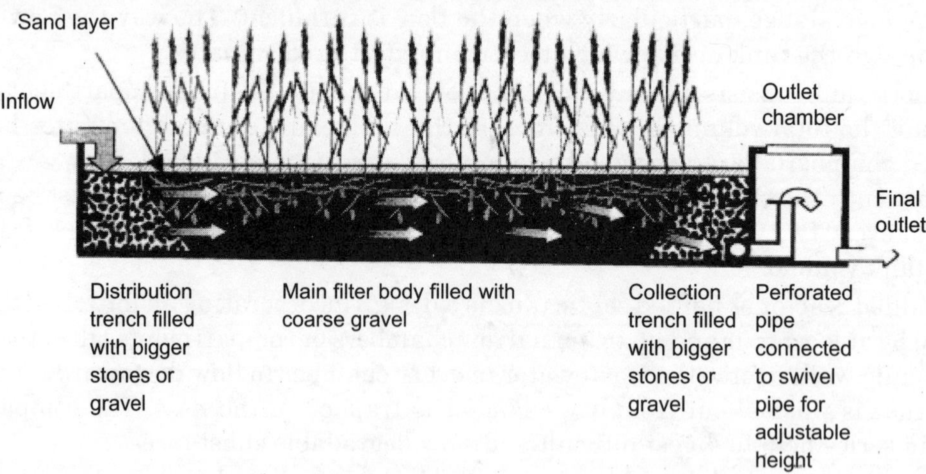

Fig.10.14: Principle of the Horizontal Planted Filter

Source: *Baetens, Tency (2000), Constructed Wetlands for Wastewater Treatment in Tropical and Sub-tropical Regions.*

The baffled reactor is suitable for all kinds of wastewaters, including domestic. Its efficiency increases with higher organic load. It is the high efficiency answer to the low efficiency septic tank, because simple and efficient operation goes along with easy construction and low cost. Treatment performance are in the range of 65 per cent to 90 per cent COD removal (70% to 90% for BOD).

Planted horizontal gravel filters are also referred to as Sub-surface Flow Wetlands (SSF), Constructed Wetlands or Root Zone Treatment Plants. The horizontal planted filter is simple by principle and requires almost no maintenance. However, design and construction require a thorough understanding of the treatment process and knowledge of the filter medium.

Planted filters are suitable for pre-treated (pre-settled) domestic or industrial wastewater of a COD content not higher than 500 mg/L. Wastewater must be pre-treated especially in respect to suspended solids, given the fact that the biggest problem in ground filters is clogging of the filter media.

The filter bed should not be deeper than the depth to which plant roots can grow (30 – 60 cm), as water will tend to flow faster below the dense bed of roots. Shallow filters are more effective compared to deeper beds of the same volume. To prevent percolation of wastewater into the ground, the bottom of the filter must be sealed. While the top part of the filter media in a planted filter is kept horizontal, the constructed bottom slopes down from inlet to outlet by preferably 1 per cent.

Planted filters are covered by suitable plantation, using any plant, which can grow on wastewater and whose roots go deep and spread wide. Phragmites australis (reed) is considered to be the best plant because its roots form horizontal rhizomes that guarantee a perfect root zone filter bed.

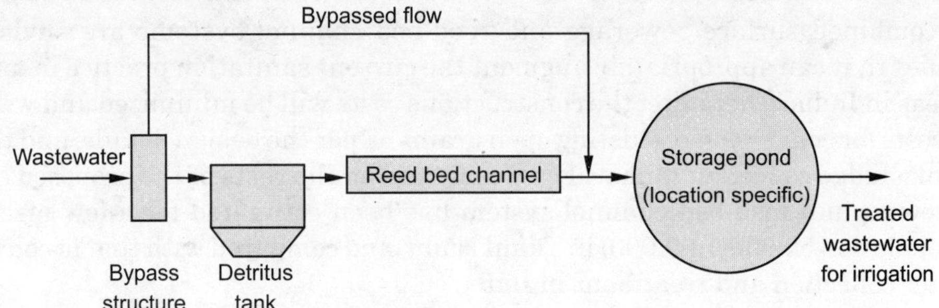

Fig.10.15: Schematic of Proposed Reed Bed System for Wastewater Reuse.

Source: *Sundarvadival M., 2002, Intermediate Technologies for Wastewater Management in India, Shelter, Vol. 5, No. 1, January.*

The elements of the reed bed channel system. It comprises:

- A storm water by-pass structure;
- A detritus tank;
- A reed bed channel; and
- A storage reservoir (location-specific).

Table 10.9: Design Details of Reed-bed Channels

Reed-bed channel:					
Average width	:	2.25 m	Porosity of the gravel bed	:	30-40%
Depth of gravel bed	:	0.6 m	Theoretical hydraulic retention time in the bed	:	2-3 days
– Bottom 122 mm graval layer	:	0.1 m	Slope of gravel bed	:	0.2-0.3%
– Middle 25 mm gravel layer	:	0.2 m	No. of plants	:	2-3 per m^2
– Top 40 mm gravel layer	:	0.3 m	Hydraulic retention time of detritus tank	:	45-60 min.

Source: *Sundarvadival M., 2002, Intermediate Technologies for Wastewater Management in India, Shelter, Vol. 5, No. 1, January.*

The combination of detritus tank and the reed-bed channel with a theoretical hydraulic retention time of 2-3 days can affect at least two log removal (from 105 to 103) or better, of faecal coliforms. With this removal rate, the recommended microbiological quality of wastewater for irrigational reuse can be achieved. Corresponding BOD removal in the reed-bed channels will be in the range of 70-80%, which also conforms the prescribed standards. By planning the combined surface sewer system to comprise two or more independent networks, each ideally draining a population of 10,000-15,000, quantity of wastewater to be treated by the reed-bed channels, and hence the area requirement at any one site can be kept at lower levels. A municipal solid waste composting system can utilize the harvested reeds from the treatment channels as an additional carbon and nutrient source, to produce quality compost. Perhaps, such a 'decentralized', non-stagnating wastewater treatment system would be more acceptable for most of the urban local authorities of India.

The combined surface sewerage and reed-bed channel systems are envisioned as technologies that can appropriately augment the current sanitation practice in majority of urban areas in India. Therefore, the constructions costs will be minimized and will include only the costs for conversion of existing open drains as per the design details, and the cost of septic tank sludge collection tanker. For a comparison, the costs of the proposed combined surface sewers and reed-bed channel system has been estimated for a few medium size towns (Population 50,000-1,00,000) in Tamil Nadu, and compared with that of conventional wastewater collection and treatment in India.

In the Indian cities water supply, sanitation/sewerage, solid waste disposal, drainage,bathing, washing, cooling and industrial activities are often mingling with each other within a building or a home. This makes the Wastewater treatment complex, for which there is a need to adopt an integrated wastewater treatment system, dealing with the various problems in a pragmatic way. This means integrating the primary, secondary and tertiary systems in a composite manner.

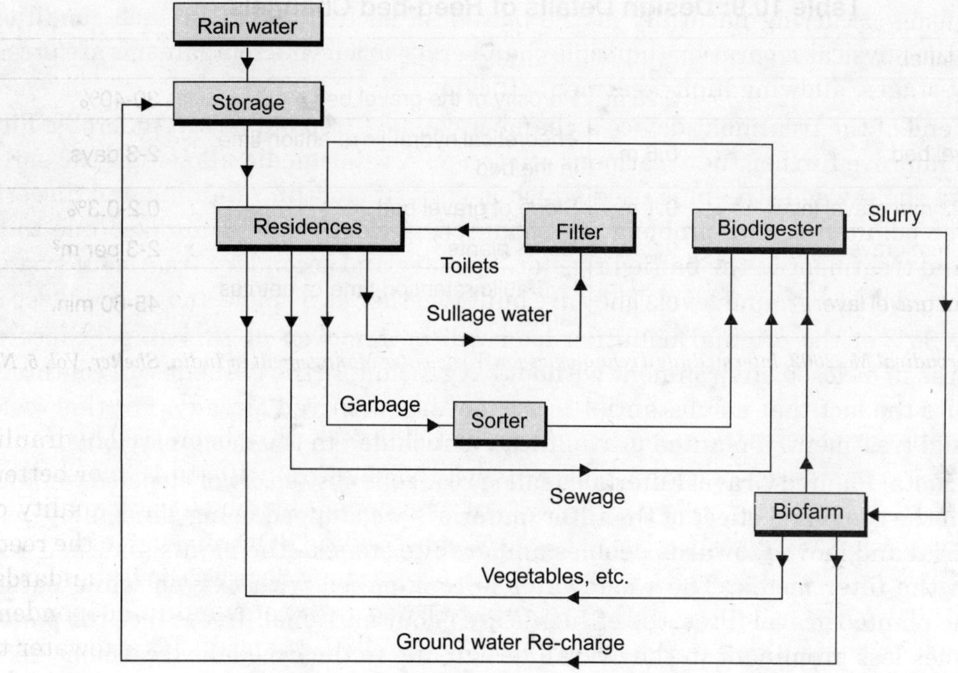

Fig.10.16: Integrated Water and Wastewater Treatment

Natural sewage treatment processes are achieved by using methods that utilize the natural occurring physical principles combined with biological activities of microorganisms. Microbes used in the treatment facility are generated from microbial populations that occur and grow naturally in the wastewater itself. Decentralized wastewater treatment systems (Dewats) are based on different natural treatment techniques, put together in different combinations according to the needs, the possibilities, the challenges and the financial implications. The different components cover the primary, secondary and tertiary treatment of wastewater.

Pre-treatment is used for screening and sedimentation process, in which the liquid part is separated from the solid matter. A device called a Septic tanks used for this phase. The septic tank is the most common, small-scale and decentralized treatment plant. It is basically a sedimentation tank in which settled sludge is stabilized by anaerobic digestion. Dissolved and suspended matter leaves the tank more or less untreated. The treatment efficiency of a septic tank is in the range of 30% COD removal. Desludging at regular intervals is absolutely necessary. A septic tank can also be incorporated into an anaerobic baffled tank as the first section as a Settler device.

In the secondary treatment phase, biological and natural chemical processes are used to digest and remove most of the organic matter A device called an Anaerobic Baffled Tank Reactors used for this phase. Several tanks (up-flow chambers) are constructed in series to digest degradable substances. Baffle walls direct the water stream between the chambers from top to bottom and up again. During the process the fresh influent is mixed and inoculated for digestion with the active blanket deposit of suspended particles and microorganisms occurring naturally at the bottom of each chamber in such conditions. Because of the physical separation (multiple chambers), various microorganisms are present at different stages, allowing high treatment efficiency.

At the end of the treatment device a chamber can be fitted out as an Anaerobic Filter in order to improve further the treatment efficiency. A filter media allowing wide spread contact with the sewage stream is used which is very efficient in retaining and digesting the left over pollutants. The problem of encountering clogging is minimized due to the digestion and treatment in the baffled reactor. Primary and secondary treatment systems are constructed below ground level. They are built together as a single and compact set-up.

Nearly 90% of the original pollution load will be removed at this stage. Since the system works in a closed environment without oxygen supply the effluent will continue to smell despite the fact that a substantial treatment has occurred already. For this reason an additional treatment, a planted gravel filter, is included in the design lay out.

A Horizontal Planted Gravel Filter(also called root-zone system or constructed wetlands) act through the combined effect of the filter material - we stopped using sand as the main filter material and moved towards pebbles and granite stones - the plants and their roots growing in the filter media. The wastewater is re supplied with oxygen while passing through the planted gravel filter; the effluents are odour and smell-free. Since the planted filter becomes less prominent in the overall design due to the excellent treatment taking place in the baffled tank reactor, this downgrading of the planted filter results in drastic cost reduction, less needed space above ground and with an additional benefit of having available larger amounts of reusable treated Wastewater, due to the decrease in the high rate of evapotranspiration normally occurring in large planted filters.

Optional Phase

The treated sewage becomes "living water again" by undergoing further biological treatment, through natural UV (ultraviolet) exposure and flowing through an open water body, a Polishing Pond. In order to function properly the pond is fitted with fish who control the mosquito larvae together with frogs and dragonflies. Different aquatic plants, water hyacinths, water lilies, lotuses, add to the overall beauty and pleasing view.

While the *E. coli* count (the contamination due to human excreta) at the outflow of a planted filter is often around 2000, the count at the outflow of a polishing pond can be well below 20. At this stage the recycled wastewater can be reused without posing any threat to humans. It is valuable for irrigation; the water is high in nutrient contents and beneficial to plant growth.

Effective Micro-organism (EM)

Micro-organisms occurring naturally in the sewage may not necessarily be the most effective and efficient microbial communities to act on the sewage during the treatment process.

The introduction of specific microbial cultures as a method for optimizing the treatment in order to increase the efficiency and effectiveness of the process is the basis for introducing EM technology in wastewater treatment facilities. Stock solutions are cheap, have to be diluted before being used in the process and are available in India.

An Integrated Treatment System

Natural sewage treatment integrates all the above mentioned devices. Both black (toilets) and grey water (sinks, bathrooms, kitchens) is used for the treatment process. Separating the different generated wastewater streams is a more sophisticated way and is being practised by industrial users. A settler, a baffled tank reactor and an anaerobic filter are integrated into a single structure, build underground. The device takes care of the primary and secondary treatment phase. The treatment efficiency is 90%. The Pollution Control Board's standards are met and the wastewater can be safely used for percolation into the soil, if required. Due to the anaerobic process the wastewater still smells when it flows out of the baffled tank reactor and anaerobic filter. If the effluent is reused, a planted filter is added which will take the odour away. With an efficient pre-treatment, the size of the planted filter can be reduced to 0.30 m^2 per person. Granite stones can be used as filter material and even plant species like cannas (Cana Indica) are able to grow in planted filters. The polishing tank is the most visible feature of the design outlay; the treated water can be pumped out and brought back into the cycle without any harm.

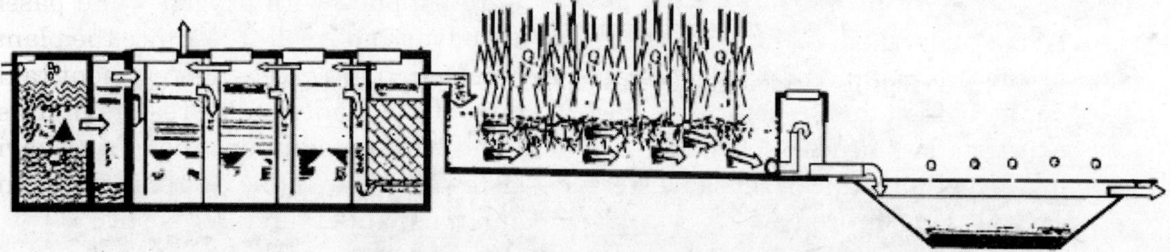

Fig.10.17: An Integrated Treatment System

Source: Baetens, Tancy, 2000

WASTE STABILISATION PONDS (WSP)

WSP systems normally comprise one or more series of anaerobic, facultative and maturation ponds. Anaerobic ponds are 2-5 m deep and receive such a high organic loading that they are devoid of oxygen. BOD removals are very high indeed: over 70 per cent at a retention time of only 1 day at 25°C.

Facultative ponds represent secondary biological treatment, but the oxygen needed by the heterotrophic bacteria is supplied through photosynthesis of the pond algae, which grow naturally and profusely in the pond, giving it its characteristic green colour. Recent research has shown that a facultative pond with a retention time of only 3-5 days at 25°C an reduce the filtered BOD to well below the 25 mg/L EU requirement for WSP effluents. (Filtered BOD excludes the BOD due to the algae.) The effluent is suitable also for restricted irrigation, provided the raw wastewater contains no more than a few hundred human intestinal nematode eggs per litre. These are sediment out in the anaerobic and facultative ponds, so that the resulting effluent contains less than 1 egg per litre—the WHO guidelines value for restricted irrigation.

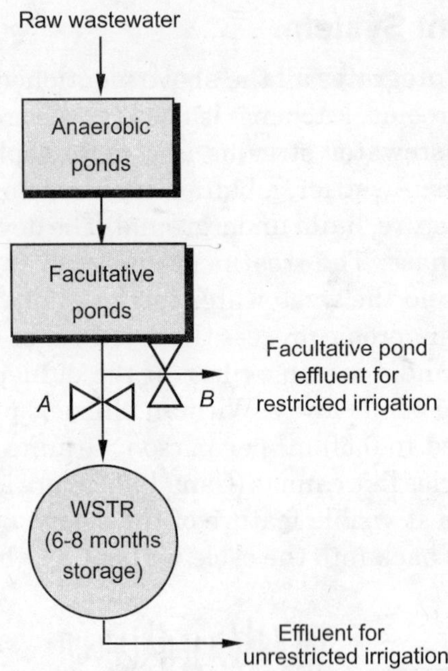

Fig.10.18: Irrigation Supplies from WSP-WSTR Hybrid System

Source: DFID

Fig.10.19: Waste Stabilisation Ponds Serving the Fortateza Industrial Area, North-East Brazil.

Maturation Ponds are primarily used to ensure the removal of fecal bacteria and viruses to safe levels so that the effluent can be used without risks to public health for crop irrigation and/or fish culture. The DFID-sponsored research in north-east Brazil has shown that bacterial pathogens such as *Vibrio cholerae, Salmonella* spp. and *Compylabacter* spp, can be completely removed in a series of WSP, and rotavirus numbers reduced to very low levels (around 3 per 10 litres).

Wastewater Storage and Treatment Reservoirs (WSTR)

In arid areas, agricultural production is limited by the amount of water, including treated wastewater, available for crop irrigation. WSTR are useful as they permit the whole year's treated wastewater to be used for irrigation, so allowing a greater area to be irrigated in the irrigation season and consequently a greater quality of food to be produced. In contrast WSP effluents can only be used during the irrigation season.

WSP-WSTR Hybrid System

The DFID-sponsored research on WSTR has shown that these reservoirs are extremely efficient in removing both fecal coliform and nematode egg numbers to the WHO guideline levels. With pre-treatment of the wastewater in an anaerobic pond a WSTR when filled with anaerobic pond effluent requires only 30 days to achieve the WHO guideline quality for unrestricted irrigation. Based on this work, and recognising that local farmers need a mix of irrigation water quality (some for restricted irrigation, some for unrestricted irrigation), the "hybrid" WSP-WSTR system shown in the figure was developed. Assuming 396 Water : A Manual for Engineers, Architects, Planners and Managers that the irrigation season lasts, for example, six months, the facultative pond effluent is used to fill the WSTR in the six months of the non-irrigation season. During the irrigation season it is used for restricted irrigation, and the WSTR contents are then also used for unrestricted irrigation.

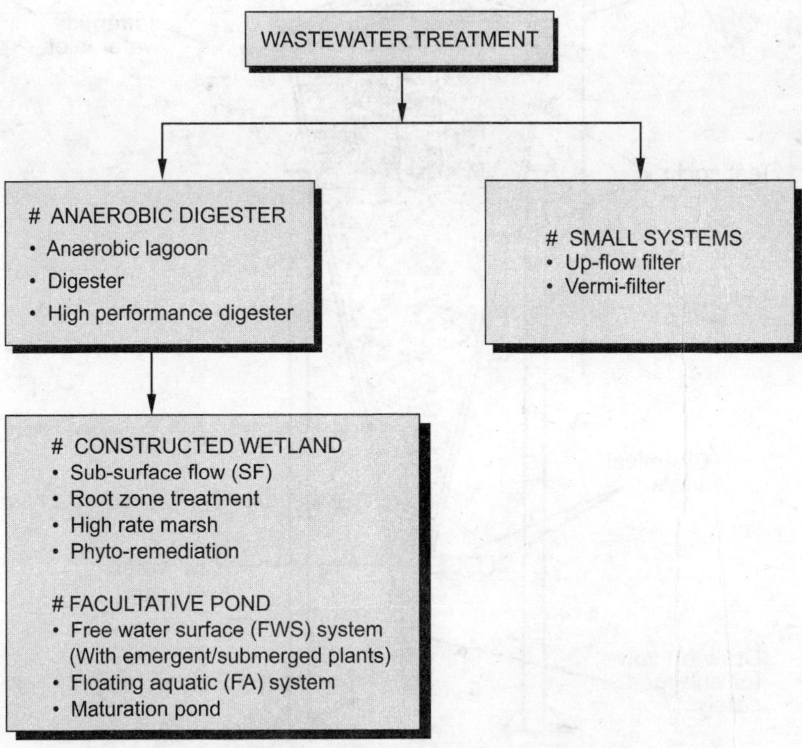

Fig.10.20: Decentralised Options for Wastewater Treatment.

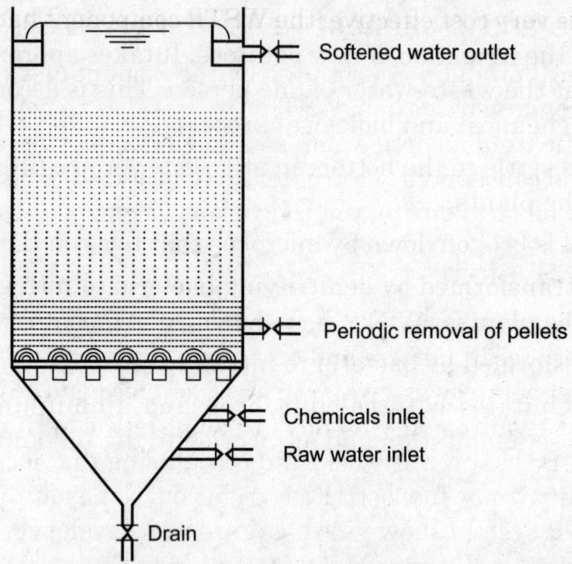

Fig.10.21: Pellet Reactor for Softening by Crystallisation.

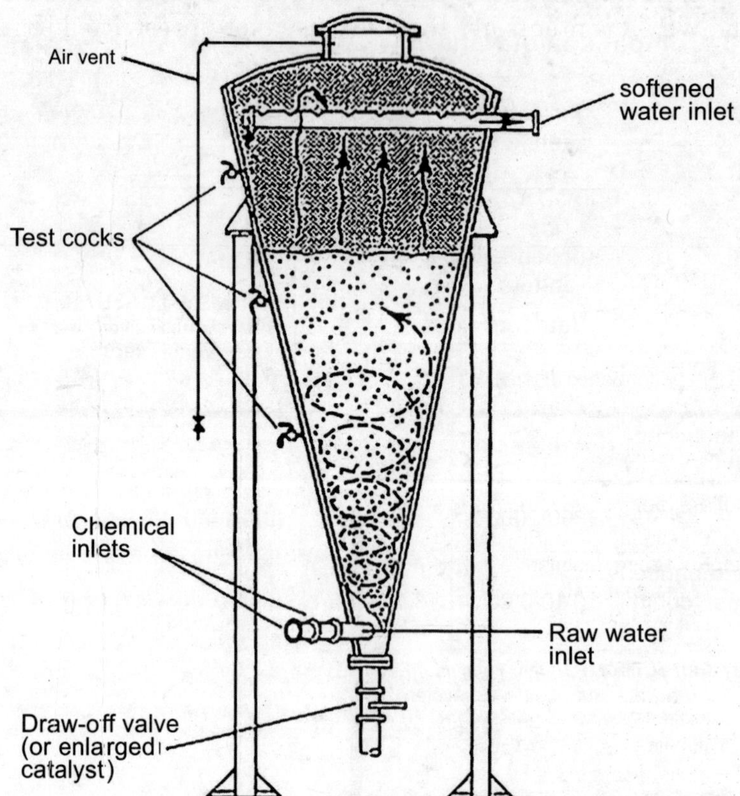

Fig.10.22: The 'Spiractor' Pellet Softener.

Source: Jan Davis, et.al (1995)

Hybrid WSP-WSTR is very cost effective; the WSTR component had an internal rate of return of around 20% for the irrigation of new potatoes. It takes approximately 1-2 square metres of wetland to treat the waste-water of one person. This is accomplished through a combination of physical, chemical and biological processes:

- Suspended solids settle to the bottom in still water or are filtered by the wetland substrates and the plants
- Organic material is broken down by microbes that lives on the roots of the plants;
- Nitrates can be transformed by denitrifying bacteria to nitrogen gas, or they can be taken up by the plants;
- Ammonia is transformed by bacteria to nitrates;
- Phosphorus precipitates with calcium, iron and aluminium compounds; it is removed both by sedimentation and adsorption to the soil and by plant uptake;
- Metals and toxic chemicals are removed by oxidation, precipitation and plant uptake;
- Pathogens die off in the inhospitable environment and are ingested by other organisms, or are killed off by antibacterial compounds (although wildlife in the wetland may introduce more).

Table. 10.10: Comparison of Natural Wastewater Treatment Technologies

	Slow-rate treatment	Rapid infiltration	Overland flow	Floating aquatic plant	Const-ructed wetland
Treatment goals	BOD TSS (total) suspended solids)	BOD TSS	BOD TSS	BOD TSS	BOD TSS
	Nutrients	—	Nutrients	Nutrients	Nutrients
Hydraulic loading (centimetres per day)	0.15-1.6	1.6-25	1-10	2-15	0.4-20
Area (hectares per thousand cubic metres per day)	6-67	0.4-6	1-10	0.7-5	0.5-20
Capital costs (US dollars per cubic metre per day)	800-2000	450-900	600-1000	500-1000	500-1000
Operation and maintenance costs (US dollars per cubic metre)	0.10-0.20	0.05-0.10	0.08-0.15	0.12-0.14	0.03-0.09

Source: Kadlee, Robert H. and Robet I, knight , 1996 Treatment Wetlands, Lewis Publishers, Boca Raton, Florida

Table. 10.11: Commonly Used Plants and Their Properties

***Phragmites australis* (common reeds) and Typha spp. (cattails):** • Flocculate colloids, eliminate pathogens ***Schoenoplectus* spp. (bulrushes):** • Take up copper, cobalt, nickel, manganese, chlorinated hydrocarbons • Eliminate pathogens • Exude antibiotics ***Scirpus* spp. (grasses):** • Break down phenols • Eliminate pathogens ***Juncus* spp. (rushes):** • Treat chlorinated hydrocarbons, cyanide compounds, phenols • Remove pathogens ***Iris psendocorus* (yellow flag):** • Remove pathogens.

Adapted from Mollison, 1988.

Zero Liquid Discharge

The 'zero liquid discharge' is a concept when the entire industrial and municipal Wastewater can be reused after recycling without discharging into the river. At present, after being treated in sewage treatment plants and common effluent treatment plants, the municipal/ industrial Wastewater goes into drains and eventually gets discharged into river. Bio-radiation process is also being adopted to clean the sewerage infested river water.

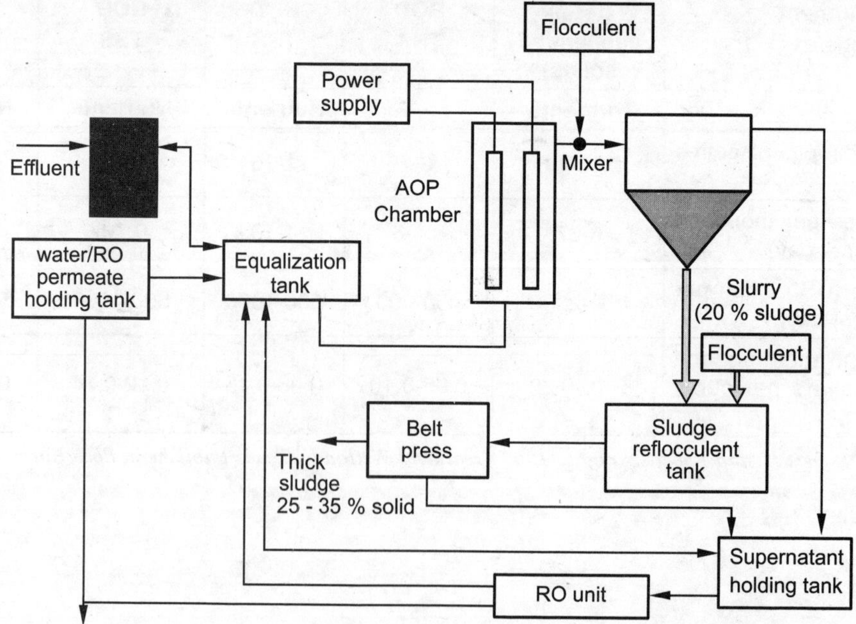

Fig.10.23: Schematic diagram of typical Zero Liquid Discharge ETP

The principle of zero discharge entails recycling of industrial wastewater, whereby wastewater is treated and reused in the process. Because of the water reuse, wastewater will not be released on the sewer system or surface water. The concept of zero discharge system is to ensure no discharge of pollutants into the, environment, and recovery of water.

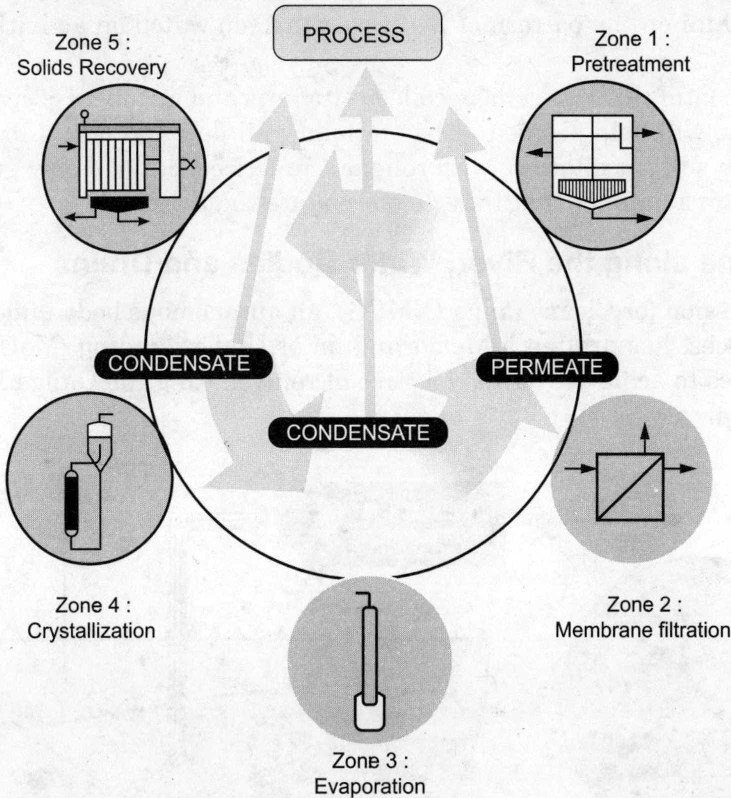

Fig.10.24: Zero Liquid Discharge (ZLD) represents the ultimate cutting-edge treatment system for the total elimination of wastewater effluent into neighbouring waterways.

Source: *Infilco Degremont Inc. (IDl), and GEA Process Engineering Inc.*

Wastewater is collected in the equalisation pond and passed through series of process equipment. Harmful organic/colloidal contents in the water are separated out in the first phase of treatment and the inorganic content in the water are separated in the second part of the process. Ultra filtration technology is used to remove harmful suspended matter in the water, which can cause damage to reverse osmosis membrane. A series of reverse osmosis units are used to recover water from the wastewater. The product water is better in terms of quality than the raw water to the plant.

The water recovered in the process is 90-95 per cent of the inlet water to the plant. Remaining 5-10 per cent wastewater is treated further for evaporation or by converting it to solid waste which can be used for land filling. Forbes Pro Water Projects use the technologies such as membrane bio-reactor to recycle the sewage and industrial Wastewater. Broadly three options are available:

1. Solar pond evaporation
2. Mechanical evaporation using solar pond
3. Thermal evaporation

Besides launching this project, the government will also set up a national volunteer force- Ganga Vahini-on the pattern of Red Cross to keep watch on activities that pollute the river.

Volunteers, mainly ex-servicemen, college students and members of civil societies and Territorial Army, will help people/tourists to follow all do's and don'ts to keep the river clean. Volunteers will not interfere with religious practices, but will only guide the people to practice them in a manner that they do not pollute the river.

Litter-free Zone along the River, Water Bodies and Drains

The National Mission for Clean Ganga (NMCG), an autonomous body under the Ministry of Water Resources, has drafted a Memorandum of Understanding (MoU) with the 118 urban local bodies to achieve overall objective of rejuvenating the Ganga River, with the following action areas:

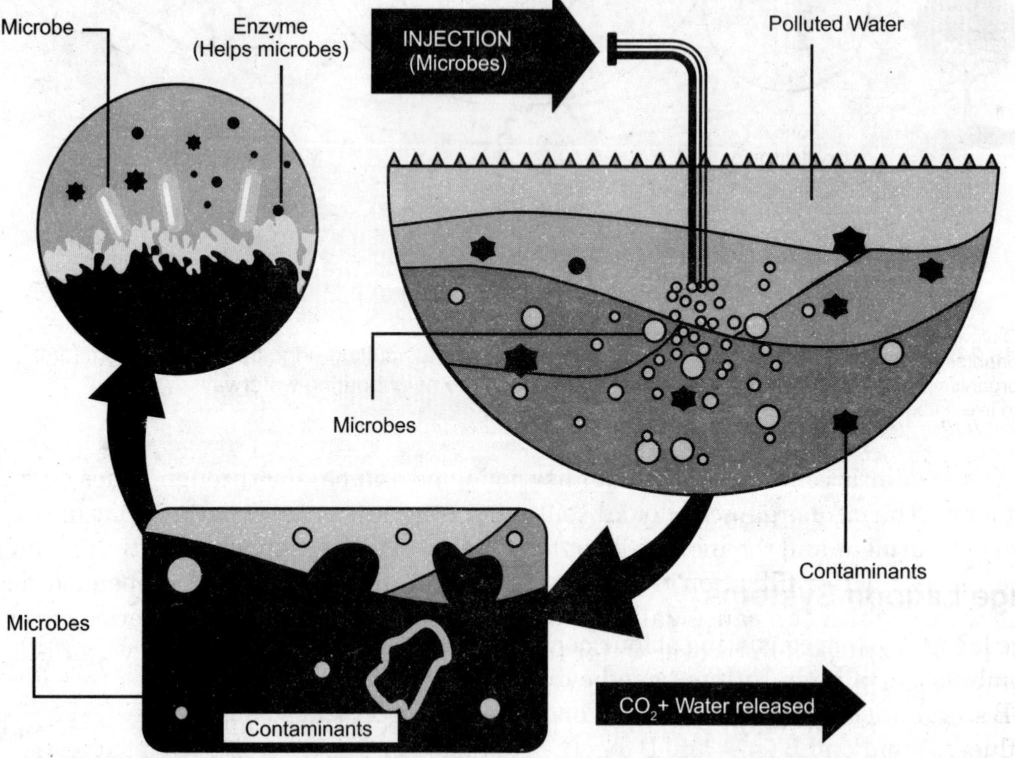

Fig.10.25: Under the bio-remediation process, bacteria are cultured in bulk and applied to the flowing sewage. Then, the microbes are activated and they multiply—with or without oxygen and food available in the form of organic matter present in sewage. Sometimes, enzymes are also added to activate the microbes.

- To maintain 500 m distance from the Ganga and connect drains as a 'litter-free zone'.
- Provide livelihood opportunities to umemployed youths to monitor litter-free zone and health of the river.
- Formation of Ganga Project Implementation Units (PIU) in various towns to enable implementation of the 'Clean Ganga' programme at town level.
- Systematic solid waste management near ghats and confluence of drains into river.
- Public outreach activities.

Natural Inversion of Pond Water

Clean-Flo's non-toxic lake, river and reservoir restoration process employs a combination of mechanical, microbial and non-toxic chemical methods. The continuous laminar flow inversion/oxygenation system, a non-toxic phosphate precipitant, water clarifier and microbes feed on organic sediments, oils, pesticides, human sewage, phosphorus and nitrogen.

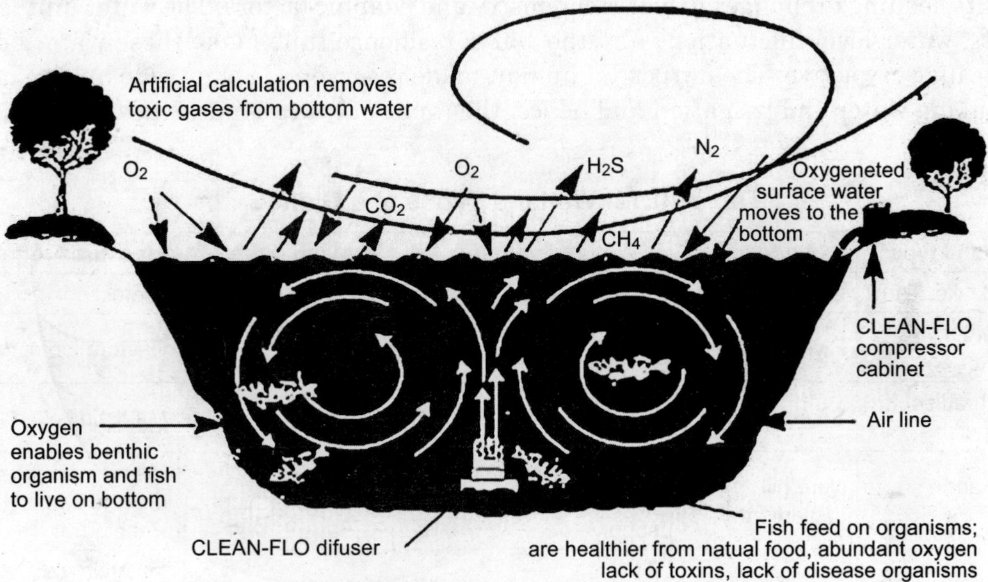

Fig.10.26: CLEAN-FLO Natural Inversion of Pond Water

Sewage Lagoon Systems

Sewage lagoon system consisting of four ponds which can be used either in series or parallel or a combination of both. Influent can be directed to flow into Pond A alone or into Ponds A and B simultaneously. Effluent from Pond A can be directed into Pond B if used in series and effluent from Pond B (or A and B together) into Ponds C and D together or successively. If treated paralleled, final effluent will be discharged from Ponds C and D. If treated in series, final effluent will be discharged from Pond D. In addition to normal valves and manholes control valves are necessary for desired flow direction. Aquatic plant treatment system depends on plant and microbial action. Aerobic bacteria in the vertical flow system nitrify ammonia to nitrate, and bacteria adapted to the anoxic conditions in the horizontal flow system denitrify nitrates to nitrites and nitrogen gas.

Wetland and Water Bodies

Swamps and marshes are often viewed as wastelands and generally filled or reclaimed to make way to increasing demand on land for other uses. Coastal and inland wet lands play a vital role in flood control, their plant species filter water, remove sediments and pollutants and yield fish and crustaceans. They are particularly valuable as repositories of many unique varieties of flora and fauna. Above all, wetlands are priceless source of food and fuel for millions of people. But as with other natural systems, India's wetlands, covering 3150 sq. km, are in imminent danger of extinction as a consequence of thoughtless and unplanned development in addition to population pressure.

On the other hand, light penetration and supply of oxygen are both adequate in shallow lakes for the development of complex food chains. These constantly react with biotic elements, leading to changes in water chemistry and edaphic factors. The water inputs and outputs, water level fluctuations and the water residence time make these water bodies biologically very active. The purpose of human-made reservoirs is to provide human needs for drinking water and irrigation and hence, the input-output regime is geared to human needs.

Table 10.12: Wetland Types at a Glance

Wetland type	Location	Wetland values	Wetland uses	Measures required
Glacial lake, Tarn	Glacier	River recharge	Conservation	Protection
Headwater wetlands	River source	High biological wildlife value	Maintenance of springs	Restoration of forest
Steep gradient river course	Source region	High biological value, flora and fauna	Fish spawning	Maintain flow, forest
Ponds and springs	Near hill settlement	High biological value, wetlands, drinking supply	Drinking water supply	Maintain springs prevent bathing, washing
Village pond	Below hill slope	Storage, percolation	Water supply, regeneration	Maintain catchment banks, prevent salutation
Barrages/ dams in hills	Valley between two hills	Storage	Water supply hydroelectricity	Maintain catchment prevent bathing, washing
Percolation tank	On gentle slopes	Groundwater recharge	Supply to wells	Maintain natural habitats, reduce siltation
Adjacent river basin	Level land below hills	Groundwater recharge, hyportheic zone maintenance	Fish, crabs river rine vegetation	Maintain flow, banks, buffers and habitats

Floodplain wetlands	In the floodplain	Groundwater recharge, flood	Fish, tourism	Prevent situation pollution, maintain habitats
Human-made reservoirs	In the plains	Storage, fish and wildlife value	Drinking water irrigation and hydroelectricity tourism	Prevent siltation and pollution, maintain habitats
River basin between two reservoirs	Plains	Groundwater recharge, maintain water quality	Biodiversity, aesthetics health requirement	Keep buffers and greenways, filters and natural habitats
Canal-side seepage	Plains	Marsh values, if property managed	Fish, crabs aquatic vegetation	Prevent pollution
Canal-fed reservóirs	Canal system	Storage, fish and wildlife value	Fish, thatch production	Prevent siltation and pollution
Deltaic wetlands	River deltas	All wetlands values	Water supply, flood and sea ingress control	Maintain quality of water habitats
Lagoons	Near the seashore	All wetlands values	Water supply flood and sea ingress control	Maintain quality of water habitats
Creeks mangroves	Near the seashore	High biological value high productivity	Fish, marine animals, fuelwood timber, fodder	Prevent pollution streamline shipping
Near-shore shallow marine areas	Near the sea	High biological value, high productivity	Fish, marine life	Maintain buffers follow coastal Zone (CRZ) rules strictly

Table 10.13 : Wetlands Beyond Wildlife Habitat: Functions, Products and Attributes Associated with Wetlands

Functions	Products	Attributes
Groundwater recharge	Agricultural crops	Active recreation
Groundwater discharge	Grazing resources	Passive recreation
Flood storage and desynchronisation	Fish	Roles in global chemical cycles
Shoreline anchoring and dissipation of erosive forces	Fuel	Maintainace of biodiversity
Sediment trapping nutrient retention and removed	Building materials	Microclimate stabilisation

Reeded Wastewater treatment is an organic method of Wastewater recycling for which parkland areas could be explored. Approximately 1-2 sm of reed bed area per person equivalent is required for correct operation. Important factors in the choice of reed bed size and type are flow rate, organic loading and the required quality of treated effluent.

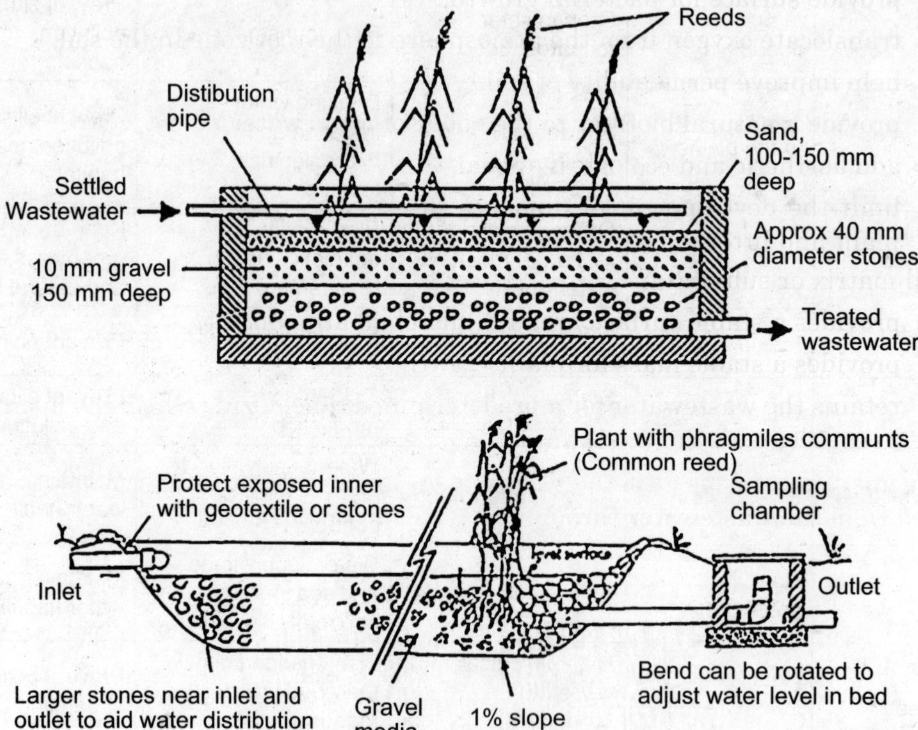

Fig.10.27: Vertical (a) and Horizontal (b) Reed-bed Waste-treatment Systems.

The topography of the site will also determine the type of reed bed to be used. Vertical reed beds require a fall along their length, and these are best sited on sloping ground to avoid pumping. A final discharge for the odourless, high-quality effluent is also required,and this could be a willow patch, a canal, a lake, or even a water feature.

A constructed wetland consist of the following basic features:

(a) A shallow bed of vary porous soil or gravel, usually 0.6 to 1.0 m deep, laid over a water proof membrane in an excavated trench. The soil may be gravel or sand or a mixture of sand and soil. This is called the substrata.

(b) An group of suitable plants, called the macrophytes or halophytes, planted over the soil.

(c) An inlet arrangement to distribute the influent uniformly across the bed.

(d) An outlet arrangement to collect the treated effluent from the wetland system.

Wastewater treatment through constructed wetland is accomplished through an integrated combination of biological, physical and chemical interactions between the plants,the substrata and the inherent microbial community. The interactions occur in the area around the roots of the plants and as such the system is called Root-Zone Treatment System (RZTS).The plants in the wastewater treatment:

(a) provide surface for bacterial growth,
(b) translocate oxygen from the atmosphere to the root-zone in the soil,
(c) help improve permeability of soil,
(d) provide a natural biofilter to the odour of wastewater,
(e) add aesthetic and ecological appeal,
(f) limit the development of nuisance insects like mosquitoes through the surface plant and litter mass.

The soil matrix or substrata:

(a) provides a stable surface area for microbial attachment,
(b) provides a stable mass for plant growth,
(c) retains the wastewater for a predetermined time during which the microbes come in contact with the wastewater and the oxygen.

The micro-organisms form the basic physiological mechanism by which pollutants are removed from the wastewater through bacterial metabolism.

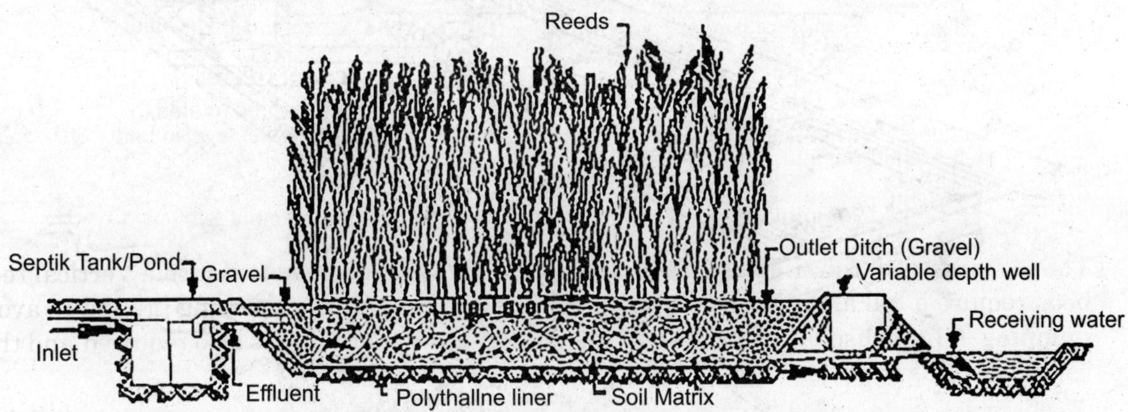

Fig.10.28: Section through a Constructed Wetland

The application of the hydrological landscape structure in the planning of a network of nature areas may result in two alternative strategies:

- The first strategy calls for an adaptation of the layout and function of land-usetypes to the units within the hydrological landscape structure. Negative influences via ground or surface water flows are avoided this way. Potentially polluting types of land use such as intensive agriculture should for instance not be located upstream of nature areas, but (relocated to those parts of the hydrological landscape structure that are dominated by local networks. This will limit the negative effects of these land-use types on water systems, both in time and space.
- A second strategy involves a direct intervention in the hydrological system, to create better conditions for the development of wetlands within the context of present land use. This strategy is applied in cases where the reallocation of land use is

considered impossible. It requires the isolation of wetlands from their surroundings by, for instance, the reconstruction or relocation of watercourses to alignments that will help prevent pollutants from upstream areas entering wetlands. The layout of buffer zones around wetlands, which help absorb differences in water tables, or strips of vegetation in which nutrients are filtered out, may also be applied.

The choice between one of the two alternative strategies depends on existing hydrological conditions and land-use types in the area concerned. In cases where natural land use is dominant, the first strategy may be adopted. In other cases, the second strategy may be preferred, sometimes in combination with the first.

The integration of urban surface water retention and drainage in the context of the landscape can be an important strategy to conservating groundwater and wetlands.

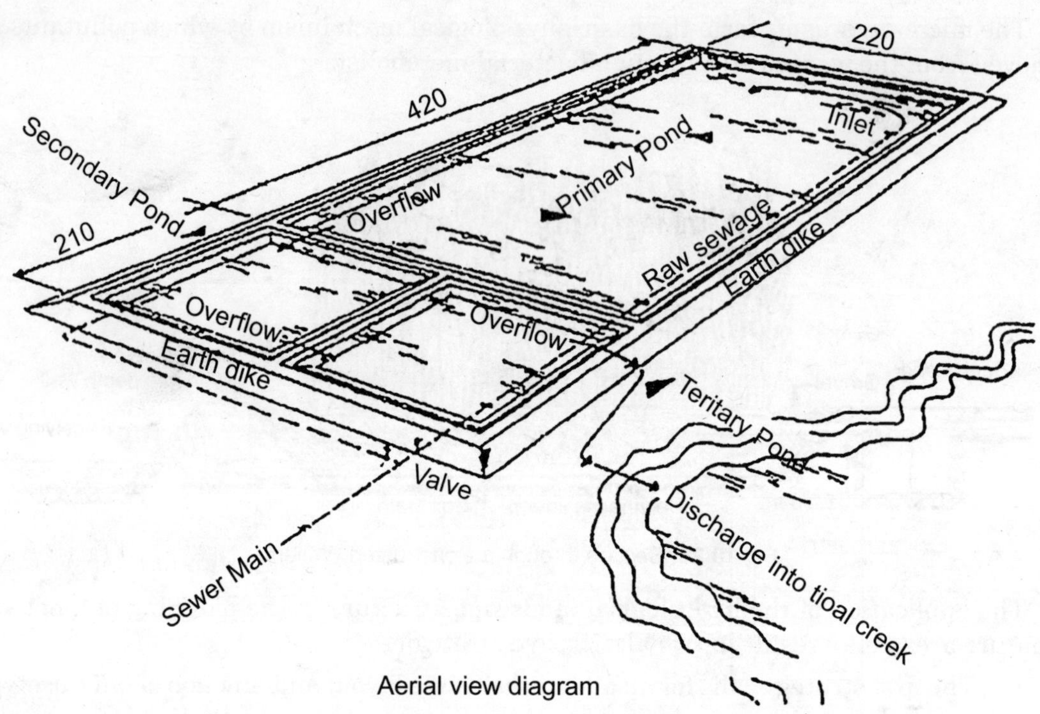

Fig.10.29: Sewage Stabilization Pond

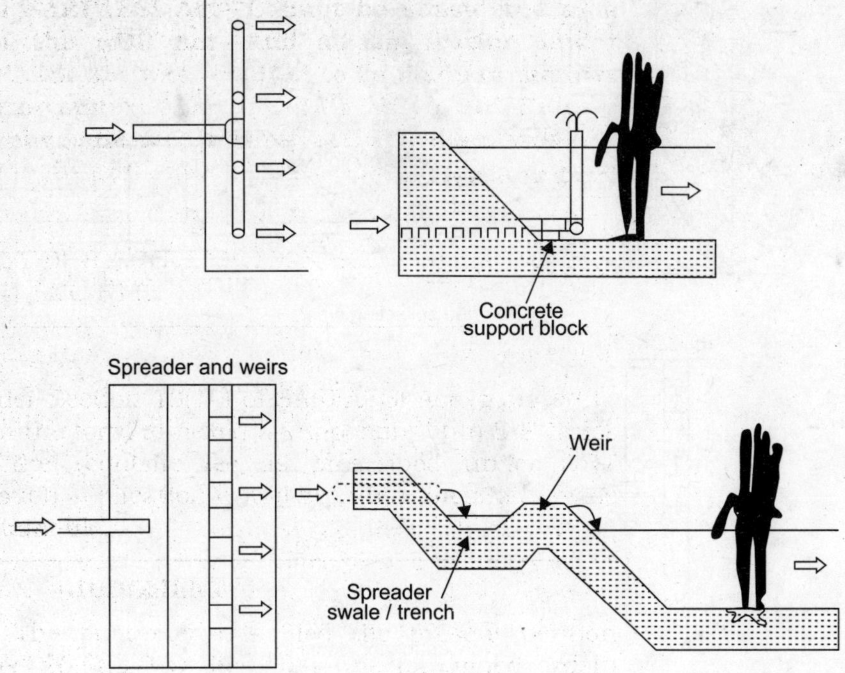

Fig.10.30: Spreader Swale/Trench

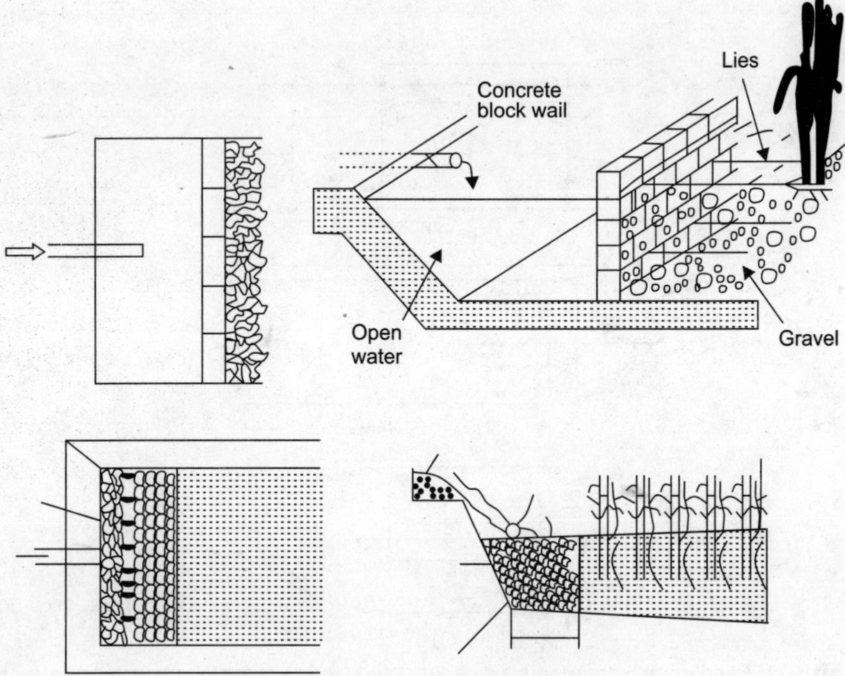

Fig.10.31: Inlet Distribution System

Source: *CPCB*

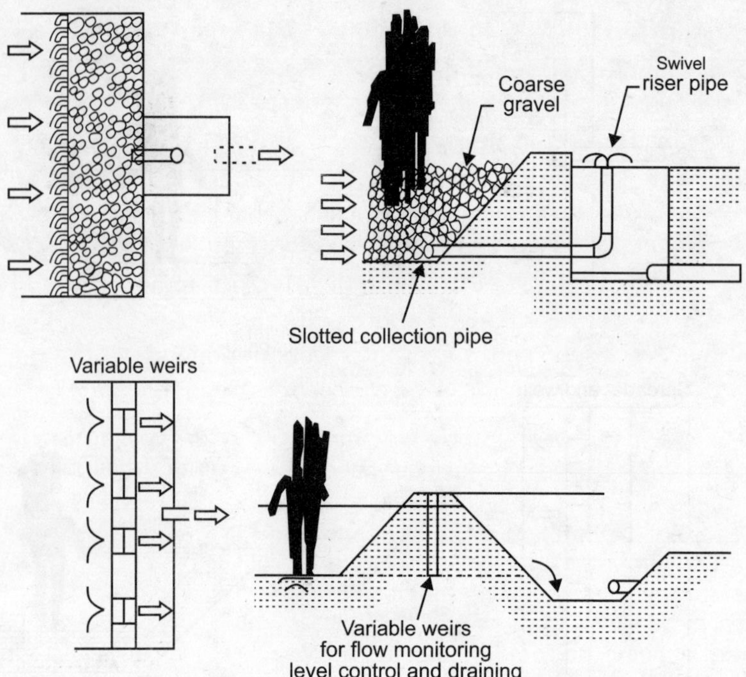

Fig.10.32: Simple Outlet Structure The outlet structures collect the effluent water without creating new dead zones in the wetland; control the depth of water in the wetland; assist in prevention of clogging and provide access for sampling and flow monitoring.

Source: *CPCB*

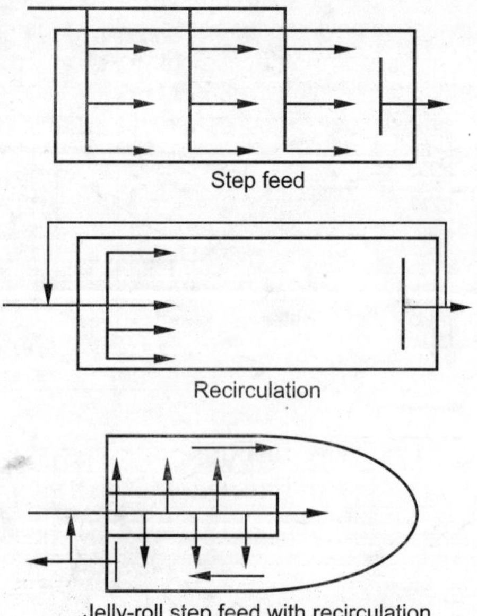

Fig.10.33: Feeding arrangements for constructed wetlands for Wastewater treatment

Source: CPCB, 2001

The simplest and the cheapest method is to use a 90° elbow with O-ring seals, which allow the end of the pipe to be raised or lowered by swivelling at the elbow. The vertical outlet pipe can comprise a series of sockets sections or telescopic sections, which can be adjusted to increase or drop the water level.

Constructed wetlands using soil as substrate will require a stone collector with an open jointed pipe running across the width of the cell at the base. It is essential to provide outlet structures in such a way to adjust the depth of water column in the bed. The water level control helps to maximise the flow path of wastewater through the reed bed.

The simplest method for distribution of flow into the wetland include saw-tooth weirs. However, sometimes saw-tooth weirs are not recommended since they are expensive to construct and maldistribution of flow is often caused by screenable material collecting on the edges. Also, the channel tends to act as a sedimentation tank and collects sludge and grit.

Source: *Central Pollution Control Board (2001), Constructed Wetlands for Wastewater Treatment, New Delhi.*

Duckweed Based Sewage Treatment System

A Duckweed Based Sewage Treatment be in the form of a single unit facultative pond, which may be used for the treatment of low strength community Wastewater. Configuration of a complete treatment system for specific situations will depend on the characteristics of the Wastewater and the objective of the treatment. Some possible situations could be as given below:

(a) **Low Strength waste-water, such as rural sullage:** Suspended solids (SS) get settled in the duckweed pond, and, the sediments at the bottom of the duckweed pond get decomposed in the anaerobic environment. Therefore, pre-sedimentation tank is not always necessary before the duckweed pond when the SS is low or moderate. Also SS is no hindrance in the growth of duckweeds. Thus in the case of community sullage, the pre-sedimentation tank will not be necessary.

(b) **Municipal Sewage:** For treatment of sewage containing high suspended solids, a pre-sedimentation tank is desirable. When the duckweed pond is to be used as the main treatment of municipal sewage of normal strengths, pre-treatment by sedimentation tank will be required for the partial removal of organics also with the objective to feed the duckweed pond with lower organic loads suitable for this kind of treatment.

(c) **Polishing of treated waste-water:** For the duckweed pond being used as a polishing pond and for nutrients removal from secondary treated effluent, it will be added in series after the conventional treatment systems.

Effluent from the Duckweed Pond may be used for pisciculture in a separate fish- pond or be discharged to the recipient body. The duckweed pond functions as an anaerobic pond except in the top layer where aerobic condition prevails. Duckweed systems are effective in treating the wastewater by supporting the right environment for the removal of pollutants by microbial metabolism and physical sedimentation and adsorption.

Duckweed Based Wastewater Treatment

Duckweed plants create an effective barrier to light transmission, thereby restricting photosynthesis by algae and sub-merged vegetation and ensuring anaerobic conditions below the top layer. Additionally, Duckweed root-lets provide a good support medium for the growth of bacteria and for filtration, and adsorption of SS as also some dissolved pollutants. Duckweed also helps in removing nutrients and heavy metals by absorbing nitrogen, phosphorus, sulfur, and trace elements.

In the Duckweed-based Sewage Treatment System, the pollutants get removed through the following mechanisms :

(a) **BOD:** Organic pollutants are decomposed microbially in the anaerobic environment, and also some get stabilized aerobically in the top layer.

(b) **Suspended Solids:** The removal of Suspended Solids (SS) is mainly due to settling of particles to the bed as sediments and by the bio-degradation of organic particles, anaerobically.

(c) **Nutrients:** Duckweed plants which grow rapidly, absorb nitrogen, phosphorus, potassium, calcium, sodium, magnesium, and also some organics from wastewater which are then removed from the system when the plants are harvested.

(d) **Metals:** Removal mechanism for metals includes plant uptake and adsorption. Duckweed Lemnoideae have been shown to possess a great ability to accumulate metals and possess considerable tolerance to high concentrations of metals.

(e) **Pathogens:** Pathogenic bacteria are removed by (i) natural die-off depending on detention time and wastewater temperature, (ii) sedimentation and (iii) adsorption. Parasites and parasite ova precipitate with other suspended solids and are trapped in the bottom sediment.

Source: *CPCB (2001) Guidelines for Duckweed Based Wastewater Treatment System*

Sand Filters

The use of sand filters is the most common form of physical treatment of water. Passing flocculated water through a sand filter strains out the floc and the particles trapped within it. Where taste and odour may be a problem (organo-leptic impacts), the sand filter may include a layer of activated carbon to remove the taste and odour. Sand filters become clogged with floc after a period in use and they are then backwashed or pressure washed to remove the floc. This backwash water is run into special settling tanks so that the floc can precipitate out and is then disposed of as waste material. In some countries this may be used as a soil conditioner.

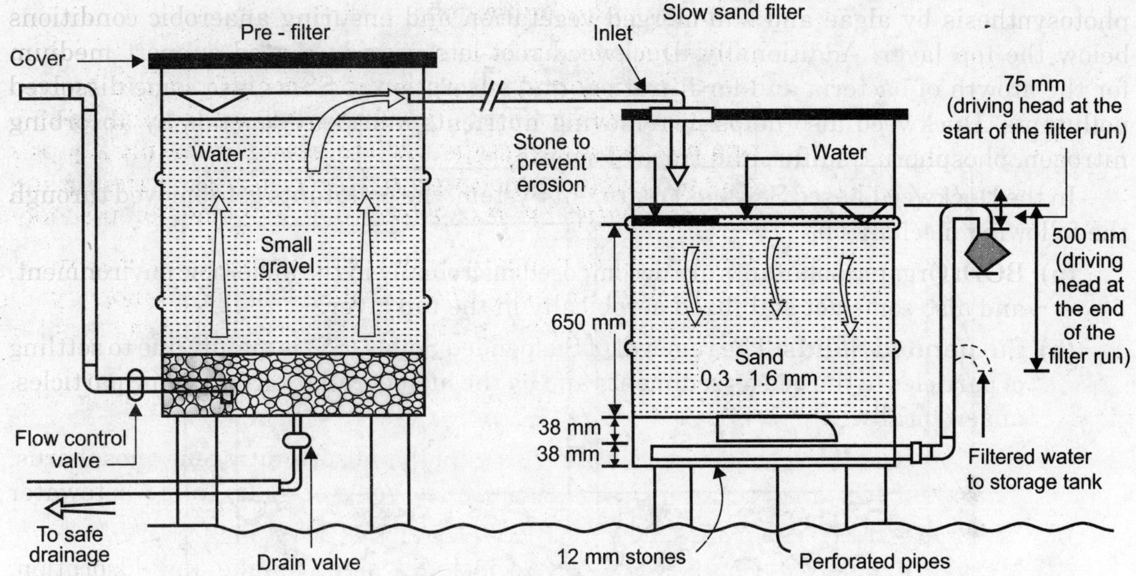

Fig.10.34: Slow sand filter process

Source: Jon Davis and Robert Lambert, 1955, Engineering in Emergencies, IT Publications, London

Source: CPCB

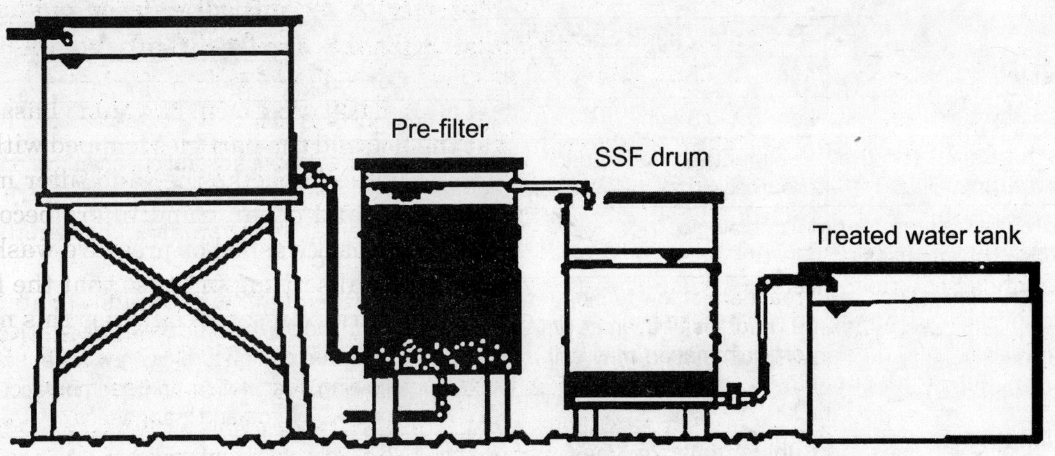

Fig.10.35: Slow Sand Drum Filter and Pre-filter

Source: Jon Davis and Robert Lambert, 1955, Engineering in Emergencies, IT Publications, London

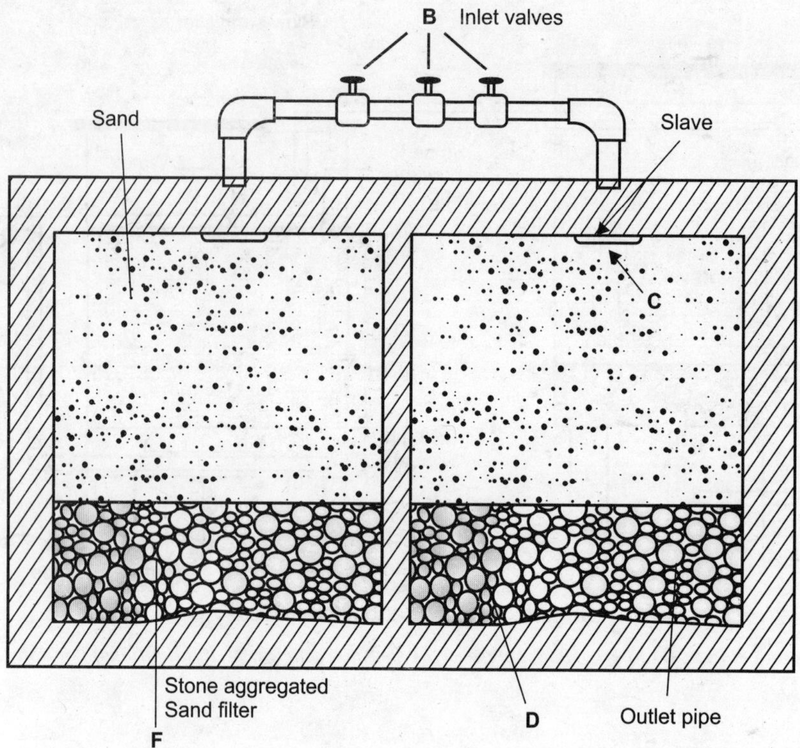

Fig.10.36: Cross-section of a Sand Filter for Treating Grey Water.

Table 10.13: Comparison of Slow Sand Filter and Rapid Filter

	Slow sand filter (SSF)	Rapid filter
Filter medium	Fire	Coarse
Filtration rate	Low: 0.1-0.3 m/h	High: 5-15 m/h
Mechanisms for the removal impurities	Sedimentation adsorption straining, chemical and micro-biological processes	Sedimentation, adsorption, chemical and microbiological processes
Main mechanism for removed of impurities	Microbiological	Physical mechanisms, especially adsorption
Principal application	Treatment of low turbidity water (<20 NTU) for distribution as potable water post-chlorination may still be carried out	Treatment of low turbidity water-chlorination is required before distribution. Treatment of high turbidity raw water and following flocculation and sedmentation. Removal of iron and manganese
Principal advantages	Substantially reduces pathogenic bacteria. potable water without further treatment. No machinery required	Relatively small and compact
Principal disadvantages	Can only/ effectively treat low turbidity water	Cannot produce a potable without further treatment. Backwashing water required to clean filter- this usually pumps.

Source: Jan Davis and Robert Lambert (1955). Engineering in Emergencies. IT Publications, London

Water Filtration Plant

There are several systems available:

- Small, high rate, pressurized sand filter with backwash. A carbon filter may also be incorporated.
- Ceramic candle filter, in which raw water is pumped through a chamber containing a number of porous ceramic candles. Ceramic candles can silt up quickly and have to be regularly cleaned, and eventually replaced.
- Diatomaceous earth (DE) filter.

The diatomaceous earth (DE) filter is a widely used system comprising stainless steel wire-wound candles enclosed in a cylindrical filter chamber. The raw water to be filtered is injected with a diatomaceous earth, or cellulose powder, which forms a thin filter coating on each candle. Water is pumped across the filter coating and exits up the inside of each candle. The flow rate gradually diminishes as the filter becomes blocked.

Wash the sand to remove fine and organic material before placing it in the filter. If the source of the sand is a river, this can be done at the river where there is a plentiful supply of water. Use a drum arrangement. Fill the drum with sand. Allow water to flow into the bottom of the drum to fluidize the sand and wash the silt away in the overflow. Continue flushing until the overflow water becomes clear.

A filter under drainage collects water which has percolated through the filter and carries that water to be outlet pipe. The Oxfam filter under drainage comprises a network of 100 mm slotted PVC pipes. Gravel above and around the under drains prevents sand migrating to the slots. Use the 'filter rule' given in Appendix 20 to size the gravel layers. The gravel should be of a uniform size for the easy passage of water and large enough not to pass through the slots in the drains. An alternative to graded layers of gravel is to interpose a layer of synthetic fabric between the filter sand and gravel surrounding the under drains. All gravel should be under laid by a 50 mm layer of sand to protect the tank lining.

The gravel and under drainage should not be placed against the wall as water which 'short-circuits' down the wall sides will not pass through the filter sand. Leave at least 600 mm of sand between the wall and under drain for water to pass through before leaving the filter.

A dug and lined slow sand filter: The design of the horizontal-flow roughing filter, as the filter chamber, an excavated basin lined with a plastic sheet. A similar approach can be adapted for the design of a slow sand filter. Chlorine residuals are of two kinds, combined residuals and free residuals. Combined residual chlorine is the proportion of the original chlorine dose which combines with ammonia and organic nitrogen compounds to form stable but less effective disinfectants than free chlorine. Free residual chlorine is that part of the chlorine dose which remains after the chlorine demand has been fully satisfied. The actual dose will depend on the condition of water. It can be expected to be in the range, 1-5 mg/L. Determine the optimum dose by trials on water samples.

To determine how much chlorine is required, it is necessary to analyse the water for the chlorine residuals. Simple colour comparator kits with reagents are available to indicate the free and combined residuals of chlorine in water. Chlorine in water will form hypochlorous acid (HClO), at low pH values or tend to dissociate into hydrogen (H+) and chlorite (ClO-) ions at higher pH values. Hypochlorous acid is the more active disinfectant and therefore chlorination is more effective in low pH (acidic) water. Chlorination is considerably less effective when the pH is greater than 8.

Fibre Filters: Fibre filters are designed to take out turbidity (suspended sediment) which contain spun cellulose or rayon. The tightly wrapped fibres form a cylinder around a tubular opening. Line pressure forces water through the wrappings to the inner opening that leads to the faucet. The fibres trap silt. Filtered water passes to the opening that leads to the faucet.

Reverse Osmosis Units: Unlike other types of devices that reduce impurities in the water, reverse osmosis removes water from the impurities. It does that by forcing water through a specially constructed, semi-permeable, nonporous synthetic membrane (usually cellulose acetate) that separates soluble and suspended particles from the water. The process removes a wide variety of substances from the water. More than 75 per cent of such minerals such as sodium, calcium and chloride may be removed from the water. Reverse osmosis also may be effective in removing fluoride, nitrate and some forms of arsenic. The membranes may last a year. Effective membrane life depends on quality of water entering the unit.

Reverse osmosis units are expensive because of the number of accessories needed to operate them. They require in-line installation and a pre-filter to remove dirt and sediment. If a backflush system is employed to prevent a build-up of contaminants on the membrane, the purchase price increases.

Ion Exchange: Most common ion exchange systems use a zeolite resin bed and simply replace unwanted Ca^{2+} and Mg^{2+} ions with benign (soap friendly) Na^+ or 1C ions. This is the common water softener. A more rigorous type of ion exchange swaps H^+ ions for unwanted calcium and hydroxide (OH–) ions for unwanted anions. The result is $H^+ + OH^- \rightarrow H_2O$. This system is recharged with hydrochloric acid and sodium hydroxide, respectively. The result is essentially deionized water.

Electro-deionization: Water is passed between a positive electrode and a negative electrode. Ion selective membranes allow the positive ions to separate from the water towards the negative electrode and the negative ions towards the positive electrode. High purity deionized water results. The water is usually passed through a reverse osmosis unit first to remove non-ionic organic contaminants.

Water Conditioning: This is a method of reducing the effects of hard water. The calcium and magnesium ions that cause lime scale are altered in a way that affects their crystallisation. This affects their adhesion to surfaces, meaning that they do not create as much limescale as untreated hard water.

Solar Disinfection (Sodis): Microbes are destroyed through temperature and UVA radiation, provided by the Sun. Water is placed in a transparent plastic bottle, which is oxygenated by shaking, followed by topping-up. It is placed on tile or metal for six hours in full Sun, which raises the temperature and gives an extended dose of solar radiation, killing any microbes that may be present. The combination of the two provides a simple method of disinfection for tropical countries.

Agitators: In the industrial effluent/sewerage/raw water treatment, mixing is an important operation which is used for preparation of solutions of chemicals like Acids/ Coagulants /Flocculants/Neutralizing agents in water or for chemicals solution mixing in the raw water/effluent/ sewerage or for blending, which is accomplished by using electrically driven agitators. The agitators are designed as per the specific content, need and function using the quality motors, gearboxes or geared motors with flat turbine blades with varying configurations.

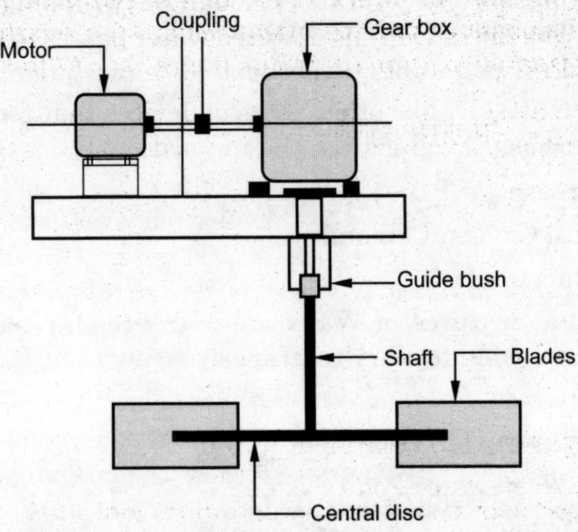

Fig.10.37: An Agitator

Solar Pond Evaporation: The objective of the system is to evaporate wastewater using solar energy. The high total dissolved solids (TDS) water is pumped into the solar evaporation pond. Due to natural evaporation process, the solids concentration takes place. Once the concentration is reached, the form of salt solids gets separated out and the salt is taken out by scrapping the solar pond.

Solar Pond with Mechanical Evaporation: The objective of the system is to concentrate the rejects or high TDS streams and make it in the solids/slurry form. The slurry/ solids can be disposed of by the hazardous solid wastes disposal option. The high TDS water is pumped into the solar evaporation pond. In the first phase of the solar ponds, the pumps re-circulate the rejects and the solids concentration takes place. Once the concentration is reached where the liquid cannot be pumped or re-circulated, the concentrated liquor is taken to the auxiliary ponds and allowed to dry naturally. In these ponds, no recirculation takes place.

Thermal Evaporation: Normally Reverse Osmosis (RO) rejected water contains minerals and salts in huge quantity. So, we cannot directly let this water go into the environment. It needs to be treated and made suitable as per environmental parameters. For that, thermal evaporation method is used to recycle the reject water.

Reject water is collected in balance tank and fed through pre-heaters (type of shell and tube heat exchangers) where it is heated up to 70-80°C with calandria chest vapours to provide the initial preheating at cheaper cost. After evaporation, liquid is separated into vapours and concentrated liquid. Vapours are collected in vapour separators and concentrated liquid is again fed into next calandria for heating. This process is repeated until final concentrated slurry is left. This slurry could be used for land filling.

DEBRIEFING QUESTIONS

1. How wastewater is a threat to the environment?
2. What are the major causes of water pollution?
3. What are the major pollutants which need to be tackled to improve the quality of wastewater?
4. Write short notes on:
   (a) Effluent Treatment Plan
   (b) Bio-chemical Oxygen Demand
   (c) Coliform Bacteria
5. Write the salient features of Wastewater treatment comprising preliminary, primary, secondary and tertiary treatment stages.
6. Write short notes:
   (a) Two Tier Wastewater Treatment
   (b) Integrated Wastewater Pond System
   (c) Root bed Channel
   (d) Floating Wetland
   (e) Bio-swale and Raingarden
7. Describe the steps involved in Decentralised Wastewater Treatment.
8. Write notes on:
   (a) Pallet Softner
   (b) Zero Liquid Discharge
   (c) Sewage Stabilisation Pond
   (d) Duckweed Based Sewage Treatment
   (e) Slow Sand Filter
   (f) Reverse Osmosis Unit
   (g) Agitator

11 Environmental Planning

Environment pollution and climate change have become the imminent reality with a rise in global temperature, changes in rainfall, floods, droughts, drastic increase in atmospheric concentrations of water vapours, carbon-dioxide, methane and nitro-oxide, and other greenhouse gases. With increasing vehicular and industrial emissions, fossil fuel usage and urban growth, coupled with growing air conditioning demand, it is projected that the urban heat can increase by 4°C. The population's health and productivity will be affected by high ambient summer temperatures. In such circumstances, action to reduce the climate change effect and environmental pollution must become a priority.

As stated by the Sustainable Development Goals (SDG 11)environmental planning aims to make cities and human settlements inclusive, safe, resilient and sustainable. Inclusive and sustainable habitat implies decoupling natural resource use and environmental impacts from economic growth, mitigation of GHG emission and also adaptation of the impact of climate change. These imply new ways of habitat planning and applications of new technologies for energy, construction, infrastructure development and transport.

The habitat is an interface between natural and built environment, as shown below:

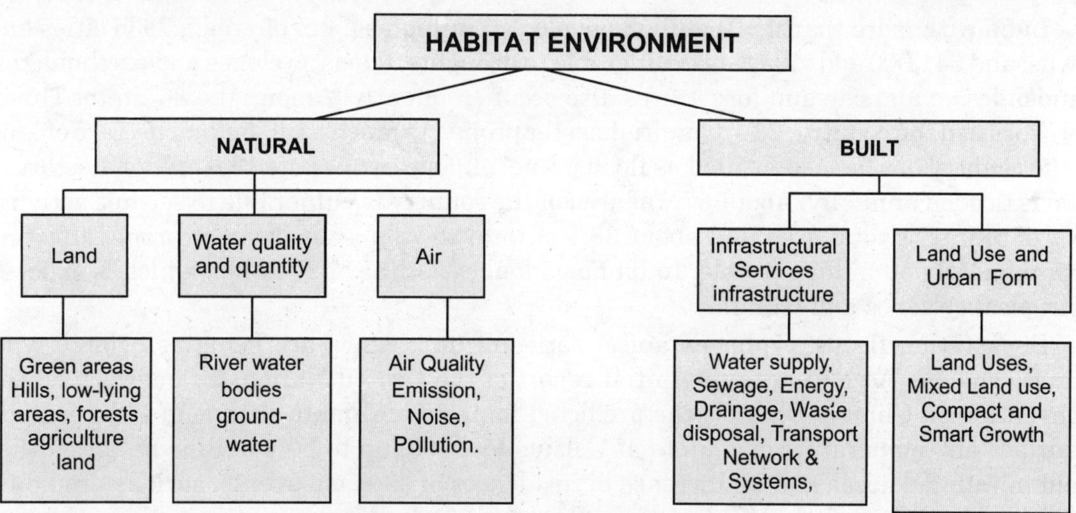

Fig.11.1: Components of urban environment

The objective of sustainable development is to provide a healthy living environment to the citizens and to manage the resources in a sustainable manner. Major issues that have to be tackled in an environmental management plan are as follows:

- Conservation of green areas, fertile land, forests, geology and hydrology
- Rejuvenation of rivers and water bodies, Groundwater augmentation
- Pollution prevention (water, air, noise, land)
- Management of pollution sources (industries, transport, households)
- Provision of efficient sustainable infrastructure (energy, water supply, sewage, sanitation, liquid and solid waste management, etc.)
- Provision of compatible land uses and compact and smart growth.

A comprehensive digitised and geo-referenced survey and a systematic investigation of 'eco determinants', which deal with relationship of humans and other forms of life to the natural and man-made environment provide the basis for a sound land use planning.

Urbanisation and Climate Change

Urbanization and economic development are changing the lifestyles of the people. Increasing production, mobility, consumption and living in comfort are enhancing the carbon emissions and climate change. Models predict an average increase in temperature of 2.3 to 4.8°C in India for the benchmark doubling of carbon dioxide scenario. Although per capita carbon emissions in India at 1.2 metric tons (mt) per year is one of the lowest in the world, which is predicted to become double within next 10 years. Already in the urban areas the people with cars and air conditioners emit 4.5 mt of carbon dioxide/greenhouse gases per year, while the low income people without car and air conditioners, emit an average of 1.1 m of CO_2/GH gases. As per UN Climate Change Panel, a benchmark of 3.0 mt per capita per year should be the upper limit.

India with more than 1210 million people, 3.3 million sq. km of area, , 7936 cities and towns, and 641,000 odd villages is vulnerable to droughts, floods, cyclones and earthquakes. Landslides, avalanche and forest fires also occur frequently. Among the 36 States/Union Territories in the country, 22 are multi-disaster prone. As much as 40 million hectare of land in the country has been identified as flood prone, and on an average 18.6 million hectare of land is flooded annually. About 57% of area of the country is vulnerable to seismic activity. 18% of country's total area and about 68% of total sown area is drought prone, affecting approximately 50 million people. India has a long coastline of 8040 km, which is exposed to tropical cyclones and tsunami.

Devastating floods, typhoons and a series of hurricanes are being associated with climate change. According to the initial report of the United Nations Framework on the Convention on Climate Change, the predicted impacts of climate change in India include a surface air temperature rise up to 4° Celsius by 2100, up to 30% decline in agriculture yield in rain-fed areas and an increase in incidences of extreme events, such as droughts, floods, cyclones, urban heat islands, earthquakes and changes in micro-climate due to shortwave radiant energy from the sun and long-wave radiant energy exchange between the surface and the atmosphere.

URBAN HEAT ISLAND (UHI)

UHI is a phenomenon where certain urban area is significantly warmer than its surrounding areas. The temperature difference usually is larger at night than during the day, and is most apparent when winds are weak. The main cause of the urban heat island is modification of the land surface by urban development, using materials which effectively retain heat and waste heat generated by transport and energy usage. As a result, there is a corresponding increase in average temperature, which is hotter than the surrounding area.

There are several causes of an urban heat island (UHI), as given below:

- The principal reason for the night time warming is due to building's surface heat radiating into the relatively cold night sky.
- The rapid rate of urbanisation, increasing concrete surface area, less green patch, dark colour roof tops.
- The urban sprawl, road building, and high-and medium-rise expansions.
- Surface radiative properties (albedo and emissivity) in the highly concrete core urban areas cause a change in the energy balance, leading to higher temperatures than surrounding rural areas.
- Lack of vegetation of urban areas, which inhibits cooling by evapotranspiration.
- The tall buildings within urban areas provide multiple surfaces for the reflection and absorption of sunlight, increasing the heat and the "urban canyon effect".
- The blocking of wind by the buildings inhibits cooling by convection.
- Waste heat from automobiles, air conditioning, industry, and other sources, pollution in urban areas also increase the UHI.

The following major issues need to be addressed by environment planning:

Land is a non-renewable resource, which is consumed to accommodate the growing population; once converted to use for habitation it becomes non-retrievable. A study of the present land use pattern in India indicates shortfall in land under forests and pasture. Lands under agricultural use have been and are increasingly getting converted and annexed for uses like expansion of settlements for human habitation. The rate of conversion shows that an additional 1.25 million hectares would be required for human settlements in future. Sacrificing agricultural land for habitation implies reduction of land for producing food for the ever growing population. Thus agricultural lands need to be protected from the process of conversion. Moreover lands that sustain biodiversity, water quality and groundwater recharge, fragile areas, sensitive areas, coastal zones, etc. need protection and conservation. Land for development is thus scarce and needs to be judiciously utilized for various uses according to the inherent capacity or the suitability of the land for the particular land use.

Water scarcity has become a persisting problem in Indian cities and towns due to rapid population growth, haphazard and unplanned development, further augmentation of the water resources seems impossible. The average annual availability of per capita water supply in the country has gone down from 5,236 cubic meters in 1951 to 2,227 cubic meters in 1991. Only 18% of the renewable water resource is being used which amounts to 2,170 cubic meters of per capita water availability *i.e.* 33% below the Asian average. India, though endowed with a good monsoon and ample rainfall uses only 10% of the annual rainfall. Thus there is a need to augment our water resources in the context of the increasing demand through recharging of groundwater resources and rainwater harvesting.

Disposal of wastewater due to the ever expanding urban fabric causes pollution of water sources, both surface water and groundwater sources. Studies conducted by the Central Pollution Control Board(CPCB) show that all the 13 major rivers of the country, which are major sources of water for the settlements are polluted to various degrees causing health hazards in urban areas. Surveys in 71 Indian cities revealed that though approximately 60% of the population is covered by sewerage system, treatment facilities are grossly inadequate. The parameters of concern are increasing coliform levels and Bio-chemical Oxygen Demand (BOD) in surface waters and increased concentration of nitrate in the groundwater. To overcome the health problems, water sources need to be protected by interception and treatment of wastewaters through appropriate technologies. A range of technical and institutional options ranging from centralized surface storage to decentralized rainwater harvesting and re-cycling, together with design of water management options are to be explored.

Air quality in Indian cities and towns is deteriorating due to vehicular and industrial emissions. Though the emissions of air pollutants in urban areas, on per capita basis, is much lower than the Asian average, the pollutant levels of SPM, CO, Pb, PAH, NO_x, etc. in many cities have exceeded the permitted maximum limits. This has led to increase in respiratory diseases. According to a survey conducted by the CPCB ambient air quality, the national capital of Delhi has reached a very critical situation of air pollution. Relatively high frequency of suspended particulate matter SPM, SO_2, NO_2, CO_2 and heavy metals including lead content in the exhaust by number of automobiles and scooters have been observed. However the recent changes in the fuel, adoption of clean technologies, and enforcement of emission norms, traffic management measures and development of mass rapid transport system have reduced the alarming pollutant levels caused due to vehicular emissions. Proper industrial location would reduce the adverse impacts on air quality.

Biodiversity is the sum total of living organisms in all types of habitats. India, being a tropical country, is one of the regions of high diversity. India is also one of the few countries in the world which have enshrined in their constitution a commitment to protect the flora and fauna. Being a signatory to the Convention on Biological Diversity, 1992 India has made a commitment to inventories and protects its biodiversity. In the pre independence years protection of flora and fauna was limited to providing protection through legislation to a few game species. This was followed by demarcating sensitive ecosystems within forest areas for conservation - Reserved and Protected forests, as per Forest Conservation Act 1927. Demarcating areas as National Parks and Sanctuaries was the next step. Setting up of Biosphere Reserves was the final step in the Protected Area Network in the Country. These efforts were targeted to protect the floral and faunal resources in areas of sparse human settlements. It is only now that it is being recognized that a wealth of living resources exist within human settlements that play a critical role in combating pollution, recharging water, and need to be protected and judiciously utilized to ensure sustainable development. The Biodiversity Action Plan, initiated in January 2000, is a step in this direction.

Energy scenario in India is characterized by increasing demand for energy growing at the rate of about three times in the last two decades making the country the fifth largest energy consumer. The post independence era of the country has experienced a shift in the bias towards commercial energy sources, which is slated to be the predominant energy source in the future. Coal, oil and gas are the primary commercial sources with coal dominating the energy supply of the country. The actual production of coal is the highest as compared to other primary energy sources. However higher reserve-production ratios of oil and gas over decades show that the rate of growth in production are higher than that of the production of coal. Power generation capacity has also increased in order to meet the rising demands. The energy consumption sectors in India include industrial, transport, commercial, agriculture and domestic sectors. The pressing problem facing the country today ' is rising consumption levels, especially oil, energy, water and food.

CAUSES AND IMPACTS

According to the Indian Network for Climate Change Assessment (2010), 58 per cent of the total emissions are caused by the energy sector, followed by the industrial sector at 22 per cent, agriculture at 17 per cent, waste at 3 per cent and other land use and forestry at 1 per cent. Of the total carbon dioxide (CO_2) emissions 85% are contributed by activities of the energy sector. Of the total methane (CH_4) and nitrous oxide (N_2O) emissions 78% are due to the agriculture sector.

Climate change can bring considerable changes in the hazard profile of the region by changing the type of hazards, changing the hazard intensities, magnitude and frequency. Many of these impacts may not be uniform across spatial and temporal scales. Devastating floods in Bangladesh, India (Chennai, Kashmir, Bangalore, Delhi, etc.), Nepal and Pakistan, typhoons in Philippines and a series of hurricanes in Florida are being associated with climate change.

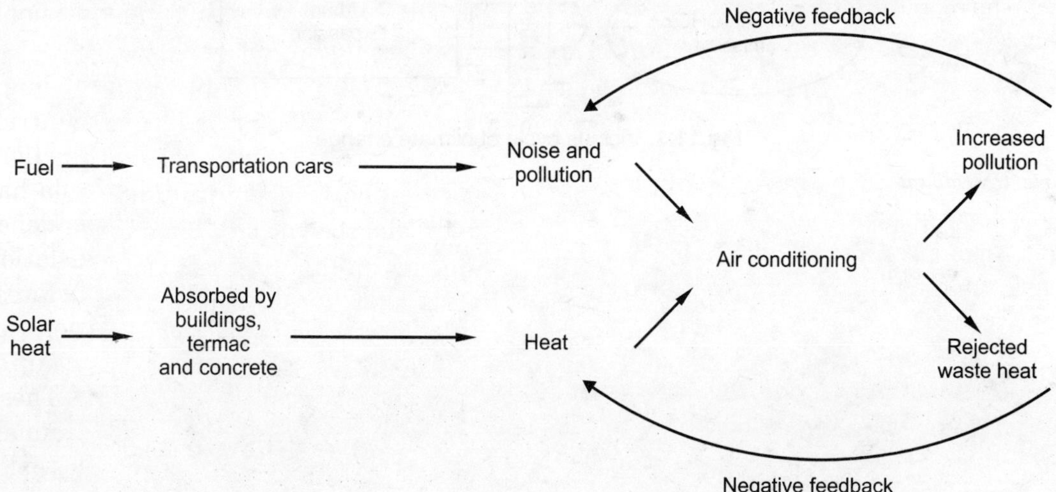

Fig.11.2: The cycle of urban pollution and rising temperature

Without the so-called greenhouse gases, including carbon dioxide, methane, nitrous oxide, and water vapour, earth would be too cold to inhabit. These gases in earth's atmosphere absorb and emit heat energy, creating the greenhouse effect that keeps our planets temperature liveable. Water vapour is the most plentiful greenhouse gas on the planet, accounting for about 60 percent of the current greenhouse effect. Even ozone helps trap some of the heat that makes life on Earth possible, but the 'ozone hole' is a separate issue not directly related to global warming. Since the industrial revolution, people have burned vast amounts of coal, petroleum, and other fossil fuels to create heat and power. This releases carbon dioxide, the most plentiful human-produced greenhouse gas, into the atmosphere. As a result more heat is trapped in earth's atmosphere instead of radiating out into space.

Because carbon dioxide lasts more than a century in the atmosphere, it is well mixed around the globe. Measurements since 1958 show a steady rise in global carbon dioxide concentrations. These have increased by 35 percent since preindustrial times. The relationship between earth's water cycle and global warming creates a well-known feedback loop. Warmer temperatures cause more evaporation from land and oceans, which produces more water vapour, which in turn contributes to warmer temperatures.

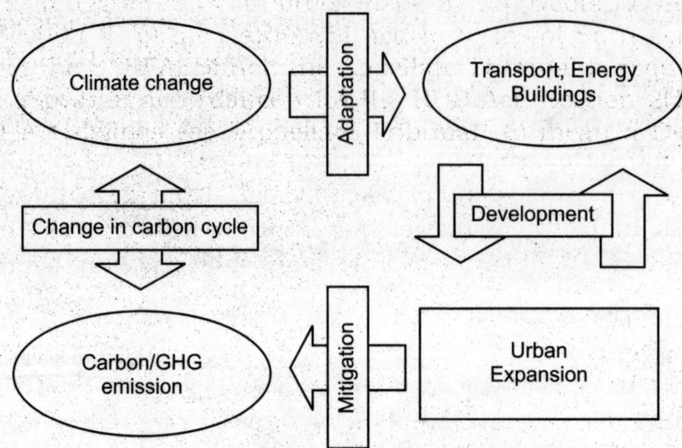

Fig.11.3: Vicious cycle of climate change

Source: UN Habitat

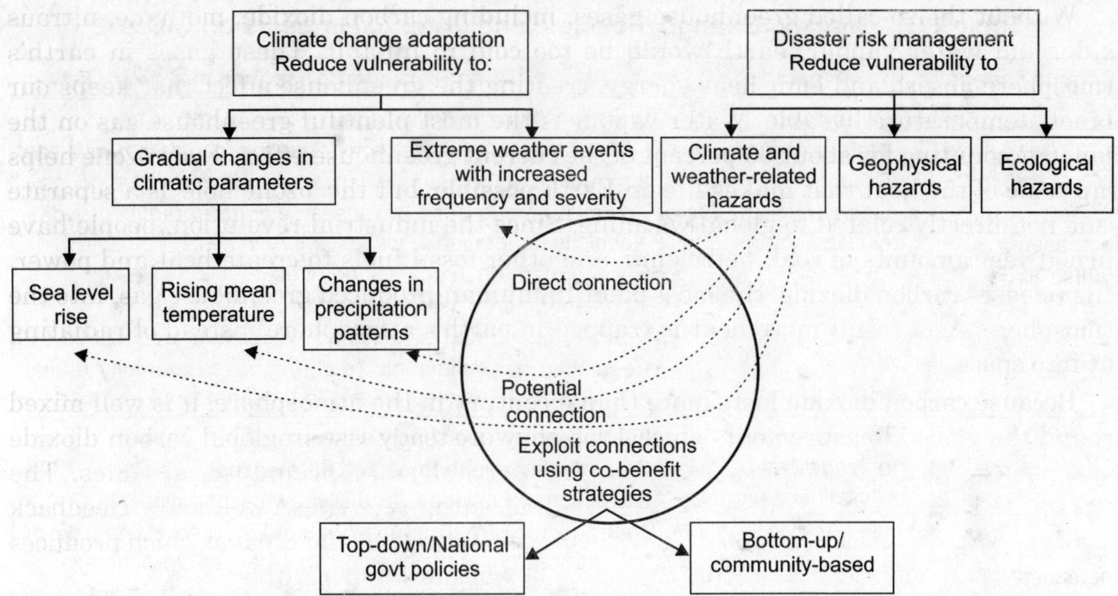

Fig.11.4: Conceptual linkages of climate change adaptation (CCA) & Disaster Risk Management (DRM)

Source: www.planetaction.org/automne_modules_files

ECOLOGICAL FOOTPRINT

Environmental analyst Mathis Wackernagel developed the concept known as ecological footprint to measure the ecological capacity and impacts. Footprint analysis measures what an economy needs from nature: the inputs that fuel it and the wastes that emerge from it. It does so using a single matrix-the number of global hectares of land and water needed by the economy. Global hectares are the area of biologically productive space (land and water with significant photosynthetic activity and biomass accumulation) with world-average of productivity. Where a nation's footprint is larger than its bio-capacity, it indicates that its economy is consuming more forests, croplands and other resources than the country can supply and thus it is overtaxing the domestic environment's capacity to absorb wastes.

According to Mathis Wackernagel, the global footprint now exceeds global bio-capacity by 40 per cent (World footprint is thus 140 per cent). By importing resources and exporting wastes, particularly carbon dioxide, the United States, Europe, Japan, India and China all live well beyond their ecological means with footprints ranging from 200 per cent to nearly 600 per cent.

The problems of urban emission and climate change pollution are inextricably linked to poverty, productivity, and broad macroeconomic performance. Problems that cross over to "green" or "social" issues, typically involve inappropriate land use, precarious housing, deficient public transportation, and road congestion. The impacts of climate change are seen in floods, heat islands, disaster, noise pollution and the degradation of environmentally fragile lands.

Table 11.1: Climatic Change, Possible and Potential Impacts on Cities

Climate change	Impact	Urban Planning Consequences
Increased temperatures	• Groundwater • Water shortage • Drought • Degraded air Quality (smog)	• Water depletion and shortage • Distress migration to cities/towns due to droughts in rural areas • Interruption of food supply networks and higher food prices • Potential energy price increases (*e.g.*, from reduced hydro-electricity generation in places where it exists) • Exaggerated urban heat island effect • Increased energy demands for cooling • Need for higher and/or additional wastewater treatment • Population health impacts (*e.g.*, increased mortality during heat waves, decreased access to food/ nutrition)
Increased precipitation	• Increased risk of landslides or mudslides on hazard slopes	• Interruption of food supply networks • Property damage (homes and businesses) • Disruption of livelihoods and city/town economies • Damage to infrastructure not designed to standards of occurrences being experienced in urban and rural areas • Displacement and population movement from informal settlements built on steep slope hazard lands, etc. • More favourable breeding grounds for pathogens (*e.g.*, mosquitoes and malaria) • Population health impacts (increased incidences of water-borne diseases like cholera)
Sea-level rise	• Coastal flooding • Salt water intrusion into groundwater supplies in coastal area increased storm surge hazard	• Displacement and population movement from Property damage (Homes and Businesses) • Damage to infrastructure not designed to standards of occurrences being experienced • Increased storm disruption of livelihoods and urban economies • Disruption of livelihoods and city/town economies • Population health impacts (injuries, increased mortality and illness) • Property damage (homes and businesses) • Population health impacts (injuries, increased mortality, distress) • Interruption of food supply networks • Increased risk of landslides or mudslides on hazard slopes

Source: Adapted from Developing Local Climate Change Plans, UN-HABITAT/International Institute for Environment & Development (2010) and Willbanks et al (2007)

Climate change has a variety of potential implications for urban areas. In terms of exposure, a key risk relates to changes in the frequency and intensity of heavy rain, storms, droughts, heat waves, wildfire and other hazard events. The urban centers more at risk are those where these events are already widespread.

Table 11.2: Projected Impacts on Urban Areas of Changes in Extreme Weather and Climate Events

Climate phenomena and their likelihood	Major projected impacts
• Warmer and fewer cold days and nights • Warmer and more frequent hot days and nights over most land areas • Warmer temperatures	• Reduced energy demand for heating • Increased demand for cooling • Declining air quality in cities • Reduced disruption to transport due to snow, ice effects on winter tourism • Changes in permafrost, damage to buildings and infrastructure
• Warm spells/heat waves. Frequency increases over most land areas	• Reduction in quality of life for people in warm areas without air conditioning; Impacts on elderly, very young and poor including significant loss of human life • Increases in energy usage for air conditioning
• Heavy precipitation events, frequency increases over most areas	• Disruption of settlements, commerce, transport and societies due to flooding • Pressures on infrastructures, potentials for use of rain in hydropower generation • Significant loss of human life, injuries; loss of and damage to property
• Areas affected by drought increases	• Water shortages for households, industries and services • Reduced hydropower generation potentials • Potential for population migration
• Intense tropical cyclone activity increases	• Disruption by flood and high winds; • Disruption of public water supply • Withdrawal of risk coverage in vulnerable areas by private insurer (at least in high income countries) • Significant loss of human life, injuries; loss of and damage to property • Potentials for population migration
• Increased incidence of extreme high sea level (excludes tsunamis)	• Costs of coastal protection versus costs of land-use relocation; • Decreased freshwater availability due to salt –water intrusion • Significant loss of human life, injuries; loss of and damage to property • Potential for movement of population and infrastructure

Source: IPCC 2007 with modifications.

In the context of growing footprints, the western countries are realising that they have carried their development too far. It is being debated whether development, which is called progress, has become a menace. Environmental problems inflict a high human and health price just 70 percent of urban dwellers are served by some form of sanitation, with only about 40 percent connected to sewers Where there is sewerage, more than 90 percent of the wastewater is discharged without treatment. The costs of these and other pollution problems alone in some developing countries are estimated to exceed 5 percent of GDP.

Poverty, economic development, and the environment inextricably linked. This linkage raises issues of equity (such as the willingness to pay for better environmental services and the issue of subsidizing basic urban services for the poor), and of the changing nature of environmental problems at different income levels. The urban poor are affected disproportionately by environmental problems and their plight exacerbates urban environmental crises. The survival of the urban poor is strongly linked to the functioning of urban labor markets and to the markets for land and housing. About a quarter of the world's urban population lives in absolute poverty—and many more live in substandard conditions. In many parts of the developing world, urban poverty has grown faster than rural poverty because of macroeconomic adjustment, inefficiencies in the urban economy, and misallocation of public resources.

The critical and most immediate problems facing cities are the health impacts of urban pollution that derive from inadequate water, sanitation, drainage and solid waste services, poor urban and industrial waste management, and air pollution, especially from particulates. Collectively dubbed the "brown agenda," this set of problems is closely linked to the poverty-environment nexus. Important underlying or related issues typically involve inappropriate land uses, precarious housing, deficient public transportation, and road congestion and accidents.

These pressing problems are also related to what may be considered more properly the "green" and the "social" issues of urban areas, such as the depletion of water and forest resources, the degradation of environmentally fragile lands, the occupation of areas prone to flooding or landslides, overcrowding, degradation or loss of historical and cultural property, noise pollution and other problems. Similarly, the emissions resulting from energy use for cooking, heating, industry, and transport contribute significantly to climate change and public health.

National Action Plan on Climate Change

The national action plan for climate change (2008) identifies measures that promote development, while also addresses climate change both adaptation as well as greenhouse mitigation measures:

The national action plan for climate change (NAPCC) is guided by the following principles:

- Protecting the poor and vulnerable sections of society through an inclusive and sustainable development strategy, sensitive to climate change.
- Achieving national growth objectives through the quantitative change in direction that enhances ecological sustainability, leading to further mitigation of greenhouse green mitigation.
- Division efficient and cost effective strategies for the demand side management.
- Developing appropriate measures for both adaptation and mitigation of greenhouse gases emissions, extensively as well as at an accelerated place.
- Engineering new and innovative forms of regulatory and voluntary mechanism to promote sustainable development.
- Effecting implementation of programmes through unique linkages, including civil society and local government institutions and through public private partnership.
- International cooperation for research, development, sharing and transfer of technologies enabled by additional funding.

The national action plan hinges on the development and use of new technologies. The implementation of the plan would be through appropriate institutional mechanism suited for effective delivery of each individual objective and includes public private partnership and civil society action.

The focus will be on promoting understanding climate change, adaptation and mitigation, energy efficiency and natural resource conservation. The following eight national missions form the core of the National Action Plan, representing multi-programmed, long term and integrated strategies for achieving the goal in context of climate change:

- National Solar Mission
- National Mission to Enhance Energy Efficiency
- National Mission on Sustainable Habitat
- National Water Mission
- National Mission for Sustaining the Himalayan Ecosystem
- National Mission for Green India
- National Mission for Sustainable Agriculture
- National Mission on Strategic Knowledge for Climate Change

The National Action Plan on Climate Change (NAPCC) and National Mission on Sustainable Habitat are important landmarks on climate change debate in India. These reflect government responses to the challenge of climate change vis-a-vis habitat. These need to link more closely with the poverty and India's variety of climatic zones, and cultures. There can not be an easy plan to climate change. India faces its own peculiar challenges for which the global approaches and priorities, such as pollution reduction, carbon emission, fuels, etc. may be inadequate.

Environment Planning Approach

The pursuit of sustainable development requires striking a balance between economic goals, human needs (social, cultural, livability and health) and environmental integrity. This involves a process of effective citizen participation in decision-making. An economic system that is self-reliant and securing ecological basis for development are the prerequisites of sustainable development. We need to relook at the fundamentals of human settlements and evolve a sustainable system whereby harmony and interrelationship between nature and life can be re-established.

Environmental Planning is basically a concern of 'design with nature'. It means development and human interventions in natural system to the extent of carrying capacity of an area. The carrying capacity of an area is the population or level of activity that can be sustained for a given length of time without depletion of the resources or breakdown of the biological natural systems.

Sustainability is a subject, which concerns everyone and cuts across various activities, disciplines and levels. It is thus an all embracing and complex agenda to plan for. Although disaggregated for analytical purpose, the ecology, landscape and cultural processes need to be tackled in an integrated manner by environmental planning and development.

The nature of environmental planning and intervention strategies depend on five factors:

1. *The natural features of urban areas*—cities tend to take their form from the nature of their site, and their environmental problems are determined in large measure by regional ecosystem characteristics. Ecosystem features vary widely, for example, whether coastal or inland, mountainous or flat, arid or humid, temperate or tropical, or some combination of these features. Each characteristic can be associated with potential problems: for example, topology affects drainage; altitude and atmospheric stability are determinants of urban ventilation and air quality.
2. *Population size and rate of growth*—the sheer magnitude and pace of population growth in cities directly affects the spatial concentration of people, industry, commerce, vehicles, energy consumption, water use, waste generation, and other environmental stresses, as well as placing pressure on the local managerial and operational capacity to respond.
3. *The level of income and economic development*—as people rise out of poverty, the nature of their environmental problems changes. For example, a low level of economic development is associated with inadequate basic services, indoor air pollution, and land degradation, whereas cities with higher levels of income and economic development suffer more from industrial and energy-related pollution of air and water and the inability to manage hazardous and toxic wastes.
4. *The diverse spatial dimensions of environmental problems*—spatial factors determine who is affected and how, the severity of impact, and the appropriate level of responsibility and decision making needed to solve problems. For example, access to basic environmental services is normally a household, neighborhood, or community issue, while air and water pollution are city- wide or regional issues.
5. *The roles of local actors*—the quality of the environment in a city is determined largely by the interaction of numerous public, private, and household actors who have an important effect on environmental problems and their solutions. The managerial and operational capacity and responsiveness of local and sectoral institutions is a critical determinant, as is the level of participation of local stakeholders.

The environmental planning approach centers around:

- Public support and participation,
- Policy interventions,
- Service delivery,
- Institutional capacity,
- The gap between urban environment and planning, and
- Sustainable development, green building and a minimalist lifestyle.

Environment and Urban Form

Several interrelated factors determine the patterns of urban sustainability, energy use and emissions. According to the studies, the size, growth, structure and density of population are key determinants of cities' carbon emissions and other environmental impacts. A negative correlation exists between population density and atmospheric emissions; for instance, a one percent increase in the density of urban areas would relate to approximately 0.7 percent decline in Carbon pollution at the city-level, with other factors held constant. Spatially compact and mixed-use urban developments have different benefits in terms of carbon emissions:

- Reduced costs for heating and cooling resulting from smaller homes and shared walls in multi-unit dwellings. For instance, households in buildings with five or more units consume 38 per cent of the energy of households in single-family homes.
- The use of energy systems covering a broader area (*e.g.* district) for cooling, heating and power generation as well as lesser line losses related to electricity transmission and distribution. The use of microgrids to meet local requirements of electricity can create efficiencies in storage and distribution.
- Reduced average daily vehicle-kilometers traveled in freight deliveries and by private motor vehicles per capita (VKT). Population density increases accessibility to such destinations as stores, employment centers and theaters. It has been found that with all other variables constant except density, "the household in a neighborhood, with 1,000 fewer units per square mile drives almost 1,200 miles more and consumes 65 more gallons of fuel per year" over a household in a higher density neighborhood.

A basic need is to create a sustainable urban structure, comprising modules of local hub/neighbourhood, community and district/zones, which allows a compact form with mixed land use with the following mandate:

- Integrated planning of land use and public transport
- Reorganisation of land use and rationalisation of density, urban restructuring for travel reduction.
- Integration of bus, MRTS, metro rail, rail corridor, cycles, walking, stations and terminals
- Integrated traffic and transit operation, control and management, and using subterranean space for transport and parking
- Telecommunication, electronic mail, video conferencing, radio paging, mobile phones, computer networking, e-mail for participatory e-governance to save travel.

The spatial model impacts the sustainability (less travel time and thus energy saving). A compact city structure should be developed to achieve the following:

- Urban Form: *(i)* High dense settlements *(ii)* Less dependence on automobile (high density) *(iii)* Clear boundary from surrounding areas;
- Spatial Characteristics: *(i)* Social fairness (high dense settlements) *(ii)* Self-sufficiency of daily life *(iii)* Independence of government (clear boundary); and
- Social Functions: There is close relationship between compact urban form and the sustainability by reduction of automobile dependence, efficient supply of social infrastructure and public services, active community relationship by high-dense habitation, and revitalization of inner city.

Compact cities can have significant implications to climate change and environment. Critical instruments for this policy option include coordination with public transport, infrastructure development, mixed land use, urban boundary, and coordination of different levels of government. Data from international cities indicates correlation of urban density with less transport energy use and car use per capita. Smart growth land use practices aim to create more accessible land use patterns, which reduce the amount of travel needed to reach goods and services.

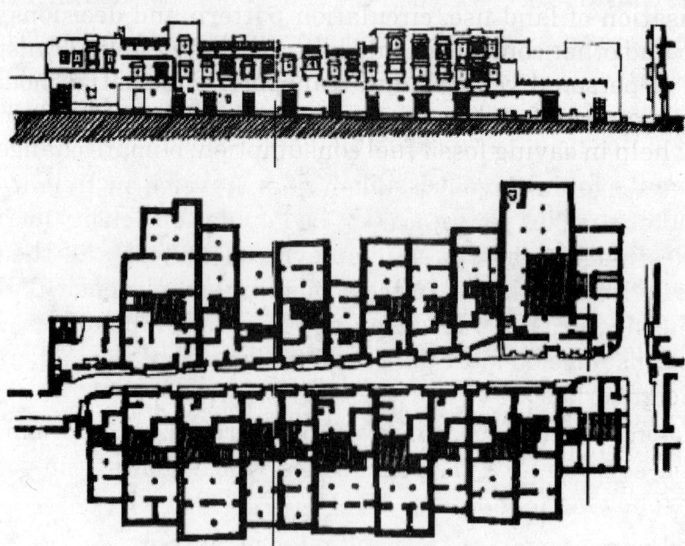

Fig.11.5: Jaisalmer: Typical residential street-compact, walkable, high density and low energy urban form (after Kulbhushan Jain)

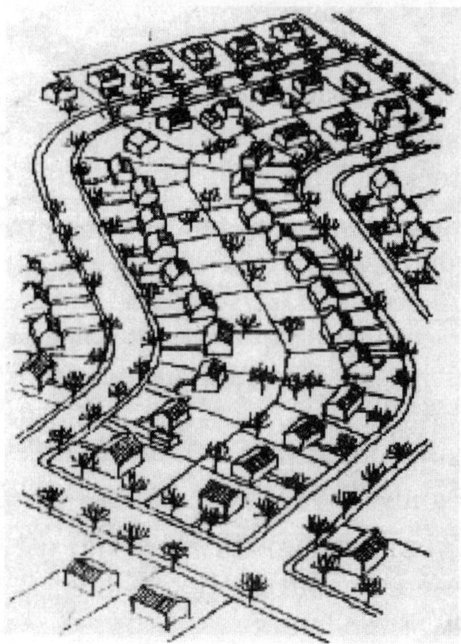

Fig.11.6: American Suburb: car – centric, high energy urban form.

Carbon neutral cities have to reduce their ecological footprint due to transport, which is responsible for 18% of carbon emissions due to use of fossil fuels. The need to travel can be reduced by mixed land use and better synergy among public transport, walking, NMTs, urban structure (form, density and floor area ratio) and by transit oriented development. An important element of policy should aim to shift private vehicle users to public transport. Apart from aspects like frequency, inter-modal integration, a possible single ticketing systems, the quality of public transport, particularly buses, would need to be significantly upgraded, inter-alia, keeping in view the element of clean and green fuel and traffic calming. Organisation of land use, circulation pattern and decisions regarding density, Floor Area Ratio and other controls should be around the Public Transport System, which reduce toe need of personal vehicles and transport. The spatial model should be based upon the principle of less travel, more energy savings. As such, a compact development can significantly help in saving fossil fuel consumption, climate change and environment. Critical instruments for more accessible urban pattern include synergy with public transport, infrastructure development, mixed land use and e-governance, which reduce the need to travel. A least-cost life cycle analysis of urban mobility options and transportation systems across arrange of carriage ways and modes should establish an optimum long term mix A mix of pedestrians, bicycles people movers, hyper- trams, hyper-buses and ferries provides safer, faster and cheaper mobility than the personal car. This call for replanning of urban form which promotes walkability and NMTs and curtail the need to travel. Transport policy and plans linked to pollution control, including noise, should be part of a City Plan. Providing a barrier free environment is a mandatory requirement for outdoor and indoor environment.

Key Indicators

The challenge of environment planning is to identify the health effects, productivity amenity, ecological values, and other key indicators and develop policy directions plan which address these issues.

- Health and Safety. In examining health and safety effects, assessment criteria include health care costs, lost working days, and higher mortality rates.
- Productivity. An assessment should judge the extent of losses in urban productivity due to external factors such as pollution from industrial areas.
- Equity. Because the negative effects of urban environmental degradation fall disproportionately on the poor, issues of equity are of critical importance. Efforts to improve universal access to basic sanitation and other essential urban services and to include the poorest of the poor in urban improvement programs will go a long way toward mitigating urban environmental degradation.
- Ecology. Ecological effects can be judged by the availability and costs of fresh water, the extent of land subsidence from overdrawn Groundwater, the vulnerability to flooding, landslides and other natural disasters, and the loss of biological diversity.
- Amenity. Effects on amenity include air and water quality, noise levels, scenic beauty, and the presence of parks and clean, open spaces.

The overriding benefit of environmental planning is that it can encompass the cumulative impacts of incremental decisions on ecosystems and their services. It examines the 'parts' to make decisions that affect the 'whole.' An effective planning framework can make the policy and planning process transparent and inclusive, assessing who benefits from which ecosystem service, helping to avoid conflicts; especially if different stakeholder groups are part of the planning process. Strategic Environmental Assessment (SEA) and Environmental Impact Assessment (EIA) can contribute to the integration of biodiversity issues and ecosystem services in local and regional planning. This safeguards livelihoods, illuminate impacts on ecosystem services and highlights the risks and opportunities associated with changes. Start locally to think globally. A good strategy considers both local and global systems and stakeholders. Spatial planning, supported by ElA and SEA, may form a basis for sustainable, economically and socially appropriate responses, to climate change. The proactive inclusion of ecosystem services allows environmental assessment to identify the economic potentials development that supports biodiversity.

Table 11.3: Policy responses integrating Eco-system Services

Policy directions	Examples of potential responses
1. Renewable energy to reduce dependence on non-renewable source	• Community energy system in Freiburg (Germany) and travel management in Calgary (Canada)
2. Carbon-neutral cities to cut and off-set carbon emissions	• Zero-carbon housing in Denmark • Urban tree and woodlands in Sacramento (USA)
3. Small-scale, distributed power and water systems with more energy-efficient service provision	• Water sensitive design that uses the complete water cycle in Hanoi (Vietnam) • Waste water agro-systems in Kolkata (India) • Local power systems and cooperatives in Malmo (Sweden)
4. Increasing photosynthetic spaces (as part of green infrastructure development) to expand renewable sources of energy and material resource needs	• Local food prevision in Devon (UK) • Biomass in Vaxjo (Sweden) • Green roofs and materials in Shanghai (China)
5. Eco-efficiency to enable the use of waste products to satisfy urban energy material resource needs	• Industries reduce waste and resource requirement by sharing waste and resources in Kalundborg (Denmark)
6. Local strategies that increase 'pride in place' by enhancing the implementation and effectiveness of innovations	• Ambitious recycling targets in Cairo (Egypt) • Maximizing urban densities in Hammar by Sjostad (Sweden):
7. Sustainable transport that reduces the adverse impacts of dependence on fossil fuels	• Participatory system that localize energy, food, materials and local production in Medeilin (Colombia) • Planning systems that capture that value of ecosystem services and creating a 'local sustainbility currency' in Curitiba (Brazil) • Urban form and density in Vancouver (Canada) Transit systems in London (UK) • Street planning and mobility management in Tokyo (Japan)
8. Development of 'cities without slums' to improve access to safe drinking water, sanitation and reduce environmental degradation	• Respecting community structure in slum resettlement in Kampung (Indonesia) • Planning for the informal economy in Somalia (UN-HABITAT initiative)

Source: Adapted from UN-HABITAT (2009)

THE ELEMENTS OF ENVIRONMENT PLANNING

There is an unprecedented challenge to re-establish the uterine relationship between nature and man, between human consumption and production. The ecological cycle, which has been disrupted by indiscriminate economic and physical demands of development needs to be revived. Looking at the damage which our cities and development have inflicted upon the environment, one of the prime agenda is to explore the possibility of creating a living environment which is self sufficient, ecologically balanced and culturally stimulating. We need to relook at the fundamentals of human settlements and evolve a sustainable system whereby harmony and interrelationship between nature and life can be re-established.

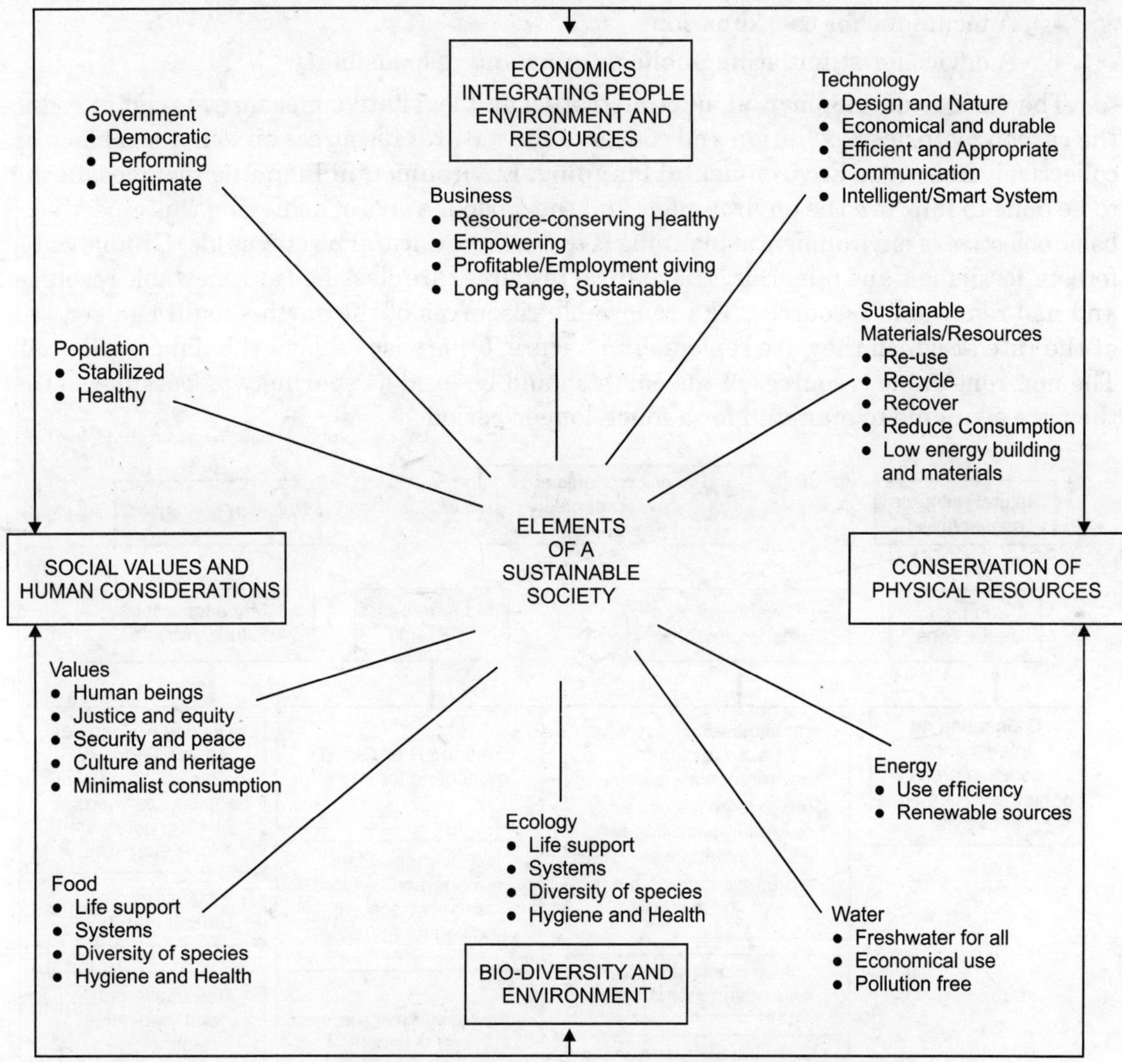

Fig.11.7: Elements of a Sustainable Society

The interaction among physical environment, natural resources, social and human factors and economic factors leads to sustainable development. The interface between physical environment and community factors leads to 'livable' development, and synthesis of community and economic factors leads to equitable development. A systematic and well planned integration of sustainability, livability and equitability would lead to a holistic, ecological development. The following are critical and essential components for the formulation of a plan:

- Sources of information
- A programme for correction
- An indicator of goals
- A technique for co-ordination
- A device for stimulating public interest and responsibility

The various developmental, administrative and legislative measures taken to reduce the carbon footprints, pollution and conserve the natural resources on long term basis are collectively known as Environmental Planning. Environmental Planning includes all that to be done to improve the environment and the various ways of achieving this objective. A basic objective of environmental planning is to conserve natural resources like Groundwater, forests, fossil fuels and minerals. The natural resources are classified as renewable resources and nao-renewable resources. The renewable resources of the earth should be used only at the rate at which they are replenish in nature, otherwise ecological balance will occur. The non-renewable resources of the earth should be used as sparingly as possible so that they are available to mankind for a much longer period.

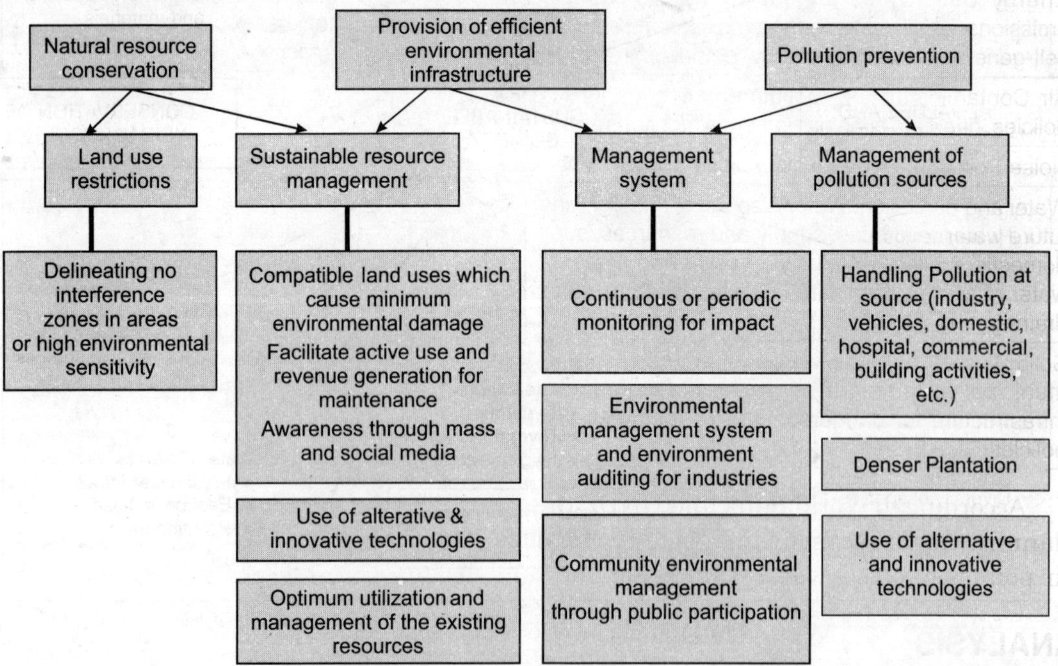

Fig.11.8: Environment Management Approach

An ecological survey is essentially an exercise in awareness. It is systematic investigation of those factors, or 'eco determinants', which deal with relationship of humans and all attendant forms of life to the natural and man-made environment. As such, it provides the basis for a sound land use planning.

Table 11.4: Objective Information for Environmental Planning

Information	Secondary Sources	Primary Sources
Socio-economic: Population, demography, income, employment, services and municipal spending	Municipality, national census	Survey
Housing Conditions: Property tenure, basic services, size and marginal units	Municipality, Ministry of Housing	Survey
Health Conditions: Basic information, mortality rates	Health services	Survey
Natural Environment: Location, type of eco-system, meterological information, drainage conditions, topography and environmental dangers	Municipality, meterological	Survey
Land Use: Urban land use, land ownership, land use law and land value	Municipality	Survey
Urban Transport: Basic information, automobile ownership, type of motorized trips, emissions, accidents, private transport restrictions (flows, stops, scheduling)	Transport infrastructure, transport Ministry/ Department, Police, Planning Department	Survey
Energy Use: Gross annual energy consumption, combustion emissions, inter-connected electric network, electric energy provision, self-generated urban electricity, domestic usage and energy prices	Electric Company, Municipality, Energy services	Survey
Air Contamination: Air pollutant concentrations, emission control policies, quality of air, environmental monitoring, environmental health	Health Services and Municipalities	Survey
Noise Pollution: Noise levels, noise pollution control	Health Services	Survey
Water and Sanitation: Water resources, underground water extraction, future water resources, supply and resources, supply and provision, domestic sanitary installations, sewerage system, industrial waste, water related contaminated policies, water quality monitoring, cost, drainage	Health Services and Municipalities, water company	Survey
Solid Waste: Total solid waste generated, municipal waste, removal, municipal expenditure on waste management, waste deposits, infrastructure for dangerous waste treatment, waste management policies	Health Services and Municipalities	Survey

According to World Bank and UNDP the objective information relevant for environment planning can be focused upon socio-economic, housing, health, land use, transport, energy, air pollution, noise, water, sanitation and solid waste.

ANALYSIS

From the information provided by the eco-determinant maps and reports, site analysis diagrams and a set of planning criteria and land use guidelines are evolved. It identifies the areas of ecological sensitivity, like aquifer, coastal areas, vegetation, bio-diversity and areas of important landscape value which summarizes the eco-survey findings.

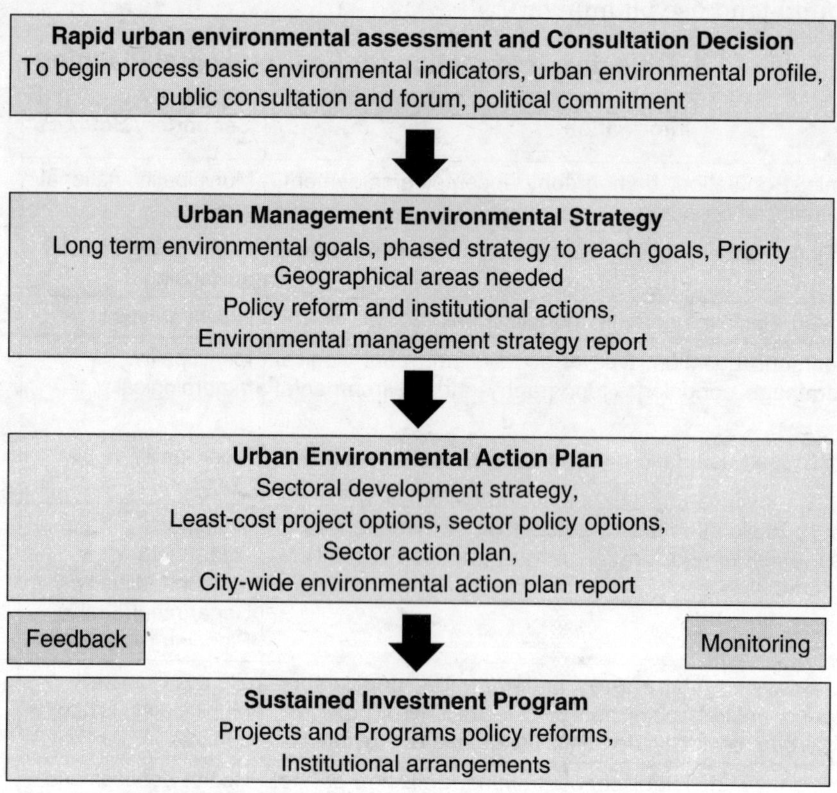

Fig.11.9: Baseline information for environmental planning

Source:UrbanManagementProgramme,WorldBank

The complete and well substantiated eco-survey maps and site analysis diagrams can be prepared on computer. Computer aided design techniques can be experimented for design conclusions and land use guidelines. If the conclusions are sound and the guidelines explicit, the entire plan would be based on a sound footing.

The maps and reports are studied in the light of the particular problem or project at hand and deduction are reached. These are normally recorded in the form of a site-analysis diagram and land use guidelines.

Table 11.5:Analytical Techniques and Applications

Technique	Application
• Spatial analysis of land use, seismic and other maps • Review of survey data and institutional analysis of maps	• Assessment of open space availability, flood prone and seismically active areas • Evaluation and functioning of sewerage and sanitation services • Association between waterborne diseases and bio-chemical oxygen demand

• Preparation and analysis of energy balance • Trend analysis • Organisational analysis • Long term marginal cost pricing • Flow chart	• Sectional concentration of polluting fuel usage and air pollution • Growth and shift in peak-hour transport • Environmental authority by level of government • Evaluation of degree of cost recovery for water and sewerage services • Occupational and environmental health data

Delineating Environmental Conservation Zone

Within most landscape a rea s there are outstanding features which should be left untouched. A strand of beach extensive beds of submerged turtle grass, a mangrove forest, a rise of sea oats, a historical run-such are the elements that give a region its essential quality. Preserve them, and the land, the environs, and life for the observers will long be the richer for it. A study of the project eco-surveys will reveal those site areas and elements, which are of highest ecological value. These should be outlined on the plans and protected in the field against the threat of damage or destruction. Often the presence of an unsuspected feature will be of real benefit to the proposed development, provided the features may be preserved, so that visually important landscape and its ecosystem are preserved.

In a real-life situation, it is necessary to identify focus areas and lay thrust on the critical issues. Although a plan should be holistic, it is not always possible to solve all the problems in a given situation, and within the available tune constraint, available finances and organisational/institutional capacity. The problem, either at the regional, local or sector level, requires planning interventions at the policy conceptual, strategic and operational levels. It should essentially deal with the following dimensions:

- Regional environment policies, conservation, revitalisation and pollution control strategies.
- Evolving standards, parameters and norms.
- Zoning and subdivision regulations, development control by design, transport and communications.
- Natural resource management—water, land, vegetation, air, etc.
- Utility system (drainage, sewage, energy etc.).
- Security, fire, police, and ambulance services.
- Pollution control strategy.
- Community, educational, recreational and cultural facilities
- Financial and investment plan.
- Action planning and program.
- Legal frame, institutions, organisation and public participation.
- Monitoring and environment audit.

Having evolved various options, the next stage is that of decision making. The decision-making process may consist of the following sequential and iterative stages- problem identification, goal/object setting, solution generation, evaluation of alternatives, choice of preferred solution, implementation and monitoring. The Environment situation study leads to development of predictions/perceptions and then to plans.

Land Suitability Matrix for Land Use Planning

Land suitability to be identified for the areas new development in order to preserve green areas, conserve fertile land and protect Groundwater recharge areas.

Land Use Existing \ Proposed	Water	Residential	Commercial	Industrial	Public	Transport	Green
Water Body	•	×	×	×	×		•
Green				×	×		•
Village			•	•	×	•	
Land Suitability GSI							
Agriculture			•	•	•	•	•
Groundwater	•				×	×	•
Green Belt	•	×	×	×	×		•
Pastures			•	•	•	•	
Forests		×	×	×	×		•
Water logged	•	×	×	×	×		

• Most Suitable × Unsuitable

Fig.11.10: Land use Suitability Matrix: Land uses that should be allocated keeping in view the environment

Environmental Stress Mapping

Environmental sensitivity mapping considers based the following parameters:

- Population affected
- Level of air pollution
- Level of noise pollution
- Level of Water pollution
- Level of land degradation
- Groundwater depletion
- Availability of green spaces
- Housing condition
- Availability and condition of civic infrastructure
- Temperature of the area (Heat Island)

Stress Mapping aims at identifying environmentally sensitive degraded areas, where there is a possibility of adverse impact on health of large amount of people. For each factor a map is prepared and the stress count, as given in the table below is calculated. The area can be divided into squares of 2.5 × 2.5 km (variable) each and each area is given a stress factor score. For example, if air in an area is highly polluted, it is identified as air pollution hotspot, and gets a score of 3 for air pollution level.

Table 11.6: Stress Count for Different Parameters

Population affected	<5000	5000-10000	>10000
	1	2	3
Distance from air pollution hotspot	In the square 3	Nearest square 2	Second square 1
Distance from noise pollution hotspot	In the square 3	Nearest square 2	Second square 1
Distance from solid waste pollution hotspot	In the square 3	Nearest square 2	Second square 1
Distance from industrial areas	In the square 3	Nearest square 2	Second square 1
Groundwater déplétion rate	0-2 m 1	2-4 m 2	>4 3
Distance from green areas	In the square 1	Nearest square 2	Second square 3
Housing condition	Slum 3	Unauthorised Urban villages 2	Planned development 1
Infrastructure condition*	All Infrastructure available 1	One of the facilities missing 2	Two or more facilities missing

** Infrastructure here means - water supply, sewage, drainage, electricity and solid waste Score*

Note: Higher the Score worst is the condition of the environment

All the stress counts of all the squares are added to give the composite stress count. This would give the most environmentally stressed areas. Hence, would require a special environmental management plan and a strong monitoring network.

Environmental Impact Assessment

The environmental impact assessment aims to predict the impact of development on the various environmental parameters and the measures that can be proposed for the minimisation of the impacts.

Environmental elements	Air quality	Surface water quality	Surface water quantity	Waste water generation	Depletion of groundwater	Groundwater pollution	Environmental infrastructure	Depletion of green areas	Loss of fertile agricultural areas
1	2	3	4	5	6	7	8	9	10
Demography									
Projected population	▲								
Increase in density	▲		▲	▲	▲		▲	▲	▲

Shelter									
Projected housing (No. of dwelling units)				▲	▲	▲	▲		▲
Redevelopment of inner city				▲	▲		▲		
Permissibility of large plots to subdivide and small plots to amalgamate to rebuild							▲	▲	
Increasing Corresponding Carrying Capacity of Social-Physical Infrastructure and Transport Networks	▲						▲		
High Rise High Density									
Development of Economic Corridors along Transport Networks				▲					
Identifying Redesignated Use Zoning for Underutilized Land Pockets	▲	▲	▲				▲		
Redeveloping Residential Area Built with Low Density				▲			▲	▲	
Adopting Land Pooling Scheme	▲	▲	▲						
Restructuring and Redevelopment of Substandard Areas,				▲			▲	▲	
Notified Slum Areas; Squatter Settlements, Resettlement Colonies	▲	▲	▲	▲			▲	▲	
Unauthorised Colonies; Villages, etc.				▲			▲	▲	
Economic Development									
Regularisation of Existing Non-conforming Industrial Clusters Qualifying the Prescribed Norms		▲		▲					
Development proposed urbanisable area along major transport corridors	▲	▲	▲	▲			▲	▲	▲
Physical Infrastructure									
Projected Additional water requited									
Projected garbage generation@ (0.68 kg pcd)							▲		
Content to be incinerated or autoclaved as per requirement (1.5 kg/bed/ day)			▲	▲	▲		▲		
Estimated power requirement									

Traffic & Transportation Increase in the vehicular traffic curb Side Parking and Increase in Road Carrying Capacity by Development of Full Cross Section of the Roads/ streets and Removal of Encroachment Decongest the existing urban transport network	▲ ▲ ▲							▲ ▲	
Mixed Land use MLU to be permitted on all floors on roads with 24 m ROW and above	▲			▲			▲	▲	
For roads less than 24 m row MLU to be permitted on ground floor only	▲			▲			▲	▲	

**Environmental Infrastructure means: water supply, sewage disposal, solid waste management*

Fig.11.11: Conceptual environment impact assessment at city level

Sustainable City

The cornerstone of making a sustainable city is to adopt an integrated approach towards ecology and the conservation of the natural resources. The composite built environment includes the environmental infrastructure—water supply, sewerage, solid waste disposal and transportation network. Planning attempt to integrate the regional, physical, environmental, transport, social, legal, management, financial and other aspects into a composite whole. It should strike a balance between conflicting demands—citizen freedom versus safeguarding community interests, commercial opportunity versus environmental sustainability, public service versus mandatory procedures. Most important is to develop a vision and ideas as the basis of urban planning. The vision for the future requires a radical transformation of infrastructure systems. Land, water, air, bio-diversity, energy and resource management should be the basis of future growth patterns along with appropriate technology, use of IT and biotechnology. The spatial scale of environmental issues need to be segregated at global, city, community and household levels.

SPATIAL SCALE	CITY/REGION (POLICY LEVEL)	COMMUNITY (LOCAL LEVEL)	HOUSEHOLD (BUILDING LEVEL)
Key Characteristics and Issues	• Acid Rain, Global Warming, Heat Islands, Ozone Layer • Sprawl & Urban Footprints • Air pollution • Water Sources, Recovery of Rivers • Power Plants & Energy	• Water • Sewerage • Garbage • Drainage • Streets/Roads/ Transport • Water	• Shelter • Water Pollution • Air Pollution/ Indoor Pollution • Drainage • Sewage/ Sanitation

	• Natural Resources Management • Health • Poverty • Industry • Highways, Roads & Transport • Waste Treatment • Landfills/Toxic Dumps • Pig. Process • Land Use • Governance and Participation	• Waste Dumping • Hooding • Noise / Stress • Natural Disasters • Micro-Climate • Energy • Safety / security	• Garbage • Ventilation • Disease Vectors • Energy / fuels • Safety / security
Kly Action Areas	• Urban Metabolism • Zero Carbon City • Zero Fossil Energy Dev. • Small Growth and Mixed Use • Public Transport • Minimum Urban Footprints • Energy Efficiency/ Alternative Sources of Energy • Regulatory and Legal Framework • Decentralised and Participator Planning • Intelligent City	• The Organisation of Space • Urban Transport & Fuels • Decentralised Services • Recycle/Reuse • Energy Efficiency / Audit • Water Conservation/ Harvesting • Conservation of Greens & Forests • Mixed Use • Zero-Run Off Drainage • Pig. Standards & Controls • Solid Waste Management	• Design with Nature • Passive Design • Water Conservation & Harvesting • Micro Climate • & Bldg. Materials • Energy Efficiency • Shelter for the Poor • Jobs Generation • Bldg. Bye Laws & Standards • Intelligent Services & Maintenance

Fig.11.12: Spatial Scale of Environmental Issues and Action Areas

The concept of energy efficiency, renewable energy and Zero-fossil Energy Development (ZED) should be incorporated in city planning and building design. In India water is emerging as the most critical sustainability constraint.

Natural Resource Management

Protection of the environmental resource base of human settlements is vital for sustainability. Starting from the Stockholm Conference (1972), the concern for the environment of settlements has been discussed in all major international and national conferences. The strategies and directives that have emerged out of these deliberations had started with initially adopting measures of pollution control, waste minimization, judicious utilization of resources. But with the rapid growth of settlements, both in number and size, alternate methods to assess the environmental status of settlements have been evolved. Assessment of the biodiversity and the energy consumption pattern and their judicious disposition and utilization are emerging strategies for resource conservation. The present paper gives an outline of the role and management of biodiversity and energy for sustainable development of settlements.

Local Responses to Climate Change

Many of the key adaptations to climate change will require individual and collective action at the community-level to build resilience and prevent harmful effects. It is widely accepted that much adaptation will be undertaken incrementally by individuals and households; and that communities and local organizations also have important roles in this. There are many examples of community-driven 'slum' upgrading that greatly reduced environmental health risks and if served with appropriate information and support, these can include attention to climate change risks (which in the next few decades are mostly increased risk levels from hazards already present). The examples show that community organizations have the capacity to build resilience and identify appropriate short- and long-term responses to climate events – if they are adequately supported by local authorities.

Table 11.7 gives examples of the kinds of local adaptation interventions needed by the key sectors drawn from the IPCC. Much of what is listed in the adaptation option/strategy will fall to local government to implement, which needs resources, policy and regulation frameworks from higher levels of government.

Table 11.7: Examples of specific adaptation interventions by sector

Sector	Adaptation option / strategy	Underlying policy framework	Key constraints and opportunities to implementation
Water	Expanded rainwater harvesting; water storage and conservation techniques; water re-use; desalination; water-use and irrigation efficiency	National water policies and integrated water resources management; water-related hazards management	Financial, human resources and physical barriers; integrated water resources management; synergies with other sectors
Infrastructure and settlements	Relocation; seawalls and storm surge barriers; dune reinforcement; land acquisition and creation of marshlands / wetlands as buffer against sea level rise and flooding; protection of existing natural barriers	Standards and regulations that integrate climate change considerations into design; land-use policies; building codes; insurance	Financial and technological barriers; availability of relocation space; integrated policies and management; synergies with sustainable development goals
Human health	Heat-health action plans; emergency medical services; improved climate-sensitive disease surveillance and control; safe water and improved sanitation	Public health policies that recognize climate risk; strengthened health services; regional and international cooperation	Limits to human tolerance (vulnerable groups); knowledge limitations; financial capacity; upgraded health services; improved quality of life

Tourism	Diversification of tourism attractions and revenues; shifting ski slopes to higher altitudes and glaciers; artificial snow-making	Integrated planning (*e.g.* carrying capacity; linkages with other sectors); financial incentives, *e.g.* subsidies and tax credits	Appeal/marketing of new attractions; financial and logistical challenges; Potential adverse impact on other sectors (*e.g.* artificial snow-making may increase energy use); Revenues from new attractions; Involvement of wider group of stakeholders
Transport	Realignment / relocation; design standards and planning for roads, rail and other infrastructure to cope with warming and drainage	Integrating climate change considerations into national transport policy; investment in research and development for special situations, *e.g.* permafrost areas	Financial and technological barriers; availability of less vulnerable routes; improved technologies and integration with key sectors (*e.g.* energy)
Energy	Strengthening of overhead transmission and distribution infrastructure; underground cabling for utilities; energy efficiency; use of renewable sources; reduced dependence on single sources of energy, increased efficiency	National energy policies, regulations, and fiscal and financial incentives to encourage use of alternative sources; incorporating climate change in design standards	Access to viable alternatives; financial and technological barriers; acceptance of new technologies; stimulation of new technologies; use of local resources

Source: Parry et al, 2007

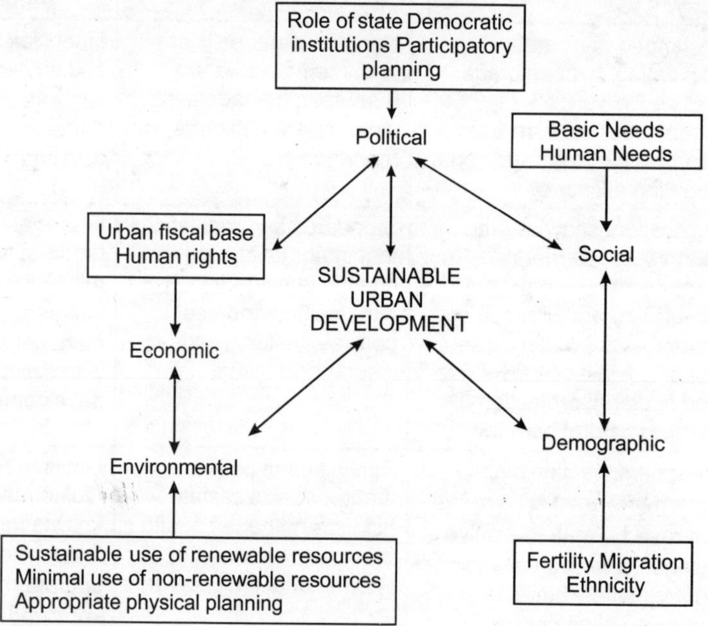

Fig.11.13: Sustainable Development Process

Source: Drakatis-Smith, Urban Studies

In contrast to emphasis on complex EIA, end of pipe pollution control measures, cleaner energy and fuels, car emission norms, etc, what most urban centres in developing countries need is not a climate change adaptation programme but a sustainable urban form that facilitates development (meeting deficits in provision for water, sanitation, drainage, electricity, tenure, healthcare, emergency services, schools, public transport, etc.) into which measures for environmental pollution control and climate-change adaptation are integrated.

Table 11.8: Conventional versus Environmental Planning Approach

Conventional Model	Sustainable Model
The Urban System	
Urban Concentrated Industrial Complexes	Regionally Dispersed Industrial Complexes
Manufacturing oriented	Community oriented
Short term economic growth emphasized	Long term development emphasized
Commodity oriented	Conservation oriented
Consumption driven	Balanced sought between consumption and conservation
Resource seen as input to production system	Resources seen as limited , vulnerable requiring stewardship
Resource intensive, governed by economic priorties	Resource conserve governed by multiple priorities
Economic costs are primary	Economic cost balanced by socio-environmental costs
The Energy System	
Fossil fuel based	Alternative energy based Conservation and renewability emphasized
Energy abundance and cheap supplies emphasised	Reduce energy intensity
Diverse sources of supply	Social/environmental cost based prices
Market based prices not reflecting social/environmental costs	
Technology focused	Conservation focused
Efficiency in economic production emphasized	Efficiency in land uses emphasized
Scale economies and technological centralization emphasized and technological decentralization sought	Modularity
The Environment System	
Human dominate the environment	Humans and environment are seen as mutually dependent
Environment as a abundant source of commodities	Exhaustibility of natural resources recognized
Environment impacts external to economic choice	Environmental impacts internal to economic choice
Rehabilitation oriented	Preservation oriented
The Technology System	
Large-scale economies sought	Moderate scale economies preferred
Centralised system emphasized	Decentralised system emphasized
Infrastructure driven technology	Users driven technology choices
Technical decision governed by economic costs	Technology decisions governed by social/ environmental costs
Environment impacts ignored	Environment sensitive design promoted

Source: Adapted from Byrne J., et al (1992) "Energy and Environmental Sustainability in East and South-east Asia, in IEEE, Technology and Society, Vol.10, No.4, page 26, reproduced in Environment and Urbanisation, Vol. 6, No.1. A

Ian McHarg in his seminal, book, Design with Nature, (1969) states that nature is process and value that has rights for continued existence: His proposition is that any place is the sum of its historic, human, biotic, abiotic, physical and biological processes, that these are dynamic, and they constitute social values, that each area has an intrinsic suitability for certain land uses and finally that certain areas lend themselves to multiple coexisting land uses. This involves superimposing maps of physiographic drainage, soils, and critical natural and cultural resources to relate with different types of human cultures.

Add Illustration X 11.14

Fig.11.14:

The basis of environment planning is the integration of greens and bio-diversity with the conservation of water resources, rivers and natural streams. The greenways and parkways along the natural drainage channels help to protect the natural drainage and also in segregating storm water from wastewater and sewerage. These ideas are reflected in planning of Delhi and Chandigarh.

While planning New Delhi, Lutyens conceived a garden city with buildings below tree height except the buildings in the monumental grid i.e. along the Central Vista. India Gate and Government Complexes were linked by a monumental grid, known as the Central Vista. The India Gate, symbolic city gate was planned on the river side, while the Capital Complex dominating the skyline was placed at the Raisina Hill. A hexagonal grid provided visual continuity of the city with the ancient monuments, like Purana Qila, Jama Masjid, Safdarjung Tomb, Ferozshah Kotla, Lodhi Garden, Humayun Tomb, etc. The new Capital was carefully placed between two natural feaures of the site, viz. River Yamuna on the East and the Ridge towards the West.

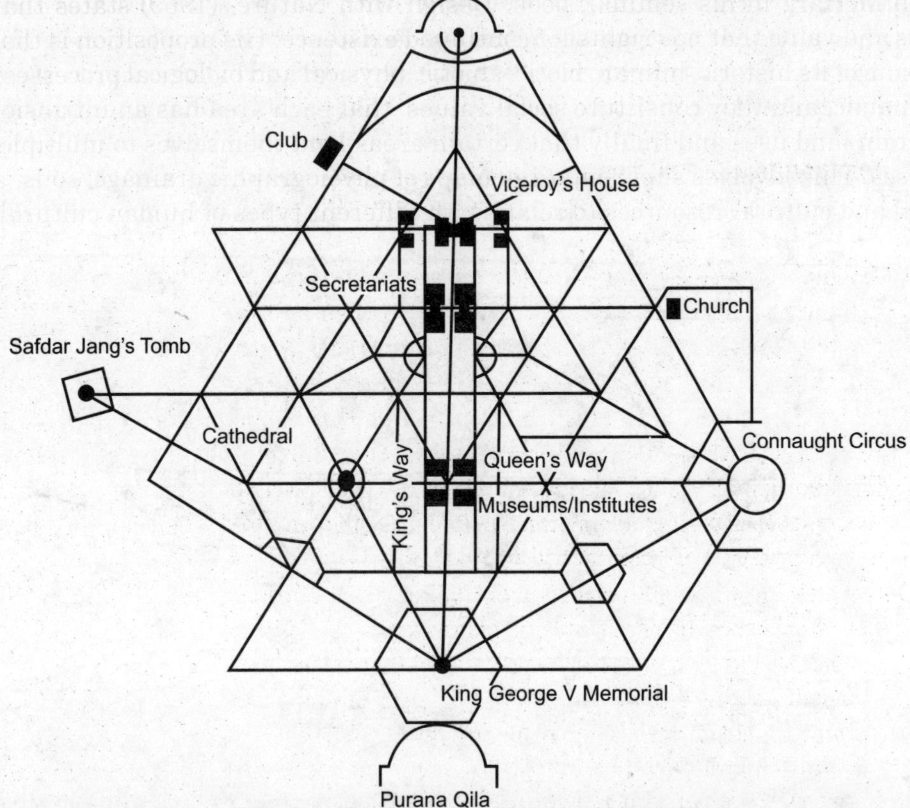

Fig.11.15: Plan of New Delhi, Edwin Lutyens

Fig.11.16: Central Vista is the centre piece of New Delhi

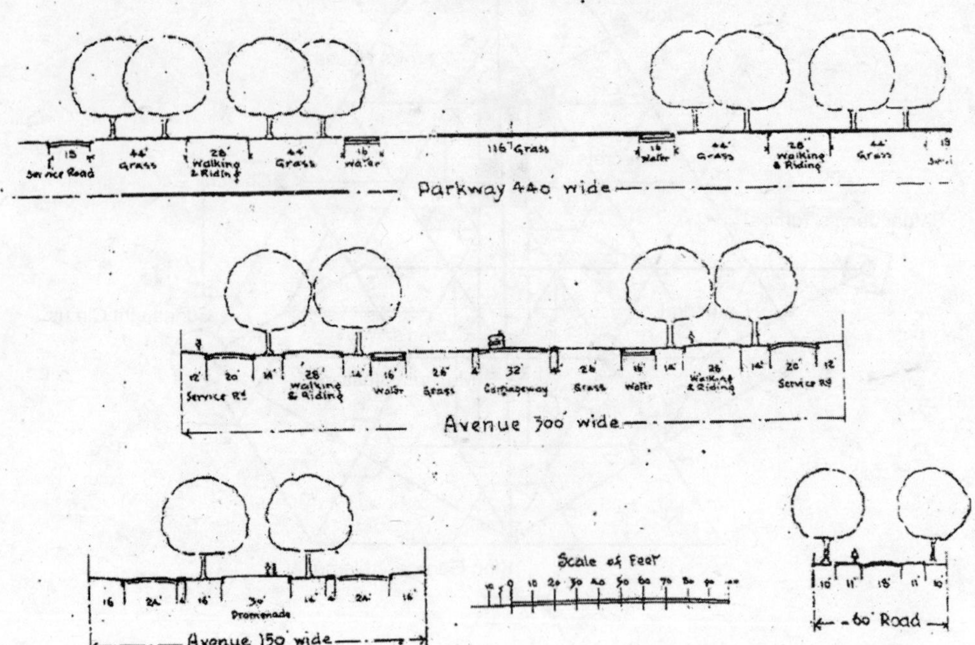

Fig.11.17: Delhi Town Planning Committee proposal for roads as parkways and avenues (1913. For a common-man, wide, comfortable and beautiful roads of New Delhi with overreaching greenery make the image of the Capital. Edwin Lutyens had a vision in providing liberal road widths, which still cater to more than 9 million vehicles in Delhi.

Source: Central PWD Archives

Chandigarh (1954), the capital of Punjab and Haryana was designed by Le Corbusier. It was based on his concept of city as a human system, comprising the head, heart and hands, with roads as the veins. It echoes Corbusier's vision of the city, its human engagement and urban landscape. Corbusier's plan for Chandigarh established a lake that impounded the water flowing from the Shivalik hills and a leisure valley along an existing stream cutting across the city. The capital complex was located at the head of the city, with the hill landscape in the backdrop. A hierarchical system of V-7 roads, streets and pathways provided mobility to its citizens. The City Centre was at the heart of the city. The industrial zone was buffered by a 'inviolable' green belt.

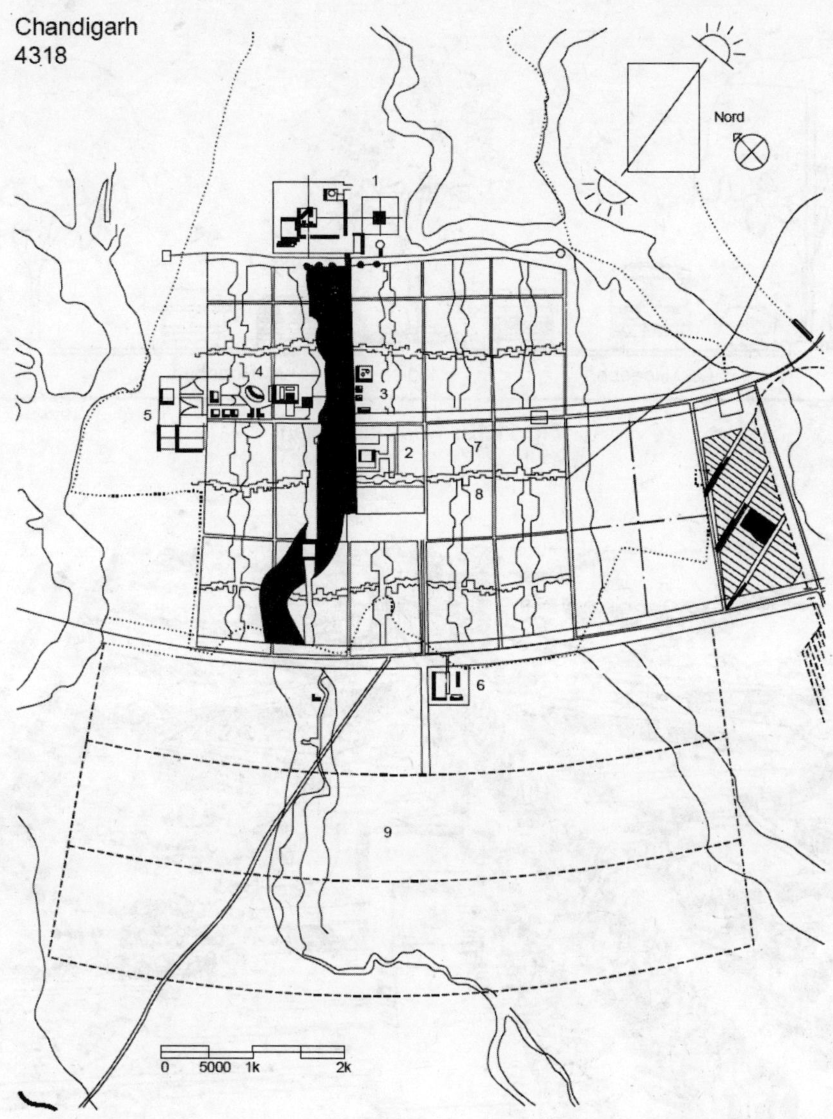

1. Capitol complex
2. Central business district
3. Hotels, restaurants and visitors' center
4. Museums, and stadium
5. University
6. Wholesale market
7. Park bands extending through residential areas
8. Shopping street (V4)
9. Area for contain expansion of the city

Fig.11.18: Plan of Chandigarh by Le Corbusier, 1951

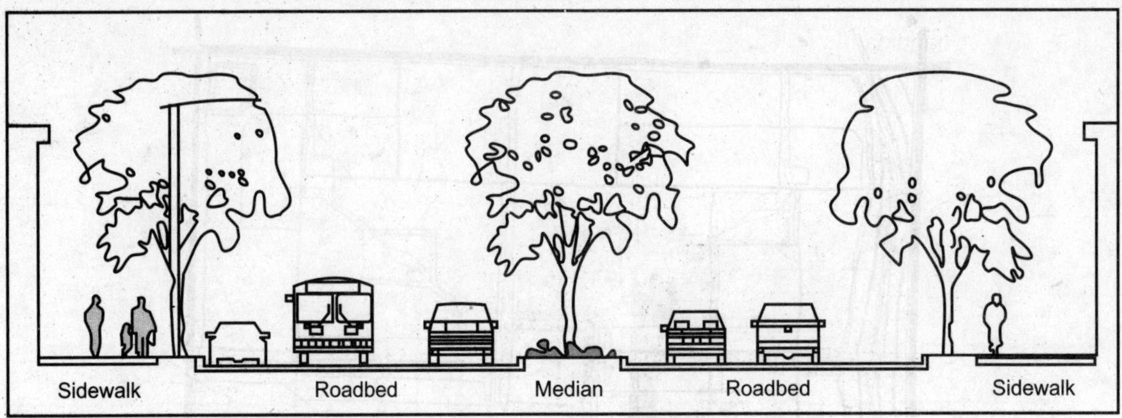

Fig.11.19: Trees in Median

Source: SMWM

Fig.11.20: Roads as Boulevards (A. K. Jain)

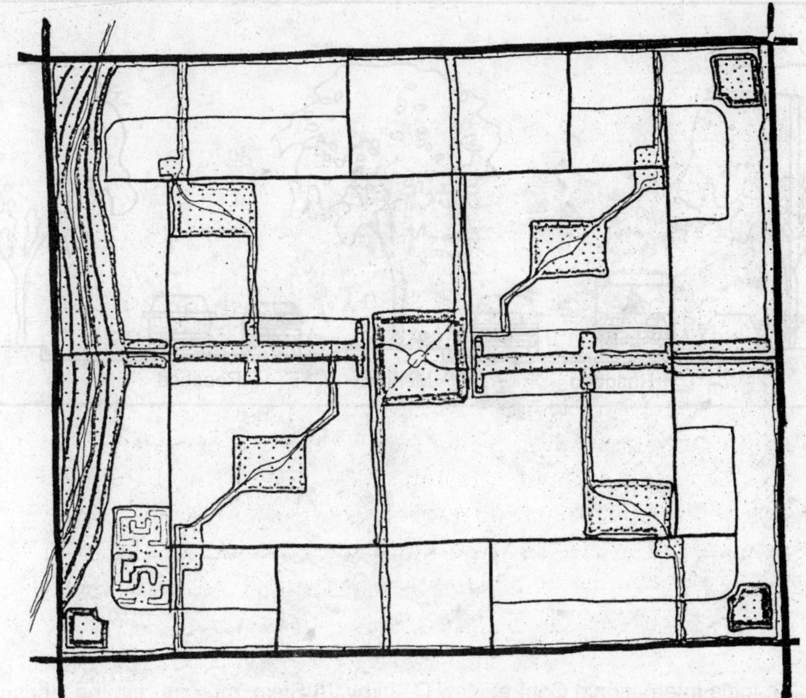

Fig.11.21: Atypical sector comprises four neighbourhoods (A. K. Jain)

Fig.11.22: Tree shaded paved surfaces

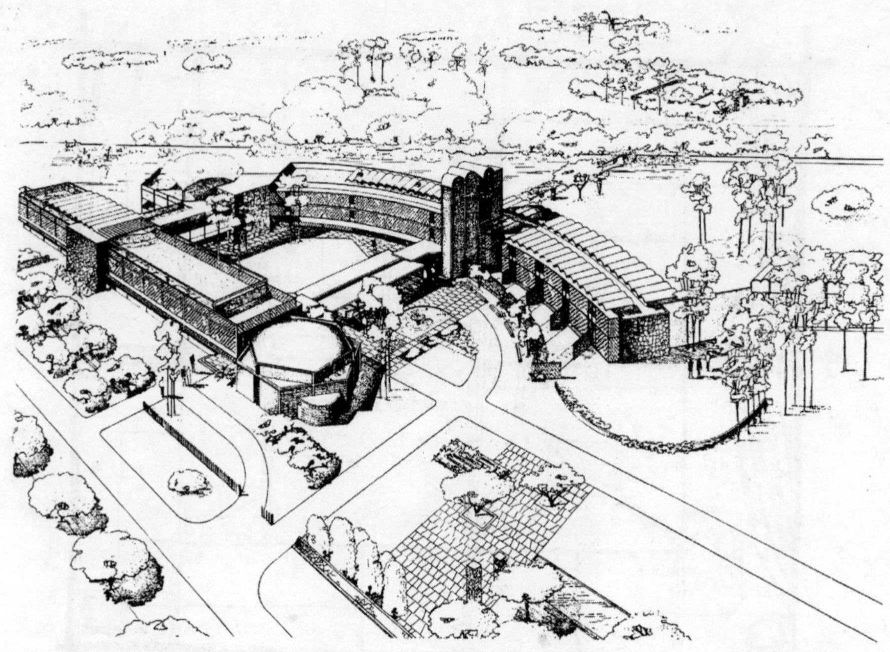

Fig.11.23: India International Centre, New Delhi by JA Stein, merges with the landscape of historic Lodi Garden

Fig.11.24: The new town square in Germany has water channels creativity designed in the landscape

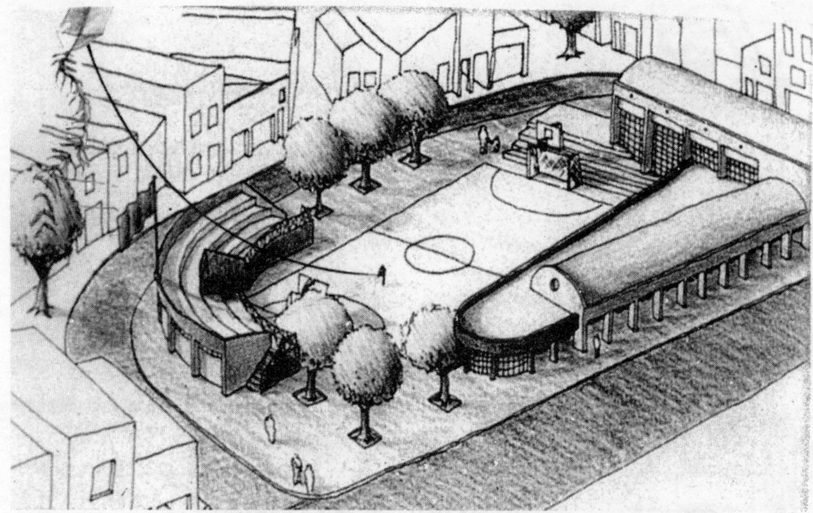

Fig.11.25: Mini-stadium in a school which also functions as an ampitheatre

Fig.11.26: A network of park and playground in a housing development

Fig.11.27: India International Centre by Joseph Allen Stein, who conceived it as 'building in the garden

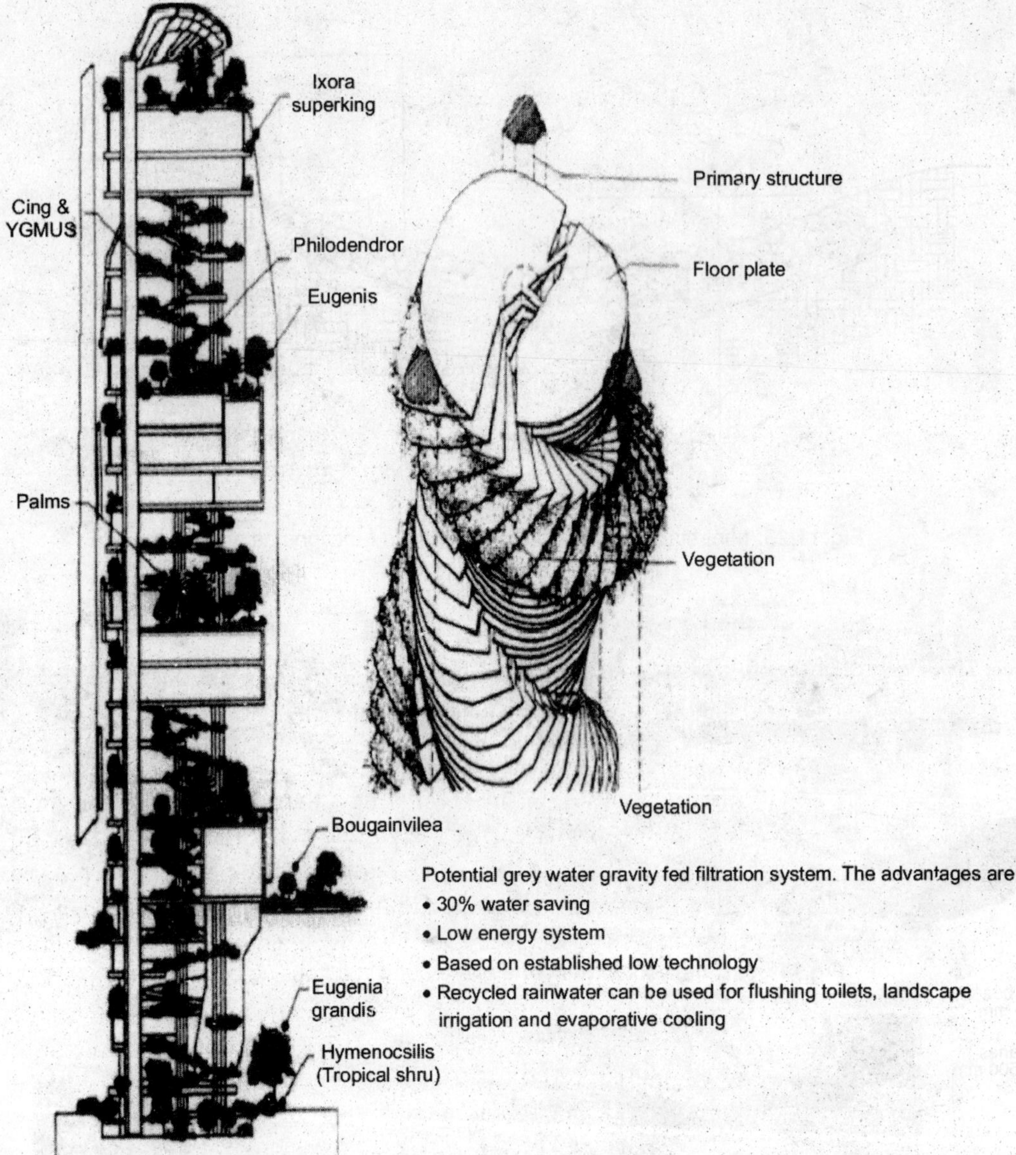

Fig.11.28: Bio-climatic skyscraper: Climate and precipitation determine the building design, its modelling of the fabric and features-Vertical, spiral/ hanging gardens, terraces, courts, atriums, louvers, tiered sun shades and shaded envelope combined with bionic controls and intelligent systems. Architect Ken Yeang

Source: *Yeang Kenneth, 1999, the Green Skyscraper, the Basis for Designing Sustainable Intensive Buildings, Prestel, New York*

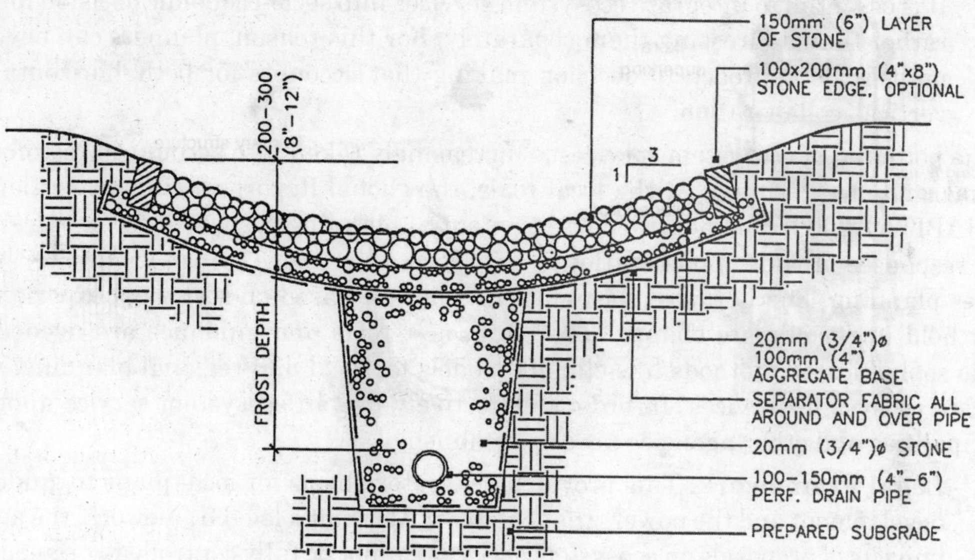

Fig.11.29: Stone lined swale with perforated drain.

Source: *Time-Saver Standards for Landscape Architecture*

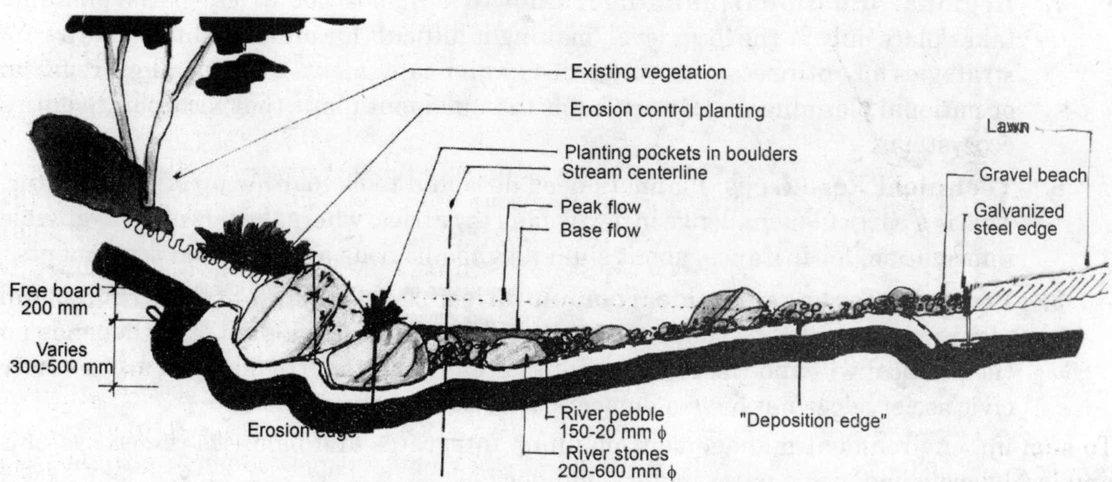

Fig.11.30: Landscape treatment of inflow channel

Source: *Greater Hyderabad Municipal Corporation*

Sustaining ecosystem is no longer just an environmental goal. It is necessary to ensure the conditions for sound economic and social development. Therefore, two key principles need to be applied if we are to integrate an ecosystem services approach into spatial planning:

- Planning must be undertaken for the functional spaces within which people live and work rather than the administrative boundaries of a single municipality or region. Ecosystems and the scales on which they deliver services should therefore be understood as the key building blocks for spatial analysis.

- It is essential to integrate ecosystem services into socio-economic decision making, rather than addressing them separately. For this reason, planners can develop a multi-scale approach to decision making that accounts for both 'horizontal' and vertical' collaboration.

The potential of ecosystem services is increasingly taken into account in regional and national land use planning. At the local scale, the Global Report on Human Settlements (UN-HABITAT 2009) has identified eight potential planning responses for urban zoning. These responses provide opportunities to incorporate the above principles into ecosystem services planning. Furthermore, assumptions that are based on historical experience no longer hold under climate change. Therefore, new tools and guidance are needed that include sophisticated methods like climate models for local and regional planning, which integrate ecosystem services. In order to effectively use an ecosystem service approach, municipalities and other agencies need to establish:

6. **Legal Framework:** This provides a statutory basis for local plans to guide both development and the powers that enforce it. Without a legal framework, the adverse impacts of proposals on ecosystem services cannot be fully controlled or remediated. Planning systems can be made more effective if local communities can design (and redesign) regulatory and legal systems to support effective development.
7. **Regional or national planning framework:** In most countries, spatial planning takes place only at the local level, making it difficult for municipalities to draw up strategies for entire ecosystems (such as water catchments). Developing a regional or national planning framework helps to implement plans that incorporate entire ecosystems.
8. **Technical Resources:** Planners need data and tools to draw up effective plans. This is a particular challenge in developing countries, where there is often negligible information, for instance, about slum neighbourhoods and informal settlements.
9. **Processes for engaging local communities:** The planning is at the core of spatial planning. Community support is essential for an effective plan. This depends on the political will and the resources of the community, particularly in areas where civic society does not have a democratic culture or institutions.

To sum up, environment management planning integrates and balances the ecosystems and local development in a participatory manner.

Principles of Environmental Planning

The Accord signed on environment day on 5th June 2005, San Francisco provides the following principles of environmental planning, which can be adapted according to local problems and priorities:

Energy: Renewable Energy, Energy Efficiency and Climate Change

Action 1: Adopt and implement a policy to increase the use of renewable energy to meet ten percent of the city's peak electrical load within seven years.

Action 2: Adopt and implement a policy to reduce the city's peak electric load by ten percent within seven years through energy efficiency, shifting the timing of energy demands, and conservation measures.

Action 3: Adopt a citywide greenhouse gas reduction plan that reduces the jurisdiction's emissions by twenty five percent by 2030, and which includes a system for accounting and auditing greenhouse gas emissions.

Waste Reduction: Zero Waste, Manufacturer Responsibility and Consumer Responsibility

Action 4: Establish a policy to achieve zero waste to landfills and incinerators by 2040.

Action 5: Adopt a citywide law that reduces the use of a disposable, toxic or non-renewable product category by at least percent.

Action 6: Implement "user-friendly" recycling and composting programs, with the goal of reducing by twenty percent per capita solid-waste disposal to landfill and incineration.

Urban Design: Green Building, Urban Planning and Slums

Action 7: Adopt a policy that mandates a green building rating system standard that applies to all new municipal buildings.

Action 8: Adopt urban planning principles that advance higher density, mixed-use, walkable, bikeable and disabled accessible neighbourhoods which coordinate land use and transportation with open space systems for recreation and ecological restoration.

Action 9: Adopt a policy or implement a program that creates environmentally beneicial jobs in slums and/or low-income neighbourhoods.

Urban Nature: Parks /Habitat Restoration / Wildlife

Action 10: Ensure that there is an accessible park or recreational open space within half-a-kilometer of every city resident.

Action 11: Conduct an inventory of existing canopy coverage in the city; and then establish a goal based on ecological and community considerations to plant and maintain canopy coverage in not less than fifty percent of all available sidewalk plating sites.

Action 12: Pass legislation that protects critical habitat corridors and other key habitat characteristics (*e.g.* water features, food bearing plants, shelter for wildlife, use of native species, etc.) from unsustainable development.

Transportation: Public Transportation /Clean Vehicles /Reducing Congestion

Action 13: Develop and implement a policy which expands affordable public transportation coverage to within half-a kilometer of all city residents in ten years.

Action 14: Pass a law or implement a program that eliminates leaded gasoline (where it is still used); and that phases down sulfur levels in diesel and gasoline fuels, concurrent with using advanced emission controls on all buses, taxis, and public fleets to reduce particulate matter and smog-forming emissions from those fleets by fifty percent.

Action 15: Implement a policy to reduce the percentage of commute trips by single occupancy vehicles by ten percent.

Environmental Health: Toxics Reduction, Healthy Food Systems and Clean Air

Action 16: Every year, identify one product, chemicals, or compounds, that is used within the city that represents the greatest risk to human health and adopt a law to provide incentives to reduce or eliminate its use by the municipal government.

Action 17: Promote the public health and environmental benefits of supporting organic foods. Ensure that twenty percent of all city facilities (including schools) serve locally grown and organic food within seven years.

Action 18: Establish an Air Quality Index (AQI) to measure the level of air pollution and set the goal of reducing by ten percent in seven years the number of days categorized in the AQI range as "unhealthy" to "hazardous."

Water: Water Access & Efficiency, Source Water Conservation and Wastewater Reduction

Action 19: Develop policies to increase adequate access to safe drinking water, aiming at access for all by 2015. For cities with potable water consumption greater than 100 liters per capita per day, adopt and implement policies to reduce consumption by ten percent by 2015.

Action 20: Protect the ecological integrity of the city's primary drinking water sources (i.e. aquifers, rivers, lakes, wetlands and associated eco-systems).

Action 21: Adopt municipal wastewater management guidelines and reduce the volume of untreated wastewater discharge by ten percent in seven years through the expanded use of recycled water and the implementation of a sustainable urban watershed planning process that includes participants of all affected communities and is based on sound economic, social, and environmental principles.

Disaggregating Environmental Planning Components

To make the cities inclusive, safe, resilient and sustainable, the following targets have been given under the UN-SDG Program:

1. Access for All to Housing and Basic Services
2. Sustainable Transport for All
3. Sustainable Planning and environmental Management
4. Safeguarding the Cultural and Natural Heritage
5. Disaster Resilience and Safety
6. Reducing the Environmental Impact and Pollution
7. Universal Access to Green Public Spaces
8. Strengthening Integrated Development Planning
9. Mitigation and Adaptation to Climate Change
10. Sustainable and Resilient Buildings and Resources

It is necessary to adopt an approach which engages with the identity of the city and meets the challenges of its population growth, housing, slums, jobs and employment, gender safety, up-gradation of old and dilapidated areas and provision of adequate infrastructure services and transport. A safe city aims to ensure safety of its citizens especially children, women and other vulnerable sections of society. A resilient city aims to ensure adjustments due to unforeseen disasters, emergencies and climate change. A sustainable involves the reversal of unprecedented air and water pollution, making urban areas clean and hygienic, conservation of natural resources, water bodies/rivers, biodiversity and heritage, along with provision of sustainable infrastructure services, urban transport and renewable energy.

1. Access for All to Housing and Basic Services

There is a need to promote and incentivize brown field development, including in-situ slum rehabilitation, regularization and redevelopment, densification and up-gradation of unauthorised colonies/villages and old residential areas. This requires the reservation of minimum 15 percent of land/FAR or 35 percent of the dwelling units for the EWS and lower income category. Compact, high rise and high density housing pattern can enable optimum use of land and other resources.

Table 11.9: Green and Brown Agendas for Urban Planning

Green agenda (ecological systems)	Brown agenda (human systems)
Ecosystems that provide green/recreation space and biodiversity protection	Waste systems that recycle and remove (solid, liquid, air) wastes from cities.
Water systems that provide a natural flow water supply and waste disposal	Energy systems that provide power, heating, cooling and lighting for city functions.
Climate and air systems that provide cities with a healthy environment	Transport systems (including fuel) that enable mobility in the city.
Agricultural and forestry systems (and that ecological services) that provide food and fiber for cities.	Building and resources/materials systems that provide the physical infrastructure of cities.

Source: Adapted from UN-HABITAT (2009).

By using land as a resource, market sale component of housing can subsidize social housing. This requires:

- Promoting a broader concept of upgrading with services, jobs and community participation,
- Variations rather than standard solutions,
- Promoting rental social housing,
- Local economic promotion and poverty reduction, and
- Security of tenure.

Inclusive housing involves the provision of mixed land use, jobs, social facilities and sustainable services, viz. drainage and flood control, sewerage, water supply, waste water recycling, rain water harvesting, solid waste management, public and individual toilets, roads, electricity, street lights, etc.

Fig.11.31: Shukhobrishti, Rajarahat, Kolkata, developed by Shapoorji Pallonji in association with the KMDA and West Bengal Housing Infrastructure Development Corporation, covers 150 acres and has 10,444 (LIG) apartments of 320 sqft, 3840 MIG apartments of 480 sqft. and 5716 apartments 690 sq.ft. subsidized by part commercial use, MIG, HIG (40%)

2. Sustainable Transport for All

Transport contributes nearly two-thirds of the total suspended particulate matter and 18 per cent of carbon emission posing a serious threat to the environment and health of the people. It is time to think and provide Integrated Transit Corridors (ITC) for BRT, Metro and trains linked with pedestrian and cycle lanes and flanked by public, semi-public, high density, high rise development. The following are the main principles of road design:

Metro, trains, sub-way and primary roads run underground for easy bike and pedestrian traffic on the grade. Besides controlling growth of private vehicles, it is necessary to explore parking space under stilts, multi-level, on roofs and in underground spaces. Seamless multimodal public transport system comprising bus rapid transit, rail based mass transport system would work only by adoption of single ticketing and restructuring land uses by transit oriented development. Subterranean garages near commuter destination reduce the need for ground parking. Digital parking meters tell mobile phone when a space opens up, reducing traffic caused by drivers trolling for space.

The concept of walk to work, transit oriented development, travel demand management and smart growth, promoting NMTs, multi-modal integration, last mile connectivity and e-governance are the pillars of sustainable urban mobility. River/water transport should be explored which is almost pollution free and most cost-effective. The concepts of cordon pricing, minimum occupancy vehicles, ceiling on new registration of private vehicles and establishment of a Unified Metropolitan Transport Authority can contribute in sustainable and coordinated urban transport.

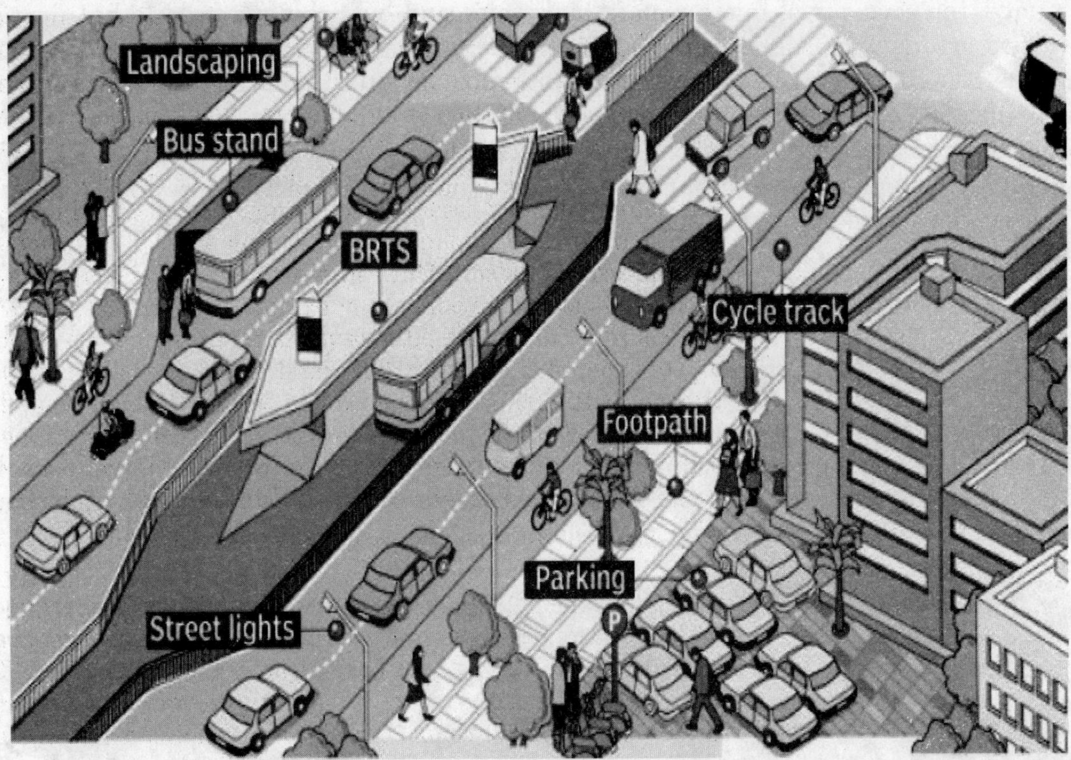

Fig.11.32: Elements of Street Design Footpaths, cycle tracks, carriageway, storm water drainage, Bus Rapid Transit System, bus stop, street vendor, pedestrian crossing, underground utility, street furniture, median, service lane, trafic calming element, parking, landscaping and streetlights.

Source: DUAC/ Amit Ghosal, Punjabi Bagh Ward No. 103, New Delhi

3. Sustainable Planning and Environmental Management

In India more than 65 million people live in slums, and unplanned settlements almost without hygienic sanitation, drinking water facility and a clean place to urinate and defecate. The public and individual toilet facility, particularly for women and children, is a critical area of concern. Intelligent, alternative and innovative ways to provide services, *i.e.* water, power, drainage and solid waste management and renewable energy have to be introduced with emphasis on servicing the informal, unplanned settlements.

The performance of present urban services and systems needs to be reassessed with reference to environment, hygiene and public health. About 50 to 60 percent area in Indian cities do not have clean drinking water supply, drainage, effluent treatment, wastewater recycling, sanitation, sewerage and solid waste management facilities, leading to serious issues in public health. Widespread method of land filling for solid waste disposal is an environmental disaster. In order to safeguard public health the urban society has to adopt re-cycling (*e.g.* water, solid waste, land), regeneration (dilapidated, old areas), recovery (wastelands, encroached areas), restructuring (networks, land use, transport), recharging (traditional areas) and rejuvenation (river, water bodies, parks etc.). Instead of conventional drainage, the concepts of micro-runoff, bio-swales and bio-drainage need to be adopted. Ponds/reservoirs and sediments traps can be planned in the catchment zones on low-lying grounds. The concepts of watershed development, rainwater harvesting and conservation and recharging of groundwater need to be institutionalized and legalised. Various options can be explored for augmentation of physical infrastructure, such as technology upgradation, dual piping system, rain water harvesting, energy audit and exploring alternative sources of energy. Several technologies, such as decentralized and compact water treatment units, solar/aerobic/oxidation and root zone systems can be employed to purify potable water from the natural sources. Various alternative technologies, like Extended Aeration Technique, Biogas production, Bubble Diffusion process, Floatation, Anaerobic Reactors, etc., are already available, which can be explored for urban sanitation. Decentralized systems based on recycling, energy generation and organic decomposing can be explored for solid waste treatment. Bio-reactor composting and vessel system (Tunnel Reactor) are new generation technologies which can be employed for treatment of urban solid waste. All the public buildings, slums, roadsides and parks, etc. need to be provided with public toilets, garbage bins, water points/public taps, etc.

Fig.11.33: Amravati, New Capital of Andhra Pradesh

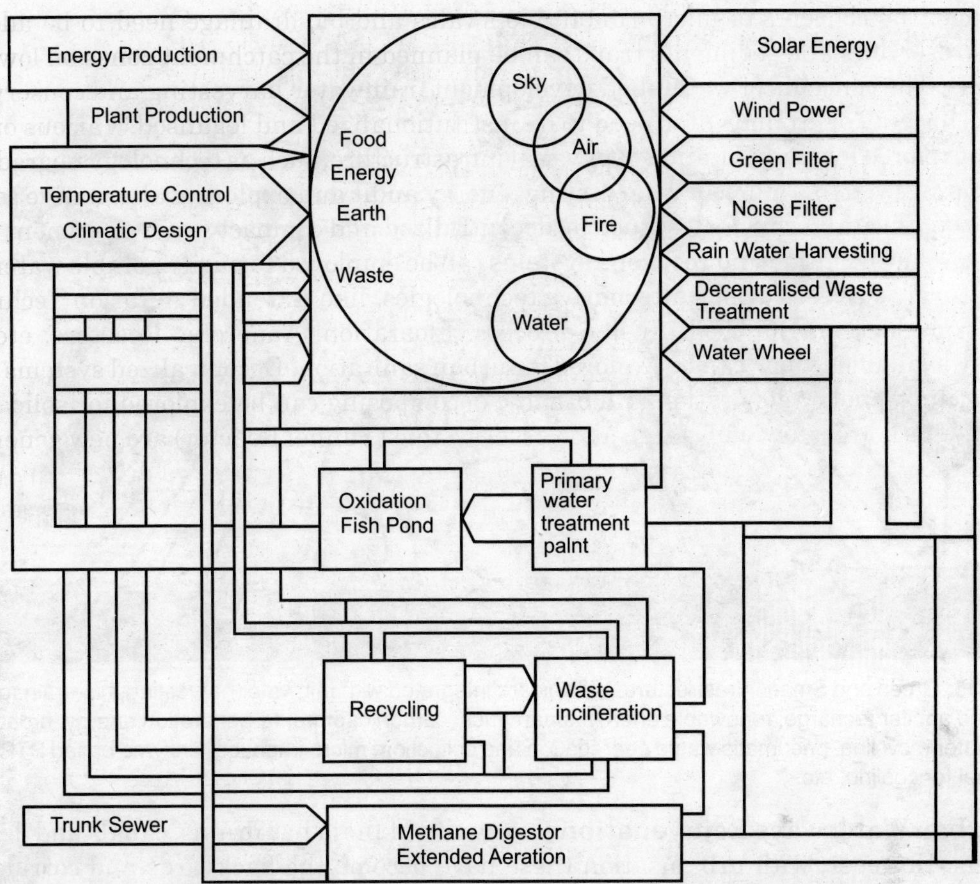

Fig.11.34: The urban process should embody the principle of circular metabolism, which means recycling wastes to produce energy, adoption of eco-technology for zero polluting energy (wind, solar and water wheel), development of green filters, noise buffers and rainwater harvesting.

Local capacity is intrinsically linked to that of the knowledge-based economy where innovation and technology are main drivers of growth, and the collective community intelligence, which underlines capacity and networks as main drivers of a community's success. This requires a planning paradigm pertinent for urban-regional development and innovation management, similar to the related concept of 'intelligent cities' (or communities, clusters, districts and multi-cluster territories). By developing sector-focused, cluster-based or more complex intelligent city strategies, territories can set in motion innovation mechanisms of global dimensions and enhance substantially their innovation systems. The critical factors those affect resource use and CO_2 emissions, such as energy use, transportation, waste prevention air quality, water quality, affordable housing, green space and transit. This implies the need to rethink and reshape the urban environment. Comprehensive transportation infrastructure, zoning, building codes, and waste management, open space and green jobs, make the cities more efficient, sustainable, and smarter.

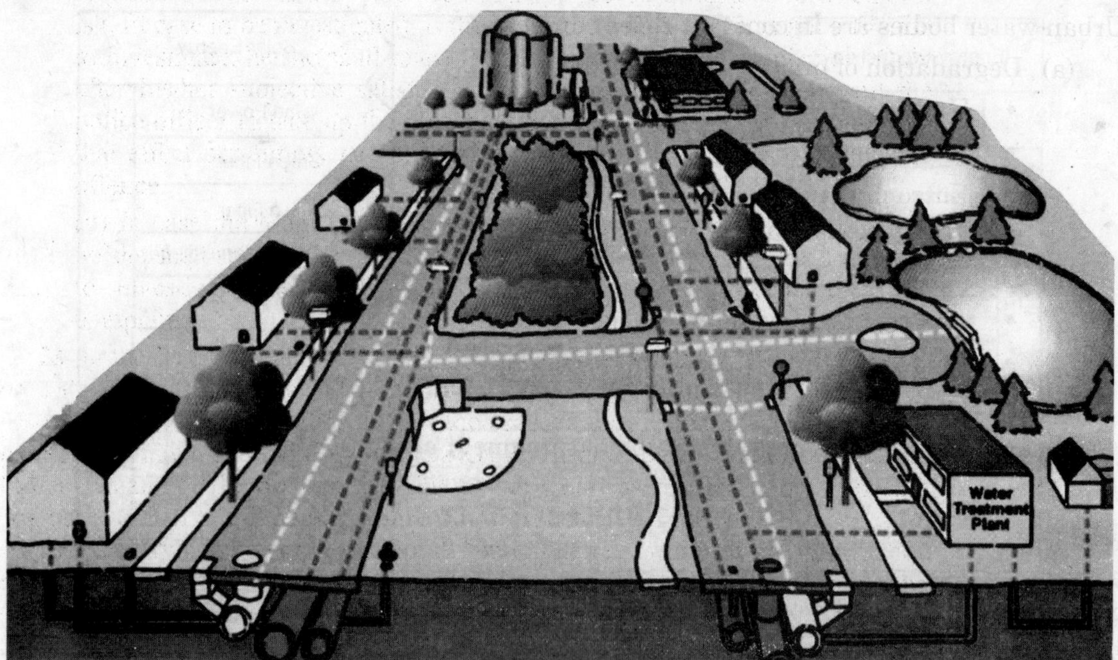

Fig.11.35: Green and Smart Infrastructure: Utilities are integrated with rainwater harvesting, bio-drainage, bio-remedied aquifer recharge, renewable energy, smart grids, carbon capture, tri-generation energy, broadband, wastewater recycling, pneumatic waste segregation and collection, micro irrigation, enzyme based STP, earth air tunnel for cooling, etc.

Urban Waterways Rejuvenation: Every city in India has drains, canals and derelict streams. However, with urbanisation these have become the backyards and corridors of insanitation, mosquitoes, encroachments and garbage dropping. These can be revived to provide valuable open space, Ghats and greenery to the people.

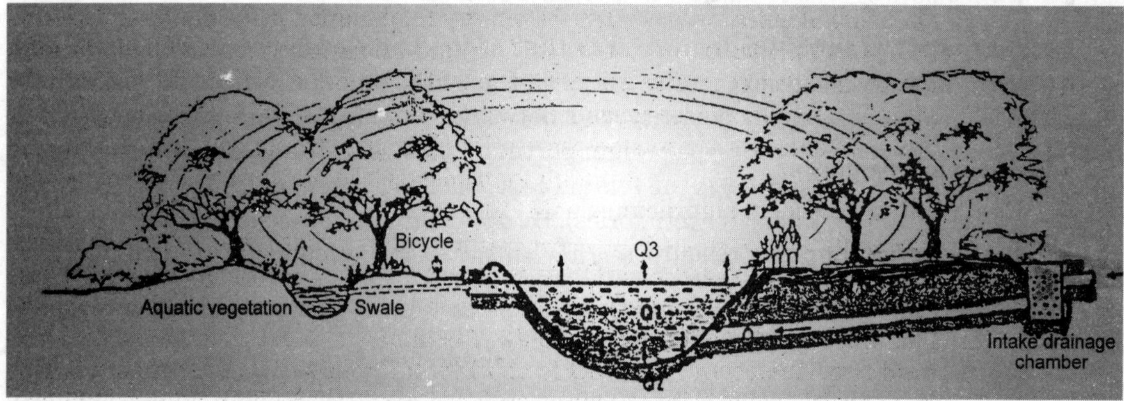

Fig.11.36: Najafgarh drain in Dwarka sub-city, New Delhi was originally Hindon River, is badly polluted. It is proposed to be recovered by pollution control, boating facilities, walk-ways, jogging and bicycle tracks, vegetation and adoption of the concept of bio-swale and balancing reservoirs to harvest the storm water

Urban water bodies are in constant threat due to :

(a) Degradation of quality
- Undertreated sewage discharge
- Solid waste disposal
- Eutrophication
- Floating vegetables

(b) Degradation of quantity
- Abstraction of water for variety of uses
- Encroachments
- Drainage and alternative land use
- Silt influx from degradation of catchment area

Urban ponds and water bodies, lakes are very important part of eco-system as they recharge groundwater, filter water is aesthetically pleasing, provide spaces for recreational activities, prevent floods and allow sediment retention. These are venues for religious and cultural activities and help to promote bio-diversity and stabilizes climate, However, urban lakes are under constant threat due to degradation of quality of waters, undertreated sewage discharge, solid-waste disposal, eutrophication, floating vegetation, There is over abstraction of water for variety of uses.

Encroachments, squatters and silt influx from degradation of catchment areas are continuously endangering the lakes. There is a need to prevent disposal of municipal and/or industrial effluent, solid-waste, idol immersion. The encroachments and over-exploitation of resources like fish, sand and water need to be checked.

Water eco-systems should aim to:

- Recharge Groundwater
- Filter water
- Aesthetically pleasing
- Recreational activities
- Prevent floods
- Sediment retention
- Religious and cultural significance
- Stabilize and improve resources like water, fish and aquatic flora

The following flow chart represents the process for lake rejuvenation from initial data collection, treatment options to planning and action.

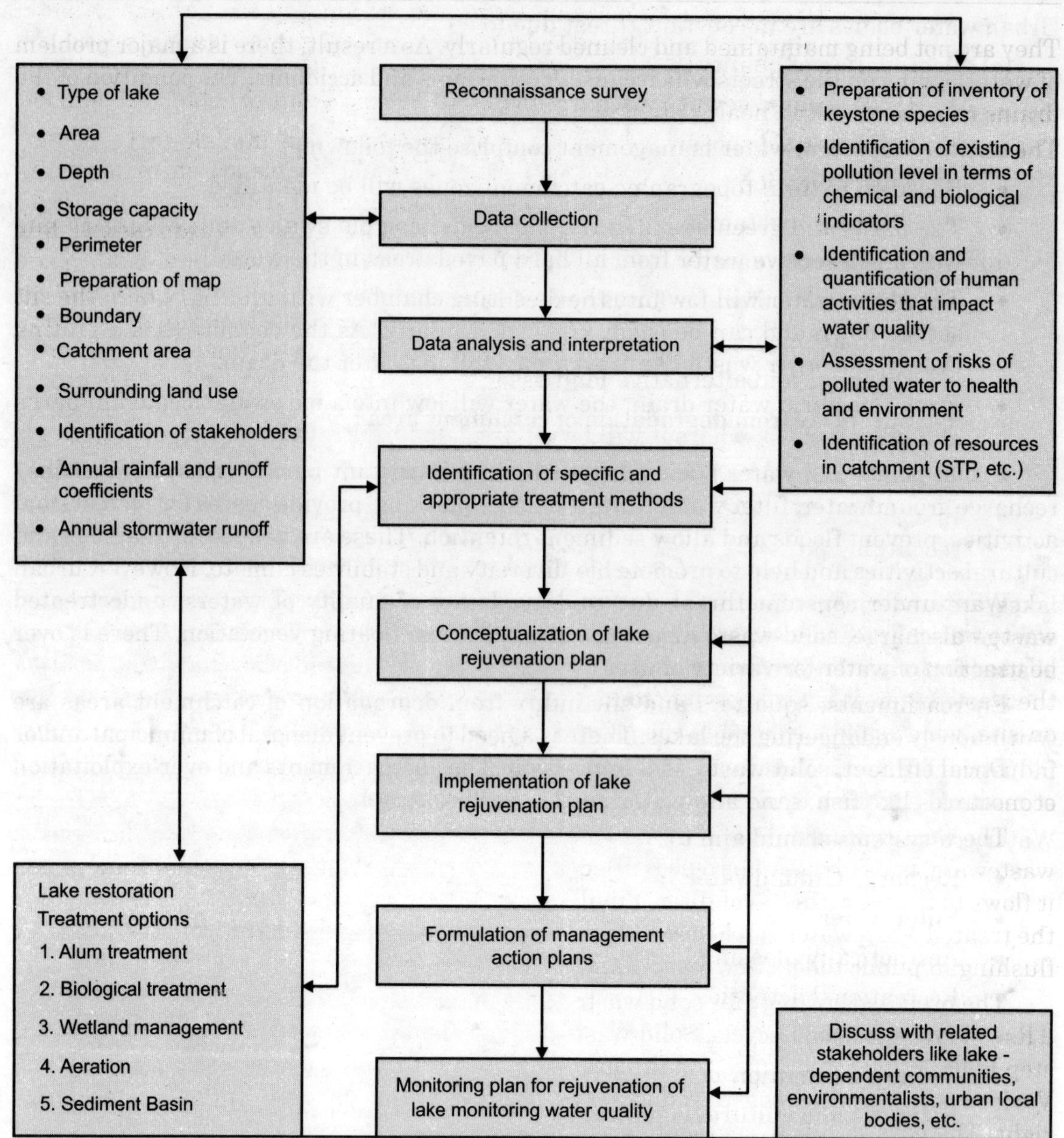

Fig.11.37: Methodology of Rejuvenation of Lakes

Source: DUAC, Amit Ghosal, 2016

Storm Water Management

Usually there are no gratings on drain inlets and de-silting chambers, due to which silt, leaves and a lot of garbage low directly into the storm water drains. These drains get clogged due to garbage and silt and results in water overflowing on to streets. Cleaning these drains is a tedious process as it involves the cleaning of the whole length of the drain.

They are not being maintained and cleaned regularly. As a result, there is a major problem of waterlogging on the streets which causes traffic jams and accidents. The condition of the drains can also result in health problems.

The strategy for storm water management comprise the following:

- Based on natural topography, catchment zones will be identified.
- The parks and green strips in the ward can have bio-swales and recharging pits which will receive water from all hard paved areas in the ward.
- The storm water will low into the de-silting chamber with grating, where the silt settles down and can be easily cleaned regularly. As the chamber has a grating over it, the other waste like leaves, etc. will not enter the drain.
- From the storm water drain, the water will low into a bio swale (located in parks and green strips), where it will be retained and infiltrated.
- This water can then be recharged into the ground.
- In case of space constraints, the water from the drain can directly low into the recharging pit.

Wastewater Treatment: The drains have become polluted due to inflow of untreated waste water. Treated wastewater that can be generated from the dwelling units which can be used for irrigation/gardening and for flushing of toilets. Instead of mechanically treating the wastewater, it is also possible to install decentralized wastewater treatment systems on-site, which can be located in the green areas or along the drains.

Decentralized Wastewater Treatment System (DEWATS) are self-sustainable, economical and easy to maintain. The working of the system is as follows:

The wastewater first goes into a two-chamber settler. After the primary treatment the wastewater goes to the Anaerobic Baffled Reactor which is a nine-chambered system. Then it flows to the planted gravel filter. Finally to the polishing pond. From the polishing pond the treated wastewater can be reused for landscaping of urban greens within the ward, for flushing in public toilets and for cleaning roads.

The proposed strategy for solid waste management should be based on the principle of 3 Rs – Reduce, Reuse, Recycle. Solid-waste management is a step by- step process and each step holds equal importance. As the first step, it is essential to reduce waste generation. Various ways in which waste reduction can be achieved is depicted in the images on the right. The next step is to segregate waste which can be achieved by keeping colour-coded segregated waste bins. Storage and collection of this waste from households should be done in segregated chambers. The waste can then be transported and treated to be reused or recycled according to its category.

The people in urban areas suffer from a high level of air borne and water borne diseases. It is estimated that 2.5 lakh children in Delhi suffer from asthma and chest ailments which are 12 time higher than the national average. Gastro enteritis and cholera break up frequently in various parts of the city. This is largely due to percolation of leachate and pathogens into the water supply system, especially sub-soil water on which large segments of population of unauthorized colonies and slums survive.

It is necessary to adopt a health oriented and environment friendly approach, particularly in respect of solid waste management, water supply, drainage and sewerage. In place of public toilets, it is necessary to encourage construction of individual toilets in the Resettlement Colonies and Slum Clusters. In case of solid waste management, adoption of recycling is necessary for improving the situation. Decentralised systems of sewerage, water supply, and waste management with community participation offer practical solutions for unauthorized colonies, villages and other areas without municipal service networks.

By proper sanitation, development and maintenance of community toilets in the localities of the poor and promoting installation of water closets (WCs) by tax reliefs, the Gandhian ideology of scavenger free society would be achieved.

To check services related health hazards, it is necessary to fix and upgrade obsolete water supply and sanitation system, replace old pipes and make the joints water tights. There has to be proper segregation of water supply and sewerage pipe lines. There should be safe distance between case pool/exudation pond/septic tanks and hand water pumps/ water storage.

The practice of disposal of waste water together with storm-water and garbage in open drains should be discontinued and waste water and rainwater disposal systems must be segregated. Proper landscaping and covering of open drains will help in achieving a better environment. To discourage breeding of mosquitoes in water bodies, fountains and regular aeration of water can be effective methods. The heat generated from sewage treatment and incineration which is presently dissipated, should be captured for generation of electricity and cooling of buildings.

The quality of water in most of the rivers and water bodies in urban areas is unfit for human consumption and a health hazard. To regenerate the rivers and control water pollution, it is necessary to develop an integrated approach towards water-sanitation-sewage and drainage. This involves inter-departmental coordination and an integrated approach.

Fig.11.38: Masdar Zero Carbon City

The concepts of watershed development and rainwater harvesting, together with recharging of underGroundwater table can substantially resolve the water crisis and improve the micro-climate. A system of connected greenways can act as a recreational and health reservoir and act as air filter for vehicular pollution. These concepts are being increasingly adopted in the planning of new areas. These have a direct bearing on improving the environment and making the city healthy and liveable.

Renewable Energy: The power demands in cities are growing exponentially, generation of which emits about one-fourth of carbon footprint. This needs action to reduce the power demand, and resorting to energy efficiency. Load management techniques should be adopted and schemes to minimise power losses should be enforced. Renewable energy sources like solar/wind energy needs mandatory incentives and instruments. Cooling, heating and lighting often accounts for bulk of the electrical consumption, which can be reduced by adopting energy efficiency practices, such as:

- Incorporating daylight design in all the buildings on a mandatory basis
- Adoption of energy efficient practices by regulation and energy labeling of lighting, air-conditioning, air-cooling products, building glass, and imposition of minimum efficiency of lighting and electrical installations.
- Utilisation of heat reflective insulation on the roofs
- Use of renewable sources of energy
- Use of technology which would consume minimum energy to maintain desired human comfort conditions. For example, replacement of bulbs by fluorescent lamps/ CFLs may bring in substantial savings in electrical consumption.
- Adoption of alternative cooling and air-conditioning systems, such as passive evaporative cooling micronisers (PECM), passive downdraft evaporative cooling (PDEC), earth air tunnel cooling, shafts and courtyards for natural cooling and ventilation.

India is endowed with abundance of sun. However, its potential in producing energy is still to be explored. In a cloudless region roof area of 100 sq. mt, receives about 500 KW per day from eight hours of sunshine. In the hot-arid regions (the greatest solar radiation intensity occurs between latitudes 15 and 35 degree north or south) more than 3,000 hours of solar radiation could be obtained and even in the warm-humid region there are about 2300 hours of solar variation. Where there is a lack of conventional materials such as coal, oil, or gas, it is that use is made of the solar energy which is generally abundant in India. Solar water heating for domestic hot water or as part of a home heating scheme has become more widely used in some countries as a response to the energy and environment crisis. As solar energy is initially captured in the form of heat, it is therefore best use of a variety of heating purposes. Solar water heaters (*e.g.* for canteens, schools, homes and hotels), solar steam cookers and water distillation units provide examples of practical applications of solar energy. It is time that solar energy is made mandatory for cooking in the establishment (hotels, restaurants, railway stations, langars, religious centers, etc) wherever more than 3000 meals are cooked per day. This has already been successfully adopted in the Brahmkumaris headquarters, at Mount Abu in Rajasthan, Auroville, Akshardham temple in New Delhi and Sai Baba Sansthan in Shirdi.

Photovoltaic power generating systems can be used in remote electric supplies for water pumps, electric fences, communications and navigation equipment; stand-alone power systems for consumer products like computers, radios, walkway lighting and auto-battery chargers. These are already in practical use, which need to be taken further as a movement.

Wind power, which is inexhaustible, and non-polluting, is a promising way out of the energy crisis. The grouping of large windmills is capable of generating 10,000 or more kilowatts of electric power. Already new type of vertical-axis wind turbines have been designed and installed in various parts of the country, which is producing electrical energy and is being utilised for grinding and pump water. They are cheaper and far more efficient than the traditional horizontal-axis wind mills.

For environmental sustainability, it is necessary to adopt energy efficiency and renewable energy at household levels. This requires adoption of green building guidelines which includes shading, renewable energy, wall insulation, green roof, energy efficient lighting and equipment, etc., which help in reducing the energy load of the building. The buildings should be compliant with energy efficiency codes and mandates.

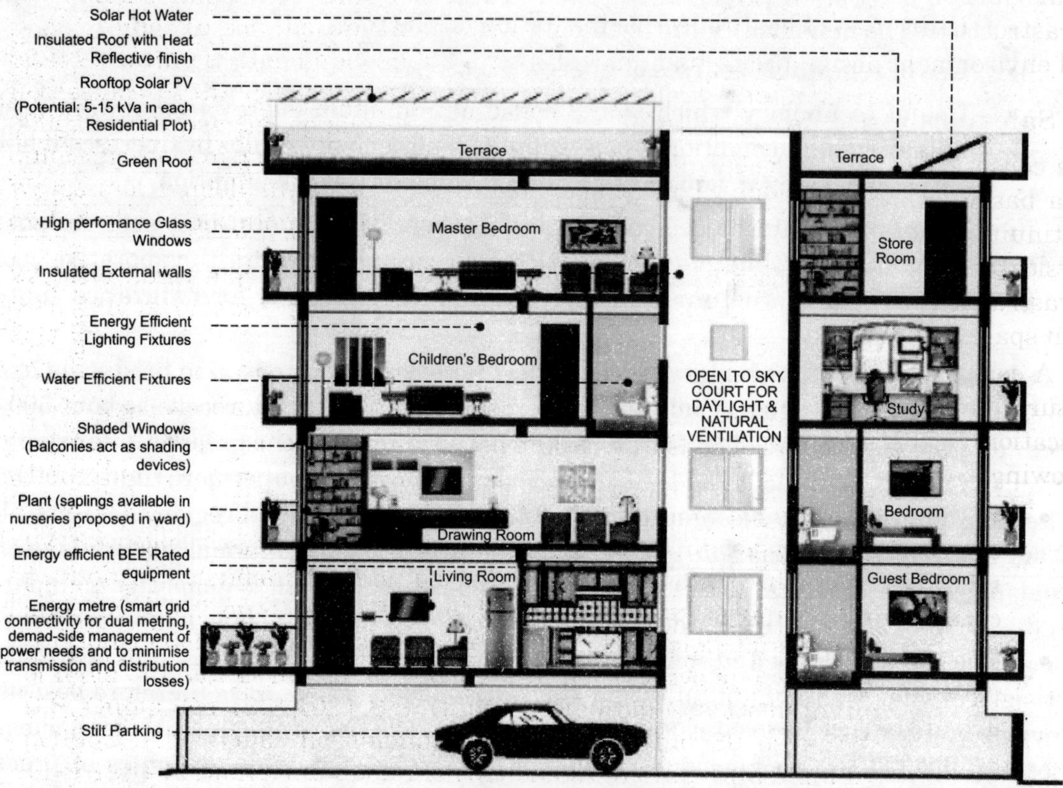

Fig.11.39: Typical Section for proposed Residential Building in Ward (Compliance with Energy Efficiency Codes and Mandate)

Source: DUAC/Amit Ghosal 2016 , Punjabi Bagh , Ward 103, New Delhi

Common utility ducts or tunnels carrying electricity, water pneumatic waste ducts, cable television and broadband internet minimize damage from traffic, road repairs, rains, etc. and make repairs easier. A series of low carbon zones across the city with co-located tri-generation energy systems (combining power, cooling and heating), dual piping for recycled water and automated, segregated waste collection and recycling would lead to bundling 'green infrastructure' together. Solid waste extracted from sewage at treatment plants is burned to make electricity. Three bins recycling is adopted with separate bins for trash, recyclable and compost. Collection charges drop as trash drops. Bio-technology, enzyme based STP, bio-remedial treatment, sludge gas/energy recovery, vermi-culture, fossilization and compositing options can be explored for waste treatment.

Rooftops painted white reflect heat, lowering a building's cooling cost and a city's heat build-up. Rooftop vegetation insulates building against heat and cold and absorbs storm-water. Passive evaporative draught cooling (PEDC) and Earth Air Tunnel (EAT) incorporating ingress of naturally cooled air 4m below ground level via clay pipes saves about 60 per cent of air conditioning load.

Taxing property owners on the volume of storm-water and waste water from kitchen and toilets that runs off their property promotes retrofits that reduce waste water volume. Zero-run off, swales, porous paving, bio-drainage and storm-surge gates in river, drains and canals can be operated when storm surges are expected to protect low lying and subterranean infrastructure. The neighbourhoods should be planned to the highest leadership in energy and environment design (LEED) standards that save energy, materials and emissions.

4. Safeguarding the Cultural and Natural Heritage

The conservation approach for monuments and listed heritage structures needs to be area based and delineate heritage zones, which have a concentration, linkages and continuity of buildings, structures of complexes united historically or aesthetically by physical development. A Heritage Zone has a range of heritage resources-housing, physical infrastructure, transport networks, listed and protected buildings, natural heritage, parks open space, water bodies, etc.

A comprehensive approach integrating conservation and restoration of natural and cultural heritage with improvement of the community in terms of livelihoods, incomes, education, health, recreation, culture and security need to be adopted. This involves the following:

- Listing and mapping of heritage/iconic buildings, parks and zones, and working out conservation guidelines and plans thereof, including taking up the repairs, retrofitting and maintenance. This requires creation of dedicated cells for conservation of cultural heritage in local bodies and planning departments.
- The rivers, drains and water bodies in urban areas usually have a high level of pollution, which is mainly from the untreated sewer and waste from adjacent urban areas. This needs control of unauthorised development and land uses which are incongruous with the environment, *e.g.* sanitary landfills, industry, thermal power station, etc. The development of greenways along drainage corridors and harvesting of rain water in balancing lakes and ponds can be a new frontier in city development.

5. Disaster and Climate Resilience

The Indian cities have become insecure, crime and danger prone, particularly for the vulnerable sections of the society – the poor, women, elderly, children and disabled. A study finds that two out of three women have faced incidents of sexual harassment more than once during the year, mostly in public spaces, roadside and public transport. Various surveys bring out the 'gender service gap' in terms of inequitable access to basic services like water and toilets for women. The lack of clean and safe public toilets for women and girls may result into sexual harassment and even violence.

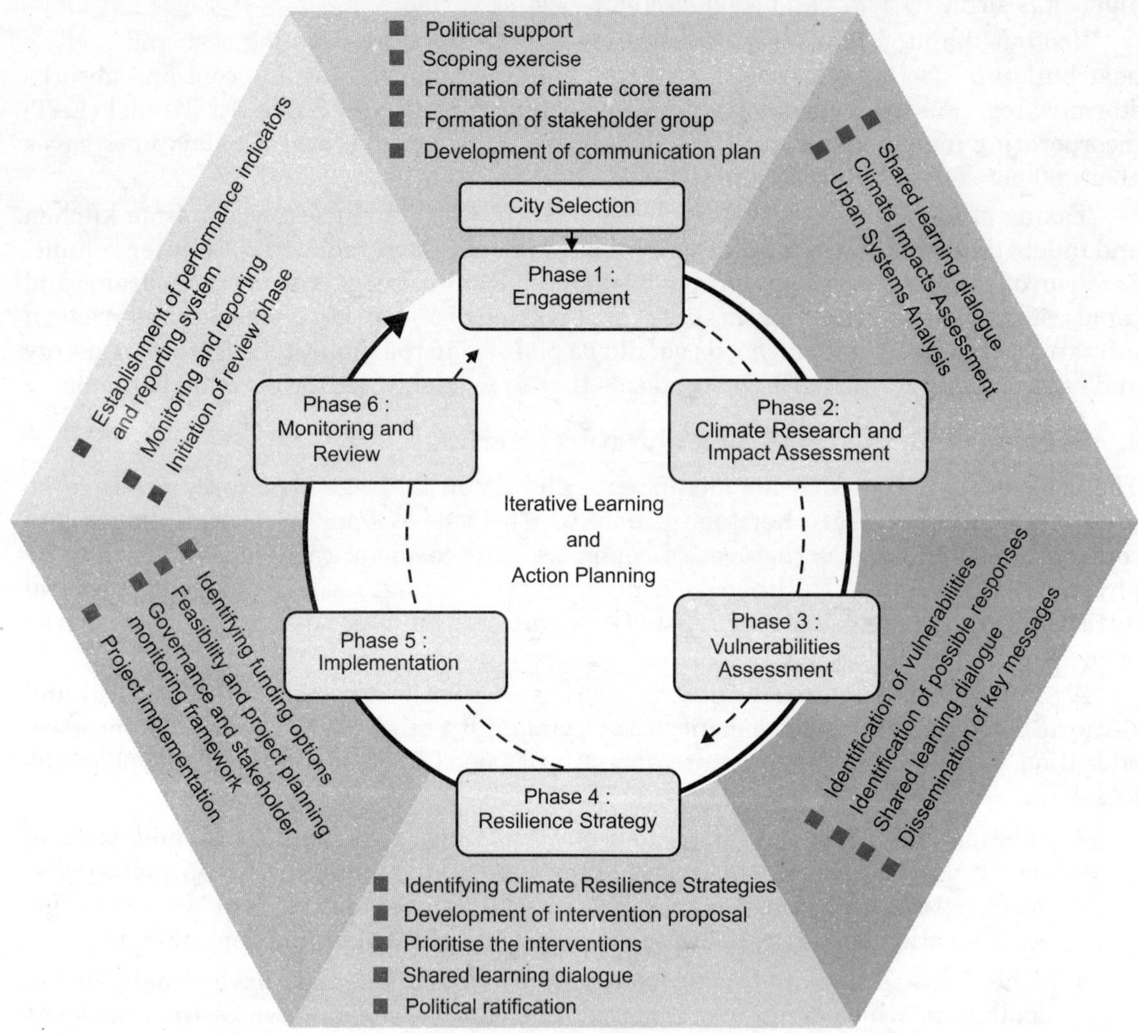

Fig.11.40: ICLEI ACCCRN Process

Source: Tewari S. ICLEI ACCCRN Process, presented at the International Conference on Resilient Cities, Experiences from ACCCRN in Asia and Beyond, New Delhi, 29th January 2013

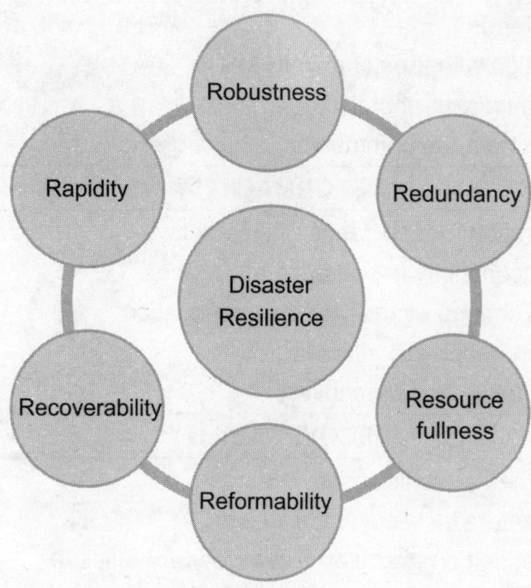

Fig.11.41: Cycle of six R's as parameter of assessment of Disaster Resilience

The Disaster Management Plan should be based on NDMA guidelines and Hyoga Convention Framework. The city can be made resilience against disasters by adopting the following 6 R's :

Table 11.10: Six Rs of Disaster Resilience

ROBUSTNESS
• Continuity in designated functions besides a level of perturbation & disruption • Clustering of services to avoid cascading failure • Contingency teams & inclusion of community • Safety parameters—accessibility, evacuation planning • Compliance of Development Controls, regulations & building bye-laws • Resources & team deployment operation of Disaster management centers • Training program for livelihood recovery, inclusion of marginalized groups • Community leadership & incentive to adopt risk reduction measures.
REDUNDANCY
• Heterogeneity/ diversity in resources for uncertain & unpredictable conditions • Ability of horizontal restructuring to avoid delays • Synchronization in distributed actions • Back up in service Provisions, transport & other infrastructural services • Segmentation in communication technology • Role of Institutions in recognizing new agents
RESOURCEFULNESS
• Reprioritization & generation of resource distribution • Availability & accessibility distributive mechanism of reserve resources • Efficiency of safety nets for support during no income period are

• Investments for structural reforms • Sharing of technical information with the community • Role of social capital & community partnership • Efficiency of collective action from the community
REFORMABILITY
• Horizontal and vertical integration for avoidance of time delays • Single window clearance for continuity in operation of services • Flexibility and adaptability to change usage, activity and location • Role rotation in planning & operation are efficient • Acceptance of new & emerging modes of recovery
RECOVERABILITY
• Distribution of roles & resources are efficient for recovery • Capacity to include diverse actors for rècover is efficient • Capacity building program for self organization & recovery are efficient • Efficiency of risk transfer mechanism/ insurance /catastrophic bonds • Accessibility and disbursement of funds to the affected communities • Emergency operation centers & available logistics • Adaptability & learning capacity of communities
RAPIDITY
• Efficiency of Immediate response on self help basis • Efficiency of Emergency responses to organize people for collective action • Readiness of Institutional framework • Emergency & evacuation warning system • Efficiency of Contingency funds

Sources: *James Smart, (2012), the Role of Post Disaster Institutions in Recovery and resilience –a comparative Study of Three Disasters-Victorian Bushfire (2009), Queensland Floods (2010-11) and Canterbury Earthquake (2010-12), Institute of Governance and Policy Studies, Victoria University, Wellington, Mehrotra, Nirmita, (2013), Disaster Resilience, Architecture-Time Space and People, Council os Architecture, New Delhi*

Comprehensive digitized and geo-referenced surveys and a systematic investigation of 'eco-determinants', which deal with relationship of humans and other forms of life to the natural and manmade environment and local climate provide the basis for a sound land use planning. Micro-zonation surveys may be referred for land use planning, layout plans and building design.

Building bye-laws may incorporate the aspects of multi-hazard safety and retrofitting. Priority should be given to public buildings such as Hospitals, Educational, Institutional, Power Stations, Infrastructure, Heritage, Monuments and lifeline structures which attract large congregation.

The safety of women is a central issue. Various initiatives include the setting of the rescue and rehabilitation centres, re-integration of victims of trafficking; the setting up of 'One Stop Crisis Centres provision of shelters, police desks, legal, medical and counselling service, and 24x7 help lines. This also needs incremental actions, such as enhancement of street lighting, establishment of 'Multi-Utility Zones'(MUZs), pedestrian safety, safer roads, provision of public toilets for women, especially in the slums and streets and safer public transport with CCTV and alarm system.

The planning of a city can contribute in crime prevention by gender sensitive design of shared, lively and defensible spaces, mixed land use and adequate shelter/tenements for the poor, together with spaces for vending and informal trades that promote livelihood and community surveillance.

In order to make the cities safer, the following strategies may be adopted:

- Development of networked emergency response and safety centres integrating police, fire, disasters, accidents, crimes, traffic and women's safety.
- Development of a network of mohalla clinics, health infrastructure, dispensaries, trauma and accident centres.
- Improve urban infrastructure – CCTV, GIS, variable and Intelligent Traffic Signage, lighting, maps, pavements, public transport, parks, drinking water points, lighting and police posts.
- Signage in all public places giving details of helpline numbers and persons to be contacted in case of emergency.
- Provision of clean and well -functioning toilets on all public roads, in schools, markets and public buildings.
- Safety of children in educational institutes, buses, playgrounds and on roads.
- Safety oriented transport planning and engineering specifications, norms and practices,
- improvement of public transport, non-motorized vehicles, sidewalks, cycle tracks, underpasses and overpasses.
- Upgrading traffic control systems, public transport, alarm system for safety of women passengers, intelligent signal control, drivers' training and management.
- Review and simplify building codes for non-engineered buildings for structure and fire safety.

6. Reducing the Environmental Impact and Pollution

The air quality in terms of pollution levels is a matter of serious concern, and is responsible for a number of respiratory diseases, heart ailments, eye irritation, asthma, etc. The three main sources of air pollution in mega-cities are vehicular emission (around 70 percent) industrial emissions (around 20 percent) with a major element of this coming from the thermal power plants, and from other sources such as diesel generator sets and domestic cooking, burning of biomass, etc.

With the phenomenal growth in the number of vehicles, almost 8-10 times in the last two decades in absolute terms the most significant aspect in the context of congestion and pollution, relates to the growth in personalized transport as compared to the availability of public transport, it has been estimated that buses, which constitute barely 1.2 percent of the total number of vehicles, cater to around only 60 percent of total transport load, while personal vehicles – cars and scooters, though almost 93 percent of the total number of vehicles, cater to around only 30 percent of the travel demand. Deficiencies in sanitation, sewerage, drainage and solid waste management for which land filling is common, add to environmental disaster.

Noise is emerging as a major pollutant and irritant in the Indian Cities, and a cause of health hazard. Against permissible level of 50 to 60 db (a), the sound level invariably exceeds 80 db (A). Faulty and leaking silencers, over-use of motor horns, screeching, tyres, rattling bodies of vehicles plying on rough roads accentuate noise level from the traffic. Besides the noise from commercial and industrial activities in residential zones. Unabated use of sound amplifiers, generator sets and fire crackers often disturb the peace in urban areas.

The need to save air, water and land form pollution is a major concern. Drinking water, sanitation, solid waste treatment and drainage are important aspects of the human environment. The performance of present sanitation systems needs to be reassessed with reference to environment, hygiene and their accessibility, The future city will be low carbon, low energy, smart, sustainable, safer and more efficient, and resilient. It will optimize the use of IT, simulation, automation and robotics that will make it smart and intelligent. Renewable, efficient energy and solar mapping would be key concerns in its planning and design. In contrast to emphasis on complex EIA, end of pipe pollution control measures, cleaner energy and fuels, car emission norms, etc. what most urban centres need is an environmental planning approach that facilitates meeting deficits in provision of basic services, *e.g.* water, sanitation, drainage, electricity, healthcare, emergency services, schools, public transport, etc.

Various alternative technologies, based on decentralized services, like Extended Aeration Technique, Bio-gas production, Bubble Diffusion Process, Flotation, Anaerobic reactors, etc. which are already in vogue, can be explored for urban sanitation.

In place of land filling for solid waste disposal, which is an environmental disaster, decentralized systems based on recycling, energy generation and organic decomposing should be explored. Bio-reactor composting and vessel system, vermi-composting, etc. are new generation technologies for waste treatment.

The concepts of energy efficiency, renewable energy and Zero-fossil Energy Development (ZED) can reduce the level of energy demand and slow down the rate at which resources are depleted. Energy efficiency involves a synergy of the various levels of planning, design, construction and maintenance.

Smart and green buildings can give energy saving up to 30%, reduce carbon emissions, provide higher efficiency and comfort with lesser energy consumption. The city and buildings have not only to be comfortable, green and efficient but also intelligent and integrated. The following measures are necessary in this regard:

- Replace high radiation absorbing surfaces such as dark roofs, asphalt streets and parking lots with highly reflecting materials.
- Adopt passive design to minimize energy use, including space heating, cooling and cooking, utilize higher efficiency air conditioning systems.
- Inclusion of green space, city forests, better spaced buildings, and better public transportation, including non-motorized transport.

- Landscape development, besides plantation, trees, vegetation, needs to ensure maintaining drainage channels, recharging of sub-soil water by porosity for absorption of water.
- Water, plantation, shade and channeling of cooling breeze are the prime elements for a comfortable micro-climate. Minimum exterior surface to maximum internal volume, correct orientation, atriums and internal courtyards reduce glare and heat. Drainage and flood control strategy may be linked with water supply and pollution control.

Besides, enforcement of Noise Pollution Rules, 2000, the urban form, land use planning, landscape and green buffers can help in reducing the noise levels. It is not desirable to locate residential area along major arteries and non-residential activities inside the residential localities that would invite traffic. By proper land use planning, mixed land use, location of public/semi-public and commercial activities along major transport arteries, a green buffer can be created for residential zone. The green buffer with thin leave trees, land formation, mounds, embankments, etc. provides effective barriers to transmission of noise.

7. Universal Access to Greens and Public Spaces

With expending urbanisation, the open spaces in cities are threatened and are being used for building, squatting, encroachments, parking, dumping of wastes, construction of roads, flyover and bridges, etc. Political and land mafia nexus looks at open as vacant space and a source of financial exploitation. This is endangering cities diverse public spaces, such as river front, green belt, ridge/forest, lakes, water bodies, square, the parks, boulevards, etc. With the expanding city, agricultural land and food production is diminishing. This is making food more expensive with longer haulage. The concept of vertical urban farming, garden and terraces can effectively add to cities green cover. Satellite control park and lawn micro-irrigation system cuts water consumption and pumping power and use of recycled wastewater. Vertical farms could reduce fertilizer and freshwater use, shorten transport and recycle grey water otherwise dumped by treatment plants. This can be a new frontier in urban landscape of Delhi.

8. Strengthening Integrated Development Planning

It is necessary to adopt an integrated approach for planning of the sub-region for economic activities, multi-modal transport system and social and physical infrastructure. The cities have limited area, for which optimum utilization of land is required so that agriculture lands and the natural features like forest, water bodies and areas of ecological importance could be conserved. The strategies have to include institutional and legal reforms and participatory local area planning, new ways of land management and implementation, monitoring and consolidation. This involves the following key actions:

- Strengthen urban planning capacity and constitute empowered Local Area/Ward Planning Committees for Local Area Planning.
- Enact/review the Town Planning and Municipal legislations
- Take up the preparation of a Sub-Regional Plan

- Constitute a Land Management Committee and bring a legislation on land pooling
- Take up Physical, Social, Environment and Financial Audits
- Bridging the gap between planning and implementation by constitution of monitoring and coordination groups at policy, strategic and operational levels, and Project Management Committees.

Fig.11.42: Vertical Urban Farm, Edible Hotel, London (Dexter Moren)

9. Mitigation and Adaptation to Climate Change

The mitigation and adaptation to climate change, disasters and risk management in the city and the safety of women, old persons, children and other vulnerable sections of the society, are key issues, which need a comprehensive approach.

The planning for climate change adaptation aims to translate and converge various sectors into a strategy. It is based upon the spatial scale of environmental problem and establishes a matrix/link between climate change and GHG/CO_2 emission and demonstrate the role of urban areas. Here, urban areas (which also include the rural) need to be defined and categorised in spatial scale (*e.g.* local community, town, city, metropolitan area, region). The spatial area can be further delineated as per the vulnerability (coastal areas, cyclone, earthquake, landslide, droughts, etc.). Climate adaptation involves a series of geographically linked and disaggregated plans, all telescopically connected. However, the focus has to be on the local action plan which is connected both way-up and down in geographical scale and also on its sides with various sectoral plans. (land, water, air, infrastructure services and settlements, shelter, public health and hygiene, transport, energy, greenery/forest, tourism, etc.).

10. Sustainable and Resilient Buildings and Resources

It is estimated that 60% of the buildings required for Indian cities by 2030 are yet to be built. This is a unique opportunity to shift towards sustainable, resilient and green buildings. For this the following principles and strategies may be adopted:

- Sustainable site planning
- Building envelope design
- Building system design HVAC (heating, ventilation and air conditioning), lighting, electrical, and water heating
- Integration of renewable energy sources to generate energy onsite.
- Selection of ecologically sustainable materials (with high recycled content, renewable resources with low emission potential)
- Passive design of building for climatic comfort
- Optimum use of energy and reducing its need by day lighting, courtyards, green roof, ventilation, etc.
- Water conservation and recycling
- Solid and liquid waste management, its treatment and reuse
- Energy efficient transport
- Efficient building systems, construction and maintenance

It is always better to specify building materials which are locally sourced, require less energy for transportation to the site. These should have low embodied energy and should be preferably natural. Promote use of recycled materials, especially, in place of high embodied energy materials like steel and aluminum. Use of reclaimed materials such as construction and demolition waste should be promoted.

Some new green materials and technologies include the following:

- Swales, bio-drainage, porous paving, rainwater harvesting, waste water recycling, dual piping.
- Aerators for water fixtures
- Waterless urinals and other water saving systems
- Micro irrigation systems, xeriscaping
- Building Integrated Solar Photovoltaic, wind energy, parabolic solar cookers
- Green roof, vertical garden, urban farms
- High COP Chillers
- Wind towers/tunnels/chimney/high albedo materials
- Geothermal system
- LED, CO_2 Sensors, Intelligent, Bionic Sensors and Controls

- Building Information System
- Rapid walls, precast, prefabricated components
- Low emitting adhesives, sealants and insulation products.

Achieving the Outcomes

The urban transformation by environment planning can't be expected without a pathway which gives a clear orientation and a series of steps to be taken in a long journey. These steps include the following:

- Setting up the benchmarks, standards and norms
- Preparation of environmental action plans at policy, strategic and operational levels
- Institutional/organizational capacity building
- Legal framework and procedural reforms
- Financial planning, outcome budgeting and harnessing the resources of government, private and community sectors.

There is a need to develop a comprehensive insight into urban problems and policies and evolve a vision with the imperatives of decentralization, localization, partnership, equity, transparency and accountability. Civic engagement and participatory development are no more subsidiary to the main strategy of 'governance', but should be its basis.

DEBRIEFING QUESTIONS

1. What is the purpose and basic elements of environmental planning?
2. Explain the components of urban environment for environment planning.
3. Write short notes on:
   (a) Sustainable Development Goal-11
   (b) Urban Heat Island
   (c) Bio-diversity
   (d) Climate Change
   (e) Ecological Footprint
   (f) National Action Plan on Climate Change
   (g) Analytical Techniques for Environmental Baseline Information
4. Explain the concept of delineation of environmental Conservation Zone.
5. Explain the process of Land Suitability Matrix for environmental conservation
6. Write short notes on:
   (a) Natural Resource Management
   (b) Environmental Infrastructure
   (c) Bio-engineering
   (d) Bio-climatic Architecture

7. Explain the concept of sustainable development vis-à-vis energy, waste reduction, urban design, green buildings, urban transport and pollution control
8. Explain the mandatory/legal framework for Environment Impact Assessment
9. What are the approaches towards planning of a climate and disaster resilient city?

Bibliography

1. Acioly, Claudio & Forbes Davidson, (1996) Density in Urban Development, Building Issues, Vol. 8, , Lund Center for Habitat Studies, Sweden
2. Action Enterprise, (2014) India Exclusion Report, BFC, Bangalore
3. ADB, (2011) Inclusive Cities, ADB, Manila
4. Adrian and Gordon Gill, (2011), Toward Zero Carbon, Image Pub., Australia
5. Aga Khan Trust for Culture, (20110, Historic City Program, Prestel, New York
6. Ahluwalia, M. S., (2011) Prospects and Policy Challenges in the Twelfth Plan, Economic and Political Weekly, Vol. 46, 21, 2011, p. 88-105.
7. Aijaz, Rumi (2015) India's Urbanisation Experiences, Operations Research Foundation, New Delhi.
8. AMDA,(2009) Land Policy for NCTD-2021, New Delhi: Association of Municipal and Development Authorities,
9. Anurag, R. (2007) Green is not just a colour: architecture towards a sustainable development. Architecture + Design 24(10) 92-100.
10. Archaeological Survey of India, Legislations http://asi.nic.in/asi legislations. asp 2007
11. Archer, R W. (1993) Urban redevelopment with landowner participation using the land pooling/readjustment technique. Bangkok: Asian Institute of Technology.
12. Arvind, K., Baker, N., Yannas, S. &Szokolay, S. V. (ed,) 2001). Climatic Responsive Architecture, Tata McGraw Hill, New Delhi
13. Baker, Lee and David Sims 91991) India Public/ Private Partnerships in Land Development.
14. Baker, N. V., (1987) Passive and Low Energy Building Design for Tropical Island Climates, Commonwealth Secretariat, London
15. Bandhopadhya, Dr. Bibek, (1998), Solar Water Heating, IREDA News, Jan.-March, 1998
16. Bangalore Development. Authority (BDA) (2007) gReport on Comprehensive Development Plan (Revised) Bangalore, 2007.
17. Bansal, A., and A.K. Jain. "Acquisition, Development and Colonisation of Land in Gurgaon: The Impact of Strategies and Programmes." Research Study for Indian Human Settlement Program, New Delhi,1986.

18. Bansal, N. K. &Minke, G. (eds) (1995), Climatic Zones and Rural Housing in India,
19. Banshree, Mitra. 1987, Lands Supply for Low Income Housing in Delhi, Seminar IHS Rotterdam
20. Barbara Harris-White, (ed) 2016, Middle India and Urban-Rural Development, Four Decades of Change, Springer, New Delhi/London
21. Barton, H., Davis, G. and Guise, R., (1995), Sustainable Settlements: A Guide for Planners,
22. BEE, (2007), Energy Conservation Building Code, Bureau of Energy Efficiency, Ministry of
23. BEE, (2009), Energy Conservation Building Code User Guide, Bureau of Energy Efficiency,
24. Bell, Keith Cliford, World Bank, Land Governance, in Coordinates, December 2011, pp. 18-20
25. Benninger, Christopher, (2011) Letter to a Young Architect, CCBA Pvt. Ltd., Pune
26. Bertaud, Alain, (2014) "Converting land into affordable housing floor space," Policy Research Working Paper Series 6870, The World Bank.
27. Bhatia, Gautam, (2014) For the New Prime Minister, Times of India, 25th May 2014, New Delhi
28. Billand, Charles J. Delhi Case Study: Formal Serviced Land Delivery. Washington D.C.: Prepared for A.I.D. PADCO,1990.
29. BIS, (2005), National Building Code of India SP 7, Bureau of Indian Standards, New Delhi
30. Boesiger, W (Ed.) (1999) Le Corbusier-Complete Works, Birkhauser Publishers, Bese
31. Brown, A., 2015, Planning for Sustainable and Inclusive Cities in the Global South, GIZ/DFID
32. Bunyard Peter, (1978), Sewage Treatment in a Swedish Sculpture Garden, Ecologist, No.1, Jan-Feb
33. Calvita, Nico and Alan Mallach, 2010, Inclusionary Housing in International Perspective: Affordable Housing, Social Inclusion and Land Value Recapture, Cambridge, MA: Lincoln Institute of Land Policy
34. Census of India (2011), Primary Census Abstract – Data Highlights, India, Series 1, New Delhi: Office of the Registrar General and Census Commissioner, India.
35. Census of India (2012), H-Series Tables on Census Houses, Household Amenities and Assets, 2011, Registrar General and Census Commissioner, India, Ministry of Home Affairs, New Delhi.
36. Census of India, (2011), Census Report, Government of India, New Delhi
37. Center for Policy Research, 2012, Governing Indian Mega Cities, New Delhi
38. Centre for Science and Environment, (2012) Excreta Matters, 71 Cities Survey, CSE, New Delhi,

39. Chen, M Skinner, C., 2014, The Urban Informal Economy: Enhanced Knowledge, Appropriate Policies and Effective Organisation, Routledge, New York
40. Chennai Metropolitan Development. Authority, Master Plan for Chenai, CMDA, Chennai
41. Christ, Emanual and Christoph Ganenbein, 2015, Typology, review No. III, ETH, Zurich
42. Cities on Water, (1991), Impact of Sea Level Rise on Cities and Regions, Proceedings of the
43. Commonwealth Science Council, London
44. Courtney, J. 1978, Urban Land-Uses Regulation, in Urban Land Policy: Issues and Opportunities, edited by H. Dunkerly and C.M Whitehead, World Bank, washington.
45. Crowe, Timothy D., (2000), Crime Prevention through Environmental Design, 2nd Edition, Stoneham, MA Butterworth, Heinemann
46. CRY (2012), Annual Survey report 2012-13, New Delhi
47. Davis, John Emmeus (ed) 2010, The Community Land Trust Reader, Lincoln Institute of Land Policy. May
48. DDA, Disposal and Developmen t of Nazul Land Rules,1981.
49. DDA, Master Plan for Delhi : 2021. New Delhi: Delhi Development Authority, 2007.
50. De Soto, H., (2000), The Mystery of Capital: Why Capitalism Triumphs in the West and Fails Everywhere Else, Basic Books, New York
51. Delhi Development Authority (DDA) Guidelines on Land Management. New Delhi: Unpublished reports, 1992 and 2014.
52. Deshpande, Subhash. Vacuum Sewerage System, IPT Journal, Jan-Feb 2006.
53. Designers and Developers, the Local Government Management Board, London
54. DPU & HIC, 2015, Habitat III National Reporting Processes: Locating Right to the City and the Role of Civil Society, University College, London
55. Drakatis Smith, (2001), Handbook of Urban Studies, Sage Publication, London
56. Dreze, Jean, and Amartya Sen (1996) India, Economic Development and Social Opportunity, New Delhi: Oxford University Press
57. DUEIIP, 2001, Status Report of Delhi 21, Delhi Urban Environment and Infrastructure Improvement Project, Government of India, Ministry of Environment and Forest and Government of NCT of Delhi Planning Department, New Delhi
58. Dupont, V. (2008), Slum Demolition in Delhi Since the 1990s: An Appraisal, Economic and Political Weekly, Vol. 43, 28, (2008), p. 79-8 Economic Growth, Mumbai,
59. Erpenstein, Annette, (2012), Back to the Future: The Rebirth of Public Place in Seoul, Trialog
60. Faetanini, M et al, 2011, Urban Policies and the Right to the City in India, CSH and UNESCO, New Delhi

61. First International Meeting of Cities on Water (R. Frassetto, ed.), Venice, 11-13 December 1989
62. Fonesca, Rory, (1976), The Walled City of Old Delhi in Shelter and Society ForschungszentrumJulich, (Germany).
63. Frampton, Kenneth (1996), Charles Correa, Perennial Press, Mumbai
64. garwal, Anil & Sunita Narain, (1997), Dying Wisdom (India's Traditional Water Harvesting System), Centre for Science & Environment, New Delhi
65. GEAG, (2009), Vulnerability Analysis, Gorakhpur City, GEAG, Gorakhpur
66. Gehl Jan, 2010, Cities for People, Island Press, Washington/London
67. Ghertner, D Asher, 2015, Rule by Aesthetics- World Class City in Making in Delhi, Oxford University Press, New Delhi
68. Girardet, Herbert, (1997), Sustainable Cities – A Contradiction in Terms, Architectural Design, London
69. GOI, (1956), Slum Areas (Improvement and Clearance) Act, Government of India
70. GOI, (2006), Integrated Energy Policy, Expert Committee Report, Planning Commission, Government of India. http://planningcommission.nic.in/reports/genrep/rep_intengy.pdf.
71. GOI, Ministry of Urban Development (website), JNNURM Toolkit, Guidelines and Progress (2006-13), New Delhi
72. GOI, Tendulkar Committee Report (2009), Planning Commission, Government of India, New Delhi.
73. Goodall, Christ, (2007) How to Live a Low Carbon Life, Earthscan, London
74. Gotsch Peter, (2012), IDEAN Principles-Ingredients of Sustainable and Resilient Neighbourhoods
75. Government of Delhi, (2013), Delhi Human Development Report, New Delhi.
76. Government of India (2007): The Eleventh Five Year Plan, the Planning Commission, New Delhi.
77. Government of India (2007): Urban National Housing and Habitat Policy, Ministry of Urban Affairs and Employment, New Delhi
78. Government of India (2008) PM's council on climate change, National Action Plan for climate Change, New Delhi
79. Government of India (2009), National Mission on Sustainable Habitat, MOUD, New Delhi
80. Government of India, (2012) Task Force Report on Urban Housing, Ministry of Urban Affairs and Employment, New Delhi
81. Government of India, (2013), Twelfth Five Year Plan, Planning Commission, New Delhi.
82. Government of NCT Delhi, Delhi Human Development Report, 2013, Oxford, New Delhi

83. Govt. of India, (2011): Report of the High Powered Committee on Urban Infrastructure, (Isher Ahluwalia Report), Ministry of Urban Development, New Delhi.
84. GTZ, Capacity Building for Ecological Sanitation, Paris: UNESCO, 2006.
85. Habitat, Planning Sustainable Cities, Global Report on Human Settlements, Earthscan,
86. Hardoy, I. E., Mitlin, D., Satterthwaite, D., 1992: Environmental Problems in Tnird World Cities. Earthscan Publications Ltd., London
87. Hough, Michael, (1995), Cities & Natural Process, Routeledge, London
88. IBM Institute for Business Value Analysis, How City System can be Made Smarter with Instrumentation, Interconnection and Intelligence
89. ICLEI (2007), Preparing for Climate Change: A Guidebook for Local, Regional and State Government, Washington
90. ICLEI, (1997), Local Agenda 21 Survey, UN Department for Policy Coordination and Sustainable
91. IEA (2008). India Energy Efficiency Policies and Measures: Energy Conservation Building Code, http:// www.iea.org/textbase/pm?mode=pm&id=4162&action=detail
92. IEA, (2007). Energy balances of Non-OECD countries, International Energy Agency
93. IGBC, (2007), LEED - India, Green Building Rating System, abridged reference guide for new construction & major renovations (LEED-India NC), version 1.0, Indian Green Building Council, Hyderabad. http://www.igbc.in/site/mmbase/attachments/48344/LEED.Abrid_Ver.pdf
94. INTACH, Delhi: The Built Heritage (2 vols), New Delhi–1999.
95. INTACH, Special Issue on Heritage, Architecture + Design, May-June 1994.
96. IPCC (2007), Climate Change, Impacts Adaptation and Vulnerability, IPCC, Cambridge and New York
97. Jacobs, Jane, 1961, The Death and Life of American Cities, Random House, USA
98. Jacobus, Ricks, 2015, Inclusionary Housing Creating and Maintaining Equitable Communities, Cambridge MA, Lincoln Institute of Land Policy
99. Jaggi, I.R., (1991). Readymade Housing, Indian Architect and Builder, October 1991
100. Jain A. K., (2009), Urban Land Policy and Public Private Partnership for Real Estate and Infrastructure Projects, Readworthy Publications, New Delhi.
101. Jain A.K (2013), Dillinama, Synergy books New Delhi
102. Jain A.K (2013), Sustainable Urban Transport and System, Khanna Publishers, New Delhi
103. Jain A.K (2014), Revisiting Land Acquisition and Urban Process, Readworthy Publications, New Delhi
104. Jain A.K (2014), Right to toilet, Readworthy Publication, New Delhi
105. Jain A.K, (2016), Urban Transformation, Discovery Publishing House, New Delhi

106. Jain A.K. (2010) Urban Planning and Governance, Bookwell Publishers, New Delhi
107. Jain A.K., (2009), Planning for inclusive Community and Social Infrastructure, Readworthy Publication, New Delhi.
108. Jain A.K., (2015) Transforming Delhi, Bookwell, New Delhi
109. Jain A.K., (2016), Building Twenty Million Homes, Discovery Publishing House, New Delhi
110. Jain A.K., (2016), Landscape Architecture- Planning, Engineering and Horticulture, Discovery Publishing House, new Delhi
111. Jain, A K., (2010), Urban Housing and Slums, Readworthy Publications, New Delhi.
112. Jain, A. K., (2011), Making Infrastructure Work, Discovery Publishers, New Delhi,
113. Jain, A. K., (2015), Conservation of Cultural Heritage, Discovery Publishers, New Delhi,
114. Jain, A. K., (2015), Regeneration of Rivers and Water Resources, Discovery Publishers, New Delhi
115. Jain, A. K., (2015), Smart Cities: Vision and Action, Discovery Publishers, New Delhi,
116. Jain, A. K., (2015), Towards Cleaner and Hygienic Cities and Villages, Discovery Publishers, New Delhi
117. Jain, A. K., (2011), the Informal City, Readworthy Publishers, New Delhi,
118. Jain, A.K. (2007), Water – A Manual for Architects, Engineers, and Managers, Daya Publishers, New Delhi.
119. Jain, A.K., (2001): Ecology and Natural Resource Management, Management Publishing Co. Dehradun-
120. Jain, A.K., (2005): "Emergency Planning and Shelter for Safety", Inst. of Engineers Seminar, New Delhi.
121. Jain, A.K., (2007), Handbook on Disaster Management, Pragun, Delhi.
122. Jain, A.K., (2009), Low Carbon Cities-Policies, Planning and Practice, Discovery Publishers, New Delhi
123. Jain, A.K., (2015), The Idea of Green Building, Khanna Publishers, New Delhi.
124. Jain, A.K., (2015), The Idea of Green Building, Khanna Publishers, New Delhi.
125. Jain, Kulbhushan and Meenakshi Jain, (2000), Architecture of the Indian Desert, AADI Centre, Ahmedabad and the University of Virginia, US
126. Jain, Shikha, DRONAH, Presentation on "Heritage Legislation in India", an overview, IHCN, second biennialconference, Hyderabad, AP, 18-20 Dec. 2008.
127. Jain, Shikha, Havelis: A Living Tradition of Rajasthan, Shubi Pub, Gurgaon, 2004.
128. Jodidio, Phillip, 2015, Heritage of the Mughal World, Aga Khan Historic City Programme, Prestel, Munich
129. Kabre, C. (2010)., A new thermal performance index for dwelling roofs in the warm humi tropics, Building and Environment, 45: 727-738

130. Kachru, Upendra (1994), A Novel Method of Cooling and Heating of Buildings-Envelope Conditioning, Proceedings on Conference -Excellence in Housing, Feb-24th and 25th Dallas, Texas, Energy Efficient Building Association, USA
131. Kenworthy and F.B. Laube (1999) An International Sourcebook of Automobile Dependence in Cities, University Press of Colorado, Boulder
132. Kenworthy, J. (2011) 'An International Comparative Perspective on Fast Rising Motorization and Automobile Dependence' in Urban Transport in the Developing World: A Handbook of Policy and Practice, edited by Harry T. Dimitriou and Ralph Gakenheimer, Edward Elgar Publishing Ltd. , Cheltenham, UK
133. Klaus, Daniels, (1994), the Technology of Ecological Building, Birkhauser, Verlg, Berlin
134. Koenigsberger, Ingersoll, Mayhew, Szokilay; (1975) Manual of Tropical Housing and Building; Orient Longman Private Limited,
135. Kundu, Amitabh, 2006, Trend and Patterns of Urbanisation and their Economic Implications in India, Infrastructure Report 2006, Infrastructure Development Finance Company, Oxford University Press, New Delhi
136. Landsberg, H. E., (1981), the Urban Climate. Academic Press, New York
137. Laul, Anil, Green is Red, ASHRAY, Anangpur, Haryana
138. Livemint, The Evolution of Land Acquisition Act, New Delhi, May 22, 2015.
139. Lowry, W. P. (1988), Atmospheric Ecology for Designers ana Planners. Peavine Press, McMinnville, Oregon.
140. Manda, Vikas, Land Acquisition Law in India: A Historical Perspective, International Journal of Innovative Research and Studies (May, 2014)
141. Martin Khor and Lim Li Lin, eds, 2001, Provision of Public Housing in Singapore in Good Practices and innovative Experiences in the South, Vol. 2, Social Policies, Indigenous knowledge and appropriate technology
142. McHarg, Ian, (1969), Design with Nature, New York, Natural History Press
143. McHarg, Ian, (1969), Design with Nature, New York, Natural History Press,
144. McKinsey Global Institute (2014), "A Blueprint for addressing the global affordable housing challenge", New York.
145. McKinsey Global Institute, (2010), India's Urban Awakening: Building Inclusive Cities, Sustaining
146. Mehrotra, Nirmita, (2013), Disaster Resilience, Architecture-Time Space and People, Council os Architecture, New Delhi
147. Mehta, Dinesh and Mira Mehta. Land for Shelter Delivery of Serviced Land in Metropolitan India. Ahmedabad, India: Prepared for USAID Centre for Environmental Planning,1991, Metropolitan Area—2011; 1995.
148. Ministry of Housing and Poverty Alleviation, (2015) Pradhan Mantri Awas Yojana, Guidelines , Government of India, New Delhi,.

149. Ministry of Rural Development, (1999) Total Sanitation Guidelines Government of India,

150. Ministry of Urban Development Govternment of India, New Delhi, (2001), the Constitutional (Seventy-Fourth) Amendment Act, 1992, Power to the People: The Nagarpalika Act.

151. Ministry of Urban Development, (2011), Report of the High Power Committee on Urban Infrastructure, New Delhi.

152. Ministry of Urban Development, (2014): Swachh Bharat (Clean India) Campaign, Government of India, New Delhi

153. Ministry of Urban Development, (2015) AMRUT, (Atal Mission for Rejuvenation of Urban Transformation) Government of India, New Delhi

154. Ministry of Urban Development, (2015): HRIDAY (Heritage and Redevelopment for Integrated Development), Government of India, New Delhi

155. MOHUPA, (NBO) 2012, Urban Housing Shortage, 2012-17, New Delhi.

156. Ministry of Urban Development, (2015): Smart City Mission Guidelines, Government of India, New Delhi

157. Misra, B. (1990), A Comparative Study of Urban Land Management (ULM) in Selected Developing Asian Countries and Japan in Economic Region Planning Research, No 93. Nagoya Japan: Economic Research Center Nagoya University, Nagoya,Japan.

158. MNRE (2010), Green Rating for Integrated Habitat Assessment, Ministry of New and Renewable Energy, Government of India, New Delhi, http://www.grihaindia.org/index.php.

159. MNRE (2014), Report on New and Renewable Energy, Ministry of New and Renewable Energy, Government of India, New Delhi, http://www.grihaindia.org/index.php.

160. Modi, Narendra, 2010, Convinient Action, Mcmillan, New Delhi

161. MoEF (2006). Environmental Clearances, Ministry of Environment, Forests and Climate Change, Governmen of India, New Delhi, http://moef.nic.in/modules/project-clearances/ehvironment-clearances

162. MOHUPA (Ministry of Husing and Urban Poverty Alleviation), 2013, Inclusive Urban Planning, ban Poor Reportv, Government of India, New Delhi

163. MOHUPA (Ministry of Husing and Urban Poverty Alleviation), 2015, Model Tenancy Act, Government of India, New Delhi

164. MOHUPA (Ministry of Husing and Urban Poverty Alleviation), 2015, Pradhan Mantri Awas Yojana -Housing for All, (Urban), Sceme Guidelines, Government of India, New Delhi

165. MOHUPA (Ministry of Husing and Urban Poverty Alleviation), 2015, National Urban Rental Housing Policy (Draft), Government of India, New Delhi

166. MOHUPA, (NBO) 2013, State of Housing in India: A Statistical Compendium, New Delhi.

167. MOHUPA, (NBO) 2013, State of Slums in India, A Statistical Compendium, New Delhi.

168. MoUD, (2012). "Improving Urban Water Supply & Sanitation Services, Advisory Note", Ministry of Urban Development, Government of India, April 2012.

169. MOUD/ Wilbur Smith Associates, (2008), Traffic and Transport Policies and Strategies in Urban Areas in India, Ministry of Urban Development, New Delhi,

170. Narain, Sunita. "The Flush Toilet is Ecologically Mindless", Down To Earth, New Delhi, Feb. 2002.

171. National Council of Applied Economic Research (1995), India Infrastructure Report, New Delhi.

172. National Council of Applied Economic Research (2012), Land Pooling Policy for Delhi, New Delhi (unpublished).

173. National Council of Applied Economic Research (2015), Affordable Housing Sector in India, New Delhi (unpublished).

174. National Sample Survey Office (2010), Migration in India, NSS 64th Round (July 2007 to June 2008), Ministry of Statistics and Programme Implementation, New Delhi.

175. Nations Conference on Human Settlements, Intanbul, Turkey, 3-14 June 1996

176. NCAER (2013), Land Pooling Policy for Delhi, unpublished report, New Delhi.

177. Newman, Peter, R Kenworthy, (1999), Sustainability and Cities: Overcoming Automobile Independence, the Centre for Resource Economics, Washington DC, USA.

178. OECD, 2010, Organising Local Economic Development, the Role of Development Agencies and Companies, OECD

179. Pandya, Y. (2007). Towards symbiosis of nature and architecture, Architecture + Design

180. Parry, M., N. Arnell, P. Berry, D. Dodman, S. Fankhauser, C. Hope, S. Kovats, R. Nicholls, D Satterthwaite, R. Tiffin and T. Wheeler (2009) Assessing the costs of adaptation to climate change: a review of the UNFCCC and other recent estimates, International Institute for Environment and Development / Grantham Institute for Climate Change, London

181. Plummer Jenelle, (2012), Focusing Partnerships, Earthscan, London Power, Government of India, New Delhi

182. Punjab Government, 1988, A Gazetteer of Delhi 1883-4, Vipin Jain, Gurgaon

183. Rai, Markandey, (2012), Creating Safe and Secure Metropolitan City: the UN Perspective, UN Habitat, Nairobi

184. Rao P.K., (1991), Safeguards for Sound Concrete, Indian Architects and Builders, October 1991

185. Rees, William E., (1997) Ecological Footprints and Urban Transportation, Velocity, Barcelona.

186. Replogle, M. (1993), Bicycle and cycle rickshaws in Asian cities, 6th conference on urban transport in developing countries, CODATU-VI, Tunis, 1993.

187. Report of High Powered Committee on Urban Infrastructure (Isher Ahluwalia Report), (2011) Ministry of Urban Development, Government of India, New Delhi,.

188. Ribeiro, E.F.N. (1985), Improved Sanitation and Environmental Health Condition with Evaluation of Sulabh International Low Cost Sanitation in Bihar, Patna: SulabhInternational,.

189. RICS (2010), "Making Affordable Housing Work in India", RICS, CBRE and LEVVEL.

190. RTPI, 2014, Fostering Growth-Understanding and Strengthening the Economic Benefit of Planning, RTPI, London

191. Sampat, Preeti 2013, Limits to Absolute Power: Eminent Domain and the Right to Land in India, EP Weekly, May 11, pp.40-52.

192. Sanjay, P. (1992). Energy conscious architecture: an endless quest. Architecture + Design

193. Satterthwaite, D. (2007) the Transition to a Predominantly Urban World and its Underpinnings, Human Settlements Discussion Paper, IIED, London

194. Satterthwaite, D. (2008). Cities' contribution to global warming: Notes on the allocation of greenhouse gas emissions. Environment and Urbanization, 20 (2).

195. Satterthwaite, D., S. Huq, H. Reid, M. Pelling and P. Romero Lankao (2007) Adapting to climate change in urban areas: the possibilities and constraints in low- and middle-income nations, IIED Human Settlements Discussion Paper Series, Climate Change and Cities, London

196. Saxena, K. B. 2011, Rehabilitation and resettlement of displaced persons. Chapter three in Development-Induced Displacement, Rehabilitation and Resettlement in India. Routledge Contemporary South Asia.

197. Scriver Peter and Amit Srivastava, 2015, India- Modern Architecture in History, ReaktionBooks, London

198. Service Report No. 376, American Planning Association, Chicago.

199. Settlements Development: Implementing Agenda 21. Nairobi.

200. Shiva, Vandana. S. Javi, S.M. Fontana, The Great Indian Land Grab, Navdanya, New Delhi,2011.

201. Shivaramakrishnan K.C., 2015, Planning for Metropolitan Region, Oxford University Press, New Delhi

202. Smart, James, (2012), the Role of Post Disaster Institutions in Recovery and resilience –a comparative Study of Three Disasters-Victorian Bushfire (2009), Queensland Floods (2010-11) and Cantebury Earthquake (2010-12), Institute of Governance and Policy Studies, Victoria University, Wellingto

203. Sorensen, Andre. Conflict, Consensus or Consent: Implications of Japanese Land Readjustment Practice for Developing Countries, Habitat International Vol. 24 pp. 51-73,2000.

204. Steinberg, Florian; 1996, Conservation and Rehabilitation of Urban Heritage in Developing Countries, HabitatInternational, June,

205. Steinberg, Florian; Concept of Evolutionary Housing Design for India, Hudco, New Delhi

206. Subrahmanyum G.V., (2009), Revised Environment Clearance Process, Prescription to the Ministry of Environment and Forests, New Delhi

207. Suburbs. Edited by P. and J., Van der Linden Edited by Paul Baross. New Delhi, India,1989.

208. Synder, Michael, (2015) The Wealth of Nation, The Caravan, New Delhi (December)

209. TCPO, (2011) Status of Town and Country Planning Departments and Act in different States and UTs in India. Spatio-Economic Development Record, Vol 18, No. 5, New Delhi,2011

210. Tewari, S., (2013), ICLEI ACCCRN Process, Presented at the International Conference on Resilient Cities, Experiences from ACCCRN in Asia and Beyond, New Delhi

211. Thurow, C., (1983): Improving Street Climate through Urban Design Planning. Advisory

212. Uchida, Hirotsugu and Andrew Nelson (2008), 'Agglomeration Index – Towards a New Measure of Urban Concentration', Background Paper Prepared for World Development Report 2009.

213. UN Global Compact and Oxfam, 2015, overty Footprint: Overview, Newyork/London

214. UN Habitat, (2003), The Challenge of Slums - Global Report on Human Settlements, London,

215. UN Habitat, (2007), Enhancing Urban Safety and Security, Nairobi

216. UN Habitat, 2013, States of the World's Cities 2012/2013: Prosperity of Cities, Nairobi

217. UN Habitat, 2014, A Practical gude to Designing, Planning and Executing City-wide Slum Upgrading Programmes, Nairobi

218. UN Habitat, 2015, Habitat III Issue Paper, Nairobi

219. UN, 2013, A New Global Partnership: Eradicate Poverty and Transform Economies through Sustainable Development, New York

220. UNDP and UN Solution Exchange (2012) Social Inclusion of Manual Scavengers. A Report of National Round Table Discussion Organised by United Nations Development Programme and

221. UNEP, 2015, District Energy in Cities: Unlocking the Potential of Energy Efficiency and Renewable Energy, New York

222. UNESCO, 2010 Historic Districts for All, New Delhi, UNESCO

223. UN-Habitat (2011),Cities and Climate Change,Global Report on Human Settlements, Earthscan, UK/USA/Nairobi.
224. UN-Habitat (2011), Cities and Climate Change, Global Report on Human Settlements, Earthscan, UK/USA/Nairobi.
225. UN-Habitat (2013), Planning and Design for Sustainable Urban Mobility,Global Report on Human Settlements, Earthscan, UK/USA/Nairobi.
226. UN-Habitat, (2010), The State of the World's Cities: Bridging the Urban Divide, London, Washington D.C., Earthscan
227. UN-Habitat/International Institute for Environment & Development (2010), Developing Local Climate Change Plans, UK/USA/Nairobi, 2009
228. Unified Traffic and Transport Infrastructure Planning and Engineering Centre (UTTIPEC) (2012) Draft Transit Oriented Development Guidelines, New Delhi
229. UNISDR, 2015, United Nations Global Assessment Report on Disaster Risks, UN Office for Disaster Risk Reduction, Geneva
230. USAID, 1991Constraints Affecting the Efficiency of Urban Land Markets in India, New Delhi
231. Volwahsen, Andreas, 2002 Imperial Delhi, Prestel Verlag, Munich
232. Vuchie, R.V. (1981), Urban Public Transport System and Technology, Prentice Hall, Eaglewood Cliffs, New Jersey
233. Watson, V., 2009, The Planned City Sweeps Away the Poor: urban Planning and 21st Century Urbanisation, Elsevier, Progress in planning 72 (2009) p 151-193
234. Wilbur Smith Associates, 2008, ,12 Indian Cities, Trnsport Indicators, MOUD, Government of India, New Delhi
235. William E. Soper, (1979) Surface Application of Sewage Effluent and Sludge, Madison, Wiscosin, USA,
236. World Bank (2013), Voices of the Poor, Washington, USA.
237. World Bank (2015), Leveraging Urbanisation in South Asia, Washington, USA.
238. World Bank, (2002), India's Transport Sector, the Challenges Ahead, Washington, D.C.
239. World Bank, (2009), Reshaping Economic Geography: The World Development Report 2009 Washington DC.
240. World Health Organisation, (2004), World Report on Road Traffic Injury Prevention, Geneva
241. Yeang Kenneth, (1999), the Green Skyscraper, the Basis for Designing Sustainable Intensive Buildings, Prestel, New York
242. www.healthplaces.org.in
243. www.igbc.in
244. www.marimatic.fi
245. www.planetaction.org
246. www.windowfarms.org